BASIC DRUG
DISCOVERY DEVELOPMENT

D1145607

Academic Press is an imprint of Elsevier
32 Jamestown Road, London NW1 7BY, UK
525 B Street, Suite 1800, San Diego, CA 92101-4495, USA
225 Wyman Street, Waltham, MA 02451, USA
The Boulevard, Langford Lane, Kidlington, Oxford OX5 1GB, UK

Notices
Knowledge and best practice in this field are constantly changing. As new research and
experience broaden our understanding, changes in research methods, professional practices, or medical treatment may become necessary.

Practitioners and researchers must always rely on their own experience and knowledge
in evaluating and using any information, methods, compounds, or experiments described
herein. In using such information or methods they should be mindful of their own safety
and the safety of others, including parties for whom they have a professional responsibility.

To the fullest extent of the law, neither the Publisher nor the authors, contributors, or
editors, assume any liability for any injury and/or damage to persons or property as a
matter of products liability, negligence or otherwise, or from any use or operation of any
methods, products, instructions, or ideas contained in the material herein.

ISBN: 978-0-12-411508-8

British Library Cataloguing in Publication Data
A catalogue record for this book is available from the British Library

Library of Congress Cataloging-in-Publication Data
A catalog record for this book is available from the Library of Congress

For information on all Academic Press publications
visit our website at http://store.elsevier.com/

Working together
to grow libraries in
developing countries

www.elsevier.com • www.bookaid.org

BASIC PRINCIPLES OF DRUG DISCOVERY AND DEVELOPMENT

BENJAMIN E. BLASS

Temple University School of Pharmacy
Moulder Center for Drug Discovery Research
North Broad Street
Philadelphia, PA, USA

AMSTERDAM • BOSTON • HEIDELBERG • LONDON
NEWYORK • OXFORD • PARIS • SAN DIEGO
SAN FRANCISCO • SINGAPORE • SYDNEY • TOKYO

Academic Press is an imprint of Elsevier

Dedication

Sir Isaac Newton, one of the greatest scientists of his time, wrote "If I have seen further it is by standing on ye shoulders of Giants." Although he was almost certainly referring to his scientific achievements, the underling concept of learning from our forbearer is true in any endeavor. Indeed, this concept can be further extended to include those who are there in the present day, supporting the activities of an individual as he or she attempts to accomplish that which they view as important. With this thought in mind, I have dedicated this book to the scientists who came before me, those who mentored me, and those who work with me on a daily basis. In addition, and perhaps more importantly, this text is dedicated to the loving and supportive family that has helped me become the person that I am today. Special thanks are offered to my mother, father, sister, brother, my three children, and of course, my wife Kathleen. These are the giants on whose shoulders I have stood upon.

Contents

Companion site for this book can be accessed at
    http://booksite.elsevier.com/9780124115088

Foreword

The last three decades have witnessed a revolution in the drug discovery and development process. Medicinal chemistry and in vitro screening that were once major bottlenecks in the process of identifying novel therapeutics have been dramatically accelerated through the incorporation of automation and the development of enabling technologies such as recombinant DNA and transfection technology. High-throughput screening (HTS), parallel synthesis, and combinatorial chemistry have facilitated the synthesis and biological evaluation of large numbers of potentially useful compounds. These activities, in turn, have generated vast amounts of data that can be analyzed to develop structure–activity relationships and structure–property relationships useful for the optimization of lead compounds. At the same time, new techniques, technological advances, and a greater understanding of the importance of pharmacokinetics, animal models, and safety studies have dramatically altered how new molecules are selected for clinical study. Design strategies of clinical trials, biomarkers, translational medicine, the regulatory landscape, intellectual property rights, and the business environment have also changed dramatically over the course of the last 30 years.

The complexities of the drug discovery and development process cannot be overstated, nor can the wide range of expertise required for the successful development of new, marketable therapeutics. In order to thrive in this every changing landscape, individuals interested in a career in the pharmaceutical industry or related fields must be more than simply experts in their chosen field of study. They must also have an understanding of the numerous, overlapping fields of their colleagues. *Basic Principles in Drug Discovery and Development* has captured the critical information on the disparate processes, technologies, and expertise required for modern drug discovery and development and presents it in a logical and concise manner for students, practicing scientists, and nonscientists with an interest in the pharmaceutical industry. Dr Benjamin E. Blass, an experienced educator and scientist with foundational knowledge in medicinal chemistry, drug design, biological targets, and over 20 years of experience in industrial and academic drug discovery and development, provides a comprehensive account of the many functions involved in drug discovery and development, from initial medicinal chemistry conceptualization and in vitro biological evaluation to clinical trials and beyond.

There are many aspects of this book that will help practicing scientists, graduate students, and future drug discovery researchers to develop a strong foundation in the concepts that govern the multidisciplinary process of drug discovery. Through this unique text, they will acquire an understanding of key aspects of drug discovery and development. The organization of the subject material was chosen to allow the readers to incrementally increase their knowledge in the wide range of disciplines required to identify new, marketable therapeutic agents. The book is thoroughly written and includes 13 chapters with over 300 figures and 900 references. Throughout the text, the reader will become familiar with more than 100 drugs and clinical candidates that exemplify important theories and practices.

Each chapter contains examples of drugs pertaining to the material in the chapter. The opening chapter provides an overview of drug discovery and development. This serves as the foundation for the following 10 chapters which describe the various functions involved in drug discovery and development. The early phases of drug discovery are described in detail through discussions of important topics such as target identification, target validation, lead identification, multidimensional lead optimization, pharmacokinetics, preclinical pharmacodynamics, and early toxicology. This is followed by discussions of preclinical activities, clinical trial design, biomarkers, and translational medicine. Each chapter builds on the previous chapters and this approach provides the readers with an integrated view of the various, multidisciplinary functions required for the drug discovery and development process.

Chapters 11 and 12 describe two important topics essential for running an effective pharmaceutical R&D business: organizational structure and patent protection. These chapters give the reader a true understanding of the organizational structure required for the successful management of research and development organizations and the importance of protecting intellectual property to ensure a good return on investment. Patent protection is the life blood of the pharmaceutical and biotech industries, and at the same time a source of innovation for new discoveries. Patents ensure the sharing of discoveries and innovations that might otherwise be kept as trade secrets. In the final chapter, case studies demonstrating the practical application of the concepts and principles described in the previous chapters are provided. These vignettes also describe important lesson learned in each case, some of which changed the way modern drug discovery research and development programs are executed.

Although there are numerous textbooks that discuss various aspects of the drug discovery and development process, none of them provides a comprehensive view of the process. *Basic Principles in Drug Discovery and Development* is unique in its comprehensive approach to this complex endeavor. In writing this textbook, Dr Blass has provided an important new tool for the education of the next generation and a valuable resource for people with a vested interest in the identification and commercialization of novel medications.

Magid Abou-Gharbia, PhD, FRSC
Associate Dean for Research

Laura H. Carnell, Professor
Director Moulder Center for Drug Discovery Research
School of Pharmacy, Temple University
Philadelphia, PA, USA

Drug Discovery and Development: An Overview of Modern Methods and Principles

Over the course of the last two centuries, modern medicines have improved the lives of countless patients. Diseases and conditions that were once deemed incurable or fatal have been conquered with therapeutic agents designed to extend and improve quality of life. The most recent, and perhaps most notable, of these accomplishments is the transition seen in the consequences of infection with human immunodeficiency virus (HIV), the virus known to cause acquired immune deficiency syndrome (AIDS).[1] When the virus was first identified by two research groups in 1983,[2] there were few antiviral agents available, none provided effective treatment for HIV infection and infection progressed rapidly to AIDS, and finally death by opportunistic infection. By 1987, AZT® (Retrovir®, azidothymidine, Figure 1.1), the first reverse transcriptase inhibitor, was approved for clinical application for the treatment of HIV infection,[3] and additional treatment options were developed through the next three decades. The discovery of modern antiviral drugs such as Viread® (Tenofovir),[4] Zeffix® (Lamivudine)[5] (reverse transcriptase inhibitors), Viracept® (Nelfinavir),[6] Norvir® (Ritonavir),[7] and Crixivan® (Indinavir)[8] (HIV protease inhibitors) provided additional treatment alternatives (Figure 1.1). In the late 1990s, multidrug cocktail treatment regimens, also known as highly active antiretroviral therapy (HAART),[9] were introduced, further enhancing treatment options, culminating in the development of all-in-one fixed combination medications such as Complera®[10] and Stribild®.[11] While additional progress is still required, it is clear that modern drug discovery and development changed the course of the AIDS epidemic in less than three decades, allowing patients who were once given a death sentence to lead productive lives.[12]

Cancer treatment has seen a similar transition, as survival rates for many types of cancer have dramatically improved as a result of the discovery and

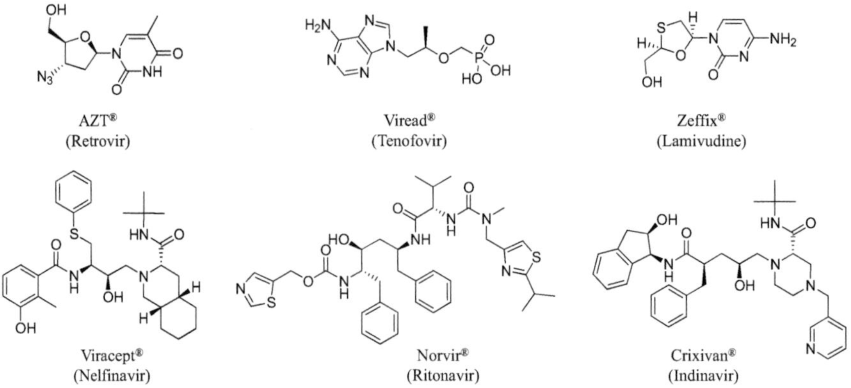

AZT®
(Retrovir)

Viread®
(Tenofovir)

Zeffix®
(Lamivudine)

Viracept®
(Nelfinavir)

Norvir®
(Ritonavir)

Crixivan®
(Indinavir)

FIGURE 1.1   Reverse transcriptase was the first enzyme successfully targeted in a drug discovery program focused on developing treatment options for HIV infection and AIDS. AZT® (Retrovir), Viread® (Tenofovir), and Zeffix® (Lamivudine) are inhibitors of this important enzyme. HIV protease, another enzyme critical to the progression of HIV and AIDS has also been the subject of intense study. The antiviral agents Viracept® (Nelfinavir), Norvir® (Ritonavir), and Crixivan® (Indinavir) are HIV protease inhibitors that were developed for the treatment of HIV infection and AIDS.

development of novel therapeutic agents. In the United States, overall cancer death rates have declined 11.4% between 1950 and 2009, and progress against some specific cancer types has been substantial. Breast cancer, prostate cancer, and melanoma, for example, have seen significant increases in their 5-year survival rates over the same period. The 5-year survival rate for breast cancer increased from 60% to 91%, while the survival rate for prostate cancer increased from 43% to over 99%, and melanoma survival rose from 49% to 93%.[13] A significant portion of the improved clinical outcomes in cancer can be attributed to improved chemotherapeutic agents. The identification of antitumor natural products and natural product analogs such as Taxol® (Paclitaxel),[14] Velban® (Vinblastine),[15] Adriamycin® (Doxorubicin),[16] and Hycamtin® (Topotecan)[17] has clearly demonstrated the importance of natural products in modern medicine, while the development of small molecule kinase inhibitors such as Gleevac® (Imatinib),[18] Tasigna® (Nilotinib),[19] and Tarceva® (Erlotinib)[20] provide clear evidence of the power of modern drug discovery techniques (Figure 1.2).

The treatment of cardiovascular disease has also seen dramatic improvements in the wake of the discovery of a multitude of therapeutic agents designed to mitigate symptoms or prevent the underlying causes of the disease. A myriad of treatments have been developed to address hypertension, also known as "the silent killer" because of its asymptomatic nature, leading to improvements in both the quality of life and life expectancy of patients. Diuretics (e.g., Midamor® (Amiloride),[21] Lozol® (Indapamide)[22]), β-blockers (e.g., Tenoretic® (Atenolol),[23] Inderal® (Propranolol)[24]), and

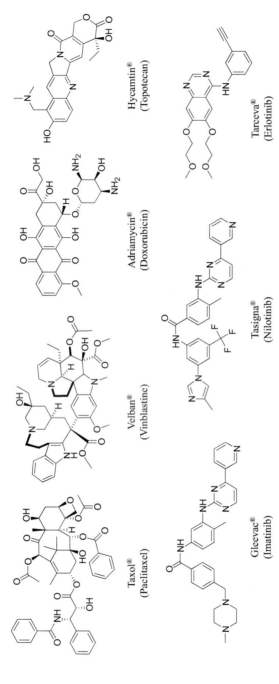

FIGURE 1.2 The natural products Taxol® (Paclitaxel), Velban® (Vinblastine), Adriamycin® (Doxorubicin), and Hycamtin® (Topotecan) are exemplary natural products that have been developed for the treatment of cancer, while Gleevac® (Imatinib), Tasigna® (Nilotinib), and Tarceva® (Erlotinib) were developed for the treatment of cancer through the application of modern drug discovery programs.

angiotensin-converting enzyme (ACE) inhibitors (e.g., Capoten® (Captopril),[25] Vasotec® (Enalapril)[26]) are just a few of the types of treatments currently available to lower blood pressure and keep cardiovascular disease at bay. Revolutionary changes occurred in the prevention of cardiovascular disease with the introduction of HMG-CoA reductase inhibitors, also known as statins.[27] Lipitor® (Atorvastatin),[28] Zocor® (Simvastatin),[29] and a number of related compounds have demonstrated remarkable capacity to lower cholesterol levels, a major risk factor associated with cardiovascular disease (Figure 1.3).[30]

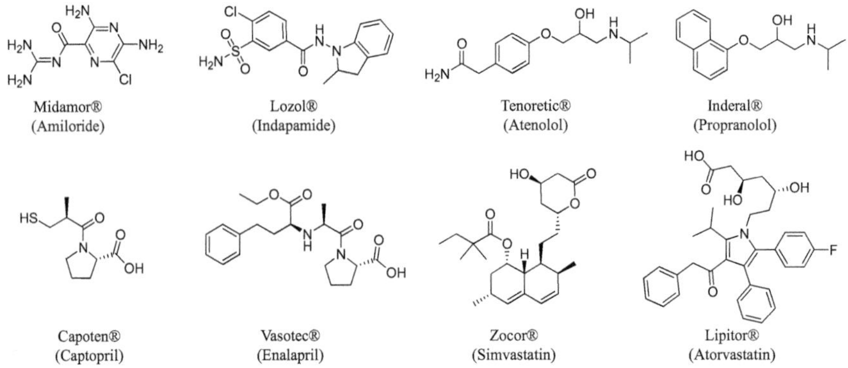

| Midamor® | Lozol® | Tenoretic® | Inderal® |
| (Amiloride) | (Indapamide) | (Atenolol) | (Propranolol) |

| Capoten® | Vasotec® | Zocor® | Lipitor® |
| (Captopril) | (Enalapril) | (Simvastatin) | (Atorvastatin) |

FIGURE 1.3 The diuretics Midamor® (Amiloride) and Lozol® (Indapamide), the β-blockers Tenoretic® (Atenolol) and Inderal® (Propranolol), the ACE inhibitors Capoten® (Captopril), Vasotec® (Enalapril), and the HMG-CoA reductase inhibitors Lipitor® (Atorvastatin) and Zocor®, (Simvastatin) have significantly improved the treatment of cardiovascular disease.

Similar improvements in disease management, symptomatic relief, and improvements in the quality of life through the development of novel chemotherapeutics could be described across a wide range of health issues. It is clear that the treatment of infectious disease, pain management, respiratory disease, and many other conditions has been profoundly and positively impacted by the identification of novel therapies.[31] There are, however, many challenging areas of health care that remain in need of improved medicine and advances in current therapy. Alzheimer's disease, for example, is the most common form of dementia, and was originally described by German psychiatrist and neuropathologist Alois Alzheimer in 1906. Over 100 years later, treatment options for this disease remain limited at best, despite the enormous amount of effort and research funding dedicated to identifying novel treatments. Potential drug targets such as β-secretase (BACE), γ-secretase, glycogen synthase kinase 3β (GSK3β), and cyclin-dependent kinase-5 (CDK5)[32] have been extensively studied, but clinically effective, disease modifying agents have as yet to be identified.

Additional challenges also exist in areas once thought to have been conquered by modern science. In the 1960s and 1970s, for example, it was widely believed that modern medicine had all but conquered infectious disease and that the major classes of antibacterial agents, β-lactams, quinolone, tetracyclines, and macrolide antibiotics (Figure 1.4) would provide all

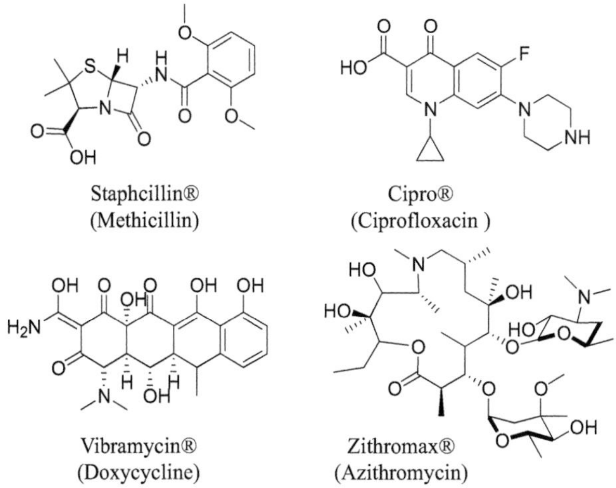

Staphcillin® (Methicillin)

Cipro® (Ciprofloxacin)

Vibramycin® (Doxycycline)

Zithromax® (Azithromycin)

FIGURE 1.4 Staphcillin® (Methicillin), Cipro® (Ciprofloxacin), Vibramycin® (Doxycycline), and Zithromax® (Azithromycin) are representative examples of β-lactam, quinolone, tetracycline, and macrolide antibiotics respectively.

of the tools necessary to protect humanity. The rise of methicillin-resistant *Staphylococcus aureus* (MRSA) in the 1980s and 1990s, however, has made it clear that additional tools will be required in order to maintain the upper hand in the war against bacterial infection. Methicillin (Staphcillin®) was introduced in 1959 as a means of treating penicillin-resistant infections, but less than two years later, resistant strains were identified in European hospitals. By the 1980s, MRSA had spread throughout the globe, and as of 2009, MRSA infections cost the US health system $3 billion to $4 billion annually.[33]

There is no doubt that the discovery of new therapeutic agents has a positive impact on society, but to the casual observer, it is not clear how this goal is achieved. On the surface, providing a drug necessary to solve a medical problem would seem to be a relatively simple task; identify the cause of the disease or malady and design a drug that will fix or eliminate the problem. In the case of infectious disease, eliminate the infectious agent, whether bacterial or viral, and the health problem is solved. This is, of course, a very simplistic view, as there are many factors to consider beyond killing the offending organism. There are millions of compounds that will kill an infectious organism, but how many of these compounds can do so without negatively impacting the host? How many of the remaining compounds can be delivered as safe and effective therapeutic agents? How does one determine which of the nearly infinite possibilities are useful and which ones are not? Of the useful compounds, which ones will be of interest to the companies that manufacture drugs and which ones will not? These issues are exceptionally complex, and become even more so, when the health issue is something other than an invading organism. In considering chronic pain management, for example, a drug provided to a patient should alleviate the chronic pain without interfering with the pain associated with protective instincts, such as withdrawing one's hand from a hot stove. This added complexity is a common feature of the vast majority of disease states and must be addressed in order to successfully develop any new medication.

Given the large number of complex issues associated with drug design and development, it should be abundantly clear that no one individual could possibly conquer all of the tasks required to discover, develop, and successfully bring to market a new therapeutic entity. The process is a multidimensional one and as such requires the coordinated effort of individuals with a wide array of expertise such as medicinal chemistry, *in vitro* biology, drug metabolism, animal pharmacology, formulations science, process chemistry, clinical research, intellectual property, and many other fields. Enabling technologies, such as high throughput screening, molecular modeling, pharmaceutical profiling, and biomarker studies, also play key roles in modern drug research. It is critical that anyone interested in pursuing a career in the development of pharmaceutically useful agents, whether in an industry setting or an academic institution, must be willing and able to participate in collaborative research effort over a significant period of time. In addition, it is important that any participant in this field understands the magnitude of the costs associated with the pursuit of new drugs. The rewards for those who are successful can be substantial, as indicated by the success of compounds such as Lipitor® (Atorvastatin), which had peak annual sales of over $13 billion,[34] Prozac® (Fluoxetine, peak sales $2.8 billion),[35] and Singulair® (Monteluclast, 2011 sales $5.5 billion),[36] but the cost in time and resources is substantial (Figure 1.5). As indicated in Figure 1.6, it has been estimated that the identification of a single marketed drug can require an initial examination of over 100,000 candidate compounds, hundreds of preclinical animal studies, and numerous clinical trials involving thousands of patients. A recent analysis of clinical trial success rates has indicated that only 1 out of every 10 clinical candidates will successfully traverse clinical trials and reach the market. This represents a success rate of less than 0.001% if measured by the number of compounds examined at the outset of the process. If measured according to the number of programs required to advance a single drug to market, program attrition rates indicate that only 1 in 24 programs is successful.

The cost associated with the identification of useful and marketable therapeutic entities is also staggering. As of 2011, it is estimated that a

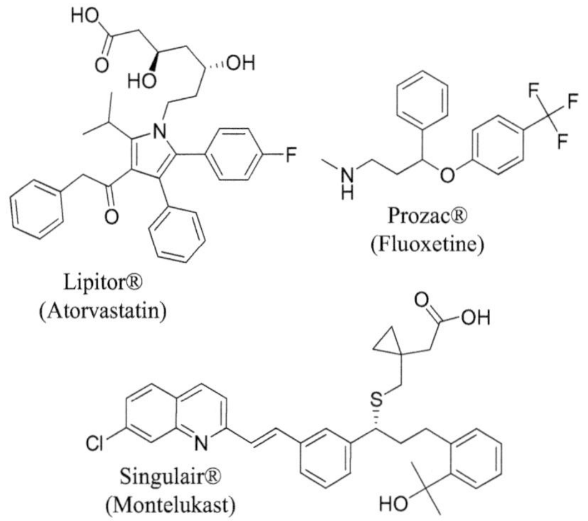

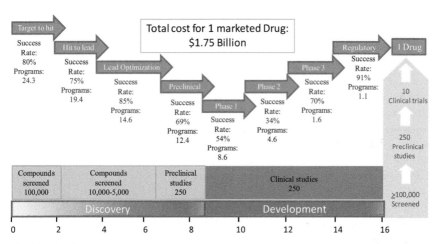

FIGURE 1.5 Lipitor® (Atorvastatin), Prozac® (Fluoxetine), and Singular® (Monteluclast) are some of the most successful drugs in the history of the drug industry. Each has produced multibillion dollar franchises, providing their owners with ample resources for the pursuit of additional novel therapeutic entities.

FIGURE 1.6 An analysis of the various stages of the drug discovery and development process provides an indication of the success rate of each stage of the process. Based on these estimates, only 1 out of every 24 early stage programs (Target to hit stage) will produce a marketed therapy. The cost to develop a single new drug must also account for the costs associated with all of the programs that are unsuccessful. The total cost is estimated to be $1.75 billion.

single new drug costs over $1.75 billion to discover and develop.[37] As a measure of comparison, the same amount of money could be used to buy 17 Boeing 737 jet aircrafts (based on 2012 prices on Boeing's Web site), purchase approximately 7000 homes (assuming $250,000 per home), 70,000 automobiles (average price $25,000), or provide for the raising 7000 children born in 2010 to the age of 18. The costs and complexity of drug discovery and development is staggering.

DRUG DISCOVERY AND DEVELOPMENT FROM 20,000 FEET (FIGURE 1.7)

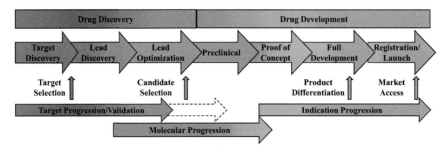

FIGURE 1.7   The drug discovery and development process viewed from "20,000 feet."

Fortunately, like most complex processes, drug discovery and development can be broken down into many smaller tasks and functions. At the highest level, the process can be divided into two major stages. The first, referred to as drug discovery, includes all of the experimentation and studies designed to move a program from the initial identification of a biological target and associated disease state to the identification of single compound with the potential to be clinically relevant. The drug discovery stage may be further broken down into three distinct phases: target discovery, lead discovery, and lead optimization. Each phase of drug discovery is designed to establish a scientific link between a biological target (e.g., an enzyme, G-protein-coupled receptor, ion channel, etc.) and a disease state model designed to mimic the human disease state. This process, often referred to as target progression and target validation, is accomplished through the use of molecular probes designed to identify multiple series of compounds that will modulate the activity of the biological target of interest. In many cases, known compounds are employed to facilitate target selection, and are eventually transitioned into novel compounds through the processes of lead discovery and lead optimization. In lead discovery phase, sets of structurally related compounds with the desired biological activity are identified (lead discovery) through biological screening of large numbers of compounds. Once a candidate series has been identified, the lead optimization phase begins. In this phase, structural analogs within a lead series are studied to identify a single compound that may be progressed into the drug development stage. Typically, multiple lead series are identified in both the lead discovery and lead optimization phases through iterative rounds of experimentation. In many cases, the lead discovery and lead optimization phases overlap, as a typical drug discovery program will produce multiple sets of related compounds with the potential for identification of candidates that might progress into drug development. This approach is required for success, as it is often difficult, if not impossible, to identify the lead series that will contain the final lead candidate in the early phases of the drug discovery process. Parallel operations of this type mitigate the risk of failure of any one series of compounds. The lead discovery phase typically concludes with the successful demonstration of *in vivo* efficacy in an appropriate animal model employing a compound that possess physical and chemical properties consistent with eventual clinical study in the drug development stage.

The second major stage, drug development, typically begins once a single compound has been identified, which is then progressed through various studies designed to support its approval for sale by the appropriate regulatory bodies. The first step in this process is the submission of an Investigational New Drug (IND) Application that requests permission to move a clinical candidate into human study. This document provides regulatory agencies with detailed preclinical data describing animal pharmacology and toxicology studies, chemical manufacturing information (including formulation, stability studies, and quality control measures), and, of course, detailed clinical protocols that describe how the clinical compounds will be studied in human populations if the studies are approved.

While clinical trial designs can vary substantially from one candidate to another, the general goals of phase I, II, III, and IV are the same. Chapter 9 will provide a more detailed review of clinical trials, but the basic tenants of clinical trials are as follows. In phase I clinical trials, safety and tolerability of an investigational new drug is examined in a small number of healthy individuals, typically 20 to 100 people, with the goal of determining if safety margins are suitable for further progression in the clinical trial process. Pharmacokinetic and pharmacodynamic aspects of the candidate are closely monitored, and the drug candidate is typically administered first in a single ascending dose (SAD) study, followed by a multiple ascending dose (MAD) study. In the SAD study, the drug is given to a group of subjects once and they are monitored to determine the impact of the drug. If there are no adverse effects, then a second group is treated with a single higher dose of the drug candidate and monitored as before. The cycle is repeated until intolerable side effects appear in order to determine the maximum tolerated dose (MTD). MAD studies are similar, but each group of subjects is provided with multiple low doses of a candidate compound over a set time. As in the SAD studies, the manifestation of clinically intolerable side effects defines the MTD for the MAD studies. The data developed through the course of the phase I studies are used to determine the doses that will be used in phase II and phase III clinical trials.

Phase II typically involves 100 to 300 patients and are designed to determine whether or not the clinical candidate provides the desired biological impact. Safety studies also continue through phase II trials. In the first part of phase II trials, referred to as phase IIA, the goal is to determine the dose required to provide the desired therapeutic impact or endpoint for the clinical candidate. Once the proper dose levels are determined, phase IIB studies can be initiated. The goal of phase IIB studies is to determine the overall efficacy of candidate compounds in a limited population of subjects. The majority of clinical drug candidates fail in phase II studies due to safety issues or lack of efficacy. As of 2011, only 34% of phase II clinical candidates successfully reach phase III studies.

The effectiveness of new drug candidates in larger patient population are determined in phase III clinical trial. These studies are typically randomized and involve hundreds to thousands of patients at multiple clinical trial sites and are designed to determine the efficacy of the candidate compound relative to the current standard of care. The cost and time associated with this phase of clinical study can vary dramatically depending on the clinical endpoint under investigation. Clinical trials for new, acute treatments, such as novel antibacterial agents, are shorter and involve far fewer patients than clinical trials for chronic conditions such as osteoarthritis. Patients are also closely monitored for adverse side effects, as the larger patient pools can identify safety issues that did not become apparent in smaller phase II trials. The number of subjects, time requirements, and complex design of phase III clinical trials (especially in chronic medical conditions) dictate that they are the most expensive aspect of drug discovery and development. Upon completion of phase III trials, a New Drug Application is submitted to the appropriate regulatory body. This document typically contains comprehensive details of both animal and human studies, all safety findings (adverse and side effects), manufacturing procedures (including methods of analysis to ensure drug quality), detailed formulation information for all dosing methods studied, and storage conditions. Regulatory reviews can lead to requests for additional information regarding the submission, or even additional clinical trials to further establish either safety or efficacy. Ideally, these reviews lead to regulatory approval, including labeling requirements, and approval to market the new drug.[38]

Approval of regulatory bodies does not, however, signal the end of clinical trials. In many cases, regulatory agencies will require additional follow-up studies, often referred to as phase IV trials or postmarketing surveillance. In general, these studies are designed to detect rare adverse effects across a much larger population of patients than could be supported in phase III trials or long-term adverse effects that might be outside of the scope of phase

III trial durations. The impact of phase IV studies can include alterations to labeling based on safety results, contraindication for use of the new drug in combination with other medications, or even the withdrawal of marketing approval if the findings are severe enough. The COX-2 selective non-steriodal anti-inflammatory agent Vioxx® (Rofecoxib),[39] for example, was removed from the market after phase IV studies indicated that it increased the risk of ischemic events in patients (Figure 1.8). In a similar fashion, Baycol® (cerivastatin),[40] a 3-hydroxy-3-methylglutaryl-coenzyme A (HMG-CoA) reductase inhibitor marketed by Bayer AG for the treatment of high cholesterol and cardiovascular disease, was voluntarily removed from the market after reports of fatal rhabdomyolysis (Figure 1.8).

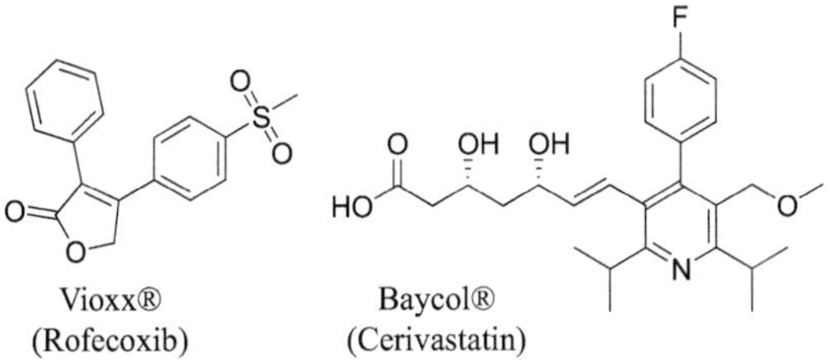

Vioxx®
(Rofecoxib)

Baycol®
(Cerivastatin)

FIGURE 1.8 Vioxx® (Rofecoxib) and Baycol® (Cerivastatin) were removed from the market as a result of an increased risk of ischemic events and fatal rhabdomyolysis respectively.

It is should be noted that safety studies are not the only reason for phase IV clinical trials. Companies often use the data provided in postmarketing surveillance and additional clinical studies to identify competitive advantages, new markets, and new indication for their products. There is some level of risk associated with conducting trials designed to identify clinical superiority, as the results of competitive trials are often difficult to predict. In some cases, a company's plan to demonstrate that their compound is superior to a competitor's drugs backfires, and they prove the opposite.

TARGET SELECTION: THE FIRST STEP FORWARD

The process of identifying a new drug candidate begins with identifying a disease state or condition that can be addressed or modified through the application of a suitable chemotherapeutic intervention. In theory, the most pressing medical needs would have the highest priority in order to ensure improvement of the overall quality of life for patients. In practice, however, there are many factors that contribute to the decision of which disease or condition to attempt to address through drug discovery programs. First, the pathway to develop a therapeutic intervention may not be clear for a particular disease or condition even though the medical need is urgent. For example, while it is clear that Alzheimer's disease is pressing medical need,[41] to date there are no disease modifying therapies available, despite the extraordinary amount of capital expended in an effort to identify useful therapies. This is in part due to the lack of proven targets for Alzheimer's disease. Similarly, while there is a clear and pressing need for additional treatments for schizophrenia,[42] the current level of understanding of the disease state and lack of sufficient animal models[43] is a hindrance to progress in this important area.

There are also drug targets that have a theoretical connection to a particular disease state, but have as yet to be proven relevant to the human condition through the application of an appropriate chemotherapeutic agent. Cholesteryl ester transfer protein (CETP), for example, plays a key role in the interconversion of high density lipoproteins (HDL) and low density lipoproteins (LDL), and it has been suggested that inhibitors of this enzyme would have a positive impact on patients suffering from hypercholesterolemia.[44] While potent CETP inhibitors have been identified, such as Torcetrapib (CP-529,414)[45] and Dalcetrapib (JTT-705)[46] (Figure 1.9), none have

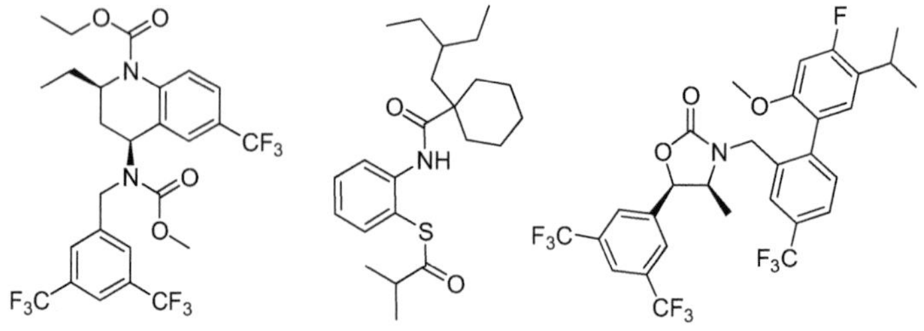

FIGURE 1.9 Torcetrapib (CP-529,414), Dalcetrapib (JTT-705), and Anacetrapib (MK-0859) are cholesteryl ester transfer protein (CETP) inhibitors that have been clinically studied as potential treatments for hypercholesterolemia. Torcetrapib increased HDL levels and decreased LDL levels, but increased mortality rates, while Dalcetrapib was not efficacious in clinical trials. Anacetrapib increased HDL levels and decreased LDL levels, and did not negatively impact mortality rates.

been approved for marketing as the clinical candidates examined to date failed to demonstrate statistically significant beneficial effects in patients. It is possible that these results are an indication that CETP inhibition is not a viable drug target for the treatment of cardiovascular disease. It is also possible, however, that the clinical candidates examined to date are flawed in ways unrelated to the CETP that prevented them from functioning in the desired manner (e.g., off-target effects, pharmacokinetic issues).

In the case of Torcetrapib (CP-529,414), clinical trial data demonstrated that the drug candidate increased HDL levels and decreased LDL levels, indicating that clinical efficacy could be achieved.[47] Unfortunately, the clinical candidate also caused increase in blood pressure and mortality rates, leading to the termination of clinical development in 2006.[48] Dalcetrapib (JTT-705) clinical studies were terminated by Roche in 2012 due to lack of efficacy.[49] On the other hand, clinical trials with Anacetrapib (MK-0859, Figure 1.9),[50] which also targets CETP, have successfully demonstrated that this compound can increase HDL levels and decrease LDL levels without increase in blood pressure or increased risk of cardiovascular disease-related deaths or events.[51] As of the writing of this text, the value of CETP as a drug target remains an unanswered question.

A similar scenario has surrounded γ-secretase inhibition. While it is known that γ-secretase plays a key role in the formation and deposition of amyloid plaques during the progression of Alzheimer's,[52] inhibitors of this key enzyme have failed to provide the clinically beneficial results expected. Semagacestat (LY450139, Figure 1.10), a compound developed

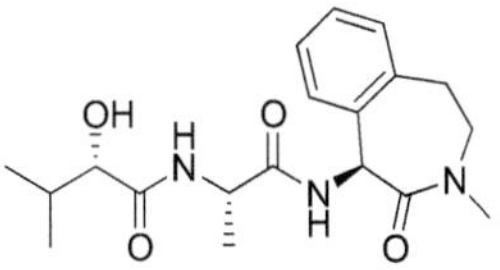

FIGURE 1.10 The γ-secretase inhibitor Semagacestat (LY450139) failed to improve cognitive function in Alzheimer's patients, despite the fact that it lowered amyloid plaque formation.

Semagacestat
(LY450139)

by Eli Lilly inhibits γ-secretase, shows a dose-dependent lowering of amyloid plaque formation in humans, but did not improve cognitive function in patients. In fact, Semagacestat produced statistically significant declines in cognitive function compared to the placebo group in clinical trials.[53] Once again, this raises the question as to whether the target pathway is a dead end for treatment of the disease in question or if the compound in question is flawed in some unforeseen manner. In the case of Semagacestat (LY450139), it is possible that unexpected off-target activity may be clouding the clinical results. Semagacestat (LY450139) also interferes with Notch signaling, which plays a key role in cognitive function,[54] and it is not unreasonable to suggest that Notch signaling modulation masked potential positive effects that might have been observed if this off-target activity was absent.

The unanswered question raised by the failures of clinical candidates such as Torcetrapib (CP-529,414) and Semagacestat (LY450139) highlights the risks associated with choosing a target that is not clinically proven, as well as the potential for negative clinical results to be clouded by factors not related to the targeted mechanism. There are, however, substantial financial incentives to attempt to develop a "first-in-class" therapeutic agent. Prior to the introduction of the statins (or HMG-CoA reductase inhibitors),[55] there was no clear pathway forward to inhibit cholesterol synthesis. The companies that took the leap of faith that inhibition of HMG-CoA reductase would provide therapeutic relief for the prevention of cardiovascular disease received significant financial rewards in the marketplace as indicated by the success of drugs like Mevacor® (Lovastatin)[56] and Lipotor® (Atorvastatin).[57]

There are, of course, many biological targets with proven clinical utility that an organization might choose to focus on, such as phosphodiesterase-5 (PDE-5),[58] β-adrenergic receptors,[59] and 5-hydroxytryptamine (5-HT) receptors.[60] In considering whether or not to pursue known drug targets, one must keep in mind both the benefits and potential pitfalls related to previously defined targets. On the positive sides, a wealth of research and development information will be available in the literature, as companies and research institutions (universities, non-profit research institutions, etc.) patent and publish their research in order to garner support for their marketed products and research programs. The availability of research tools such as biological assays, reference compounds, and clinical trial data can provide an excellent springboard for a drug discovery program. On the other hand, the availability of this kind of information presents a significant hurdle to the development of new therapeutic agents, as any new compound or biological agent will be required to demonstrate clinical superiority to the current standard of care. Also, scientific disclosures in the literature will be available as prior art and could prevent an organization from gaining patent protection for their research (this topic will be covered in more detail in Chapter 12). If, however, one is successful in developing a new therapeutic entity based on clinically proven targets, substantial benefits can be available. Sepracor, now a division of Sunovia, for example, took the risk of developing a new antihistamine at a time when the market was dominated by Seldane® (Terfenadine).[61] They were able to demonstrate that Allegra® (Fexofenadine), a metabolite of Seldane® (Terfenadine), is significantly safer than its predecessor and quickly took over the antihistamine market (Figure 1.11).[62]

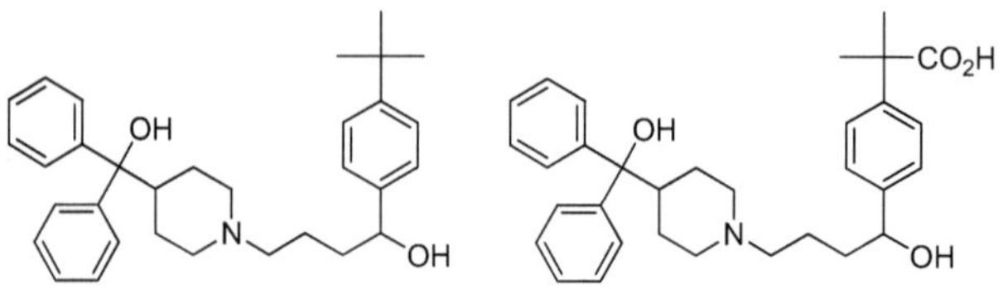

Seldane®   Allegra®
(Terfenadine)   (Fexofenadine)

FIGURE 1.11   Seldane® (Terfenadine), the first non-sedating antihistamine, dominated the market until serious safety issues were identified. It was replaced by a Allegra® (Fexofenadine), an active metabolite that is safer than the original.

Financial considerations also play a major factor in the determination of which diseases and potential drug targets are examined and which are not. Clearly, the amount of money and time available to pursue new therapeutic entities is limited, so not every target or disease state can be addressed. In the corporate world, disease state and target selection is generally driven by the ability to generate profitable products whose sale will support future research programs. On the surface, this would seem to dictate that only diseases or conditions with large numbers of patients would be of interest to corporate entities, but this is not the case. Chronic diseases such as osteoporosis, hypertension, hypercholestremia, and arthritis clearly have a large patient population that creates opportunities for corporations. Rare diseases, however, also present significant opportunities and a pathway for growth. Amyotrophic lateral sclerosis (ALS), for example, is a disease with a small, but consistent patient population with significant unmet medical needs. At any given time, there are only 20,000–30,000 ALS patients in the United States whose life expectancy is only 3–5 years, and there are no life extending therapies currently available.[63] This would appear to be a very small market that is unlikely to provide the kind of profitability required to sustain a corporation. However, it is important to realize that if a suitable treatment were available, this terminal condition would become a chronic condition wherein patients would be treated for the disease throughout the course of an otherwise normal life span. In addition, increased survival time for ALS patients would lead to a larger patient pool, providing additional revenue to a company that develops a life-extending treatment for ALS.

The selection of targets and disease states of interest sets the course for all future aspect of a research program, so the importance of this decision cannot be understated. Once the biological target is selected, the process of identifying a clinical candidate can begin.

HIT IDENTIFICATION: FINDING A STARTING POINT

Once a target of interest has been identified, the remainder of the research program is essentially a quest to identify a single compound that is suitable for use in a clinical setting. Of course, this relatively simple statement is actually a representation of an exceptionally complex and multifaceted problem. Currently, there are over 70 million compounds registered in the Chemical Abstract Service database,[64] and the total number of possible compounds to consider as drug candidates is nearly infinite, so the question of where to start the process is significant. Fortunately, there are some guidelines that have been developed in order to provide some guidance as to where one might begin to look for biologically useful molecules. Lipinski's rule of 5,[65] for example, suggests that the majority of druglike compounds exist within a limited portion of chemical space. This concept will be discussed in greater detail in Chapter 5, but for the purposes of this discussion, Lipinski's rules suggest that "druglike" compounds will have (1) a molecular weight lower than 500, (2) a logP below 5, (3) less than 5 hydrogen bond donors, (4) less than 10 hydrogen bond acceptors, and (5) less than 10 rotatable bonds. While there are exceptions to these rules (most notably in the natural products arena), their application to chemical space can be useful in that it provides a framework for further movement towards a manageable number of compounds for consideration.

However, these limitations still leave an enormous expanse of chemical space that could be mined in an effort to identify compounds that interact with a biological target of interest. This issue is further complicated by the fact that drugs interacting at the same target may have very little structural overlap. There are, for example, clear similarities between the HMG-CoA reductase inhibitors Lipitor® (Atorvastatin),[28] Lescol® (Fluvastatin),[67] and Crestor® (Rosuvastatin)[68] (Figure 1.12). They each contain a para-fluorbenzene

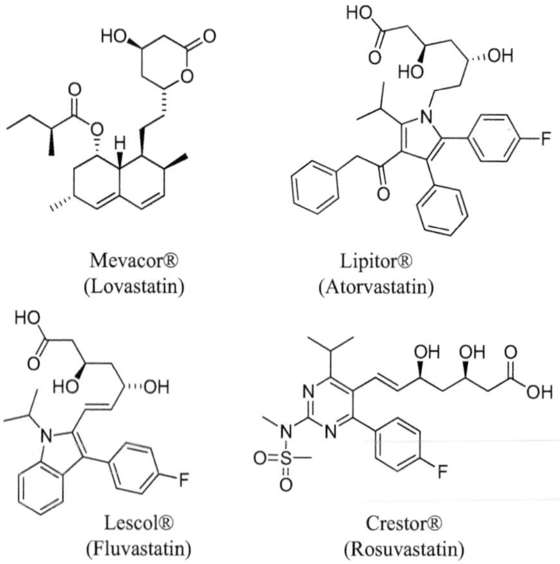

Mevacor®
(Lovastatin)

Lipitor®
(Atorvastatin)

Lescol®
(Fluvastatin)

Crestor®
(Rosuvastatin)

FIGURE 1.12   The HMG-CoA reductase inhibitors Mevacor® (lovastatin), Lipitor® (atorvastatin), Lescol® (Fluvastatin), and Crestor® (rosuvastatin) have some structural similarities, but there are a number of differences that make each unique.

ring and 1,3-diol-carboxlyic acid side chain displayed in a similar orientation, but the remainder of the three compounds are substantially different from each other. Mevacor® (Lovastatin, Figure 1.12),[66] which also inhibits HMG-CoA reductase, is from an entirely separate structural class, and it is not clear to the naked eye how this compound is related to the previously mentioned drugs. Similarly, Viagra® (Sildenafil)[69] and Cialis®

(Tadalafil)[70] are both PDE-5 inhibitors, but structurally, they are quite different (Figure 1.13). It is not immediately clear how these two compounds

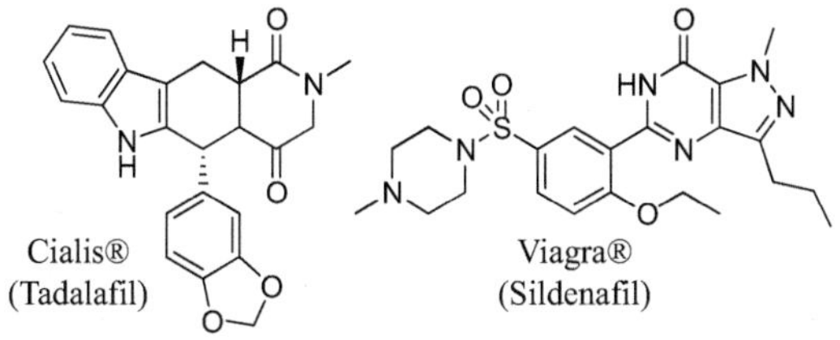

Cialis®
(Tadalafil)

Viagra®
(Sildenafil)

FIGURE 1.13   The PDE-5 inhibitors Cialis® (tadalafil) and Viagra® (sildenafil) are structurally dissimilar even though they interact with the same macromolecular target.

are related or why they would serve the same biological function. The same is true of morphine,[71] Demerol® (Meperidine)[72] and Duragesic® (Fentanyl) (Figure 1.14).[73] While all of these compounds are μ-opioid receptor agonists,

Morphine

Demerol®
(Meperidine)

Duragesic®
(Fentanyl)

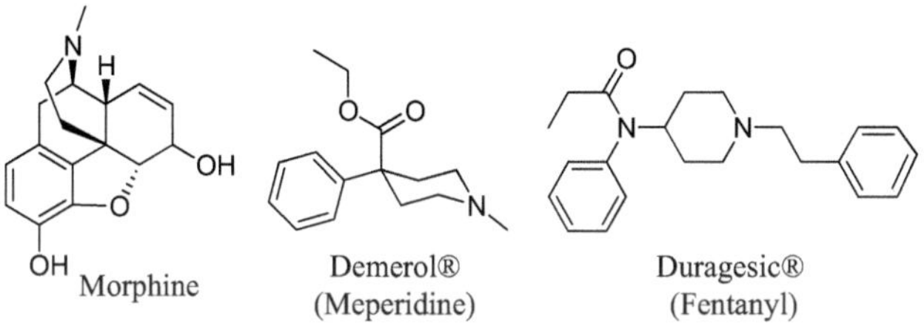

FIGURE 1.14   Morphine, Demerol® (Meperidine), and Duragesic® (Fentanyl) are all μ-opioid receptor agonists.

it would not be obvious to a casual observer that they share a common biological target. The selective serotonin reuptake inhibitors (SSRIs) Zoloft® (Sertraline),[74] Zelmid® (Zimeldine),[75] Celexa® (Citalopram),[76] Prozac® (Fluoxetine),[77] and Paxil® (Paroxetine)[78] represent distinct chemical classes that are not clearly related to each other (Figure 1.15). Given the breadth of

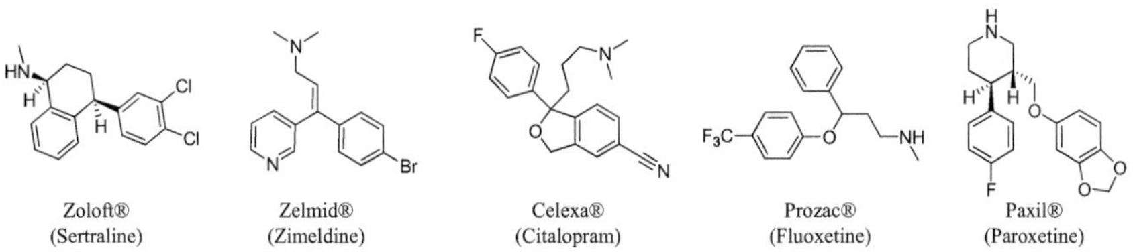

Zoloft®
(Sertraline)

Zelmid®
(Zimeldine)

Celexa®
(Citalopram)

Prozac®
(Fluoxetine)

Paxil®
(Paroxetine)

FIGURE 1.15   Zoloft® (Sertraline), Zelmid® (Zimeldine), Celexa® (Citalopram), Prozac® (Fluoxetine), and Paxil® (Paroxetine) are all selective serotonin reuptake inhibitors (SSRIs) useful for the treatment of depression, but structural similarities are limited.

structural diversity that can be employed for any single molecular target, it is clear that even the process of finding an initial chemical lead for a drug discovery program can be challenging.

Fortunately, a number of tools and methods have been developed to address the simple and yet very complex question of identifying a molecular starting point for a drug discovery program. Essentially, there are two general methods utilized in modern drug discovery programs, physical high throughput screening (HTS) methods[79] and virtual high throughput screening methods.[80] There is some degree of overlap between the two categories, and the use of one set of tools does not preclude the use of the other. In point of fact, they are often employed in tandem in order to improve the likelihood of success. Physical high throughput screening approaches depend on the ability to screen large compound libraries containing hundreds, thousands, or even millions of samples. Large libraries are often designed to be chemically diverse in order to cover as much of the "drug-like" chemical space as possible, but focused libraries designed to target specific types of biological targets (e.g., kinases, phosphatases) have also been employed. Physical samples for screening are available from commercial sources (e.g., Maybridge, Enamine, Aldrich, etc.), and pharmaceutical companies generally maintain an internal compound collection of proprietary compounds.

Physical HTS techniques also require sophisticated, fully automated systems capable of manipulating reagents and 96-, 384-, or even 1496-well microtiter plates, as well as reagent distribution, data acquisition, and waste disposal for thousands of samples per hour. Automated data analysis is also required in order to handle the volume of information generated in a typical high throughput screening run.

There are some key points that one must consider in evaluating the data provided by an HTS screen. First and foremost is the possibility of false positives and false negative results. In physical screening methods, the sheer number of manipulations involved leaves open the possibility that an error may occur during some facet of reagent handling (such as a clogged pipette tip). There is also the possibility that the screening sample may have degraded over time, creating "ghost samples" within the chemical library (in other words, a sample whose structure no longer matches the material originally entered into the library). In order to ensure that programs are directed towards authentically active compounds, the chemical integrity of "hit" samples is generally assessed using High-performance liquid chromatography/Mass spectroscopy (HPLC/MS) methods. In addition, biological screening is often repeated with the "hit" compounds in order to validate the HTS results.

As an alternative to physical HTS methods, it is also possible to perform virtual high throughput screening (also referred to as *in silico* screening). In this scenario, advanced molecular modeling techniques are combined with virtual chemical libraries (data files containing detailed structural information on millions of compounds) and structural data on the biological target in order to assess a compound's ability to interact with the target of interest. Virtual chemical libraries are often freely available from commercial vendors (the largest of which is the ZINC database; http://zinc.docking.org/) and, as with physical samples, pharmaceutical companies generally maintain virtual libraries of their proprietary compounds for internal use. Structural information on biological targets may be available through X-ray crystallography, as a large number of protein crystal structures are available through the Research Collaboratory for Structural Bioinformatics (RCSB) Protein Data Bank (http://www.rcsb.org/pdb/home/home.do). If a structure is not readily available, it may be possible to create a homology model of the biological target using crystal structure data of a closely related macromolecules.[81] In either case, the individual compounds of the chemical libraries can then be "docked" in a hypothetical binding site in the target of interest to determine a relative rank order for the entire set of compounds. Automated data analysis tools are then employed that organize the predictions provided by the "docking" of the chemical libraries to the hypothetical binding sites of the biological targets. The predictions can then be used to select a smaller subset of a large library for physical biological screening as potential starting points.

Much like physical HTS, there are some important limitations that must be considered in evaluating virtual screening data. First and foremost, virtual screening results are predictions based on model system and not actual data on physical compounds. As such, the quality of the results will depend on the quality of the model. *In silico* models based on X-ray crystal structures tend to be stronger models than homology models built on related biological structures, but it is important to realize that there are limitations associated with X-ray crystal-based models as well. Crystal structures can provide exceptionally detailed structural information, with resolution as low as 1.5 Å, but by definition, X-ray crystal structures are solid state version of the desired target. It is possible that the structure provided by X-ray crystallography matches the biologically active form of the biological target of interest, but it may not. In "real-life" situations, biological targets are either dissolved in water or membrane bound, and it is possible that they may have a different configuration in these situations as compared to the close-packed structure of a crystal form. Also, in many cases, sections or a macromolecule must be altered or removed in order to generate a crystallizable form of the biological target (with or without a ligand). Given these limitations, virtual "hits" should also be physically validated in biological screening efforts in order to confirm that the predictions provided by molecular modeling are representative of the real system.

Irrespective of the initial screening method employed (physical or virtual), a successful screening effort will produce a subset of potential "hit" compounds that will need to be examined in order to determine whether or not follow-up efforts are warranted. This determination, also referred to as "lead discovery," (Figure 1.7) can be quite complex in itself depending on the number and nature of the "hits" identified. If 500,000 compounds are screened and only 0.1% of the compounds in the library provide interesting biological results, this still leaves 500 compounds to be evaluated. Ideally, the initial "hits" will belong to a relatively small number of structural classes, and then each structural class can be independently analyzed to determine if further effort in the class is warranted. Small groups of related compounds demonstrating the desired biological activity can provide a significant advantage in further efforts, as structure–activity relationship data may become apparent at an early stage. (The concept of structure–activity relationships will be covered in more detail in Chapter 5.) Also, the preparation of additional analogs may be simplified, as synthetic methods may already be available. On the other hand, the presence of set of related compounds within a library suggests that they may have been prepared for a project with a different biological target. Intellectual property issues may also exist, as patent rights and ownership could become a serious question, especially if the compounds were part of a set that has been previously patented, previously published, or purchased from a commercial vendor. Intellectual property consideration will be explored in more detail in Chapter 12.

In some instances, "hit" compounds may be singletons. Isolated compounds can be more difficult to follow up on, as the original HTS data set will not provide any additional guidance on how to proceed. It is, however, still possible to generate more data on related compounds that may be available from outside of the original compound library through either commercial sources or additional synthetic efforts.

Once the initial "hit" compounds have been identified, confirmed, and a compound class (or perhaps more than one compound class, depending on the available resources) has been selected for further study, an iterative process of compound acquisition/synthesis, biological screening, and data evaluation begins with the goal of improving the potency of the compounds (Figure 1.16). In each cycle of the "lead optimization" process,

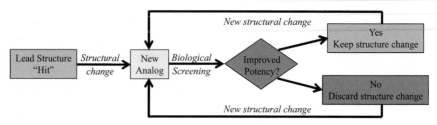

FIGURE 1.16 The lead optimization cycle begins with the identification of a lead structure ("hit") in a relevant biological assay. New analogs with structural modifications are prepared and screened in the biological assay. If the assay results improve, then the changes are kept and the cycle is repeated. If the changes are detrimental, then the changes are discarded and the cycle is repeated. This process continues until a candidate compound with the desired properties is identified.

new data are produced as changes in the molecular structure are made to the "lead" compounds, and these data are used to design the next generation of compounds. This cycle of generating structure–activity relationship data continues until a compound suitable for clinical evaluation has been identified. The nature of this process and the associated medicinal chemistry will be discussed in greater detail in Chapter 5.

IDENTIFY A CLINICAL CANDIDATE: JUGGLING THE PROPERTIES

Simply identifying a compound that is potent at the target of interest is a difficult task to begin with, but shear potency is not enough to allow a compound to be considered for clinical development and ultimately commercialization. The process of discovering a suitable drug candidate is, in many ways, a juggling act performed by drug discovery scientists (Figure 1.17). As programs progress from hit and lead identification to

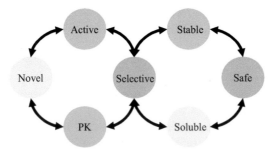

FIGURE 1.17 The identification of a clinical candidate requires consideration of a variety of properties beyond activity at the biological target of interest. Drug discovery and development programs seek to optimize as many of these properties as possible in order to identify the best opportunity for success.

lead optimization and an eventual clinical study, hundreds if not thousands of compounds will be examined. It is the drug discovery scientist's responsibility to identify a compound that will not only modulate the target of interest but also possesses the correct balance of properties required to create a usable drug. Potency at a biological target is only the beginning of a long series of screening processes that must be performed in order to demonstrate that a compound will survive the rigors of a discovery program. The specific strategies employed are different for each program, but in general they can be mapped in a screening cascade (Figure 1.18) that sets gating guidelines for each level of the screening process, from initial activity screening through *in vivo* animal efficacy studies. The screening cascade is designed to decrease the number of

FIGURE 1.18 A screening tree is designed to identify lead compounds by establishing a series of qualifications or "gates" that a compound must surpass in order to advance through the process.

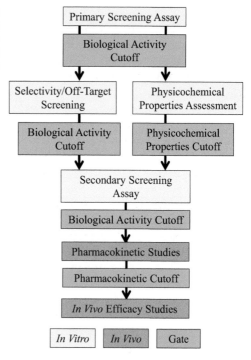

compounds examined at each level in order to ensure that compounds with flaws are removed as early as possible. The cascade, also referred to as a screening tree, begins with *in vitro* profiling and then transitions into *in vivo* studies designed to determine a compound's pharmacokinetic profile and demonstrate efficacy in an appropriate animal model.

At the top of the cascade, compounds are screened for activity against the biological target and a threshold of interest is generally set to determine if compounds are active enough to warrant further investigation. Potency is, of course, an important issue, as dosing requirements are lower for compounds that are more potent. All other things being equal, compounds with higher potency can be dosed at lower levels, decreasing the likelihood of side effects. A compound with target potency of 5 nM in theory could be provided to a patient at a significantly lower dose than a compound serving the same function but with a potency of 5 μM.

Once a compound has satisfied the potency criteria, selectivity and physicochemical properties criteria are typically examined. Nature has developed exquisite systems to accomplish very specific tasks with highly selective systems, but many of these systems overlap structurally, and this can have a significant impact on the biological properties of a given test compound. Thus, the next biological screening step in a typical screening cascade is often an assessment of a compound's potency at biological systems that are closely related to the target of interest. The Kv1.5 channel, for example, is a voltage-gated potassium channel that has been the target of research programs atrial arrhythmia, and many compounds have been identified that can block this channel with a high level of potency.[82] There are, however, over 70 other voltage-gated potassium channels with varying degrees of similarity to the Kv1.5 channel, and undesired activity at any of these related channels could create unwanted side effects in human or animal studies. For example, the Kv1.5 channel is closely related to the hERG channel. Blockade of the hERG channel has been linked to torsade de pointes and sudden cardiac death,[83] so any compound moving forward in this area would need to be counterscreened for hERG activity in order to ensure that advancing compounds do not present a risk of sudden cardiac death in a clinical setting. This is a rather extreme example of the importance of proper selectivity, but it should be clear that failure to achieve proper target selectivity in this area represents a significant barrier to moving a program forward.

Similarly, there are over 500 known kinase enzymes,[84] and any drug discovery program designed to target a single kinase, or even a family of kinases, has an associated risk of identifying compounds that are active at multiple members of this large family of related enzymes. In order to mitigate this risk, kinase programs routinely screen test compounds against panels of related kinases in order to understand the risks associated with off-target activity.

In general, compounds that are potent at the target of interest, but are also potent at a variety of other targets ("promiscuous compounds"), do not move forward in a drug discovery program, as the risk of undesired (or unpredicted) side effects is too high. The level of selectivity required, however, is dependent on the program, the nature of the potential side effect presented by off-target activity (off-target activity leading to excessive hair growth might be tolerable, whereas sudden cardiac death through poor hERG selectivity is not), the target patient population, duration of treatment (some side effects only appear upon extended exposure to a drug), and a variety of other factors. Overall, target selectivity is a major factor to consider.

An active and suitably selective compound, however, is not necessarily a good drug candidate. Physicochemical properties also play a major role in determining whether or not a compound is suitable for further investigation. *In vitro* screens designed to predict absorption, distribution, metabolism, and excretion (*in vitro* ADME) are generally performed early in a program in order to ensure that candidates reaching the drug development pipeline are "druglike" in nature (Figure 1.19). Compounds that have poor aqueous solubility, for instance, are often difficult to develop as drugs. In order for a drug to exert an influence on a biological target, it must be soluble in biological fluid at a level consistent with its potency. Thus, the level of

FIGURE 1.19   An *in vitro* ADME profile can be used to identify compounds that have "druglike" properties. Assays are available to determine metabolic stability, plasma stability, aqueous solubility, Pgp efflux susceptibility, solution stability, Cyp450 inhibition, bioassay solubility, blood–brain barrier penetration, and permeability.

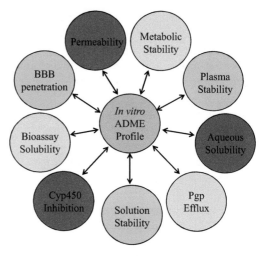

solubility required for a given compound is directly linked to its potency. As potency increases, the requisite solubility decreases, as less drug is required to provide the intended effect. Solubility also has a direct impact on absorption, as a compound must be soluble in biological fluids in order for it to successfully pass through a biological membrane and reach its intended target.

The ability of a compound to penetrate cellular membrane (its permeability) can also be a determining factor in the success or failure of a given candidate compound. If a compound is potent, selective, and soluble, but unable to pass through a biological membrane, it may not be able to reach the target of interest and fail to demonstrate the desired efficacy. Orally active drugs must be absorbed in the gastrointestinal tract, and drugs that target intracellular system must also pass through the cell membrane in order to reach their intended targets. Extracellular targets, of course, do not face this added issue, but CNS drug candidates face the added complexity of required permeability through the blood–brain barrier (BBB). Additionally, there are efflux pumps (e.g., P-glycoproteins (Pgps))[85] designed to remove xenobiotic material that can limit permeability, preventing efficacy. The inability of a compound to penetrate a cell membrane represents a significant issue that could prevent further investigation of the candidate compound.

Metabolic and chemical stability are also important considerations. If all of the previously mentioned criteria are met, and a compound is able to enter the body, but is immediately metabolized, efficacy studies will fail to show the desired results. However, the relative rate of metabolism that can be allowed for a successful compound depends on the goals of the project. If, for example, the goal is to develop a new antibacterial agent, then high-metabolic stability will likely be desired so that the potential drug candidate will be available in the circulation long enough to kill the invading organism. If, however, the goal is to develop a new surgical anesthetic, metabolic stability may be less of an issue, as it may be desirable for drug efficacy to fade rapidly upon termination of dosing regimens.

In a related sense, compounds that have chemical stability issues may be problematic as drugs. Special packaging systems, some as simple as amber bottles for light-sensitive compounds or cold storage, may be required in order for the drug to be available commercially when a patient is in need. While these kinds of issues are not insurmountable, generally speaking, more chemically stable compounds are preferred.

It is also important to consider how candidate compounds may impact the normal metabolic processes, potentially altering the metabolism of drug products used in tandem with the candidate compound. Inhibition of key metabolic enzymes in the liver, such as Cyp3A4, Cyp2D6, and Cyp2C9, members of the cytochrome P450 (Cyp450) family of metabolic enzymes, are often studied using *in vitro* screening methods (liver microsomes) in order to determine the risk of drug–drug interactions.[86] A compound that meets all other *in vitro* criteria and demonstrates efficacy, may still fail as a drug candidate if it is determined that there is significant risk of drug–drug interactions. The withdrawal of Seldane[61,87] from market is a classic example of the risks associated with unintended inhibition of the normal metabolic processes, and is discussed in greater detail in Chapter 13.

Positive results through the *in vitro* screening portion of a discovery program represent a significant accomplishment, but are still not necessarily indicative of success. The pharmacokinetic properties (PK) of a candidate compound must be determined in order to answer key questions about the *in vivo* fate of the potential drug candidate. For example, if the candidate compound is dosed orally, what percentage of the oral dose actually reaches the systemic circulation? How rapidly is the candidate compound excreted or metabolized? Does the compound reach systemic concentrations high enough to suggest that *in vivo* efficacy should be expected in an animal model? Is the compound freely distributed through the body, or does it concentrate in a particular organ or tissue type? The answer to these and a number of similar questions will have a significant impact on the ability of any given compound to provide the desired *in vivo* efficacy in a given animal model. Irrespective of the positive results of *in vitro* screening, compounds with poor PK profiles are not likely to be successful drugs.

Compounds found to possess suitable PK profiles must, of course, demonstrate activity in key animal efficacy trials before they can be considered for clinical study. The type of efficacy studies required is based on the desired biological endpoint (disease state), and a full discussion of *in vivo* efficacy models is well beyond the scope of this text. Some examples are given in chapter 7, but it should be clear that the ultimate goal of a discovery program is to identify compounds that meet all of the aforementioned *in vitro* criteria, demonstrate efficacy in the appropriate animal model, and have PK properties consistent with the desired dosing regimen.

Safety and side-effect profiles are also major concerns, and there are a number of *in vitro* and *in vivo* screens that can be used to assess the risks associated with a compound (e.g., *in vitro* hERG screening,[88] Ames mutagenicity screening,[89] dog cardiovascular safety assessment[90]). The nature and scope of safety studies is well beyond the scope of this text, but should always be a major concern in the minds of drug discovery scientist as any project moves forward. It is also important to realize that the side-effect profiles for a potential clinical candidate are somewhat dependent on the intended use. For example, compounds used to prevent life-threatening illness, such as cancer, AIDS, and ALS, may be given more latitude with their side-effect profiles, given the severity of the illness. On the other hand, treatments designed for chronic use or non-life-threatening conditions, such as osteoarthritis or neuropathic pain, must be scrutinized for possible safety issues or side effects. The concern for safety bridges through all aspects of discovery and development of novel therapeutics. It is impossible to guarantee that compounds entering clinical development will be safe, but compounds with "red flags" in safety screens are generally not pursued as drugs.

Finally, the ability to identify patentable compounds will also gate the progress of any drug discovery program. As mentioned earlier, drug discovery and development is an exceptionally expensive endeavor. Market exclusivity through patent protection provides the necessary financial incentive required for companies to invest in new drug development. Compounds and compound classes that cannot be protected through the issuance of patents are unlikely to be pursued by private organizations, as recouping the significant investments required to move a compound into clinical use becomes challenging. Further information regarding the importance of patent protection in the pharmaceutical industry is provided in Chapter 12.

Drug discovery scientists walk on the edge of several precarious slopes in attempting to identify potential new therapeutic entities. Balancing the needs of potency, selectivity, solubility, stability, pharmacokinetics, safety, and novelty is critical to the success of any project, and failure to deliver in any one of these areas can terminate the forward progression of a test compound.

QUESTIONS

1. What are the three major phases of drug discovery?
2. What are the four major phases of drug development?
3. Describe the lead optimization cycle.
4. What is a screening cascade (also referred to as a screening tree)?
5. Why is compound selectivity an important aspect of drug discovery?

6. What does the term *in vitro* ADME refer to?

7. Name five properties that are a part of a compound's *in vitro* ADME profile.

References

1. a. Weiss, R. A. How Does HIV Cause AIDS? *Science* **1993,** *260* (5112), 1273–1279.
   b. Douek, D. C.; Roederer, M.; Koup, R. A. Emerging Concepts in the Immunopathogenesis of AIDS. *Annu. Rev. Med.* **2009,** *60,* 471–484.
2. a. Gallo, R. C.; Sarin, P. S.; Gelmann, E. P.; Robert-Guroff, M.; Richardson, E.; Kalyanaraman, V. S.; Mann, D.; Sidhu, G. D.; Stahl, R. E.; Zolla-Pazner, S.; et al. Isolation of Human T-cell Leukemia Virus in Acquired Immune Deficiency Syndrome (AIDS). *Science* **1983,** *220* (4599), 865–867.
   b. Barre-Sinoussi, F.; Chermann, J.; Rey, F.; Nugeyre, M.; Chamaret, S.; Gruest, J.; Dauguet, C.; Axler-Blin, C. Isolation of a T-lymphotropic Retrovirus from a Patient at Risk for Acquired Immune Deficiency Syndrome (AIDS). *Science* **1983,** *220* (4599), 868–871.
3. a. Nakashima, H.; Matsui, T.; Harada, S.; Kobayashi, N.; Matsuda, A.; Ueda, T.; Yamamoto, N. Inhibition of Replication and Cytopathic Effect of Human T Cell Lymphotropic Virus Type III/lymphadenopathy-associated Virus by 3'-azido-3'-deoxythymidine *In vitro*. *Antimicrob. Agents Chemother.* **1986,** *30* (6), 933–937.
   b. Birnbaum, G. I.; Giziewicz, J.; Gabe, E. J.; Lin, T. S.; Prusoff, W. H. Structure and Conformation of 3'-azido-3'-deoxythymidine (AZT), an Inhibitor of the HIV (AIDS) Virus. *Can. J. Chem.* **1987,** *65* (9), 2135–2139.
   c. Global HIV/AIDS Timeline, Kaiser Family Foundation, http://www.kff.org/hivaids/timeline/hivtimeline.cfm.
4. a. Holy, A.; Dvorakova, H.; Declercq, E. D. A.; Balzarini, J. M. R. *Preparation of Antiretroviral Enantiomeric Nucleotide Analogs;* 1994. WO9403467.
   b. Deeks, S. G.; Barditch-Crovo, P.; Lietman, P. S.; Hwang, F.; Cundy, K. C.; Rooney, J. F.; Hellmann, N. S.; Safrin, S.; Kahn, J. O. Safety, Pharmacokinetics, and Antiretroviral Activity of Intravenous 9-[2-(R)-(phosphonomethoxy)propyl]adenine, a Novel Anti-human Immunodeficiency Virus (HIV) Therapy, in HIV-infected Adults. *Antimicrob. Agents Chemother.* **1998,** *42* (9), 2380–2384.
5. a. Soudeyns, H.; Yao, X. I.; Gao, Q.; Belleau, B.; Kraus, J. L.; Nguyen-Ba, N.; Spira, B.; Wainberg, M. A. Anti-human Immunodeficiency Virus Type 1 Activity and *In vitro* Toxicity of 2'-deoxy-3'-thiacytidine (BCH-189), a Novel Heterocyclic Nucleoside Analog. *Antimicrob. Agents Chemother.* **1991,** *35* (7), 1386–1390.
   b. Coates, J. A. V.; Mutton, I. M.; Penn, C. R.; Storer, R.; Williamson, C. *Preparation of 1,3-oxathiolane Nucleoside Analogs and Pharmaceutical Compositions Containing Them;* 1991. WO9117159.
6. a. Jungheim, L. N.; Shepherd, T. A. *Preparation of HIV Protease Inhibitors and Their (Aminohydroxyalkyl)piperazine Intermediates;* , 1995. WO9521164.
   b. Kaldor, S. W.; Kalish, V. J.; Davies, J. F., II; Shetty, B. V.; Fritz, J. E.; Appelt, K.; Burgess, J. A.; Campanale, K. M.; Chirgadze, N. Y.; Clawson, D. K.; et al. Viracept (Nelfinavir Mesylate, AG1343): A Potent, Orally Bioavailable Inhibitor of HIV-1 Protease. *J. Med. Chem.* **1997,** *40* (24), 3979–3985.
7. a. Kempf, D. J.; Norbeck, D. W.; Sham, H. L.; Zhao, C.; Sowin, T. J.; Reno, D. S.; Haight, A. R.; Cooper, A. J. *Preparation of Peptide Analogs as Retroviral Protease Inhibitors;* 1994. WO9414436.
   b. Markowitz, M.; Mo, H.; Kempf, D. J.; Norbeck, D. W.; Bhat, T. N.; Erickson, J. W.; Ho, D. D. Selection and Analysis of Human Immunodeficiency Virus Type 1 Variants with Increased Resistance to ABT-538, a Novel Protease Inhibitor. *J. Virol.* **1995,** *69* (2), 701–706.

8. Dorsey, B. D.; Levin, J. R. B.; McDaniel, S. L.; Vacca, J. P.; Guare, J. P.; Darke, P. L.; Zugay, J. A.; Emini, E. A.; Schleif, W. A.; Quintero, J. C.; Lin, J. H.; Chen, W.; Holloway, M. K.; Fitzgerald, P. M. D.; Axel, M. G.; Ostovic, D.; Anderson, P. S.; Huff, J. R. "L-735,524: The Design of a Potent and Orally Bioavailable HIV Protease Inhibitor". *J. Med. Chem.* **1994,** 37 (21), 3443–3451.

9. a. *Antiretroviral Therapy for HIV Infection in Adults and Adolescents: Recommendations for a Public Health Approach;* World Health Organization, 2010.

b. Autran, B.; Carcelain, G.; Li, T. S.; Blanc, C.; Mathez, D.; Tubiana, R.; Katlama, C.; Debré, P.; Leibowitch, J. Positive Effects of Combined Antiretroviral Therapy on CD4+ T Cell Homeostasis and Function in Advanced HIV Disease. *Science* **1997,** 277 (5322), 112–116.

c. Kroon, F. P.; Rimmelzwaan, G. F.; Roos, M. T.; Osterhaus, A. D.; Hamann, D.; Miedema, F.; van Dissel, J. T. Restored Humoral Immune Response to Influenza Vaccination in HIV-infected Adults Treated with Highly Active Antiretroviral Therapy. *AIDS* **1998,** 12 (17), F217–F223.

10. O'Neal, R. Rilpivirine and Complera: New First-line Treatment Options. *BETA Bull. Exp. Treat. AIDS Publ. San Francisco AIDS Found.* **2011,** 23 (4), 14–18.

11. U.S. FDA Approves Gilead's Stribild™, a Complete Once-daily Single Tablet Regimen for Treatment-naïve Adults with HIV-1 Infection, Gilead Sciences, Inc. Press Release 1728981. http://investors.gilead.com/phoenix.zhtml?c=69964&p=irol-newsArticle&ID=1728981

12. May, M. T.; Ingle, S. M. Life Expectancy of HIV-positive Adults: A Review. *Sex. Health* **2011,** 8 (4), 526–533.

13. Howlader, N.; Noone, A. M.; Krapcho, M.; Neyman, N.; Aminou, R.; Altekruse, S. F.; Kosary, C. L.; Ruhl, J.; Tatalovich, Z.; Cho, H.; et al. SEER Cancer Statistics Review, 1975-2009, Vintage 2009 Populations. *Natl. Cancer Inst.* **2012.** Bethesda, MD.

14. a. Manfredi, J. J.; Horwitz, S. B. Taxol: An Antimitotic Agent with a New Mechanism of Action. *Pharmacol. Ther.* **1984,** 25 (1), 83–125.

b. Donehower, R. C.; Rowinsky, E. K. An Overview of Experience with TAXOL (Paclitaxel) in the U.S.A. *Cancer Treat. Rev.* **1993,** 19 (Suppl. C), 63–78.

15. Johnson, I. S.; Armstrong, J. G.; Gorman, M.; Burnett, J. P., Jr. The Vinca Alkaloids: A New Class of Oncolytic Agents. *Cancer Res.* **1963,** 23, 1390–1427.

16. Tan, C.; Etcubanas, E.; Wollner, N.; Rosen, G.; Gilladoga, A.; Showel, J.; Murphy, M. L.; Krakoff, I. H. Adriamycin – An Antitumor Antibiotic in the Treatment of Neoplastic Diseases. *Cancer* **1973,** 32 (1), 9–17.

17. a. Boehm, J. C.; Johnson, R. K.; Hecht, S. M.; Kingsbury, W. D.; Holden, K. G. *Preparation, Testing, and Formulation of Water Soluble Camptothecin Analogs as Antitumor Agents;* 1989. EP321122.

b. Kingsbury, W. D.; Boehm, J. C.; Jakas, D. R.; Holden, K. G.; Hecht, S. M.; Gallagher, G.; Caranfa, M. J.; McCabe, F. L.; Faucette, L. F.; Johnson, R. K.; et al. Synthesis of Water-soluble (Aminoalkyl)camptothecin Analogs: Inhibition of Topoisomerase I and Antitumor Activity. *J. Med. Chem.* **1991,** 34 (1), 98–107.

18. a. Zimmermann, J.; Buchdunger, E.; Mett, H.; Meyer, T.; Lydon, N. B. Potent and Selective Inhibitors of the ABL-kinase: Phenylaminopyrimidine (PAP) Derivatives. *Bioorg. Med. Chem. Lett.* **1997,** 7 (2), 187–192.

b. Druker, B. J.; Lydon, N. B. Lessons Learned from the Development of an Abl Tyrosine Kinase Inhibitor for Chronic Myelogenous Leukemia. *J. Clin. Invest.* **2000,** 105 (1), 3–7.

19. a. O'Hare, T.; Walters, D. K.; Deininger, M. W. N.; Druker, B. J. AMN107: Tightening the Grip of Imatinib. *Cancer Cell* **2005,** 7 (2), 117–119.

b. Breitenstein, W.; Furet, P.; Jacob, S.; Manley, P. W. *Preparation of Pyrimidinylaminobenzamides as Inhibitors of Protein Kinases, in Particular Tyrosine Kinases for Treating Neoplasm, Especially Leukemia;* 2004. WO2004005281.

20. a. Kim, T. E.; Murren, J. R. Erlotinib (OSI/Roche/Genentech). *Curr. Opin. Invest. Drugs* **2002,** 3 (9), 1385–1395.

b. Schnur, R. C.; Arnold, L. D. *Preparation of N-phenylquinazoline-4-amines as Neoplasm Inhibitors;* 1996. WO9630347.

21. Gombos, E. A.; Freis, E. D.; Moghadam, A. Effects of MK-870 in Normal Subjects and Hypertensive Patients. *N. Engl. J. Med.* **1966,** *275* (22), 1215–1220.

22. Campbell, D. B.; Phillips, E. M. Short Term Effects and Urinary Excretion of the New Diuretic, Indapamide, in Normal Subjects. *Eur. J. Clin. Pharmacol.* **1974,** *7* (6), 407–414.

23. Hansson, L.; Aberg, H.; Jameson, S.; Karlberg, B.; Malmcrona, R. Initial Clinical Experience with I.C.I. 66.082 [4-(2-hydroxy-3-isopropylaminopropoxy)phenylacetamide], a New β-adrenergic Blocking Agent, in Hypertension. *Acta Med. Scand.* **1973,** *194* (6), 549–550.

24. Shanks, R. G.; Wood, T. M.; Dornhorst, A. C.; Clark, M. L. Some Pharmacological Properties of a New Adrenergic Beta-receptor Antagonist. *Nature* **1966,** *212* (5057), 88–90.

25. Ondetti, M. A.; Rubin, B.; Cushman, D. W. Design of Specific Inhibitors of Angiotensin-Converting Enzyme: New Class of Orally Active Antihypertensive Agents. *Science* **1977,** *196* (4288), 441–444.

26. Patchett, A. A.; Harris, E.; Tristram, E. W.; Wyvratt, M. J.; Wu, M. T.; Taub, D.; Peterson, E. R.; Ikeler, T. J.; Ten Broeke, J.; Payne, L. G.; et al. A New Class of Angiotensin-Converting Enzyme Inhibitors. *Nature* **1980,** *288* (5788), 280–283.

27. Ray, K. K.; Seshasai, S. R. K.; Erqou, S.; Sever, P.; Jukema, J. W.; Ford, I.; Sattar, N. Statins and All-cause Mortality in High-risk Primary Prevention: A Meta-analysis of 11 Randomized Controlled Trials Involving 65,229 Participants. *Archives Intern. Med.* **2010,** *170* (12), 1024–1031.

28. Roth, B. D. *Preparation of Anticholesteremic (R-(R*R*))-2-(4-fluorophenyl)-β,γ-dihydroxy-5-(1-methylethyl-3-phenyl-4((phenylamino)carbonyl)-1H-pyrrolyl-1-heptanoic Acid, its Lactone Form and Salts Thereof;* 1991. EP409281.

29. Olsson, A. G.; Molgaard, J.; von Schenk, H. Synvinolin in Hypercholesterolaemia. *Lancet* **1986,** *2* (8503), 390–391.

30. Lewington, S.; Whitlock, G.; Clarke, R.; Sherliker, P.; Emberson, J.; Halsey, J.; Qizilbash, N.; Peto, R.; Collins, R. Blood Cholesterol and Vascular Mortality by Age, Sex, and Blood Pressure: A Meta-analysis of Individual Data from 61 Prospective Studies with 55,000 Vascular Deaths. *Lancet* **2007,** *370* (9602), 1829–1839.

31. Add Notation for Medicinal Chemistry Text.

32. a. Salomone, S.; Caraci, F.; Leggio, G. M.; Fedotova, J.; Drago, F. New Pharmacological Strategies for Treatment of Alzheimer's Disease: Focus on Disease Modifying Drugs. *Br. J. Clin. Pharmacol.* **2012,** *73* (4), 504–517.

b. Mondragon-Rodriguez, S.; Perry, G.; Zhu, X.; Boehm, J. Amyloid Beta and Tau Proteins as Therapeutic Targets for Alzheimer's Disease Treatment: Rethinking the Current Strategy. *Int. J. Alzheimer's Dis.* **2012,** 1–7.

33. Walsh, C. T.; Fischbach, M. A. New Ways to Squash Superbugs. *Sci. Am.* **2009,** *301* (1), 44–51.

34. Johnsson, L. *With New Generic Rivals, Lipitor's Sales Halved.* The Associated Press, December 19, 2011. http://www.businessweek.com/ap/financialnews/D9RNTN0O0.htm.

35. Wong, D. T.; Perry, K. W.; Bymaster, F. P. The Discovery of Fluoxetine Hydrochloride (Prozac). *Nat. Rev. Drug Discov.* **2005,** *4,* 764–774.

36. The Associated Press. *Merck Shares Climb after Drug Trial Ends Early;* July 12, 2012. http://www.businessweek.com/ap/2012-07-12/merck-shares-climb-after-drug-trial-ends-early.

37. Paul, S. M.; Mytelka, D. S.; Dunwiddie, C. T.; Persinger, C. C.; Munos, B. H.; Lindborg, S. R.; Schacht, A. L. How to Improve R&D Productivity: The Pharmaceutical Industry's Grand Challenge. *Nat. Rev. Drug Discov.* **2010,** *9,* 203–214.

38. Meinert, C. L. *Clinical Trials: Design, Conduct, and Analysis;* Oxford University Press: New York, 2012.

39. Karha, J.; Topol, E. J. The Sad Story of Vioxx, and What We Should Learn from It. *Cleveland Clin. J. Med.* **2004,** *71* (12), 933–939.

40. a. Furberg, C. D.; Pitt, B. Withdrawal of Cerivastatin from the World Market. *Curr. Controlled Trials Cardiovasc. Med.* **2001,** *2*, 205–207.

   b. Psaty, B. M.; Furberg, C. D.; Ray, W. A.; Weiss, N. S. Potential for Conflict of Interest in the Evaluation of Suspected Adverse Drug Reactions: Use of Cerivastatin and Risk of Rhabdomyolysis. *J. Am. Med. Assoc.* **2004,** *292* (21), 2622–2631.

41. Minati, L.; Edginton, T.; Bruzzone, M. G.; Giaccone, G. Current Concepts in Alzheimer's Disease: A Multidisciplinary Review. *Am. J. Alzheimer's Dis. Other Dementias* **2009,** *24* (2), 95–121.

42. a. Labrie, V.; Roder, J. C. The Involvement of the NMDA Receptor D-serine/glycine Site in the Pathophysiology and Treatment of Schizophrenia. *Neurosci. Biobehav. Rev.* **2010,** *34* (3), 351–372.

   b. Ibrahim, H. M.; Tamming, C. A. Schizophrenia: Treatment Targets beyond Monoamine Systems. *Annu. Rev. Pharmacol. Toxicol.* **2011,** *51,* 189–209.

   c. Conn, P. J.; Lindsley, C. W.; Jones, C. K. Activation of Metabotropic Glutamate Receptors as a Novel Approach for the Treatment of Schizophrenia. *Trends Pharmacol. Sci.* **2009,** *30* (1), 25–31.

43. a. Marcotte, E. R.; Pearson, D. M.; Srivastava, L. K. Animal Models of Schizophrenia: A Critical Review. *J. Psychiatry Neurosci.* **2001,** *26* (5), 395–410.

   b. Jones, C. A.; Watson, D. J. G.; Fone, K. C. F. Animal Models of Schizophrenia. *Br. J. Pharmacol.* **2011,** *164* (4), 1162–1194.

44. Barter, P. J.; Brewer, H. B., Jr.; Chapman, M. J.; Hennekens, C. H.; Rader, D. J.; Tall, A. R. Cholesteryl Ester Transfer Protein: A Novel Target for Raising HDL and Inhibiting Atherosclerosis. *Arterioscler. Thromb. Vasc. Biol.* **2003,** *23* (2), 160–167.

45. Nissen, S. E.; Tardif, J. C.; Nicholls, S. J.; Revkin, J. H.; Shear, C. L.; Duggan, W. T.; Ruzyllo, W.; Bachinsky, W. B.; Lasala, G. P.; Tuzcu, E. M. Effect of Torcetrapib on the Progression of Coronary Atherosclerosis. *N. Engl. J. Med.* **2007,** *356* (13), 1304–1316.

46. Huang, Z.; Inazu, A.; Nohara, A.; Higashikata, T.; Mabuchi, H. Cholesteryl Ester Transfer Protein Inhibitor (JTT-705) and the Development of Atherosclerosis in Rabbits with Severe Hypercholesterolaemia. *Clin. Sci.* **2002,** *103* (6), 587–594.

47. a. Clark, R. W.; Sutfin, T. A.; Ruggeri, R. B.; Willauer, A. T.; Sugarman, E. D.; Magnus-Aryitey, G.; Cosgrove, P. G.; Sand, T. M.; Wester, R. T.; Williams, J. A.; et al. Raising High-density Lipoprotein in Humans through Inhibition of Cholesteryl Ester Transfer Protein: An Initial Multidose Study of Torcetrapib. *Arterioscler. Thromb. Vasc. Biol.* **2004,** *24* (3), 490–497.

   b. http://clinicaltrials.gov/ct2/show/NCT00139061 Phase III Assess HDL-C Increase and Non-HDL Lowering Effect of Torcetrapib/Atorvastatin vs Fenofibrate.

   c. http://clinicaltrials.gov/ct2/show/NCT00134511 Phase III Study to Evaluate the Effect of Torcetrapib/Atorvastatin in Patients with Genetic High Cholesterol Disorder.

   d. http://clinicaltrials.gov/ct2/show/NCT00134485 Phase III Study to Evaluate the Safety and Efficacy of Torcetrapib/Atorvastatin in Subjects with Familial Hypercholerolemia.

   e. http://clinicaltrials.gov/ct2/show/NCT00134498 Phase III Study Comparing the Efficacy & Safety of Torcetrapib/Atorvastatin and Atorvastatin in Subjects with High Triglycerides.

   f. http://clinicaltrials.gov/ct2/show/NCT00267254 Phase III Clinical Trial Comparing Torcetrapib/Atorvastatin to Simvastatin in Subjects with High Cholesterol.

48. Berenson, A. Pfizer Ends Studies on Drug for Heart Disease. *N.Y. Times* **December 3, 2006.**

49. a. Bennett, S.; Kresge, N. Bloomberg, June 7, 2012. http://www.bloomberg.com/news/2012-05-07/roche-halts-testing-on-dalcetrapib-cholesterol-treatment-1-.html.

   b. Michelle Fay Cortez, M. F. *Roche's Good Cholesterol Drug Shows Negative Side Effects;* November 5, 2012. http://www.businessweek.com/news/2012-11-05/roche-s-good-cholesterol-drug-shows-negative-side-effects. Bloomberg Businessweek.

50. Gutstein, D. E.; Krishna, R.; Johns, D.; Surks, H. K.; Dansky, H. M.; Shah, S.; Mitchel, Y. B.; Arena, J.; Wagner, J. A. Anacetrapib, a Novel CETP Inhibitor: Pursuing a New Approach to Cardiovascular Risk Reduction. *Clin. Pharmacol. Ther.* **2011,** *91* (1), 109–122.

51. Cannon, C. P.; Shah, S.; Dansky, H. M.; Davidson, M.; Brinton, E. A.; Gotto, A. M., Jr.; Stepanavage, M.; Liu, S. X.; Gibbons, P.; Ashraf, T. B.; et al. Safety of Anacetrapib in Patients with or at High Risk for Coronary Heart Disease. *N. Engl. J. Med.* **2010,** *363* (25), 2406–2415.

52. Kaether, C.; Haass, C.; Steiner, H. Assembly, Trafficking and Function of Gamma-secretase. *Neurodegener. Dis.* **2006,** *3* (4–5), 275–283.

53. Extance, A. Alzheimer's Failure Raises Questions about Disease-modifying Strategies. *Nat. Rev. Drug Discov.* **2010,** *9*, 749–751.

54. a. Schor, N. F. What the Halted Phase III Γ-secretase Inhibitor Trial May (Or May Not) Be Telling Us. *Ann. Neurol.* **2011,** *69* (2), 237–239.

    b. Hopkins, C. R. ACS Chemical Neuroscience Molecule Spotlight on Semagacestat (LY450139). *ACS Chem. Neurosci.* **2010,** *1*, 533–534.

55. Farmer, J. A. Aggressive Lipid Therapy in the Statin Era. *Prog. Cardiovasc. Dis.* **1998,** *41* (2), 71–94.

56. Krukemyer, J. J.; Talbert, R. L. Lovastatin: A New Cholesterol-Lowering Agent. *Pharmacotherapy* **1987,** *7* (6), 198–210.

57. Chong, P. H.; Seeger, J. D. Atorvastatin Calcium: An Addition to HMG-coa Reductase Inhibitors. *Pharmacotherapy* **1997,** *17* (6), 1157–1177.

58. a. Haning, H.; Niewohner, U.; Schenke, T.; Es-Sayed, M.; Schmidt, G.; Lampe, T.; Bischoff, E. Imidazo[5,1-f][1,2,4]triazin-4(3H)-ones, a New Class of Potent PDE 5 Inhibitors. *Bioorg. Med. Chem. Lett.* **2002,** *12*, 865–868.

    b. Kukreja, R. C.; Ockaili, R.; Salloum, F.; Yin, C.; Hawkins, J.; Das, A.; Xi, L. *J. Mol. Cell. Cardiol.* **2004,** *36* (2), 165–173.

59. Krauseneck, T.; Padberg, F.; Roozendaal, B.; Grathwohl, M.; Weis, F.; Hauer, D.; Kaufmann, I.; Schmoeckel, M.; Schelling, G. A B-adrenergic Antagonist Reduces Traumatic Memories and PTSD Symptoms in Female but Not in Male Patients after Cardiac Surgery. *Psychol. Med.* **2010,** *40*, 861–869.

60. a. Hoyer, D.; Clarke, D. E.; Fozard, J. R.; Hartig, P. R.; Martin, G. R.; Mylecharane, E. J.; Saxena, P. R.; Humphrey, P. P. International Union of Pharmacology Classification of Receptors for 5-hydroxytryptamine (Serotonin). *Pharmacol. Rev.* **1994,** *46* (2), 157–203.

    b. Nichols, D. E.; Nichols, C. D. Serotonin Receptors. *Chem. Rev.* **2008,** *108* (5), 1614–1641.

61. a. Sorkin, E. M.; Heel, R. C. Terfenadine. A Review of its Pharmacodynamic Properties and Therapeutic Efficacy. *Drugs* **1985,** *29* (1), 34–56.

    b. Thompson, D.; Oster, G. Use of Terfenadine and Contraindicated Drugs. *J. Am. Med. Assoc.* **1996,** *275* (17), 1339–1341.

62. a. Bernstein, D. ,I.; Schoenwetter, W. F.; Nathan, R. A.; Storms, W.; Ahlbrandt, R.; Mason, J. Efficacy and Safety of Fexofenadine Hydrochloride for Treatment of Seasonal Allergic Rhinitis. *Ann. Allergy Asthma Immunol. Off. Publ. Am. Coll. Allergy Asthma Immunol.* **1997,** *79* (5), 443–448.

    b. Meltzer, E. O.; Casale, T. B.; Nathan, R. A.; Thompson, A. K. Once-daily Fexofenadine HCl Improves Quality of Life and Reduces Work and Activity Impairment in Patients with Seasonal Allergic Rhinitis. *Ann. Allergy Asthma Immunol. Off. Publ. Am. Coll. Allergy Asthma Immunol.* **1999,** *83* (4), 311–317.

63. Amyotrophic Lateral Sclerosis (ALS) Fact Sheet. *National Institute of Neurological Disorders and Stroke;* 2012. http://www.ninds.nih.gov/disorders/amyotrophiclateralsclerosis/detail_ALS.htm.

64. Chemical Abstracts Services. A Division of the American Chemical Society. website: http://www.cas.org/content/chemical-substances.

65. a. Lipinski, C. A.; Lombardo, F.; Dominy, B. W.; Feeney, P. J. Experimental and Computational Approaches to Estimate Solubility and Permeability in Drug Discovery and Development Settings. *Adv. Drug Deliv. Rev.* **2001,** 46.

    b. Roth, B. D. *Preparation of Anticholesteremic (R-(R*R*))-2-(4-fluorophenyl)-β,γ-dihydroxy-5-(1-methylethyl-3-phenyl-4((phenylamino)carbonyl)-1H-pyrrolyl-1-heptanoic Acid, Its Lactone Form and Salts Thereof;* , 1991. EP409281, 3–26.

66. Yamamoto, A.; Sudo, H.; Endo, A. Therapeutic Effects of ML-236B in Primary Hypercholesterolemia. *Atherosclerosis* **1980,** *35* (3), 259–266.

67. Bader, T.; Fazili, J.; Madhoun, M.; Aston, C.; Hughes, D.; Rizvi, S.; Seres, K.; Hasan, M. Fluvastatin Inhibits Hepatitis C Replication in Humans. *Am. J. Gastroenterol.* **2008,** *103* (6), 1383–1389.

68. McTaggart, F.; Buckett, L.; Davidson, R.; Holdgate, G.; McCormick, A.; Schneck, D.; Smith, G.; Warwick, M. Preclinical and Clinical Pharmacology of Rosuvastatin, a New 3-hydroxy-3-methylglutaryl Coenzyme a Reductase Inhibitor. *Am. J. Cardiol.* **2001,** *87* (5A), 28B–32B.

69. Boolell, M.; Allen, M. J.; Ballard, S. A.; Gepi-Attee, S.; Muirhead, G. J.; Naylor, A. M.; Osterloh, I. H.; Gingell, C. Sildenafil: An Orally Active Type 5 Cyclic GMP-specific Phosphodiesterase Inhibitor for the Treatment of Penile Erectile Dysfunction. *Int. J. Impotence Res.* **1996,** *8* (2), 47–52.

70. Daugan, A.; Grondin, P.; Ruault, C.; Le Monnier de Gouville, A. C.; Coste, H.; Kirilovsky, J.; Hyafil, F.; Labaudinière, R. The Discovery of Tadalafil: A Novel and Highly Selective PDE5 Inhibitor. 1: 5,6,11,11a-tetrahydro-1H-imidazo[1',5':1,6]pyrido[3,4-b]indole-1,3(2H)-dione Analogues. *J. Med. Chem.* **2003,** *46* (21), 4525–4532.

71. Novak, B. H.; Hudlicky, T.; Reed, J. W.; Mulzer, J.; Trauner, D. Morphine Synthesis and Biosynthesis—An Update. *Curr. Org. Chem.* **2000,** *4,* 343–362.

72. Lomenzo, S.; Izenwasser, S.; Gerdes, R. M.; Katz, J. L.; Kopajtic, T.; Trudell, M. L. Synthesis, Dopamine and Serotonin Transporter Binding Affinities of Novel Analogues of Meperidine. *Bioorg. Med. Chem. Lett.* **1999,** *9* (23), 3273–3276.

73. Stanley, T. H. The History and Development of the Fentanyl Series. *J. Pain Symptom Manage.* **1992,** *7* (3 Suppl. l), S3–S7.

74. Owens, M. J.; Morgan, W. N.; Plott, S. J.; Nemeroff, C. B. Neurotransmitter Receptor and Transporter Binding Profile of Antidepressants and Their Metabolites. *J. Pharmacol. Exp. Ther.* **1997,** *283* (3), 1305–1322.

75. Coppen, A.; Rao, V. A. R.; Swade, C.; Wood, K. Inhibition of 5-hydroxytryptamine Reuptake by Amitriptyline and Zimelidine and Its Relationship to Their Therapeutic Action. *Psychopharmacology* **1979,** *63* (2), 125–129.

76. Keller, M. B. Citalopram Therapy for Depression: A Review of 10 Years of European Experience and Data from U.S. Clinical Trials. *J. Clin. Psychiatry* **2000,** *61* (12), 896–908.

77. Wong, D. T.; Horng, J. S.; Bymaster, F. P.; Hauser, K. L.; Molloy, B. B. A Selective Inhibitor of Serotonin Uptake: Lilly 110140, 3-(p-trifluoromethylphenoxy)-n-methyl-3-phenylpropylamine. *Life Sci.* **1974,** *15* (3), 471–479.

78. Dechant, K. L.; Clissold, S. P. Paroxetine. A Review of its Pharmacodynamic and Pharmacokinetic Properties, and Therapeutic Potential in Depressive Illness. *Drugs* **1991,** *41* (2), 225–253.

79. a. Hann, M. M.; Oprea, T. I. Pursuing the Leadlikeness Concept in Pharmaceutical Research. *Curr. Opin. Chem. Biol.* **June 2004,** *8* (3), 255–263.

    b. Howe, D.; Costanzo, M.; Fey, P.; Gojobori, T.; Hannick, L.; Hide, W.; Hill, D. P.; Kania, R.; Schaeffer, M.; Pierre, S. S.; et al. Big Data: The Future of Biocuration. *Nature* **2008,** *455* (7209), 47–50.

    c. In *High-Throughput Screening in Drug Discovery;* Hüser, J., Ed. , 2006. (Mannhold, R. (Series Editor); Kubinyi, H. (Series Editor); Folkers, G. (Series Editor). Wiley-VCH.

80. a. Rester, U. From Virtuality to Reality – Virtual Screening in Lead Discovery and Lead Optimization: A Medicinal Chemistry Perspective. *Curr. Opin. Drug Discovery Dev.* **2008,** *11* (4), 559–568.

b. Rollinger, J. M.; Stuppner, H.; Langer, T. Virtual Screening for the Discovery of Bioactive Natural Products. *Prog. Drug Res.* **2008**, *65* (211), 213–249.

c. Walters, W. P.; Stahl, M. T.; Murcko, M. A. Virtual Screening – An Overview. *Drug Discovery Today* **1998**, *3* (4), 160–178.

81. a. Kaczanowski, S.; Zielenkiewicz, P. Why Similar Protein Sequences Encode Similar Three-dimensional Structures? *Theor. Chem. Accounts* **2010**, *2010* (125), 543–550.

b. Zhang, Y. Progress and Challenges in Protein Structure Prediction. *Curr. Opin. Struct. Biol.* **2008**, *18* (3), 342–348.

c. Capener, C. E.; Shrivastava, I. H.; Ranatunga, K. M.; Forrest, L. R.; Smith, G. R.; Sansom, M. S. P. Homology Modeling and Molecular Dynamics Simulation Studies of an Inward Rectifier Potassium Channel. *Biophys. J.* **2000**, *78* (6), 2929–2942.

d. Ogawa, H.; Toyoshima, C. Homology Modeling of the Cation Binding Sites of Na^+K^+-atpase. *Proc. Natl. Acad. Sci. U.S.A.* **2002**, *99* (25), 15977–15982.

82. a. Wang, Z.; Fermini, B.; Nattel, S. Evidence for a Novel Delayed Rectifier K^+ Current Similar to Kv1.5 Cloned Channel Currents. *Circ. Res.* **1993**, *73*, 1061–1076.

b. Fedida, D.; Wible, B.; Wang, Z.; Fermini, B.; Faust, F.; Nattel, S.; Brown, A. M. Identity of a Novel Delayed Rectifier Current from Human Heart with a Cloned Potassium Channel Current. *Circ. Res.* **1993**, *73*, 210–216.

c. Brendel, J.; Peukert, S. Blockers of the Kv1.5 Channel for the Treatment of Atrial Arrhythmias. *Expert Opin. Ther. Pat.* **2002**, *12* (11), 1589–1598.

83. a. Taglialatela, M.; Castaldo, P.; Pannaccione, A. Human Ether-a-gogo Related Gene (HERG) K Channels as Pharmacological Targets: Present and Future Implications. *Biochem. Pharmacol.* **1998**, *55* (11), 1741–1746.

b. Vaz, R. J.; Li, Y.; Rampe, D. Human Ether-a-go-go Related Gene (HERG): A Chemist's Perspective. *Prog. Med. Chem.* **2005**, *43*, 1–18.

c. Kang, J.; Wang, L.; Chen, X. L.; Triggle, D. J.; Rampe, D. Interactions of a Series of Fluoroquinolone Antibacterial Drugs with the Human Cardiac K^+ Channel HERG. *Mol. Pharmacol.* **2001**, *59*, 122–126.

84. Manning, G.; Whyte, D. B.; Martinez, R.; Hunter, T.; Sudarsanam, S. The Protein Kinase Complement of the Human Genome. *Science* **2002**, *298*, 1912–1934.

85. a. Hennessy, M.; Spiers, J. P. A Primer on the Mechanism of P-glycoprotein the Multidrug Transporter. *Pharmacol. Res.* **2007**, *55*, 1–15.

b. Aller, S. G.; Yu, J.; Ward, A.; Weng, Y.; Chittaboina, S.; Zhuo, R.; Harrell, P. M.; Trinh, Y. T.; Zhang, Q.; Urbatsch, I. L.; et al. Structure of P-glycoprotein Reveals a Molecular Basis for Poly-specific Drug Binding. *Science* **2009**, *323* (5922), 1718–1722.

c. Liu, X.; Chen, C. Strategies to Optimize Brain Penetration in Drug Discovery. *Curr. Opin. Drug Discovery Dev.* **2005**, *8*, 505–512.

d. Schinkel, A. H. P-Glycoprotein, a Gatekeeper in the Blood–Brain Barrier. *Adv. Drug Deliv. Rev.* **1999**, *36*, 179–194.

86. a. Rodrigues, A. D. *Drug-Drug Interactions;* Marcel Dekker: New York, 2002.

b. Shimada, T.; Yamazaki, H.; Mimura, M.; Inui, Y.; Guengerich, F. P. Interindividual Variations in Human Liver Cytochrome P-450 Enzymes Involved in the Oxidation of Drugs, Carcinogens and Toxic Chemicals: Studies with Liver Microsomes of 30 Japanese and 30 Caucasians. *J. Pharmacol. Exp. Ther.* **1994**, *270* (1), 414–423.

87. a. Stinson, S. C. Uncertain Climate for Antihistamines. *Chem. Eng. News* **1997**, *75* (10), 43–45.

88. a. Dorn, A.; Hermann, F.; Ebneth, A.; Bothmann, H.; Trube, G.; Christensen, K.; Apfel, C. Evaluation of a High-Throughput Fluorescence Assay Method for HERG Potassium Channel Inhibition. *J. Biomol. Screening* **2005**, *10* (4), 339–347.

b. Redfern, W. S.; Carlsson, L.; Davis, A. S.; Lynch, W. G.; MacKenzie, I.; Palethorpe, S.; Siegl, P. K.; Strang, I.; Sullivan, A. T.; Wallis, R.; et al. Relationships between Preclinical Cardiac Electrophysiology, Clinical QT Interval Prolongation and Torsade De Pointes for a Broad Range of Drugs: Evidence for a Provisional Safety Margin in Drug Development. *Cardiovasc. Res.* **April 1, 2003**, *58* (1), 32–45.

89. a. Mortelmans, K.; Zeiger, E. The Ames Salmonella/Microsome Mutagenicity Assay. *Mutat. Res.* **2000,** *455* (1–2), 29–60.

b. McCann, J.; Choi, E.; Yamasaki, E.; Ames, B. N. Detection of Carcinogens as Mutagens in the Salmonella/microsome Test: Assay of 300 Chemicals. *Proc. Natl. Acad. Sci. U.S.A.* **1975,** *72* (12), 5135–5139.

c. Hakura, A.; Suzuki, S.; Satoh, T. Advantage of the Use of Human Liver S9 in the Ames Test. *Mutat. Res.* **1999,** *438* (1), 29–36.

90. Guth, B. D. Preclinical Cardiovascular Risk Assessment in Modern Drug Development. *Toxicol. Sci.* **2007,** *97* (1), 4–20.

2

The Drug Discovery Process: From Ancient Times to the Present Day

Throughout the course of history, there has been a near constant need for therapeutic intervention for the treatment of disease. Efforts to provide for this need can be traced to prehistoric times as evidenced in cave drawings from 7000 to 5000 BC that are suggestive of the use of hallucinogenic mushrooms. The concept that curing diseases or alleviating symptoms could be accomplished by eating, drinking, or applying substances to the body is ancient, but the methods used to discover therapeutic agents has changed dramatically over the course of human history. In its earliest form, from ancient times until the mid-nineteenth century, the identification of new drugs was primarily the result of serendipity, as the foundational science required for the systematic study of potential new therapeutic entities had not yet been established. Modern methods of drug discovery have evolved over the last two centuries, however, as a result of advances in basic science (e.g., chemistry, biology, pharmacology) and applied science (e.g., transgenic animal models, molecular modeling, robotics) leading to a process that is far less dependent upon the serendipitous identification of therapeutic agents. A third factor, governmental and regulatory oversight, which focuses primarily on ensuring the safety and efficacy of new medications, has also had a major impact on modern drug discovery over the last century. This chapter will review the evolution of the drug discovery process from ancient times to the modern age, focusing on key scientific advances and the regulatory environment that changed the way in which new drugs are identified.

Basic Principles of Drug Discovery and Development
http://dx.doi.org/10.1016/B978-0-12-411508-8.00002-5

THE AGE OF BOTANICALS: PREINDUSTRIAL DRUG DISCOVERY

The search for effective methods and medications designed to improve the quality and length of life predates the age of modern discovery by several thousand years. While it is unclear exactly when humanity began to understand that ingestion of specific materials (i.e., drugs) could influence physiology, disease-related or otherwise, there is evidence suggesting that these concepts were beginning to evolve as early as prehistoric times. Plant remains from between 7000 and 5500 BC found in the Spirit Caves of north-western Thailand included seeds of the betel nut, a mildly psychoactive agent, indirectly suggesting its use in the Neolithic period.[1] Human consumption, perhaps for the alteration of perception, is also suggested by the presence of skeletal remains from 2680 BC in the presence of lime-containing betel nut shells found in the Duyong cave of the Philippines. Although not conclusive evidence of consumption, the presence of the lime and betel nuts is consistent with practices designed to aid in the absorption of the active ingredient (arecoline) while chewing that are still in practice in modern India.[2] The prehistoric use of hallucinogenic mushrooms is also implicated by Saharan cave drawings (c.7000–5000 BC),[3] suggesting that humanity became aware of the potency of these plants long before recorded history.

Direct evidence of the early identification of the most frequently consumed drug in history, alcohol, is far easier to obtain. Although it is unclear how the fermentation of alcoholic beverages was discovered, there is ample evidence indicating that its discovery occurred early in human history. Strong evidence exists indicating that alcoholic beverages were developed as early as the Neolithic period, and that its use was common across the ancient world.[4] Given that the effects of alcohol consumption occurs rapidly upon ingestion, it is not surprising that various alcoholic beverages were among the first drugs to be widely consumed for either recreational or medicinal purposes.

When humanity began to recognize the medicinal properties of various plants and chemicals is also an open question. It is clear, however, that the pursuit of treatments for diseases and symptom relief is not a phenomenon of the modern world. The Mesopotamians documented their medical methods and prescriptions on stone tablets. One of the oldest and largest collections from this civilization consists of a series of 40 tablets from around 1700 BC that are collectively known as "Treatise of Medical Diagnosis and Prognoses." Included among the writings are some of the earliest recorded uses of drugs for medicinal purposes (Figure 2.1(a)).[5] In a similar fashion, the Ebers Papyrus was written by the ancient Egyptians around 1550 BC and contains several hundred "prescriptions" for the treatment of disease or symptomatic

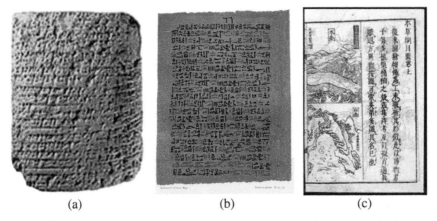

(a)   (b)   (c)

FIGURE 2.1 (a) Cuneiform clay tablets unearthed from the library of King Ashurbanipal at Nineveh describe Mesopotamian medical practices. It is estimated that they originated between 1900 and 1700 BC. *(Image from The Schoyen Collection, MS2670,* http://www.sch oyencollection.com/smallercollect_files/ms2670.jpg.*).* (b) A section of the Ebers Papyrus, a compilation of Egyptian medical knowledge composed around 1550 BC containing over 800 prescriptions for various conditions *(Image from the NIH U.S. National Library of Medicine Archives,* http://www.nlm.nih.gov/archive/20120918/hmd/breath/breath_exhibit /MindBodySpirit/IIBa18.html.*).* (c) A page from the Pen-tsao Kang-mu, a compilation of traditional Chinese medicines written by Li Shih-chen. The completed text contains over 1800 Chinese medicines and 11,000 prescriptions *(Image from the U.S. National Library of Medicine, History of Medicine Division* http://www.nlm.nih.gov/exhibition/chinesemedici ne/images/017c.jpg.*).*

relief (Figure 2.1(b)).[6] The origin of traditional Chinese medicine is largely unknown, but it is estimated that the practices and methods are at least 2000 years old. The herbalist and acupuncturist Li Shih-chen completed the first draft of Pen-tsao Kang-mu, which is widely considered the most comprehensive text on traditional Chinese medicine, in 1587. The text describes hundreds of distinct herbs and thousands of combinations useful for treating disease and alleviating symptoms (Figure 2.1(c)).[7]

There are some commons threads that run through all preindustrial drug discovery efforts, irrespective of their country or region of origin. First, they depended almost exclusively on plants, plant-derived mixtures, or plant extracts, as the ability of preindustrial society to isolate or prepare pure chemicals with medicinal value was limited. Second, medications developed in the preindustrial ages were identified using empirical observation of the presence or absence of symptoms in patients, rather than an understanding of the disease or condition afflicting the patient. Third, and perhaps most importantly, all of the efforts to develop new medication in the preindustrial age of drug discovery did so in the absence of the vast majority of the fundamental knowledge required to understand even the basic principles of disease progression. This almost certainly led to the use of any number of concoctions with little true medicinal value and some that were actually detrimental to the patients' well-being.

Despite these facts, there are a number of medications that were identified prior to the advent of modern drug discovery that still play an important role in modern medicine. The treatment of malaria caused by *Plasmodium falciparum*, for example, was revolutionized by the discovery of quinine, an alkaloid found in cinchona bark (Figure 2.2). Agostino Salumbrino

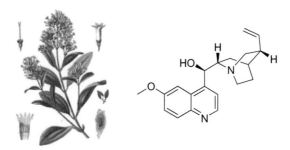

FIGURE 2.2   The cinchona tree (left) and quinine (right) were important players in the treatment of malaria infection for over 300 years. *Source:* http://upload.wikimedia.org/wikipedia/commons/e/e8/Koeh-179-cropped.jpg.

(1561–1642), a Jesuit living in Lima, Peru, observed the Quechua people chewing the bark from the cinchona tree in an effort to relieve shivering and fevers. Although Salumbrino certainly had no knowledge of the causative malaria parasite, he did recognize that the symptoms of the febrile phase of malaria might be positively impacted by the cinchona bark and arranged for a sample to be shipped to Rome for evaluation as a treatment for malaria. The cinchona bark, also known as Jesuit's bark or Peruvian bark, and the Quechua people thus became the source of the first successful antimalarial agent, a drug that was a first line treatment for malaria infection until 2006.[8]

In a similar fashion, cardiac glycosides were identified as an important treatment for congestive heart failure via the foxglove plant, which contains high levels of several cardiac glycosides in the leaves (Figure 2.3). The use of the foxglove plant as part of an herbal remedy for dropsy, swelling, and fatigue, all of which are symptoms of congestive heart failure, can be traced to medieval Europe. Although the structure and mechanism of action were clearly not known at the time, William Withering deduced that the foxglove plant was the source of the active ingredient of the herbal remedy in 1785 and unknowingly provided a primary treatment for congestive heart failure, digoxin that is still routinely used in modern medicine.[9]

FIGURE 2.3   The common foxglove plant (*Digitalis purpurea*, left) contains cardiac glyco-sides such as Digoxin (right), which are known to increase cardiac contractility via inhibition of myocardial sodium/potassium ATPase. *Source: Kurt Stüber* http://en.wikipedia.org/wiki/Foxglove#mediaviewer/File:Digitalis_purpurea2.jpg.

PAUL EHRLICH: THE FATHER OF MODERN DRUG DISCOVERY[10]

There are many additional examples of useful drugs that were discov-ered in the preindustrial era, such as morphine,[11] cocaine,[12] and aspirin,[13] but it was Paul Ehrlich's (Figure 2.4) efforts that are most often cited as the

FIGURE 2.4   Paul Ehrlich (1854–1915), the founder of modern drug discovery was a physician and scientist noted for his discover-ies in the fields of hematology, immunology, and chemotherapy. In 1908, he received the Nobel Prize in Physiology or Medicine in recognition of his work. *Source: NIH U.S. National Library of Medicine* http://ihm.nlm.gov/luna/servlet/view/search?q=B07744.

starting point for modern drug discovery methods. Ehrlich's early observations of differential affinities of biological tissues for various dyes, such as Trypan red, Trypan blue and methylene blue (Figure 2.5), lead him to

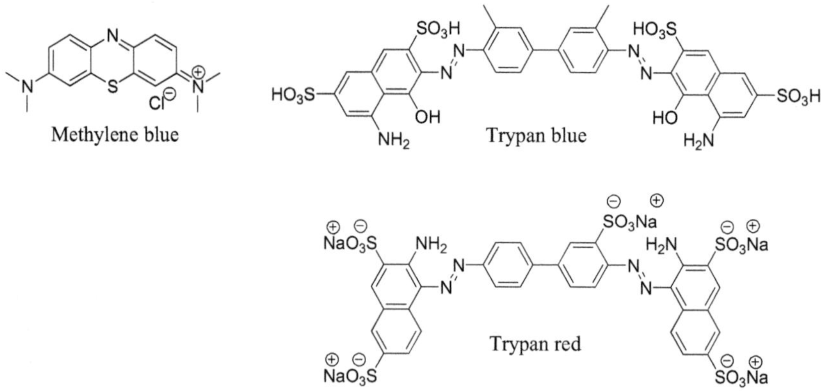

FIGURE 2.5 Methylene blue is used in Wright's stain, Jenner's stain, and northern blotting experiments. Trypan blue and Trypan red are commonly employed as stains that distinguish between viable and non-viable cells.

postulate the existence of "chemoreceptors" that influence the interaction of cells with the chemicals around them to produced a biological effect. He further theorized that "chemoreceptors" of infectious organisms or cancer cells would be different from those of the host and that the differences could be exploited to produce a therapeutic benefit (the "magic bullet" theory). These concepts, along with his hypothesis that the chemical composition of drugs controlled their mode of action in an organism, formed the basis of modern chemotherapy. His initial successful treatment of two malaria patients using methylene blue lead him to conclude that this dye possessed a clear affinity for the malaria parasite over the host, and that compounds previously used only as dyes might have therapeutic value. In an effort to capitalize on these theories, Ehrlich and his colleague began a systematic evaluation of hundreds of commercial synthetic dyes in mice infected with *Trypanosoma equinum*, also known as sleeping sickness. In 1904, these first attempts to develop structure–activity relationships (see Chapter 5 for a full discussion of this concept), led to the identification of Trypan red as an agent capable of killing this infections in mice. Unfortunately, resistant strains of the organism developed and eventually killed the mice, as well as rats and dogs that were also studied, marking setbacks in the research efforts, but also prompting Ehrlich to hypothesize the development of resistant organisms. More importantly, however, these efforts marked the first concerted effort to discern the relationship between chemical structure and biological activity in an effort to develop new therapeutic agents in conjunction with a chemical manufacturing company, also known as a pharmaceutical product pipeline.

Full validation of Ehrlich's methods came with the identification of Salvarsan, the first successful synthetic chemotherapeutic drug, and the first truly effective treatment of syphilis (Figure 2.6). Prior to the identification

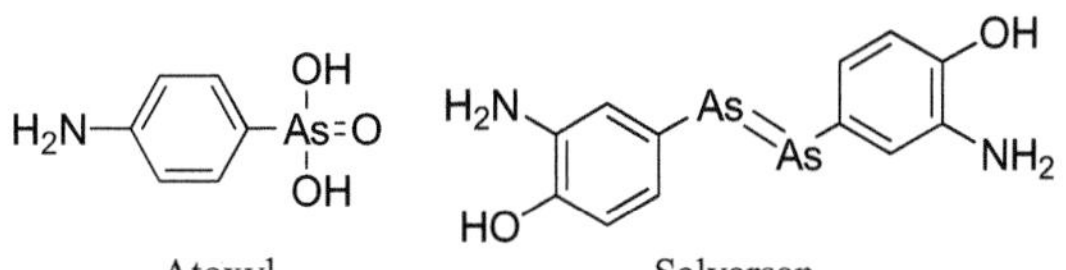

Atoxyl   Salvarsan

FIGURE 2.6   The discovery of Salvarsan from Atoxyl is one of the earliest examples of an effective drug discovery program. Salvarsan was the drug of choice for the treatment of syphilis until the mid-1940s, when it was replaced by penicillin.

of the causative agent of syphilis, *Treponema pallidum*, Ehrlich and his colleagues had prepared and tested a number of phenyl arsenide analogs in an attempt to improve upon the drug Atoxyl, a treatment for African sleeping sickness (African trypanosomiasis) with a high risk of blindness. Erich Hoffman (1868–1959), a contemporary scientist of Ehrlich's, noted the similarities between the causative agents of the two diseases, and at Hoffman's urging, Ehrlich reexamined the phenyl arsenide analogs in a rabbit model of syphilis developed by Sahachiro Hata (1873–1938). These efforts lead to the identification of arsphenamine in 1909 as a lead compound for the treatment of syphilis. Clinical results demonstrating its efficacy in patients were presented at the 1910 Congress for Internal Medicine, and Hoechst marketed the drug as Salvarsan. Thus, the age of modern drug discovery was born.

MILESTONES IN DRUG DISCOVERY

Ehrlich's research and methods provided much of the foundation of what eventually became modern drug discovery, but his efforts did not provide many of the important tools that are now commonplace. When Paul Ehrlich unknowingly launched the age of modern drug discovery, the ability to prepare, analyze, and screen compounds for biological activity was in its infancy. Over the course of the next 100 years, critical tools required to efficiently identify biologically active compounds and understand how they function, both in a whole organism and in isolated systems, were developed. The fields of animal modeling, X-ray crystallography, molecular modeling, high throughput screening, and high throughput chemistry, as well as biotechnology tools such as recombinant DNA and transfection technology grew as knowledge in basic sciences such as biology and chemistry expanded. In many cases, the development of new technology in one field led to advances in a related field. The advent of transfection technology, for example, provided the tools necessary to generate transgenic and knockout animal models. Advances in

X-ray crystallography led to advances in molecular modeling and computational chemistry, and the combination of increased computer capacity, automation science, and *in vitro* screening techniques led to the introduction of high throughput screening. There can be no question that a wide range of scientific disciplines influenced the development of modern drug discovery science. It is well beyond the scope of this text to provide a complete history of the various important fields that influenced the evolution of the drug discovery process. The history and growth of synthetic organic chemistry or *in vitro* biology, for example, would require many texts unto themselves. There are, however, some scientific advances that had foundational impact on drug discovery. An understanding of their history provides insight into how the process developed to its current status, and perhaps some guidance as to where the field may be going in the future.

Milestones in Animal Models: Breeding a Better Model

The Wistar Rat

While modern drug discovery research is performed using a wide assortment of standardized animals from any number of different species, this was not the case at the beginning of the twentieth century. Up until 1906, there were no standardized animal models available and the common house mouse, *Mus musculus*, was used for laboratory research. This changed in 1906, however, with the introduction of the Wistar rat (Figure 2.7),[14] a strain

FIGURE 2.7 The Wistar rat is a product of research lead by Milton Greenman and Henry Donaldson at the Wistar Institute, the first independent biomedical research facility in the United States, which was founded in 1892. © istock. com/VseBogd

of albino rats belonging to the species *Rattus norvegicus*, which marked the first effort to develop a "pure strain" animal as a model organism for medical research. It is estimated that over 50% of all laboratory rat strains are descendants of the original colony established at Wistar Institute and it remains one of the most commonly employed rat strains in modern medical research. While a full listing of rat models developed using the Wistar rat is well beyond the scope of this text, there is little doubt as to the importance of this watershed animal model. Wistar rats models include spontaneously diabetic rats,[15] spontaneous tumor formers (the Rochester strain),[16] high anxiety behavior rats, low anxiety behavior rats,[17] the Lobund-Wistar rat model of prostate cancer,[18] Wistar Kyoto rats (an important model of attention deficit disorder),[19] myelin deficient rats,[20] and the spontaneously hypertensive rat (SHR),[21] the most widely studied model of hypertension.

Immunocompromised Mice

The Nude Mouse[22]

The availability of the Wistar rat and the concept of using standardized animal strains led other research teams to examine their animal colonies more closely in an attempt to identify useful subpopulations. Thousands of useful animal models across a range of different species have been identified in the intervening time period, but few have had the impact of the nude mouse and the severe combined immune deficient mouse. Prior to the development of these two animal models, the ability to study human tumor progression in animals was limited by T-lymphocyte-mediated rejection of implanted human tumors. The nude mouse (Figure 2.8) was

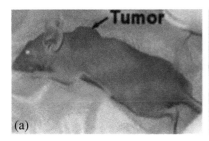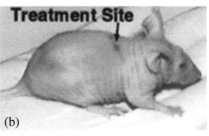

FIGURE 2.8   The nude mouse: Disruption of the FOXN1 gene in mice produces a strain of mice with a severely inhibited immune system due to the absence of the thymus, a major source of T-cells. The nickname "nude mouse" is based on the most obvious physical feature resulting from the disruption of the FOXN1 gene, a distinct lack of body hair. Nude mouse bearing subcutaneous tumor before (a) and after (b) high intensity focused ultrasound treatment. *Source: Reprinted from Vaezy, S.; Fujimoto, V. Y.; Walker, C.; Martin, R. W.; Chi, E. Y.; Crum, L. A. Treatment of uterine fibroid tumors in a nude mouse model using high-intensity focused ultrasound. Am. J. Obstet. Gynecol., 183 (1), 6–11, copyright 2000 with permission from Elsevier.*

originally identified in the Virus Laboratory, Ruchill Hospital, Glasgow in 1962,[23] and it was subsequently demonstrated that they were congenitally athymic.[24] In the absence of the thymus, nude mice are unable to generate mature T-lymphocytes, which severely limits their ability to mount an immune response. In the absence of a pathogen, the nude mice have a similar life span to their normal counterparts, but they are unable to reject transplanted tissues such as human tumors. Both primary and metastatic tumors of human origin can be grown and studied in the nude mouse. As such, they were rapidly accepted as a major model for the study of cancer progression and therapeutic intervention. The nude mouse also facilitated the study of infectious disease, as it became possible to study pathogen progression and potential therapies in the absence of a full immune response.

The SCID Mouse

The development of immune compromised models was further advanced in 1983 with the introduction of the severe combined immune deficient (SCID) mouse.[25] An autosomal recessive mutation in mice was identified at the Fox Chase Cancer Center that, when homozygous, leads to animals that are severely deficient in B- and T-lymphocytes. This leaves them highly susceptible to infectious disease, irrespective of the nature of the pathogen, and, similar to nude mice, unable to reject transplanted tissues. The introduction of the SCID mouse model, and the variations that were developed as a result of its identification, provided an additional platform for the study of cancer and infectious disease that was previously unavailable to the research community.

Transgenic Animal Models

Up until 1974, the ability to develop new animal models was limited to selective breeding and depended on the natural occurrence of mutations, such as the nude mouse, to provide improved models for research. Direct manipulation of an animal's genetic codes was not possible. This changed, however, with the introduction of transgenic science. The initial breakthrough in this area was provided by Rudolf Jaenisch, who successfully inserted simian virus 40 DNA sequences into mice.[26] Although the genes were not passed onto offspring, these efforts marked the first successful transfer of foreign DNA into an animal suitable for drug discovery research. Subsequent efforts by Frank Ruddle (Yale),[27] Frank Constantini (Oxford), and Elizabeth Lacy (Oxford)[28] demonstrated that the addition of foreign DNA to single cell mouse embryos provided incorporation of the foreign DNA, and the new genes were passed on to subsequent generations (Figure 2.9). These efforts marked the beginning of a new era in both animal modeling and drug discovery. It was now possible to insert disease-related genes into animals that did not normally demonstrate the pathology in question. Mouse models of Alzheimer's disease, for example, were created by inserting DNA that induced the production of Aβ42 plaques, a hallmark of this disease, providing a new platform for the study of this important malady.[29] Similarly, models of human obesity have been generated in mice through the transgenic methods, providing significant insight into the mechanism of obesity.[30] The pathogenesis of viral infections such as HIV, hepatitis (B and C), polio, and measles

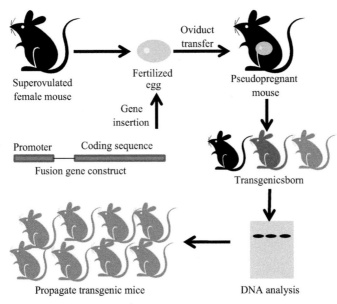

Oviduct transfer

Superovulated female mouse

Fertilized egg

Pseudopregnant mouse

Gene insertion

Promoter   Coding sequence

Fusion gene construct

Transgenicsborn

Propagate transgenic mice   DNA analysis

FIGURE 2.9   Transgenic animal models are developed through a combination of selective breeding and genetic manipulation. A gene construct suitable for insertion into an organism's DNA is prepared and then inserted into a fertilized egg via microinjection. The altered embryos are then implanted into a suitably pseudo-pregnant female and carried to term. After birth, genetic profiling is employed to identify offspring that are carriers of the transgene. Identification of transgene positive progeny is then followed by selective breeding to further the germ line.

have all been studied through the development of transgenic models through expression of either the human receptor for the virus or viral proteins important for pathogenesis.[31] Production of therapeutically relevant biomolecules has also been accomplished through the generation of transgenic animals.[32] Human antithrombin,[33] fibrinogen,[34] and monoclonal antibodies[35] have all been produced via transgenic science. It is well beyond the scope of this text to describe the wide array of transgenic animal models that has been developed since these initial experiments, but the impact of transgenic animals has been significant (Figure 2.10).

Knockout Animal Models

The advent of transgenic technology in animal models opened the door to knockout animal models. By the late 1980s, it had been well established that new animal models could be developed through the insertion and expression of foreign DNA in animal models. The next logical step, the suppression of normal gene function, was addressed by Capecchi, Evans, and Smithies in 1989 when they introduced the first knockout mouse.[36]

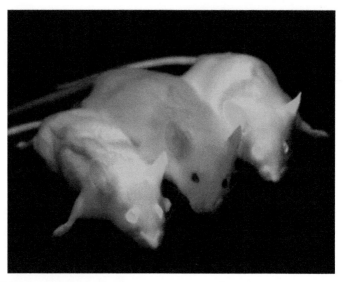

FIGURE 2.10   Transgenic insertion of the gene responsible for the production of green fluorescent protein (GFP) results in mice that fluoresce when exposed to ultraviolet light. The GFP gene has been successfully expressed in bacteria, fungi, plants, insects, and mammalian cells. Martin Chalfie, Osamu Shimomura, and Roger Y. Tsien were awarded the Nobel Prize in Chemistry in 2008 in acknowledgement of their work on GFP technology. *Source: Moen, I.; Jevne, C.; Wang, J.; Kalland, K. H.; Chekenya, M; Akslen, L. A.; Sleire, L.; Enger, P.; Reed, R. K.; Yan, A. M.; Stuh, L. E. B. Gene expression in tumor cells and stroma in dsRed 4T1 tumors in eGFP-expressing mice with and without enhanced oxygenation. BMC Cancer 2012, 12:21.* http://doi:10.1186/1471-2407-12-21.

In their seminal experiments, they were able to eliminate functional hypoxanthine-guanine phosphoribosyl transferase genes (*hprt*) in mouse embryonic stems cells using either a sequence replacement targeting vector or a sequence insertion targeting vector (Figure 2.11). In both cases, the insertion of foreign DNA into the otherwise functional DNA segment led to the suppression of the *hprt* gene in viable embryonic stem cells, which were then implanted into the uterus of a healthy mouse and progressed to birth. In the following years, thousands of knockout mouse models have been developed to study a wide range of disease states. The p53 knockout mouse, for example, has been an important model in the study of cancer progression and therapy. The absence of functional p53 tumor suppressors, encoded by the TP53 gene, provides a mouse strain that mimics Li-Fraumeni syndrome. The resulting mice are far more susceptible to tumor formation.[37] Additional examples include the Fmr1 knockout mouse that serves as a model for Fragile X-related mental retardation,[38] the nescient helix loop helix 2 (Nhlh2) knockout mouse that decreases the levels of α-melanocyte-stimulating hormone and thyrotropin-releasing hormone, providing a model for the study of obesity,[39] and the ApoE knockout mouse in which the expression of Apolipoprotein E is suppressed, leading to the formation of vascular plaques similar to those found in humans suffering

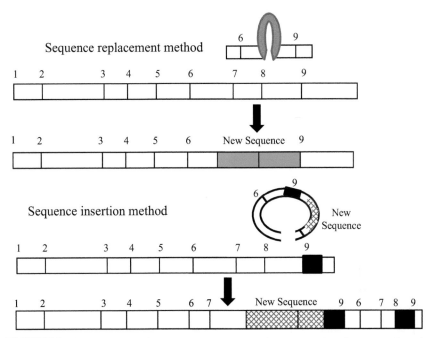

FIGURE 2.11 In the sequence replacement method of producing knockout animal models, gene disruption is accomplished by replacing a portion of the target DNA with a new sequence. An alternative method of producing knockout animals, the sequence insertion method, inserts a new sequence of DNA that is a repeat of a portion of the original DNA sequence. In both instances, the DNA is no longer capable of producing the gene product. Genetic screening of offspring animals followed by selective breeding can then be used to establish the germ line.

from hypercholsteroemia.[40] Since the introduction of knockout technology, thousands of knockout mice have been created in an effort to better understand gene function and disease progression. The importance of this technology was recognized in 2007 with the awarding of the Nobel Prize to Capecchi, Evans, and Smithies for their pioneering work in this area.[41]

Milestones in Molecular Science

While advances in animal models were providing more and more information into the physiological outcomes of potential therapies, they provided little, if any, knowledge as to the molecular interaction required for biological activity. Elucidating the mechanistic aspects of drug action or disease progression at a molecular level requires the ability to prepare molecules suitable for testing, an understanding of the structure of the target (e.g., enzymes, receptors, etc. See Chapter 3), and the ability to screen for biological activity in isolated systems (e.g., *in vitro* screening. See Chapter 4). In the intervening time between Paul Ehrlich's pioneering efforts and the present day, substantial progress has been made in preparing novel compounds through advanced organic synthesis, elucidating the molecular structure of biological targets, understanding the interaction of the aforementioned targets with biologically relevant molecules, and increasing the pace at which the science is explored through the application of robotics, automation, and computer technology. Advances in one of these overlapping fields often provided support for new discoveries or technological advancements in related areas. The growth of X-ray crystallographic knowledge, for example, had a tremendous impact on the science of molecular modeling and computational chemistry, and both of these fields relied heavily on advances in computer technology, an area totally outside of drug discovery, to increase capabilities and capacity. While it is not possible to describe the complete history of the development of the full range of tools employed to understand the molecular basis of disease processes and drug action, an examination of the history of some of the key technologies developed for this purpose provides a wealth of insight into how modern drug discovery systems developed over the course of the last century.

X-ray Crystallography

Understanding the molecular structure of biological targets and associated ligands is a critical aspect of modern drug discovery. At the beginning of the twentieth century, however, modern analytical methods were just beginning to be developed. The field of X-ray crystallography was still in its infancy when Paul Ehrlich launched the research that eventually led to the identification of Salvarsan. In fact, the existence of X-rays themselves had only recently been discovered by Wilhelm Conrad Röntgen in 1895,[42] and the concept that crystalline materials could diffract an X-ray beam and the resulting scattering pattern was related to the molecular structure of the material was still a novel one at the turn of the twentieth century.[43]

The first successful application of this technology to an organic compound was reported in 1923 by Raymond and Dickinson, who elucidated the structure of hexamethylenetetramine,[44] but it was Dorothy Crowfoot Hodgkin[45] who propelled the field into the world of biomolecules and drug discovery. She was among the first to realize the potential for the application of X-ray crystallographic techniques to organic compounds and biomolecules. If Paul Ehrlich is the father of drug discovery, then Dorothy Crowfoot Hodgkin is the mother of protein crystallography. Her accomplishments include the first diffraction pattern of a crystalline protein, pepsin,[46] as well as diffraction pattern images of a host of important proteins including lactoglobulin[47] and insulin.[48] In 1969, 34 years after Hodgkin took her first X-ray diffraction photographs of insulin, she

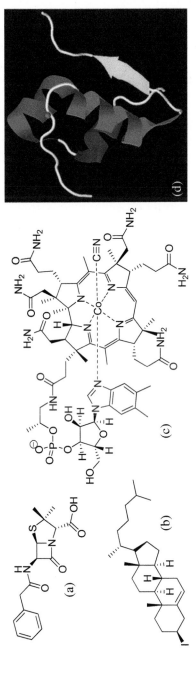

FIGURE 2.12 Dorothy Crowfoot Hodgkin (1910–1994), a graduate of the University of Cambridge, was an earlier pioneer in the field of X-ray crystallography, especially with respect to its application to biomolecules. She is credited with providing definitive structures for a variety of important molecules including (a) benzylpenicillin, (b) cholesteryl iodide, (c) vitamin B_{12}, and (d) insulin (RCSB 4INS).

and her colleagues reported the crystal structure of rhombohedral 2 zinc insulin at 2.8 Å resolution, providing an atomic model for the protein.[49] In the intervening years, she revolutionized the field of X-ray crystallography by solving numerous atomic structures of compounds such as cholesteryl iodide,[50] establishing for the first time the relative stereochemistry of steroids, benzylpenicillin salts,[51] identifying the β-lactam substructure for the first time, and vitamin B_{12},[52] the first naturally occurring organometallic compound with biological significance (Figure 2.12). In 1964, Hodgkin received the Nobel Prize in chemistry for her contributions to the field.[53]

The remarkable work of Dorothy Crowfoot Hodgkin and the scientists that followed in her footsteps provided the scientific community with their first clear pictures of structures of biomolecules. Thousands of protein structures, both in the presence and absence of a ligand have been reported, and the information embedded within these structures has provided a detailed understanding of how drugs interact with their target proteins. The Protein Data Bank (http://www.rcsb.org/pdb/home/home.do), first established in 1971 with 7 structures, contains over 82,000 protein structures as of 2012.[54] Nucleic acid X-ray structures, the most famous of which is the Watson and Crick DNA structure introduced in 1953,[55] have also been exceptionally valuable tools in determining the molecular interaction required for normal, pathological, and drug-mediated biology. The Nucleic Acid Database (http://ndbserver.rutgers.edu/index.html), a more recently created publicly available database, was established in 1992 to provide the scientific community with access to three dimensional structures of nucleic acids, and contains over 6300 solved structures as of 2012.[56] Finally, The Cambridge Structural Database (http://www.ccdc.cam.ac.uk/products/csd/), founded in 1965,[57] focuses on small molecule crystal structures, and contains structural information on nearly 600,000 small molecules as of 2012.[58]

Molecular Modeling and Computational Chemistry

Although Heisenberg's 1925 paper on quantum mechanics[59] is widely considered to be the first publication in the field of computational chemistry and molecular modeling, it would take an additional 36 years for the concept of using computers to calculate and predict chemical properties and interactions to arrive. In 1961, James Hendrickson calculated the conformational energies of cycloheptanes using an IBM 709 computer (Figure 2.13) that was capable of "8000 additions/subtractions, 4000 multiplications/divisions, or 500 complex functions per second."[60] In essence, he launched the field of molecular modeling with a computer that had fewer capabilities and less capacity than most cellular telephones.

FIGURE 2.13 The IBM 709 computer, introduced in 1958, had less computer power than modern cellular phones. *Source: IBM 709 front panel at the Computer History Museum by Arnold Reinhold* http://en.wikipedia.org/wiki/IBM_709#mediaviewer/File:IBM_709_front_panel_at_CHM.agr.jpg.

A few years later (1966), Cyrus Levinthal described his efforts to combine computer simulations with molecular graphics to visualize and study the structures of proteins and nucleic acids,[61] marking the dawn of computer-aided drug design.

The impact of molecular modeling and computational chemistry grew as the computer industry became more and more sophisticated, but the overall premise of the field remained the same. Computers and software could be used to understand the relationship between structural features and physical/chemical properties, including those that were critical to drug function. In addition, knowledge of these relationships could be used to alter or improve the physical and chemical properties of compounds, such as biological activity, solubility, and metabolic stability. By the late 1970s, independent commercial ventures based on computer-assisted modeling were beginning to appear. Molecular Design Limited and Tripos (Figure 2.14) were the first of many organizations built to exploit the ever-growing understanding of molecular interaction with the goal of designing better molecules *in silico*. In 1984, computing capabilities and molecular modeling capabilities had grown to the point where protein simulation was possible and BioDesign launched the first commercial program designed for this purpose. Continued growth in computer power and changes in the drug discovery industry led to the development of additional software tools between 1984 and the present day. Tools designed to assess molecular diversity, design compound libraries, create screening sets based on molecular similarity, and automate the docking

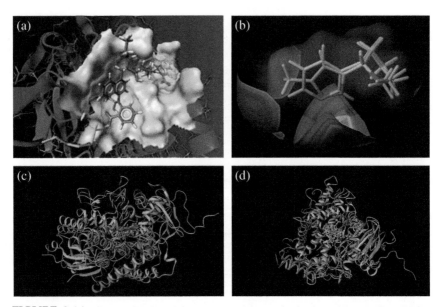

FIGURE 2.14   Tripos, the first independent company focused on the use of computer aided drug design, was founded in 1979. One of its products, Benchware 3D Explorer, provides drug discovery scientists with the ability to visualize and manipulate protein-ligand structures on a desktop computer. Ligands can be modified within the context of a protein in order to gain insight into the impact of structural changes on the potential binding energy of a new proposed ligand. Image (a) shows the PDB structure of protein tyrosine phosphatase 1B (RCSB 1NNY) with a potent inhibitor, which can be readily visualized and manipulated by non-experts in molecular modeling using the Tripos Benchware 3D Explorer software. Important aspects of binding, such as hydrogen bonding, hydrophobic interactions, and structural compatibility between the ligand and protein are readily identified. The surface of the binding site, highlighted in light blue (Connolly surfaces), enables the user to see the shape complementarity of the ligand and the protein. The Sybyl-X software system, also a product of Tripos, offers more advanced capabilities, such as virtual high throughput screening in which potentially millions of compounds are docked into a target protein's binding site and scored to provide an estimate of their relative binding energy at the target of interest. Pharmacophore-based virtual high throughput screening, a method of overlaying and comparing a compound of interest with potentially millions of compounds to determine their similarity, and therefore, potential for binding at a macromolecular target, is also possible with Sybyl-X. Image (b) shows an overlay of nicotine and an oxazole derivative, comparing their overall molecular architecture. The grey, translucent surface provides visualization of the molecular volume of the aligned molecules, the red area represents significant differences in hydrophobic surfaces between the two compounds, and the blue/green surface indicates a high degree of electrostatic potential overlap in the two structures. Comparisons of this type can be automated, scored, and sorted in order to facilitate the identification of potentially interesting molecules based on their similarity to known compounds of interest using Sybyl-X. Comparison of macromolecular structures is also facilitated with Sybyl-X. Panels (c) and (d) provide different views of an overlay of steroid 17-alpha-monooxygenase (Cyp17A1, RCSB 3RUK), a key enzyme in steroidogenesis, and cholesterol 7-alpha-monooxygenase (CYP7a1, RCSB 3DAX), the rate limiting enzyme in the synthesis of bile acid from cholesterol. Key differences in the binding sites in the two related enzymes can be exploited by drug discovery scientists to create compounds that are highly selective for one enzyme over the other.

of large compound libraries into biological targets are now common-place in the pharmaceutical industry.[62] Computer-driven predictions of chemical and physical properties are also commonplace, as are homology models[63] designed to provide a better understanding of molecular interactions when X-ray crystal structures are not available. The ability to employ computer-aided design will continue to grow as computer science advances and additional structural details become available.

High Throughput Technology: Chemical Synthesis and Screening Science

While advances in animal models, X-ray crystallography, and molecular modeling had a substantial impact on the course of drug discovery, they did not address the two key bottlenecks in the process, chemical synthesis and screening science. In fact, for the majority of the twentieth century, these issues remained unresolved. Prior to the development of high throughput technologies, drugs were discovered primarily using endogenous ligands, natural products, or marketed drugs as starting points in an animal model. Chemical modifications to improve efficacy was followed by additional *in vivo* screening to chart a path forward.[64] By the 1980s, most pharmaceutical companies' compound collections consisted of only a few thousand compounds acquired through historical projects and screening programs remained primarily a manual process, heavily dependent on low throughput assays and animal models.[65] The situation changed, however, over the last two decades of the twentieth century with the creation of the fields of high throughput chemistry and high throughput screening. Although it is not clear when the concepts for each field were developed, there were significant technological hurdles to overcome in order to accomplish the end goal of increased efficiency in both chemical synthesis and biological screening.

In the case of high throughput chemistry, also referred to as combinatorial chemistry or parallel synthesis, the groundwork that provides the basis for much of the modern methods can be traced back to earlier synthetic efforts that were not originally geared towards increasing efficiency. The preparation of small, druglike compounds on polymer-based material, for example, was first reported by Robert B. Merrifield in 1963 when he described the synthesis of a short peptide sequence on a polystyrene resin (also known as solid phase peptide synthesis).[66] Shortly thereafter, Merrifield reported the preparation of the biologically active peptides bradykinin,[67] bovine insulin,[68] and deaminooxytocin,[69] thereby validating the approach. As interesting as these efforts may have been at the time, the utility of preparing compounds on solid support was met with some degree of skepticism, as indicated by Rappaport and Crowley's 1976 publication entitled "Solid Phase Organic Synthesis: Novelty or

Fundamental Concept?" The article focused on the "ambiguous limitations of non-peptide solid-phase chemistry whose resolution is required if the process is to mature from publishable novelty to fundamental methodology."[70] By the mid-1980s, however, advances in polymer science, automation, and chemical synthesis paved the way for explosive growth in the field of high throughput synthesis, beginning with the independent work of Richard Houghten[71] and H. Mario Geysen.[72] Houghten and Geysen separately described methods for the synthesis of large arrays of small peptides using solid support and successfully applied them to identify biologically active peptides. The practice of high throughput chemistry transitioned out of peptides and into druglike space by the early 1990s with the nearly simultaneous disclosure of the synthesis of arrays of functionalized 1,4-benzodiazepines on solid support by Jonathan A. Ellman[73] and S. Hobbs DeWitt (Figure 2.15).[74]

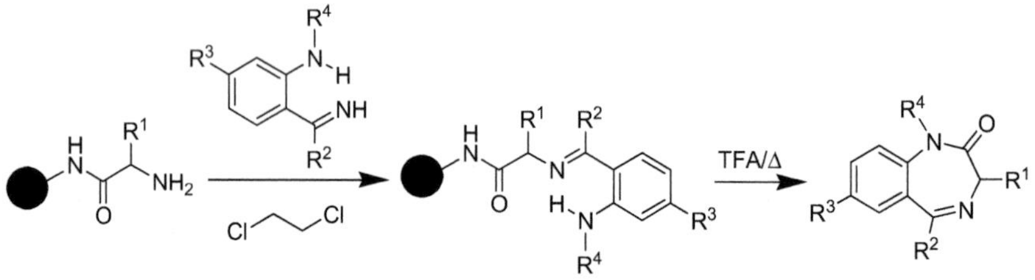

FIGURE 2.15   In 1992, professor Jonathan Ellman and his colleagues demonstrated that 1,4-benzodiazepine derivatives could be prepared on solid support. The application of solid phase chemistry to produce analogs of market drugs such as Valium® (diazepam), Ativan® (lorazepam), and Rivotril® (clonazepam) demonstrated that drug-like compounds could be prepared in this manner.

After these seminal reports, pharmaceutical companies began to incorporate the concepts and practices of solid phase synthesis and high throughput chemistry into their research programs. Resin-bound synthesis continued to progress into the small molecule arena,[75] but at the same time, older techniques were reexamined and new technologies were developed with the goal of increasing the synthetic output of medicinal chemists. Multicomponent reactions designed to incorporate multiple elements of diversity in a single step, such as the Ugi reaction,[76] the Biginelli reaction,[77] and the Passerini reaction[78] were revisited and employed to generate libraries of druglike compounds (Figure 2.16). New equipment dedicated to the rapid synthesis of hundreds, if not thousands, of compounds was developed, along with the technology necessary to purify, store, and retrieve hundreds of thousands of compounds. By the end of the twentieth century, compound collections at most major pharmaceutical companies had eclipsed 500,000 compounds,[65] and by 2013, the number of commercially available screening compounds exceeded 21 million.[79]

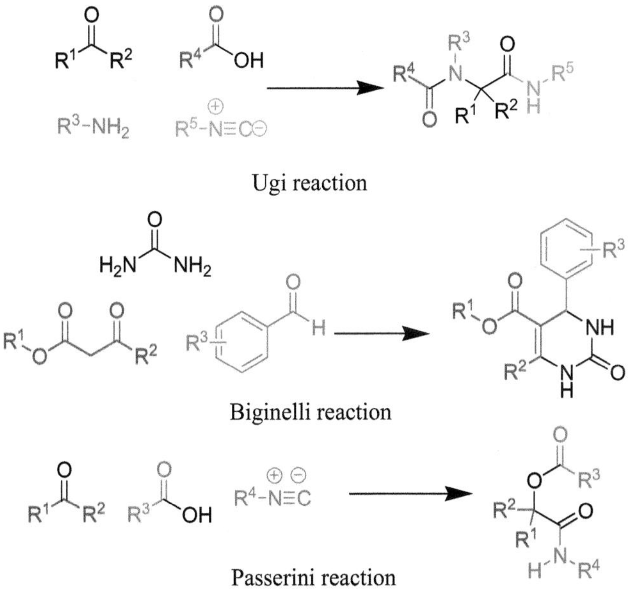

FIGURE 2.16 The Ugi reaction was discovered in 1959 by Karl Ugi, the Biginelli reaction was reported in 1891 by Pietro Biginelli, and the Passerini reaction was discovered in 1921 by Mario Passerini. These reactions have been repurposed for the preparation of large compound libraries suitable for HTS screening.

The development of high throughput screening occurred almost in parallel with high throughput chemistry, although a different set of technological advances were required. Through the 1950s, 1960s, and the 1970s, the pharmaceutical industry moved more and more towards a paradigm of screening compounds in cellular assays and isolated enzyme assays prior to animal testing in an effort to decrease costs and increase efficiency. An increasing understanding of the biochemical basis of disease provided the foundation for new biochemical assays, but the capacity to screen natural product extracts and compound collections was limited by the technology of the time. Prior to the mid-1970s and earlier 1980s, conventional methods of protein isolation and purification severely limited the amount of protein available for any given screen, thus driving up the costs. In addition, cellular assays were limited to using naturally occurring cell lines that could be grown in a reliable fashion.

The biotechnology revolution and the rise of robotics and automation, however, profoundly altered the landscape of compound screening. By the mid-1980s, major advances in biochemistry and molecular biology opened new pathways to the production of large quantities of proteins and "designer" cell lines. Technological breakthroughs, such as recombinant DNA, transfection science, polymerase chain reaction (PCR), and cloning made it possible for scientists to generate cell lines that overexpressed targeted biomolecules, essentially eliminating the supply limitations of the past. Recombinant proteins could be harvested from cellular factories, providing ample quantities of target proteins. Alternatively, custom cell lines could be designed to incorporate biomolecular targets to support cellular screening assays. At the same time, advances in the fields of computer science, robotics, and automation led to the development of robotic platforms capable of performing repetitive motion tasks previously handled by humans, increasing accuracy and efficiency of any number of tasks in multiple fields (Figure 2.17).

FIGURE 2.17   The automated uHTS system at Bristol-Myers Squibb. Integral components and subsystems are shown; (1) Compound store, (2) Hit-picking robot, (3) 3456 reagent dispensing robot, (4) Transport, (5) Incubators, (6) Piezo-electric distribution robot, (7) Topology compensating plate reader, (8) 1536 reagent dispensing robot, (9) Automated plate replicating system, (10) High-capacity stacking system. *Source: Reprinted from Cacace, A.; Banks, M.; Spicer, T.; Civoli, F.; Watson, J. An ultra-HTS process for the identification of small molecule modulators of orphan G-protein-coupled receptors. Drug Discovery Today, 8 (17), 785–792, copyright 2003, with permission from Elsevier.*

While it is not clear exactly when and where automation technology merged with the field of drug discovery, it is clear that by the end of the twentieth century nearly all pharmaceutical companies had transitioned to high throughput screening methods. Initially, screening assays were performed in 96-well microplates (standardized by the Society for Biomolecular Screening and the American National Standards Institute), but the drive for increased efficiency and lower costs eventually lead to the development of 384-, 1536-, and even 3456-well plate technology (Figure 2.18). The miniaturization of screening technologies also spawned advances in micro-fluidics and signal detection methods, as the increased plate density required decreased solution volumes and smaller signal windows. A standard 96-deep well plate could hold up to 1.0 mL of fluid per well, while the

FIGURE 2.18   The typical *in vitro* screening assay employs 96 (left), 384 (middle), or 1536 (right) well plates. As the plates increase in well number (density), the well volume decreases and reagent requirements drop accordingly. The cost saving associated with higher density plates can be substantial.

corresponding 3456-well plate would be limited to a much smaller fluid volume per well. In addition, the density of signals from a 3456-well plate is much higher than that of a 96-well plate (a 3456-well plate contains the same number of wells as 36 96-well plates in the same space), requiring the development of more sophisticated data acquisition tools. By the end of the twentieth century, compound libraries containing hundreds of thousands of compounds could be screened for activity against multiple targets in a matter of days, a feat that would be impossible if attempted manually.

The combination of high throughput chemistry and high throughput screening, however, led to a massive increase in the amount of data produced in any given research program. It quickly became apparent that the bottlenecks of chemical synthesis and biological screening had been replaced by a new bottleneck, data analysis. If, for example, a single enzyme target was screened against a compound library containing 500,000 compounds at a single concentration in triplicate to ensure accuracy, this would produce 1.5 million data points that would need to be associated with the compound library. If one assumes that the hit rate for this hypothetical library of compounds is 0.2%, then 1000 compounds would be identified for follow-up screening to determine their potency (i.e., their IC_{50}). In addition, the majority of drug discovery programs have multiple screening targets for the purposes of determining selectivity, so millions of more data points would become available on compounds of interests across multiple biological targets. The addition of high throughput screening assays to determine physical properties, such as solubility, and druglike properties, such as microsomal stability and permeability, add even more data for analysis and correlation.

Clearly, the level of data available rapidly exceeded the human capacity to evaluate in the absence of computer-driven support. Efforts to address this growing issue led to the development of complex database software systems designed to capture data from a variety of sources (i.e., robotic screening platforms), link the data to a specific chemical structure within the database, and convert the data to a human readable form. Advances in molecular modeling and computational chemistry were also leveraged to increase efficiency, leading to the incorporation of structural data into

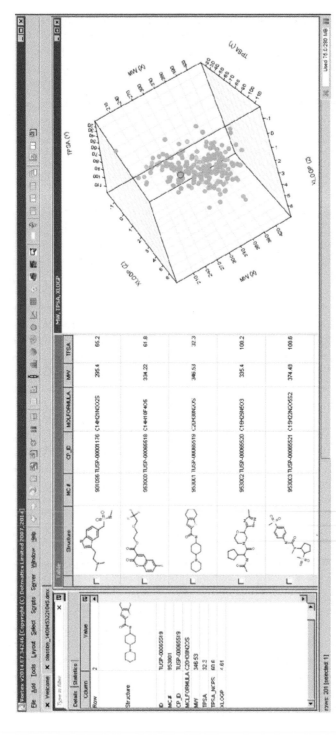

FIGURE 2.19 Cheminformatics software platforms provide scientist with the ability to link compound structures to physicochemical properties (e.g., molecular formula, molecular weight, Topological Polar Surface Area (TPSA), solubility, etc.) and screening data from multiple sources in a searchable database. Groups of structurally related compounds can be identified using sub-structure searching tools, and multidimensional analysis of compound associated data can be used to design next generation compounds with properties consistent with program goals. In this example, a series of compounds are analyzed using the Dotmatics software suite and three dimensional plot has been created to compare changes in molecular weight, TPSA, and cLogP.

modern database software and the birth of the field of cheminformatics. Originally defined by F.K. Brown in 1998,[80] cheminformatics has been applied in drug discovery to store, index, and search information related to individual compounds or groups of compounds. Specialized software provided by companies such as Chemaxon, Core Informatics, Tripos, and Dotmatics is now common place in drug discovery, and allows scientists to evaluate millions of data points with the click of a mouse (Figure 2.19).

Milestones in Biotechnology

Although there were many remarkable discoveries made in the first 70 years of the twentieth century, such as penicillin antibiotics,[81] benzodiazepine central nervous system (CNS) drugs,[82] and macrolide antibiotics,[83] drug discovery scientists of this age were limited in their ability to identify and interrogate targets of interest. The generation of new animal models was restricted to selective breeding of naturally occurring mutations (e.g., the nude mouse) and protein production was limited by the expression levels of proteins in naturally occurring cells. Similarly, the development of cellular assays was dependent on naturally occurring cell lines. The dawn of the age of biotechnology, however, ushered in a new era of drug discovery and disease understanding. Beginning in the 1970s, the restrictions imposed by natural evolution and selection were lifted as scientists began to develop technologies that allowed them to manipulate the DNA of living organisms. Initial experiments in the early 1970s designed to demonstrate that non-native DNA could be prepared (recombinant DNA) and transferred into living cells (transfection technology) were quickly followed by the application of similar technology to generate animals with non-native DNA (transgenic and knockout animal models). In 1975, monoclonal antibodies were introduced, adding further fuel to the biotech fire, and by 1980 companies such as Genentech and Amgen were founded to harness the new techniques for therapeutic purposes. Continued scientific advances, such as polymerase chain reaction (PCR) technology, successful macromolecular therapeutics (recombinant proteins, monoclonal antibodies, and receptor construct/fusion proteins), and the Human Genome Project, further expanded the reach of biotechnology. By the end of the twentieth century, less than 30 years after the initial experiment that launched the field, biotechnology had transformed the process of drug discovery and created a multi-billion dollar industry of its own. In early 2009, Genentech was purchased by Roche for over $46 billion,[84] and as of the end of 2013, Amgen had grown into a $90 billion company.[85] These examples clearly demonstrate the importance of the biotechnology revolution and the profound impact it had on the pharmaceutical industry.

Recombinant DNA and Transfection Technology

Watson and Crick's 1953 discovery of the three dimensional structure of DNA provided an understanding of its physical structure, but this knowledge did not provide the tools necessary to manipulate DNA. It would take another 20 years to develop this technology. The first step in this process was the identification of the enzymes involved in DNA production, modification, and degradation. Significant progress was achieved in the 1950s and 1960s. In 1956, DNA polymerase I, an enzyme capable of copying DNA template strands, was identified by Arthur Kornberg.[86] This was the first of many enzymes identified as acting on polynucleotide sequences that would lay a critical foundation for the experiments that led to the development of the technology necessary to not only manipulate the DNA of a species, but also to transfer functional DNA between species. The time period between 1956 and 1975 witnessed the identification of DNA active enzymes such as the DNA ligases, the enzymes responsible for joining DNA strands end to end,[87] exonucleases, which remove nucleotides from DNA chains,[88] and terminal transferases (also known as terminal deoxynucleotidyl transferases), enzymes capable of adding nucleotides to the 3' end of DNA.[89] Reverse transcriptases,[90] enzymes capable of converting RNA into DNA, were also identified in this time period. The identification of the restriction enzymes (also known as restriction endonucleases), however, was the key to unlocking the puzzle. This class of enzymes, capable of creating two incisions across a double stranded DNA chain, provided DNA duplex segments with complimentary single stranded ends (also referred to as "sticky ends" or "cohesive ends").[91] Essentially, this provided scientists with the ability to carve out specific segments of duplex DNA strands, the nature of which are dictated by the selectivity of the particular restriction enzymes employed. DNA strands with complimentary "stick ends" could then be stitched together with the appropriate enzymes, thereby creating synthetic DNA, also referred to as recombinant DNA (Figure 2.20).

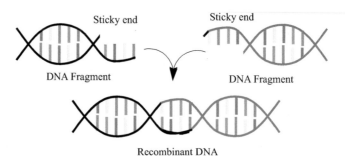

FIGURE 2.20 The identification of enzymes responsible for building, degrading and modifying DNA changes was critical to the development of recombinant DNA technology. Once these enzymes became available, DNA chains with complementary "sticky ends" could be stitched together to form "designer" DNA strands.

While the molecular biology of nucleic acid synthesis was being unraveled, scientists were also developing an understanding of virus form and function. The concept of infectious agents smaller then bacteria originated with French microbiologist Charles Chamberland in 1884. His studies of the infectious agent using filtration methods designed to remove bacterial organisms clearly demonstrated that a non-bacterial agent (eventually identified as the tobacco mosaic virus) was responsible for an infection present in tobacco plants. Over the course of the next several decades, methods to grow, isolate, and examine viruses evolved. Bacteriophages, viruses that infect bacteria which eventually became powerful tools in the study of DNA transfer, activation, and inactivation, were identified through the separate work of Frederick Twort[92] and Félix d'Herelle[93] at the turn of the twentieth century. In 1931, the cultivation and isolation of influenza and a number of other viruses using fertilized chicken eggs was reported by Ernest William Goodpasture,[94] opening the doorway to mass production of virus particles for scientific study. Further improvements in methods for the production and study of viruses continued between 1930 and 1970, setting in place another piece of the puzzle that eventually became recombinant DNA technology. The knowledge developed through the study of virus biology would eventually be utilized to develop the delivery vehicles necessary to move the science of recombinant DNA forward.

By the end of the 1960s, all of the tools necessary for the manipulation of genetic material in living organisms were in place, and in 1969, Peter Lobban, a graduate student working in the Biochemistry Department at Stanford University Medical School took the first steps down the path. His Ph.D. project proposal, presented to his research review committee as part of his progress towards his degree, suggested the merging of DNA modification technology and viral biology to provide a method of artificially transferring genetic material from one species to another.[95] His theories were quickly validated with the first publication of this new technology appearing in 1972 in which David Jackson et al. described methods of inserting new DNA into simian virus 40 (SV40).[96] By 1973, scientists at Stanford University had published methods for the end to end joining of DNA molecules[97] and the construction of biologically functional bacterial plasmids.[98]

Then, in 1974, the labs of Stanley N. Cohen and Herbert W. Boyer at Stanford University fundamentally altered the landscape of the pharmaceutical industry with a patent application (serial number 520,961) that described methods

> for genetically transforming microorganisms, particularly bacteria, to provide diverse genotypical capability and producing recombinant plasmids... which is used to transform a susceptible and compatible microorganism.... The newly functionalized microorganism may then be used to carry out their new function; for example, by producing proteins which are the desired end products, or metabolites of enzymatic conversions or be lysed and the desired nucleic acids or proteins recovered.[99]

The work of Cohen and Boyer provided methods to produce proteins and other cellular products by simply creating a stable microorganism that could be tailor made to produce the material and used as a factory (Figure 2.21). Cell lines that overexpressed cell surface receptors were also

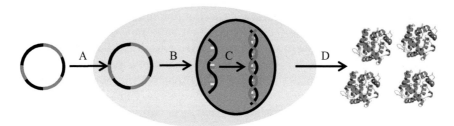

FIGURE 2.21   Stable cell lines capable of producing large amounts of a desired protein such as insulin can be prepared by (A) introducing a gene of interest into a cell (B) that subsequently enters the nucleus where it is (C) incorporated into the cell's chromosomal DNA. A stable cell line can be grown that will (D) express the desired protein that can be harvested from the growth media.

eventually developed as a result of this work, enabling the detailed study of a host of cellular targets that were previously difficult to examine due to low levels of expression. Inserting the proper DNA sequences into a suitable cell line would provide cells that overexpressed the desired target or protein (an overexpressing cell line), greatly amplifying the presence of the biological target of interest. Transfection technology is nearly omnipresent in the modern drug discovery lab. In the decades that followed Cohen and Boyer's initial patent, thousands of new cell lines have been developed using recombinant DNA and transfection technology. This seminal work also became one of the major underpinnings for the biotechnology industry. Recombinant human insulin, developed at Genentech and licensed to Eli Lilly,[100] was the first recombinant protein to gain market approval and many others have followed. As of 2012, less than 45 years after the initial suggestions of Peter Lobban, the biotechnology industry has grown from a series of lab experiments into a $300 billion industry[101] and has become an integral part of the modern drug discovery process.

Polymerase Chain Reaction (PCR) Technology

One of the early limitations of the biotechnology industry was the ability to prepare and analyze DNA. Although the tools to manipulate and analyze DNA (i.e., enzymes acting upon DNA) had been identified in between 1950 and 1970, the process was slow and manual. The utility of the science itself had been clearly demonstrated by the early

1980s, but the ability to generate large quantities of DNA remained lacking. The first attempt to replicate DNA using the enzymatic tools developed in the previous decades was reported by Kleppe and his coworkers in 1971,[102] but the process was far from optimal. DNA replication required that the double stranded helix be separated into the two parent strands, which can be accomplished by heating the sample to a high enough temperature (DNA melting). Upon cooling and in the presence of complimentary DNA primers (oligonucleotide starting points for DNA synthesis), nucleotide building blocks, and a suitable DNA polymerase, the DNA is replicated, providing copies of the original for study. Repeating the cycle provides additional copies of identical DNA with an exponential growth rate (Figure 2.22). The key limitation prior to the advent of modern PCR

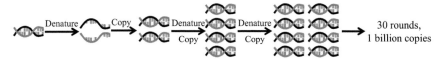

FIGURE 2.22   In the polymerase chain reaction process, each cycle of the denaturing and copying process doubles the number of copies of DNA present. Three copies cycles provides eight copies of the DNA chain. After 30 rounds of the sequence, the DNA amplification exceeds one billion copies.

technology, however, was the melting step. Like most enzymes, most DNA polymerases denature at the higher temperatures required for separation of duplex DNA, so each cycle of heating and cooling required the addition of fresh DNA polymerases, making the process of DNA replication both time consuming and expensive.

The landscape changed in 1976 with the discovery of Taq polymerase. This particular variant of DNA polymerase was isolated from *Thermus aquaticus*, a member of a family of unusual bacteria, thermophilic bacteria, which can survive in temperatures of up to 80 °C (175 °F). Prior to the discovery of these microbes in the geysers of the Yellowstone National Park,[103] it was generally believed that life could not be sustained above 55 °C, but clearly these organisms and other like them disagreed. In order to survive the harsh conditions of the geysers, *T. aquaticus* had developed biological systems that did not break down at the elevated temperatures of its environment. This included a variation of DNA polymerase, Taq polymerase,[104] which could function at elevated temperatures without denaturing. This paved the way for automation of DNA amplification. With the addition of this tool to the growing biotechnological toolbox, DNA amplification could be accomplished without adding additional DNA polymerase at the end of each cycle, greatly simplifying the process. In 1983, Kerry Mullis and his colleagues at Cetus Corporation, an early player in the biotechnology field, were the first to harness these tools to create automated PCR equipment based on thermocycling, an alternating series of heating and cooling steps.[105]

With the advent of this technology, it became possible to generate millions of copies of a DNA strand using an efficient, automated process. This, in turn, facilitated rapid advances in numerous areas including genetic sequencing of organisms, cloning, diagnosis of hereditary diseases, and the detection of infectious agents. Genetic finger printing as a means to determine paternity and the use of genetic material in forensic sciences also developed as a result of the development of PCR technology. Multiple variations of PCR technology have been developed since the original reports, and much like recombinant DNA technology, PCR has become common place in modern drug discovery. The importance of this work was recognized with the awarding of the Nobel Prize in chemistry to Kerry Mullis in 1993. (The prize was shared with Michael Smith, who focused on site-directed mutagenesis.)

Monoclonal Antibody and Hybridoma Technology

While the development of monoclonal antibody technology is certainly a product of the biotechnological revolution of the twentieth century, it is clear that the importance of antibodies in general was recognized decades earlier. Paul Ehrlich was the first to suggest the term antibodies in 1891, and in 1897, he introduced the "side chain theory" of antibody/antigen interaction, which suggested that receptors on the surface of a cell could bind to antigens and stimulate the production of antibodies.[106] At the time, Ehrlich did not have the tools necessary to definitively test his theory, and it was almost 50 years later that Astrid Fagreaus determined that B-cells were the source of Ehrlich's antibodies.[107] The concept of monoclonal antibodies is also significantly older than the technology itself, as it was originally suggested by F. M. Burnet and his colleagues in the 1950s.[108] In brief, Burnet's theory, which proved to be correct, stated that upon full differentiation, antibody-producing B-cells (and their progeny) produce only a single type of antibody that would bind to only one target molecule. He further suggested that the polyclonal nature of the immune response observed in animals upon exposure to an antigen was the result of multiple lines of B-cells producing different antibodies targeting the same antigen, but through different structural features of the antigen (antigenic determinants). Although Burnet's theories clearly pointed to the concept that monoclonal antibodies could be produced from a uniform B-cell line, the technology to create a stable, antibody-producing cell line was not available at the time.

In a somewhat ironic twist of nature and science, the solution to the problem of generating stable cell lines capable of producing monoclonal antibodies evolved from the disease that is the target of numerous monoclonal antibody-based therapies. Cancer, specifically multiple myeloma, was the key to unlocking this puzzle. By the early 1970s, multiple myeloma had been recognized as a malignant disorder of antibody-producing cells, and in 1973 Jerrold Schwaber and Edward Cohen reported the fusion of antibody-secreting mouse myeloma cells with human peripheral blood lymphocytes. The resulting hybrid cell line could be grown continuously and more importantly, produced human antibodies along with the mouse antibodies.[109] These milestone experiments were followed by the 1975 report of Georges Köhler and César Milstein[110] of stable cell lines generated by the fusion of mouse myeloma cells with antibody producing mouse B-cells. Each fusion cell, more commonly referred to as a hybridoma cell, produced a single antibody and, upon isolation and cloning, provided access to a stable cell line that produced a single antibody (Figure 2.23). This new process opened the

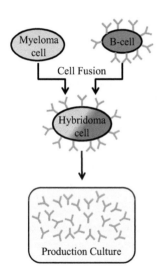

FIGURE 2.23 Monoclonal antibody producing cells can be prepared by fusing an antibody producing b-cell with a myeloma cell. The resulting hybridoma cell can be isolated, and cloned to provide a stable cell line capable of producing a single antibody (a monoclonal antibody).

doorway to generating large amounts of tailor-made, highly specific, monoclonal antibodies, creating immense opportunities for the fledgling biotechnology industry. In recognition of these groundbreaking achievements, Georges Köhler and César Milstein shared the 1984 Nobel Prize in Physiology or Medicine with Niels Kaj Jerne, a Danish immunologist, "for theories concerning the specificity in development and control of the immune system and the discovery of the principle for production of monoclonal antibodies."

The ability to prepare significant quantities of monoclonal antibodies targeting virtually any macromolecular target had a profound impact on the drug discovery industry. It was quickly realized that these new tools could be used to enhance screening technology, study cell surface proteins in more detail, purify proteins, and perhaps most importantly, they had the potential to be exquisitely specific therapies. For example, a monoclonal antibody could be designed to target a specific cell type, such as cancer cells, through an antigen unique to the target, such as a cell surface protein. Binding of the monoclonal antibody to the target would then prompt an immune system response targeting the cells for destruction.

The allure of treating patients with such highly specific drugs led to an immense level of research in the area. By the late 1980s, the humanization of monoclonal antibodies[111] had been achieved and in 1986, the first monoclonal antibody therapy, Orthoclone OKT3® (Muromonab-CD3), was approved by the FDA for the prevention of transplant rejection.[112] By the beginning of the twenty-first century, monoclonal antibody therapy had established itself as a significant player in the pharmaceuticals industry. Drugs such as Herceptin® (Trastuzumab, breast cancer) and Remicade® (Infliximab, arthritis) had achieved blockbuster status, with multibillion dollar annual sales, and every major pharmaceutical company was investing in this technology.

THE RISE OF BIOLOGICS AND MACROMOLECULAR THERAPEUTICS

The importance of the biotechnological revolution that began in the 1970s cannot be understated. Modern drug discovery would be a far more difficult task if it were not for the ground breaking technologies that were developed during this time period. High throughput screening, for example, depends on the production of large quantities of proteins and antibodies, neither of which would be available without recombinant DNA, PCR, and hybridomas. The identification of novel, druggable targets would be far more difficult, and programs designed to fully map the genome of a given species, such as the Human Genome Project,[113] would be all but impossible. Transgenic and knockout animal models that have provided significant insights into disease mechanisms and drug therapy would not be available if not for the pioneering efforts of the scientists that drove biotechnology forward.

The most important impact of this era, however, has been the development of groundbreaking therapeutic agents. Recombinant human insulin is probably the most well-known, but dozens of other important therapeutic agents have changed the lives of patients around the world. Recombinant human proteins such as Activase® (Alteplase tissue plasminogen activator, Genentech),[114] Epogen® (erythropoietin, Amgen)[115] and many others provide treatments for conditions that might otherwise be impossible to treat. Monoclonal antibodies such as Remicade® (Infliximab, Janssen

Biotech)[116] and Herceptin® (Trastuzumab, Genentech)[117] have revolutionized the treatment of arthritis and cancer, respectively, providing therapeutic relief to millions.

Hybrid technologies have also emerged. Receptor construct fusion proteins, for example, have emerged as a new tool that employs both antibody and protein technology. Therapies such as Orencia® (Abatacept, arthritis),[118] AmeviveE® (Alefacept, psoriasis),[119] and Eylea® (Aflibercept, wet macular degeneration)[120] are each composed of a protein receptor segment and an immunoglobulin structure. The protein receptor portion provides selectivity for the target of interest, while the immunoglobulin structure provides metabolic stability.

Major companies such as Genentech and Amgen emerged as key early player in the commercialization of biotechnology, and in the modern world of drug discovery, every major pharmaceutical company is vying for a piece of the biotechnology pie. While it is clear that the therapeutic utility of macromolecules is of vital importance to the drug industry, it is well beyond the scope of this text to describe the methods used to discover them. Those interested in the specific details of the discovery of each class of biotherapeutics are encouraged to consult texts that are focused on those particular areas. Sections of this text covering aspects of drug discovery such as pharmacokinetics, clinical trials, patent law, animal models, and translation medicine are, however, just as relevant to biotherapeutics as they are to small molecule therapies.

SOCIETAL AND GOVERNMENTAL IMPACTS

While scientific advances certainly played a major role in defining the drug discovery process that has evolved over time, it is abundantly clear that societal forces also helped shape the modern drug discovery process. It is well established that humanity has been searching for treatments to alleviate suffering and disease for hundreds, if not thousands, of years, and in many cases, the identification of new therapeutic agents can be directly linked to a major public health need of a particular time period. Agostino Salumbrino's observations that eventually led to the discovery of quinine,[8] for example, were driven by societal needs to deal with malaria, a major public health issue. In the same sense, but in a more modern setting, the development of modern cardiovascular drugs was driven by the realization that cardiovascular disease is a major mortality risk. Similarly, advances in antiviral drug technology were primarily driven by the societal impact of the AIDS epidemic.

Of course, many drugs have been brought to market by pharmaceutical companies in the absence of social pressures, but the appearance of the drugs on the market led to societal demand for additional treatments. It is hard to argue, for example, that the development of Rogaine® (Minoxidil)[121] for the treatment and prevention of hair loss was prompted by a major health issue surrounding baldness. However, this drug was originally developed as an antihypertensive agent. The company that developed it, Upjohn Corporation, simply capitalized on an interesting side-effect observed in patients being treated for high blood pressure to create a new market for one of its products. Since the introduction of Rogaine, many companies have spent millions of dollars to break into the hair growth market. The social pressures of vanity and an aging population, as well as a corporate desire for increased profits, clearly influenced the choice to pursue Rogaine and drugs with similar utility.

Apart from influencing which disease states and conditions would be addressed over time, it is also clear that societal forces, often through governmental intervention and regulation, have played an important role in determining how drugs are developed. The modern drug discovery process is a highly regulated path to market that must be adhered to in order to successfully bring a drug to market, but this was not always the case. In fact, prior to the twentieth century, there were few, if any, laws or guidelines in place that specified what could or could not be sold as a medicine. Similarly, there were essentially no guidelines or requirements in place for determining if materials used as drugs were effective or even safe to use. While it is well beyond the scope of this text to provide a detailed accounting of the history of governmental and societal actions that contributed to the evolution of the modern drug discovery process, an examination of some of the major milestones is instructive.

The Pure Food and Drug Act of 1906[122]

At the beginning of the twentieth century, the preparation and sale of medication was virtually unregulated. In the absence of restriction by government regulation, virtually anything could be sold as a "drug" and, in many cases, chemicals that are now known to be harmful were marketed as "medicines." Syrups marketed to soothe the crying of infants and children were often laced with opiates, and addictive drugs such as cocaine and heroin were routinely part of "patent medicines." The simple requirement of an accurate listing of ingredients was not in place, and secret ingredients in medicine were part of the back drop against which the Pure Food and Drug Act of 1906 came into being. Samuel Hopkins Adams' series of articles entitled "The Great American Fraud," in which he detailed the abuses of the pharmaceutical industry of the time, is often credited with igniting the fire that lead to the passage of the first law designed to regulate the drug industry.[123]

While limited in its scope, this first attempt to ensure the safety of drugs had far reaching consequences for the pharmaceuticals industry, as it established the foundation upon which the Food and Drug Administration (FDA) would eventually be built. "Dangerous" drugs such as cocaine, heroin, alcohol, and morphine could no longer be used as secret ingredients in medicines, although they could still be included as long as they were accurately labeled. In addition, the U.S. Pharmacopeia[124] and the National Formulary[125] were put in place as the authorities for drug composition and formulation. More importantly, however, the new law provided the Bureau of Chemistry in the U.S. Department of Agriculture, the authority to establish a group of federal inspectors to enforce the laws. These inspectors were empowered to seize and destroy material found to be in violation of the new law (at the company's expense) and provide publication of all violations that occurred. Although direct financial penalties were modest, the prospect of negative publicity and physical loss of manufactured materials became a major tool in enforcing drug regulations.

The Elixir of Sulfanilamide Disaster of 1937[126]

Although the Pure Food and Drug Act of 1906 set the stage for further improvements in the regulation of the pharmaceuticals industry, the law was far from adequate. The law provided no guidance or requirements for the safety of drugs brought to market. As is often the case, a disaster of some type would be required before this would change. Thus, in 1937, the S. E. Massengill Company began marketing a new formulation for the antibiotic sulfanilamide under the name Elixir of Sulfanilamide. Sulfanilamide had been successfully employed for the treatment of streptococcal infections when provided in the form of a tablet or powder, but a liquid formulation was not available. After receiving requests from field sale agents, the company's head of chemistry and pharmacy, Harold Watkins, created a new formulation in response to this request. The new product contained three key ingredients, sulfanilamide, raspberry flavoring, and diethylene glycol (Figure 2.24). After tests for flavor, appearance, and fragrance were deemed acceptable, the new product

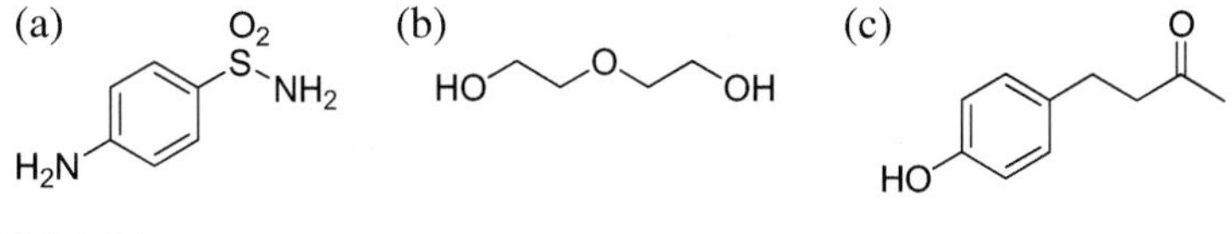

FIGURE 2.24 Elixir of Sulfanilamide marketed by S. E. Massengill contained (a) sulfanilamide, (b) diethylene glycol, and (c) Raspberry flavoring.

was prepared in bulk and distributed nationwide in September of 1937. No safety studies of any kind were performed, as none were legally required at the time.

By October of 1937, the consequence of not studying the safety of a potential new therapy became abundantly clear. In Tulsa, Oklahoma, the American Medical Association received reports from physicians that Elixir of Sulfanilamide had been linked to a number of deaths. Their laboratory quickly recognized that diethylene glycol, a compound often used as antifreeze and a deadly poison, was the cause of the deaths. The federal government was notified on October 14th, and began the process of recalling the material. Of the 240 gallons prepared and distributed for sale, 234 gallons were retrieved and destroyed, but the damage had already been done. At least 107 deaths were attributed to Elixir of Sulfanilamide poisoning. The victims, many of them children suffering from simple sore throats, died as a result of kidney failure brought on by ingestion of diethylene glycol.

Although a simple animal safety study would have quickly revealed the toxicity of diethylene glycol, there were no legal requirements for animal safety studies at the time, and none were performed. Even a brief review of the scientific literature at the time would have been sufficient to uncover the deadly nature of diethylene glycol, but even this simple precaution was not taken, and the public paid a heavy price. In fact, in 1937, there were no laws barring the sale of dangerous, untested, or even poisonous drugs. Legal authority to confiscate and destroy the Elixir of Sulfanilamide was based on misbranding the material as an "elixir" when there was no alcohol in the product, rather than the deaths that it caused, as there was no other legal basis for the recovery in 1937. The rather trivial charge of "misbranding" was the only charge that could legally be applied against the company, even though it was clearly responsible for marketing a known poison as a medicine. When pressed to admit some level of accountability for this tragedy, the company's owner, Dr. Samuel Evans Massengill, denied any responsibility for the tragedy, stating "My chemists and I deeply regret the fatal results, but there was no error in the manufacture of the product. We have been supplying a legitimate professional demand and not once could have foreseen the unlooked-for results. I do not feel that there was any responsibility on our part."

The Elixir of Sulfanilamide disaster was not the first time a dangerous drug had been brought to market, but it is widely viewed as a watershed event in the history of drug regulations. In response to this disaster, congress passed the Food, Drug, and Cosmetic Act of 1938. Under this new law, pharmaceutical companies were required to prove the safety of their new products through animal safety studies prior to receiving marketing approval. In addition, manufacturers would be required to submit an application for marketing approval to the FDA before new products could be brought to market. The New Drug Application (NDA) process had been born.

The new laws were not without flaws, however, as applications for marketing approval would be automatically approved if the FDA did not act within a set period of time. In addition, companies were not required under the new law to demonstrate that their products were effective. These issues would be addressed in later legislative and regulatory proceedings, but for all intents and purposes, the Food, Drug, and Cosmetic Act of 1938 established the framework of the modern drug approval system.[127]

The Thalidomide Story[128]

The commercialization and subsequent withdrawal from market of Thalidomide is perhaps one of the most compelling and tragic events in the history of drug discovery and development. Originally prepared in 1954 by scientists at Chemie Grünenthal GmbH, a German pharmaceutical company, Thalidomide was studied clinically soon after it was patented. By July of 1956, safety studies on animals had demonstrated that it was nearly impossible to achieve a lethal dose of the drug, so it was licensed for sale as an over-the-counter sleep aid in Germany and most of Europe. Use among pregnant woman increased significantly when it was discovered that Thalidomide was also useful as an antiemetic for the suppression of morning sickness, and the drug was marketed under as many as 37 different names worldwide.

Unfortunately, Thalidomide turned out to be far more dangerous than expected. While animal safety studies did indicate a lack of acute toxicity, other safety issues were not studied, especially those related to the effects of a drug on a developing fetus. The prevailing theory on fetal development during the 1950s was that the placenta provided perfect protection to a developing fetus, protecting it from any drugs or toxic material ingested by the mother. As such, few, if any, studies were performed to determine the safety of new drugs during pregnancy. If such testing had been done, Thalidomide would probably have never made it out of the labs at Chemie Grünenthal GmbH. In the absence of such testing, Thalidomide was used by thousands of pregnant women across the globe, but by 1959 questions began to arise about the true safety of the drug. In 1960, reports of peripheral neuropathy after long term usage began to appear in England, although the manufacturer continued to insist that the drug was safe. Frances Oldham Kelsey, an FDA physician assigned to review the Thalidomide New Drug Application, refused to provide marketing approval, however, insisting that additional safety studies were necessary before Thalidomide could be approved in the United States.

By 1961, it had become abundantly clear that Dr. Kelsey's decision to insist on more safety studies was justified. Peripheral neuropathy was just the tip of the iceberg in terms of safety issues. Less than five years after its launch as a safe drug for the treatment of morning sickness, over 10,000 children had been born with severe birth defects linked to the drug. "Thalidomide babies" were often born with misshapen or missing limbs (Figure 2.25). William McBride, an Australian obstetrician, and Widukind Lenz, a German pediatrician, independently suggested the link between Thalidomide and the birth defects that eventually lead to the revocation of marketing rights in most global markets by 1962.

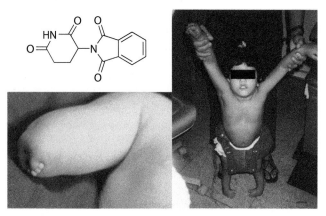

FIGURE 2.25   Thalidomide was brought to market in 1956 by Chemie Grünenthal GmbH as an over the counter treatment for morning sickness in pregnant women. By 1962, it had been withdrawn from markets across the globe as it had been definitively linked to severe birth defect in the children of woman that used it during their pregnancies. *Source: Lower left: Reprinted with permission from Davies, D. P.; Evans, D. J. R. Clinical dysmorphology: understanding congenital abnormalities. Curr. Paediatrics, 13 (4), 288–297, copyright 2003, with permission from Elsevier. Right: Reprinted from Miller, M. T.; Strömland, K.; Ventura, L.; Johansson, M.; Bandim, J. M.; Gillberg, C. Autism associated with conditions characterized by developmental errors in early embryogenesis: a mini review. Int. J. Dev. Neurosci., 23 (2-3), 201–219, copyright 2005, with permission from Elsevier.*

In response to the tragedy that had unfolded, regulatory standards for safety and efficacy testing of new drug candidates were significantly improved with the passage of the Kefauver Harris Amendment of 1962. It is also worth noting that theories on the protection provided by the placenta to an unborn child were substantially revised. The marked increase in birth defects caused by Thalidomide left little doubt that pre-natal exposure was a serious safety issue that had to be addressed. The placenta was not the perfect protection it was thought to be. Also, the importance of drug chirality was brought forward on two different levels. First, it was eventually determined that (S)-isomer of Thalidomide is a major groove binder responsible for the teratogenic nature of the drug, but the (R)-isomer does not share this liability. Prior to this realization, the possibility that single enantiomers might have different biological effects had not been widely considered, and racemic drugs were routinely studied. After the thalidomide tragedy, however, scientists began to focus more heavily on single enantiomers. In the modern era, racemic drug candidates are very rare.

In related findings, an understanding of chiral stability also began to take shape as a result of the tragedy that unfolded around Thalidomide. While it is true that the two enantiomers of Thalidomide have different biological properties, the (R)-isomer still represents a significant hazard for pregnant women. This is the result of the chiral instability of the (R)-isomer in an *in vivo* setting. At physiological pH, the (R)-isomer of Thalidomide undergoes racemization to a mixture of the two isomers (Figure 2.26),

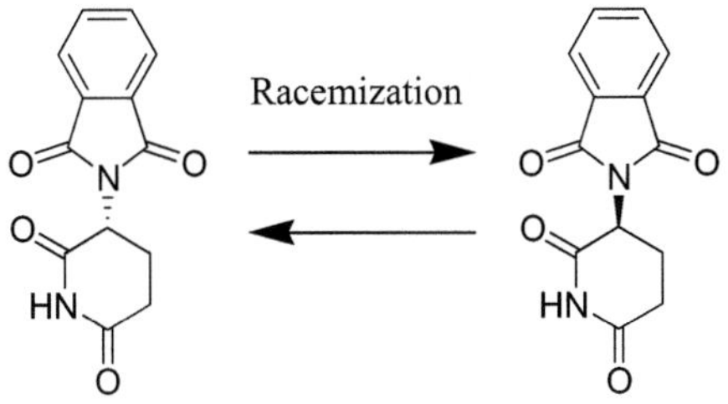

FIGURE 2.26 Thalidomide is not chirally stable *in vivo*. The (R) isomer (left) is readily converted to the (S) isomer (right). As a result, the pure (R) isomer is no safer than the originally marketed racemic material.

so even if a patient is provided only the safer (R)-isomer, the more dangerous (S)-isomer will be generated *in vivo*. In the modern drug discovery process, it is common place to check the chiral stability of possible drug candidates as a result of these findings.

REGULATORY MILESTONES

While major events such as the Thalidomide tragedy made it abundantly clear that new laws needed to be put in place to regulate the growing pharmaceutical industry, many other legislative events occurred in the absence of the substantial attention generated by a drug failure. In fact, over the course of the twentieth century, the growth of the pharmaceuticals industry has been tracked by a parallel growth in regulatory bodies responsible for oversight of the industry. The growth of the regulatory agencies was often a step behind the industry itself, as the laws granting them authority over the industry were most often reactionary in nature (i.e., developed and implemented in response to a perceived problem in the system). Over the course of the twentieth century, however, regulatory bodies such as the FDA, the European Medicines Agency (EMA), and many others across the globe have been created through legislative action with the goal of ensuring the safety of marketed drugs. Although it is well beyond the scope of this text to describe the historical context and impact of all of the laws governing the drug industry (The FDA as it exists in 2013, for example, is the result of over 200 laws passed since 1906), there are some key milestones that should be examined.

Durham–Humphrey Amendment of 1951[129]

As discussed earlier, the Pure Food and Drug Act of 1906 and the Food, Drug, and Cosmetic Act of 1938 established the authority of the FDA to act in the interest of consumers and protect the populace from dangerous drugs. In fact, the FDA used the authority granted to it in 1906 and 1938 to declare that some drugs were not safe for use in the absence of individualized medical supervision. By 1941, more than 20 drugs and drug classes, including sulfa antibiotics, barbiturates, and amphetamines, required a prescription from either a physician or dentist. Neither of these two laws, however, provided clear definitions for prescription versus nonprescription drugs, and there was no specification as to who was responsible for labeling drugs as belonging to one class or the other. In addition, there were no clear guidelines regarding refilling of prescriptions. This lack of clarity led to a number of legal battles between the FDA, the drug industry, and professional pharmacy organizations over the distribution of prescription drugs.

The Durham–Humphrey Amendment of 1951 addressed this hole in the law by firmly establishing two classes of drugs, those that required a prescription, legend drugs, and those that did not, over-the-counter (OTC) drugs. In short, under this amendment, drugs that have been proven to be safe, effective, and require little, if any, medical oversight in their use (e.g., aspirin) could be sold as OTC products. On the other hand, drugs with addictive properties (e.g., morphine) or that required medical monitoring to ensure safety (such as monitoring of liver functions required with the use of statin cholesterol-lowering drugs) could only be distributed with the consent or under the direction of a physician via prescription. The new law also codified the role of the pharmacist in ensuring that prescription drugs were provided only with a properly documented prescription and the conditions under which refills would be available.

Kefauver–Harris Amendment of 1962[130]

The tragic events that unfolded around the failure of Thalidomide led to substantial public outrage and pressure to enact stricter laws and regulations designed to ensure that public safety and well-being were at the forefront of the drug discovery process. The Kefauver–Harris Amendment of 1962 significantly broadened the FDA's regulatory authority over the pharmaceuticals industry with this goal in mind. The law required demonstration of both efficacy and safety of potential new drug candidates prior to granting marketing approval, effectively abolishing the automatic approval clause of the 1938 Food, Drug, and Cosmetic Act. The law also required that all drugs launched between 1938 and 1962 had to be proven effective in order to maintain their place in the pharmacy. The National Academy of Sciences and the FDA collaboratively studied this set of drugs, and discovered that nearly 40% were not effective. They were subsequently removed from the list of approved drugs.

Clinical trials, manufacturing processes, and even advertising of prescription drugs were also placed within the jurisdiction of the FDA. Clinical trial design had to be approved by the FDA, informed consent of study participants was required, and known side effects had to be disclosed to the public under the new law. Good manufacturing practices (GMP) and FDA access to company control and production records were also required in order to promote quality assurance. Finally, advertising for prescription drugs was placed under strict regulation. Marketing of generic drugs as new breakthrough medications was barred, and accurate disclosures of efficacy and side effects associated with drug treatment were required in all advertisements for prescription medications. In summary, the Kefauver-Harris Amendment of 1962 gave the FDA virtually complete authority over drug approval and marketing.

Hatch–Waxman Act of 1984[131]

Although the Drug Price Competition and Patent Term Restoration Act of 1984, also known as the Hatch–Waxman Act, did not have a direct impact on the drug discovery process itself, it did fundamentally alter the pharmaceutical landscape. Unlike the previously discussed laws which were focused on safety issues, the Hatch–Waxman Act was designed to encourage the growth of the generic drug industry, thereby decreasing the costs of prescription medication. Prior to the enactment of this legislation, generic drugs represented approximately 10% of the prescription market, despite the fact that many major medicines were no longer under patent protection. By 2008, the market share held by generic drugs had risen to nearly 70% of the market, providing a clear indication that the Hatch–Waxman Act was successfully increasing competition in the prescription drug market. The positive impact on health care costs in the form of cheaper, generic prescription drugs was also viewed as a positive sign to those interested in containing the rising cost of health care.

The success of the Hatch–Waxman Act was driven by a few key changes in the drug approval process and patent laws that simplified market entry for generic drugs. First, the approval process itself was simplified through the creation of the Abbreviated New Drug Applications (ANDAs). Prior to 1984, companies interested in marketing a generic version of a prescription drug were required to provide the same level of studies as the original manufacturer of the medicine in question, including animal safety studies, bioavailability studies, and human clinical trials. At the same time, generic manufacturer would not have the benefit of patent protection afforded to the "innovator" companies, making it more difficult for generic companies to recoup the substantial investments required to bring a drug to market. Under the provisions of the Hatch–Waxman Act, generic drugs could be approved for marketing based on the "innovator's" clinical and safety data. Bioequivalence studies designed to demonstrate that the generic drug provided the same bioavailability as the marketed equivalent replaced the time-consuming and expensive efficacy and safety trials, significantly lowering the cost of market entry.

Changes to the patent law and market exclusivity rules were also put into place in order to support both the "innovator" manufacturers and generic manufacturers in an effort to provide a balanced playing field for both. A "safe harbor" clause was included in the legislation that allowed generic drug companies to manufacture and study patented drugs as part of an effort to generate data necessary for an ANDA submission. In the absence of this "safe harbor" clause, generic companies could have been sued for patent infringement if their efforts to generate a generic copy of a drug occurred during the lifetime of a patent. This change coupled with the law's allowance for lawsuits by generic companies seeking to invalidate drug patents, created significant openings in the prescription drug market that have been exploited by generic drug companies. Those interested in the details of generic market entry based on lawsuits to invalidate drug patent are encourage to consult paragraph four of 35 U.S.C. 271 for additional information.

Measures to protect the "innovator" companies were also put into place, as it had become widely realized that a significant portion of a drug's patent life was being consumed by clinical trials and the FDA approval process. These processes were dramatically shortening the useful patent life of potential new drugs, thus increasing their overall cost. The ANDA provisions of the Hatch–Waxman Act threatened to further erode the useful patent life of new therapeutic agents if enacted alone. In recognition of this possible negative outcome for "innovator" companies, provisions were put in place for the extension of patent terms to compensate for time lost in both clinical trials and the approval process. In general, the length of patent protection for any drug was increased by 50% of the time spent in clinical trials and 100% of the time spent in the NDA approval process. The law also created a new class of drugs, those directed towards "orphan indications" or diseases with fewer than 200,000 patients. Companies that provide new therapies for orphan indications were granted market exclusivity for 7 years, irrespective of the patent status. This provision was designed to provide an incentive for targeting rare and neglected diseases.

Overall, the laws created under the Hatch–Waxman Act of 1984 were generally regarded as successful, although there continues to be both on-going litigation and debate regarding when and how generic drugs may be brought to market. "Innovator" companies and generic drug manu-facturers will likely continue to battle each other over the rights to either maintain or eliminate patent protection for new drugs in an attempt to maintain their profit margins, much to the delight of patent lawyers around the world.

Biologics Price Competition and Innovation Act of 2009[132]

When the Hatch–Waxman Act was being considered and eventually passed as a law, the biotechnological revolution had only just begun. Politicians concerned with the high cost of medication and health care were not aware of and did not account for the complexities of antibody therapeutics, recombinant proteins, or other macromolecular therapeu-tics. Their main concern at the time was the price of small molecule thera-peutics and the creation of a more robust generic drug market that would lower overall health care costs. As such, generic biologics were not cov-ered in the Hatch–Waxman Act, leaving generic drug companies without a regulatory pathway to gain approval of a generic equivalent of a macro-molecular therapeutic. The high price of biologics, however, made it clear very quickly that this oversight needed to be addressed. A single year's treatment with Herceptin, an antibody used in the treatment of breast can-cer, for example, can cost over $70,000.00.[133]

The substantial differences between small molecule and macromolecu-lar therapeutics made it impractical to simply apply the rules for one to the other. While it would be a relatively simple matter, for example, to ensure that the identity of a small molecule is the same in a "branded" version versus the generic version, the same cannot be said of macromolecules. Under the guidelines for small molecules, a "generic" antibody possess-ing a 99.9% overlap in structure with the "branded" version, sharing the same function and mechanism of action of the original, and with the same safety features would not be allowed on the market under the provisions of the Hatch–Waxman Act. The Biologics Price Competition and Innova-tion Act of 2009 addresses these, as well as other issues that prevented generic drug manufacturers from marketing cheaper version of biologi-cal medicines. New rules set forth in this law removed the requirement that generic macromolecules had to be identical with their branded coun-terpart. Biosimilarity replaced the identity requirement with a "highly similar" requirement, which dictated that generic macromolecules could have minor differences in the inactive components, so long as there were no clinically meaningful differences between the "branded" and "generic" versions in terms of safety, purity, and potency. Clinically, the "branded" and "generic" versions were also required to be interchangeable clinically without the aid of a health care provider and with no added safety risk upon switching between the two.

This legislation and similar laws in place across the globe will likely have a major impact on the pharmaceuticals industry. Most of the major pharmaceutical companies have shifted significant resources away from small molecules and into biologics in an attempt to maintain the high profit margins expected from drug companies, but generic companies are also entering the market. In July of 2010, the FDA approved its first generic biologic drug, a biosimilar to Lovenox® (enoxaparin sodium), the blockbuster blood-thinning drug originally brought to market by Sanofi-Aventis. It is likely additional biosimilars will be brought to market as the patent estates created in the biotech revolution come to the end of their enforcement period.

FUTURE DEVELOPMENTS IN DRUG DISCOVERY

The drug discovery process has changed significantly since the first experiments of Paul Ehrlich launched the science at the beginning of the twentieth century. Growth of the field has marched forward with advances in scientific understanding in the areas of biology, pharmacology, chemistry, and computer sciences, and it is likely that this will continue. The pace of innovation will likely quicken over time, as the technological tools currently in place allow for a far more rapid acquisition of scientific data than ever before. Where this will lead, however, is somewhat unpredictable, as the number of unanswered questions in the field of drug discovery is enormous. There is, however, some degree of certainty that the regulatory aspects of drug discovery will continue to grow with the field, although slightly out of phase with the field itself. It is, after all, difficult to create regulatory guidelines for new therapies that have as yet to be discovered.

QUESTIONS

1. Paul Ehrlich is known as the father of modern drug discovery. Between 1872 and 1874 he noted the selective affinity of dyes Trypan red, Trypan blue, and Methylene blue for biological tissues. What hypothesis did he develop based on these observations?
2. What is the significance of the Wistar rat?
3. What are SCID mice, why are they important, and how are they different from nude mice?

4. In 1974, Rudolf Jaenisch produced the first transgenic mouse. What is a transgenic animal model?
5. What is a knockout animal model?
6. Define high throughput chemistry.
7. What is recombinant DNA?
8. What technology enables the transfer of genetic material and can be used to prepare over-expression cell lines?
9. What is the general process of polymerase chain reaction (PCR) technology and why was the discovery of Taq polymerase important to the advancement of this technology?
10. Hybridoma technology was introduced by C Milstein and G. J. F. Köhler in 1975. What two cell types are fused to form a hybridoma cell line and what do the resulting cell lines produce?
11. What is a receptor construct/fusion protein and what is the function of the two components of the same?
12. What was the Elixir sulfanilamide disaster of 1937? This event led to the passage of the Food, Drug, and Cosmetic Act in 1938. What new requirements were put in place for the pharmaceutical industry?
13. First launched in 1957 as a treatment for birth defects, thalidomide was removed from the market in 1961 after 10,000 children were born with birth defects. Provide two key learnings from this event.
14. The Kefauver–Harris Amendment of 1962 placed additional requirements on new drugs entering the market. Describe them.

References

1. Gorman, C. F. Hoabinhian: A Pebble-Tool Complex with Early Plant Associations in Southeast Asia. *Science* **1969**, *163* (3868), 671–673.
2. a. Rudgley, R. *The Lost Civilizations of the Stone Age;* The Free Press: New York, 2000.
   b. Rooney, D. F. *Betel Chewing in Southeast Asia;* Centre National de la Recherche Scientifique (CNRS): Lyon, France, August 1995.
3. Samorini, G. The Oldest Representations of Hallucinogenic Mushrooms in the World (Sahara Desert, 9000–7000 B.P.). *Integr. J. Mind-Moving Plants Cult.* **1992**, *2* (3), 69–78.
4. a. McGovern, P. E.; Zhang, J.; Tang, J.; Zhang, Z.; Hall, G. R.; Moreau, R. A.; Nunez, A.; Butrym, E. D.; Richards, M. P.; Wang, Chen-shan.; et al. Fermented Beverages of Pre- and Proto-historic China. *Proc. Natl. Acad. Sci. U.S.A.* **2004**, *101* (51), 17593–17598.
   b. Homan, M. M. Beer and Its Drinkers: An Ancient Near Eastern Love Story. *Near East. Archaeol.* **2004**, *67* (2), 84–95.
   c. McGovern, Patrick E. *Ancient Wine: The Search for the Origins of Viniculture;* Princeton University Press: Princeton, New Jersey, 2007.
5. a. Kelly, K. *The History of Medicine: Early Civilizations, Prehistoric Times to 500 C.E;* Facts on File, Inc.: New York, 2009.
   b. Borchardt, J. K. The Beginning of Drug Therapy: Anceitn Mesopotamian Medicine. *Drug News Perspect.* **2002**, *15* (3), 187–192.
   c. Price, Massoume *History of Ancient Medicine in Mesopotamia & Iran;* Iran Chamber Society, October 2001. http://www.iranchamber.com/history/articles/ancient_medicine_mesopotamia_iran.php. http://www.indiana.edu/ancmed/meso.HTM.

6. *The Papyrus Ebers*(Bryan Cyril P., translator). D. Appleton and Co., 1931.
7. Read, Bernard E. *Chinese Medicinal Plants from the Pen T'Sao Kang Mu*, 3rd ed.; Peking National History Bulletin, 1936.
8. Rocco, F. *The Miraculous Fever-Tree: Malaria and the Quest for a Cure That Changed the World;* Harper Collins Publishers, Inc.: New York, 2003.
9. Aronson, J. K. *An Account of the Foxglove and Its Medical Uses 1785–1985;* Oxford University Press: New York, 1985.
10. a. Bosch, F.; Rosich, L. The Contributions of Paul Ehrlich to Pharmacology: A Tribute on the Occasion of the Centenary of His Nobel Prize. *Pharmacology* **2008,** *82,* 171–179.
    b. Drews, J. Drug Discovery: A Historical Perspective. *Science* **2000,** *287,* 1960–1964.
11. Brownstein, M. J. A Brief History of Opiate, Opioid Peptides, and Opioid Receptors. *Proc. Natl. Acad. Sci. U.S.A.* **1993,** *90,* 5391–5393.
12. a. Gay, G. R.; Inaba, D. S.; Sheppard, C. W.; Newmeyer, J. A.; Rappolt, R. T. Cocaine: History, Epidemiology, Human Pharmacology, and Treatment. A Perspective on a New Debut for an Old Girl. *Clin. Toxicol.* **1975,** *8* (2), 149–178.
    b. Karch, S. B. *A Brief History of Cocaine;* CRC Press Taylor & Francis Group: Boca Raton, Florida, 2006.
13. Jeffreys, D. *Aspirin: The Remarkable Story of a Wonder Drug;* Bloomsbury Publishing: New York, 2004.
14. a. Clause, B. T. The Wistar Rat as a Right Choice: Establishing Mammalian Standards and the Ideal of a Standardized Animal Model. *J. Hist. Biol.* **1993,** *26* (2), 329–349.
    b. Tucker, M. J. *Diseases of the Wistar Rat;* Taylor & Francis: Bristol, Pennsylvania, 1997.
15. Nakhooda, A. F.; Like, A. A.; Chappel, C. I.; Murray, F. T.; Marliss, E. B. The Spontaneously Diabetic Wistar Rat: Metabolic and Morphologic Studies. *Diabetes* **1977,** *26* (2), 100–112.
16. Crain, R. C. Spontaneous Tumors in the Rochester Strain of the Wistar Rat. *Am. J. Pathol.* **1958,** *34* (2), 311–335.
17. Liebsch, G.; Linthorst, A. C. E.; Neumann, I. D.; Reul, J. M. H.M.; Holsboer, F.; Landgraf, R. Behavioral, Physiological, and Neuroendocrine Stress Responses and Differential Sensitivity to Diazepam in Two Wistar Rat Lines Selectively Bred for High- and Low-Anxiety–Related Behavior. *Neuropsychopharmacology* **1998,** *19* (5), 381–396.
18. Pollard, M. Lobund-Wistar Rat Model of Prostate Cancer in Man. *The Prostate* **1998,** *37* (1), 1–4.
19. Drolet, G.; Proulx, K.; Pearson, D.; Rochford, J.; Deschepper, C. F. Comparisons of Behavioral and Neurochemical Characteristics between WKY, WKHA, and Wistar Rat Strains. *Neuropsychopharmacology* **2002,** *27* (3), 400–409.
20. Csiza, C. K.; de Lahunta, A. Myelin Deficiency (Md), a Neurologic Mutant in the Wistar Rat. *Am. J. Pathol.* **1979,** *95* (1), 215–224.
21. Okamoto, K.; Aoki, K. Development of a Strain of Spontaneously Hypertensive Rat. *Jpn. Circ. J.* **1963,** *27,* 282–293.
22. Giovanella, B. C.; Fogh, J. The Nude Mouse in Cancer Research. *Adv. Cancer Res.* **1985,** *44,* 70–120.
23. Flanagan, S. P. 'Nude', a New Hairless Gene with Pleiotropic Effects in the Mouse. *Genet. Res.* **1966,** *8,* 295–309.
24. Pantelouris, E. M. Absence of Thymus in a Mouse Mutant. *Nature* **1968,** *217,* 370–371.
25. Bosma, G. C.; Custer, R. P.; Bosma, M. J. A Severe Combined Immunodeficiency Mutation in the Mouse. *Nature* **1983,** *301,* 527–530.
26. Jaenisch, R.; Mintz, B. Simian Virus 40 DNA Sequences in DNA of Healthy Adult Mice Derived from Preimplantation Blastocysts Injected with Viral DNA. *Proc. Natl. Acad. Sci. U.S.A.* **1974,** *71* (4), 1250–1254.
27. Gordon, J.; Ruddle, F. Integration and Stable Germ Line Transmission of Genes Injected into Mouse Pronuclei. *Science* **1981,** *214* (4526), 1244–1246.
28. Costantini, F.; Lacy, E. Introduction of a Rabbit β-Globin Gene into the Mouse Germ Line. *Nature* **1981,** *294* (5836), 92–94.

29. Richardson, J. A.; Burns, D. K. Mouse Models of Alzheimer's Disease: A Quest for Plaques and Tangles. *Inst. Lab. Anim. Res. J.* **2002,** *43* (2), 89–99.

30. a. Gilliam, L. A. A.; Neufer, P. D. Transgenic Mouse Models Resistant to Diet-Induced Metabolic Disease: Is Energy Balance the Key? *J. Pharmacol. Exp. Ther.* **2012,** *342* (3), 631–636.

    b. Masuzaki, H.; Paterson, J.; Shinyama, H.; Morton, N. M.; Mullins, J. J.; Seckl, J. R.; Flier, J. S. A Transgenic Model of Visceral Obesity and the Metabolic Syndrome. *Science* **2001,** *294* (5549), 2166–2170.

    c. Cai, A.; Hyde, J. F. The Human Growth Hormone-Releasing Hormone Transgenic Mouse as a Model of Modest Obesity: Differential Changes in Leptin Receptor (OBR) Gene Expression in the Anterior Pituitary and Hypothalamus after Fasting and OBR Localization in Somatotrophs. *Endocrinology* **1999,** *140* (8), 3609–3614.

31. Rall, G. F.; Lawrence, D. M. P.; Patterson, C. E. The Application of Transgenic and Knockout Mouse Technology for the Study of Viral Pathogenesis. *Virology* **2000,** *271,* 220–226.

32. Houdebine, L. M. Production of Pharmaceutical Proteins by Transgenic Animals. *Comp. Immunol. Microbiol. Infect. Dis.* **2009,** *32,* 107–121.

33. Edmunds, T.; Van Patten, S. M.; Pollock, J.; Hanson, E.; Bernasconi, R.; Higgins, E.; Manavalan, P.; Ziomek, C.; Meade, H.; McPherson, J. M.; et al. Transgenically Produced Human Antithrombin: Structural and Functional Comparison to Human Plasma-Derived Antithrombin. *Blood* **1998,** *91,* 4561–4571.

34. Mccreath, G.; Udell, M. N. *Fibrinogen from Transgenic Animals.* US 20070219352, 2007.

35. a. Umana, P.; Jean-Mairet, J.; Moudry, R.; Amstutz, H.; Bailey, J. E. Engineered Glycoforms of an Antineuroblastoma IgG1 with Optimized Antibody-Dependent Cellular Cytotoxic Activity. *Nat. Biotechnol.* **1999,** *17,* 176–180.

    b. Lonberg, N. Human Monoclonal Antibodies from Transgenic Mice. *Handb. Exp. Pharmacol.* **2008,** *181* (181), 69–97.

    c. Zhu, L.; van de Lavoir, M. C.; Albanese, J.; Beenhouwer, D. O.; Cardarelli, P. M.; Cuison, S.; Deng, D. F.; Deshpande, S.; Diamond, J. H.; Green, L.; et al. Production of Human Monoclonal Antibody in Eggs of Chimeric Chickens. *Nat. Biotechnol.* **2005,** *23,* 1159–1169.

36. a. Capecchi, M. R. Altering the Genome by Homologous Recombination. *Science* **1989,** *244,* 1288–1292.

    b. Doetschman, T.; Gregg, R. G.; Maeda, N.; Hooper, M. L.; Melton, D. W.; Thompson, S.; Smithies, O. Germ-Line Transmission of a Planned Alteration Made in a Hypoxanthine Phosphoribosyltransferase Gene by Homologous Recombination in Embryonic Stem Cells. *Proc. Natl. Acad. Sci.* **1989,** *86* (22), 8927–8931.

    c. Evans, M. Embryonic Stem Cells: The Mouse Source—Vehicle for Mammalian Genetics and beyond (Nobel Lecture). *ChemBioChem* **2008,** *9,* 1690–1696.

37. a. Blackburn, A. C.; Jerry, D. J. Knockout and Transgenic Mice of Trp53: What Have We Learned about P53 in Breast Cancer? *Breast Cancer Res.* **2002,** *4* (3), 101–111.

    b. Carmichael, N. G.; Debruyne, E. L.; Bigot-Lasserre, D. The P53 Heterozygous Knockout Mouse as a Model for Chemical Carcinogenesis in Vascular Tissue. *Environ. Health Perspect.* **2000,** *108* (1), 61–65.

    c. Clarke, A. R.; Hollstein, M. Mouse Models with Modified P53 Sequences to Study Cancer and Ageing. *Cell Death Differ.* **2003,** *10,* 443–450.

38. Mientjes, E. J.; Nieuwenhuizen, I.; Kirkpatrick, L.; Zu, T.; Hoogeveen-Westerveld, M.; Severijnen, L.; Rifé, M.; Willemsen, R.; Nelson, D. L.; Oostra, B. A. The Generation of a Conditional Fmr1 Knock Out Mouse Model to Study Fmrp Function *In Vivo. Neurobiol. Dis.* **2006,** *3,* 549–555.

39. Jing, E.; Nillni, E. A.; Sanchez, V. C.; Stuart, R. C.; Good, D. J. Deletion of the Nhlh2 Transcription Factor Decreases the Levels of the Anorexigenic Peptides A Melanocyte-Stimulating Hormone and Thyrotropin-Releasing Hormone and Implicates Prohormone Convertases I and II in Obesity. *Endocrinology* **2004,** *145* (4), 1503–1513.

40. a. Bond, A. R.; Jackson, C. L. The Fat-Fed Apolipoprotein E Knockout Mouse Brachio-
cephalic Artery in the Study of Atherosclerotic Plaque Rupture. *J. Biomed. Biotechnol.*
2011, *2011,* 1–10. Article ID 379069.

   b. Zhang, S. H.; Reddick, R. L.; Piedrahita, J. A.; Maeda, N. Spontaneous Hypercholesterolemia and Arterial Lesions in Mice Lacking Apolipoprotein E. *Science* **1992,** *258*
(5081), 468–471.

41. The Nobel Prize in Physiology or Medicine 2007. http://www.nobelprize.org/nobel_p
rizes/medicine/laureates/2007/index.html.

42. Glasser, O. *Wilhelm Conrad Rontgen and the Early History of the Roentgen Rays;* Norman
Publishing, 1993.

43. a. von Laue, M. Concerning the Detection of X-ray Interferences. *Nobel Lectures, Physics*
1901–1921, Elsevier Publishing Company: Amesterdem, **1967,** pp. 348–355.

   b. Bragg, W. L.; James, R. W.; Bosanquet, C. H. The Distribution of Electrons Around the
Nucleus in the Sodium and Chlorine Atoms. *Philos. Mag.* **1922,** *44* (261), 433–449.

   c. Bragg, W. L. The Crystalline Structure of Copper. *Philos. Mag.* **1914,** *28* (165), 355–360.

44. Dickinson, R. G.; Raymond, A. L. The Crystal Structure of Hexamethylene-Tetramine. *J.*
Am. Chem. Soc. **1923,** *45,* 22–29.

45. Glusker, J. P. Dorothy Crowfoot Hodgkin (1910–1994). *Protein Sci.* **1994,** (3), 2465–2469.

46. Bernal, J. D.; Crowfoot, D. X-ray Photographs of Crystalline Pepsin. *Nature* **1934,** *133,*
794–795.

47. Crowfoot, D.; Riley, D. Crystal Structures of the Proteins an X-ray Study of Palmar's
Lactoglobulin. *Nature* **1938,** *141,* 521–522.

48. Crowfoot, D. X-ray Single-Crystal Photographs of Insulin. *Nature* **1935,** *135,* 591–592.

49. Adams, M. J.; Blundell, T. L.; Dodson, E. J.; Dodson, G. G.; Vijayan, M.; Baker, E. N.;
Harding, M. M.; Hodgkin, D.; Rimmer, B.; Sheat, S. Structure of Rhombohedral 2-Zinc
Insulin Crystals. *Nature* **1969,** *224,* 491–495.

50. Carlisle, C. H.; Crowfoot, D. The Crystal Structure of Cholesteryl Iodide. *Proc. Roy. Soc.*
Lond. Ser. A Math. Phys. Sci. **1945,** *184,* 64–83.

51. Crowfoot, D.; Bunn, C. W.; Rogers-Low, B. W.; Turner-Jones, A. X-ray Crystallo-
graphic Investigation of the Structure of Penicillin. In *Chemistry of Penicillin;* Clarke,
H. T., Johnson, J. R., Robinson, R., Eds.; Princeton University Press: Princeton, New
Jersey, 1949.

52. a. Brink, C.; Hodgkin, D. C.; Lindsey, J.; Pickworth, J.; Robertson, J. H.; White, J. G.
X-ray Crystallographic Evidence on the Structure of Vitamin B_{12}. *Nature* **1954,** *174,*
1169–1170.

   b. Hodgkin, D. C.; Pickworth, J.; Robertson, J. H.; Trueblood, K. N.; Prosen, R. J.; White,
J. G. The Crystal Structure of the Hexacarboxylic Acid Derived from B_{12} and the
Molecular Structure of the Vitamin. *Nature* **1955,** *176,* 325–328.

53. Opfell, O. S. *Lady Laureates: Women Who Have Won the Nobel Prize;* Rowman & Littlefield
Publishers, Inc.: Lanham, Maryland, 1986.

54. About the PDB Archive and the RCSB PDB. http://www.rcsb.org/pdb/static.do?p=ge
neral_information/about_pdb/index.html.

55. Watson, J. D.; Crick, F. H. C. A Structure for Deoxyribose Nucleic Acid. *Nature* **1953,** *171*
(4356), 737–738.

56. About NDB. http://ndbserver.rutgers.edu/about_ndb/index.html.

57. The Cambridge Crystallographic Data Centre (CCDC) Annual Operational Report,
2009.

58. Cambridge Structural Database Summary Statistics, 2012.

59. Heisenberg, W. Über quantentheoretische Umdeutung kinematischer und mecha-
nischer Beziehungen. *Zeitschrift für Physik* **1925,** *33,* 879–893.

60. Hendrickson, J. B. Molecular Geometry. I. Machine Computation of the Common Rings.
J. Am. Chem. Soc. **1961,** *83,* 4537–4547.

61. Levinthal, C. Molecular Model-Building by Computer. *Sci. Am.* **1966,** *214,* 42–52.

62. Richon, A. B. An Early History of the Molecular Modeling Industry. *Drug Discov. Today* **2008,** *13* (15/16), 659–664.

63. Koehl, P.; Levitt, M. A Brighter Future for Protein Structure Prediction. *Nat. Struct. Biol.* **1999,** *6* (2), 108–111.

64. Lombardino, J. G.; Lowe, J. A., III The Role of the Medicinal Chemist in Drug Discovery—Then and Now. *Nat. Rev. Drug Discov.* **2004,** *3,* 853–862.

65. Rankovic, Z.; Morphy, R. *Lead Generation Approaches in Drug Discovery;* John Wiley & Sons, Inc.: Hoboken, New Jersey, 2010.

66. Merrifield, R. B. Solid Phase Peptide Synthesis. I. The Synthesis of a Tetrapeptide. *J. Am. Chem. Soc.* **1963,** *85* (14), 2149–2154.

67. Merrifield, R. B. Solid Phase Peptide Synthesis. II. The Synthesis of Bradykinin. *J. Am. Chem. Soc.* **1964,** *86* (2), 304–305.

68. Marglin, B.; Merrifield, R. B. The Synthesis of Bovine Insulin by the Solid Phase Method. *J. Am. Chem. Soc.* **1966,** *88* (21), 5051–5052.

69. Takashima, H.; Vigneaud, V. D.; Merrifield, R. B. Synthesis of Deaminooxytocin by the Solid Phase Method. *J. Am. Chem. Soc.* **1968,** *90* (5), 1323–1325.

70. Crowley, J. I.; Rapoport, H. Solid-Phase Organic Synthesis: Novelty or Fundamental Concept? *Acc. Chem. Res.* **1976,** *9* (4), 135–144.

71. Houghten, R. A. General Method for the Rapid Solid-Phase Synthesis of Large Numbers of Peptides: Specificity of Antigen-Antibody Interaction at the Level of Individual Amino Acids. *Proc. Natl. Acad. Sci. U.S.A.* **1985,** *82,* 5131–5135.

72. Geysen, H. M.; Meloen, R. H.; Barteling, S. J. Use of Peptide Synthesis to Probe Viral Antigens for Epitopes to a Resolution of a Single Amino Acid. *Proc. Natl. Acad. Sci. U.S.A.* **1984,** *81,* 3998–4002.

73. Bunin, B. A.; Plunkett, M. J.; Ellman, J. A. The Combinatorial Synthesis and Chemical and Biological Evaluation of a 1,4-benzodiazepine Library. *Proc. Natl. Acad. Sci. U.S.A.* **1994,** *91,* 4708–4712.

74. Dewitt, S. H.; Kiely, J. S.; Stankovic, C. J.; Schroeder, M. C.; Cody, D. M. R.; Pavia, M. R. "Diversomers": An Approach to Nonpeptide, Nonoligomeric Chemical Diversity. *Proc. Natl. Acad. Sci. U.S.A.* **1993,** *90,* 6909–6913.

75. Toy, P. H.; Lam, Y. *Solid-Phase Organic Synthesis: Concepts, Strategies, and Applications;* John Wiley & Sons, Inc. Hoboken: New Jersey, 2012.

76. Ugi, I.; Heck, S. The Multicomponent Reactions and Their Libraries for Natural and Preparative Chemistry. *Comb. Chem. High Throughput Screen.* **2001,** *4* (1), 1–34.

77. Kappe, C. O. 100 Years of the Biginelli Dihydropyrimidine Synthesis. *Tetrahedron* **1993,** *49* (32), 6937–6963.

78. Dömling, A.; Ugi, I. Multicomponent Reactions with Isocyanides. *Angew. Chem. Int. Ed.* **2000,** *39,* 3168–3210.

79. https://zinc.docking.org/.

80. Brown, F. K. Chapter 35. Chemoinformatics: What Is It and How Does It Impact Drug Discovery. *Annu. Reports Med. Chem.* **1998,** *33,* 375–384.

81. Miller, E. L. The Penicillins: A Review and Update. *J. Midwifery Womens Health* **2002,** *47* (6), 426–434.

82. Wick, J. Y. The History of Benzodiazepines. *Consult. Pharm.* **2013,** *28* (9), 538–548.

83. Zuckerman, J. M.; Qamar, F.; Bono, B. R. Review of Macrolides (Azithromycin, Clarithromycin), Ketolids (Telithromycin) and Glycylcyclines (Tigecycline). *Med. Clin. North Am.* **2011,** *95,* 761–791.

84. Pollack, A. Roche Agrees to Buy Genentech for $46.8 Billion. *N. Y. Times* **March 12, 2009.**

85. Based on NYSE Stock Price on December 31st, 2013.

86. Kornberg, A.; Lehman, I. R.; Simms, E. S. Polydesoxyribonucleotide Synthesis by Enzymes from *Escherichia coli. Fed. Proc.* **1956,** *15,* 291–292.

87. Reviewed in Lehman, I. R. DNA Ligase: Structure, Mechanism, and Function. *Science* **1974,** *186* (4166), 790–797.

88. a. Klett, R. P.; Cerami, A.; Reich, E. Exonuclease VI, a New Nuclease Activity Associated with *E. coli* DNA Polymerase. *Proc. Natl. Acad. Sci. U.S.A.* **1968,** *60* (3), 943–950.

    b. Richardson, C. C.; Kornberg, A. A Deoxyribonucleic Acid Phosphatase-exonuclease from *Escherichia coli*. I. Purification of the Enzyme and Characterization of the Phosphatase Activity. *J. Biol. Chem.* **1964,** *239,* 242–250.

    c. Shevelev, I. V.; Hübscher, U. The 3′–5′ Exonucleases. *Nat. Rev. Mol. Cell. Biol.* **2002,** *3,* 364–376.

89. a. Krakow, J. S.; Coutsogeorgopoulos, C.; Canellakis, E. S. "Formation of Sedoheptulose-7-Phosphate from Enzymatically Obtained "Active Glycolic Aldehyde" and Ribose-5-Phosphate with Transketolase. *Biochem. Biophys. Res. Commun.* **1961,** *5,* 477–481.

    b. Chang, L. M. S.; Bollum, F. J. Molecular Biology of Terminal Transferase. *Crit. Rev. Biochem.* **1986,** *21* (1), 27–52.

90. a. Temin, H. M.; Mizutani, S. RNA-Dependent DNA Polymerase in Virions of Rous Sarcoma Virus. *Nature* **1970,** *226,* 1211–1213.

    b. Baltimore, D. RNA-Dependent DNA Polymerase in Virions of RNA Tumour Viruses. *Nature* **1970,** *226,* 1209–1211.

91. a. Roberts, R. J. Restriction Endonucleases. *CRC Crit. Rev. Biochem.* **1976,** *4* (2), 123–164.

    b. Meselson, M.; Yuan, R. DNA Restriction Enzyme from *E. coli*. *Nature* **1968,** *217,* 1110–1114.

    c. Dussoix, D.; Arber, W. Host Specificity of DNA Produced by *Escherichia coli*. II. Control over Acceptance of DNA from Infecting Phage Lambda. *J. Mol. Biol.* **July 1962,** *5* (1), 37–49.

92. a. Twort, F. W. An Investigation on the Nature of Ultra-Microscopic Viruses. *Lancet* **1915,** *186,* 1241–1243.

    b. Twort, F. W. The Discovery of the "Bacteriophage". *Lancet* **1925,** *205,* 845.

93. D'Herelle, F. On an Invisible Microbe Antagonistic toward Dysenteric Bacilli: Brief Note by Mr F. D'Herelle, Presented by Mr Roux. *Res. Microbiol.* **2007,** *158* (7), 553–554.

94. Goodpasture, E. W.; Woodruff, A. M.; Buddingh, G. J. The Cultivation of Vaccine and Other Viruses in the Chorioallantoic Membrane of Chick Embryos. *Science* **1931,** *74,* 371–372.

95. Lear, J. *Recombinant DNA: The Untold Story;* Crown Publishing: New York, 1978.

96. Jackson, D.; Symons, R.; Berg, P. Biochemical Method for Inserting New Genetic Information into DNA of Simian Virus 40: Circular SV40 DNA Molecules Containing Lambda Phage Genes and the Galactose Operon of *Escherichia coli*. *Proc. Natl. Acad. Sci. U.S.A.* **1972,** *69* (10), 2904–2909.

97. Lobban, P.; Kaiser, A. Enzymatic End-to End Joining of DNA Molecules. *J. Mol. Biol.* **1973,** *78* (3), 453–471.

98. Cohen, S.; Chang, A.; Boyer, H.; Helling, R. Construction of Biologically Functional Bacterial Plasmids *In Vitro*. *Proc. Natl. Acad. Sci. U.S.A.* **1973,** *70* (11), 3240–3244.

99. Cohen, S. N.; Boyer, H. W. *Process for Producing Biologically Functional Molecular Chimeras*. US 4,237,224, 1980.

100. Altman, L. K. *A New Insulin Given Approval for Use in U.S.;* New York Times, October 30, 1982.

101. *Global Biotechnology Industry Guide;* Research and Markets, Inc.: Dublin Ireland, 2013.

102. Kleppe, K.; Ohtsuka, E.; Kleppe, R.; Molineux, I.; Khorana, H. G. Studies on Polynucleotides. XCVI. Repair Replications of Short Synthetic DNA's as Catalyzed by DNA Polymerases. *J. Mol. Biol.* **1971,** *56* (2), 341–361.

103. Brock, T. D.; Freeze, H. *Thermus aquaticus,* a Nonsporulating Extreme Thermophile. *J. Bacteriol.* **1969,** *98* (1), 289–297.

104. Chien, A.; Edgar, D. B.; Trela, J. M. Deoxyribonucleic Acid Polymerase from the Extreme Thermophile *Thermus aquaticus*. *J. Bacteriol.* **1976,** *127* (3), 1550–1557.

105. a. Saiki, R.; Gelfand, D.; Stoffel, S.; Scharf, S.; Higuchi, R.; Horn, G.; Mullis, K.; Erlich, H. Primer-Directed Enzymatic Amplification of DNA with a Thermostable DNA Polymerase. *Science* **1988,** *239* (4839), 487–491.

   b. Lawyer, F.; Stoffel, S.; Saiki, R.; Chang, S.; Landre, P.; Abramson, R.; Gelfand, D. High-Level Expression, Purification, and Enzymatic Characterization of Full-Length *Thermus aquaticus* DNA Polymerase and a Truncated Form Deficient in 5′ to 3′ Exonuclease Activity. *PCR Methods Appl.* **1993,** *2* (4), 275–287.

106. Winau, F.; Westphal, O.; Winau, R. Paul Ehrlich—In Search of the Magic Bullet. *Microbes Infect.* **2004,** *6* (8), 786–789.

107. Fagraeus, A. The Plasma Cellular Reaction and its Relation to the Formation of Antibodies *In Vitro. J. Immunol.* **1948,** *58* (1), 1–13.

108. Burnet, F. M. *The Clonal Selection Theory of Acquired Immunity;* Vanderbilt University Press: Nashville, 1959.

109. Schwaber, J.; Cohen, E. P. Human X Mouse Somatic Cell Hybrid Clone Secreting Immunoglobulins of Both Parental Types. *Nature* **1973,** *244* (5416), 444–447.

110. Köhler, G.; Milstein, C. Continuous Cultures of Fused Cells Secreting Antibody of Predefined Specificity. *Nature* **1975,** *256* (5517), 495–497.

111. Riechmann, L.; Clark, M.; Waldmann, H.; Winter, G. Reshaping Human Antibodies for Therapy. *Nature* **1998,** *332* (6162), 323–327.

112. Smith, S. L. Ten Years of Orthoclone OKT3 (Muromonab-CD3): A Review. *J. Transpl. Coord.* **1996,** *6* (3), 109–119.

113. Barnhart, B. J. DOE Human Genome Program. *Hum. Genome Q.* **1989,** *1,* 1.

114. a. Anderson, C. Thrombolysis with Alteplase after Stroke: Extending Outcomes. *Lancet Neurol.* **2013,** *12* (8), 731–732.

   b. http://www.activase.com/.

115. a. Corwin, H. L.; Gettinger, A.; Fabian, T. C.; May, A.; Pearl, R. G.; Heard, S.; An, R.; Bowers, P. J.; Burton, P.; Klausner, M. A.; et al. Efficacy and Safety of Epoetin Alfa in Critically Ill Patients. *N. Engl. J. Med.* **2007,** *357* (10), 965–976.

   b. http://www.epogen.com/.

116. a. Maini, R.; St Clair, E. W.; Breedveld, F.; Furst, D.; Kalden, J.; Weisman, M.; Smolen, J.; Emery, P.; Harriman, G.; Feldmann, M.; et al. Infliximab (Chimeric Anti-tumour Necrosis Factor Alpha Monoclonal Antibody) versus Placebo in Rheumatoid Arthritis Patients Receiving Concomitant Methotrexate: a Randomised Phase III Trial. ATTRACT Study Group. *Lancet* **1999,** *354* (9194), 1932–1999.

   b. http://www.remicade.com/.

117. a. Hudis, C. A. Trastuzumab—Mechanism of Action and Use in Clinical Practice. *N. Engl. J. Med.* **2007,** *357* (1), 39–51.

   b. http://www.herceptin.com/.

118. a. Moreland, L.; Bate, G.; Kirkpatrick, P. Abatacept. *Nat. Rev. Drug Discov.* **2006,** *5* (3), 185–186.

   b. http://www.orencia.com/index.aspx.

119. Ellis, C. N.; Krueger, G. G. Treatment of Chronic Plaque Psoriasis by Selective Targeting of Memory Effector T Lymphocytes. *N. Engl. J. Med.* **2001,** *345* (4), 248–255.

120. Sorbera, L. A. Aflibercept Antiangiogenic Agent Vascular Endothelial Growth Factor Inhibitor. *Drugs Fut.* **2007,** *32* (2), 109–117.

121. Olsen, E. A.; Dunlap, F. E.; Funicella, T.; Koperski, J. A.; Swinehart, J. M.; Tschen, E. H.; Trancik, R. J. A Randomized Clinical Trial of 5% Topical Minoxidil versus 2% Topical Minoxidil and Placebo in the Treatment of Androgenetic Alopecia in Men. *J. Am. Acad. Dermatol.* **2002,** *47* (3), 377–385.

122. Barkan, I. D. Industry invites Regulation: the Passage of the Pure Food and Drug Act of 1906. *Am. J. Public Health* **1985,** *75* (1), 18–26.

123. Gadarowski, J. C., Ed. *The Great American Fraud: A Series of Articles on the Patent Medicine Evil;* CreateSpace Independent Publishing Platform: Seattle, WA, USA, January 15, 2014. (Reprinted from Collier's Weekly by Samuel Hopkins Adam (Author).

124. http://www.usp.org/.

125. http://www.usp.org/usp-nf.

126. a. Ballentine, C. *Taste of Raspberries, Taste of Death the 1937 Elixir Sulfanilamide Incident;* . FDA Consumer Magazine, June 1981.

    b. Wax, P. M. Elixirs, Diluents, and the Passage of the 1938 Federal Food, Drug and Cosmetic Act. *Ann. Intern. Med.* **March 15, 1995,** *122* (6), 456–461.

127. http://www.fda.gov/AboutFDA/WhatWeDo/History/CentennialofFDA/Centenn ial EditionofFDAConsumer/ucm093787.htm.

128. a. Kim, J. H.; Scialli, A. R. Thalidomide: The Tragedy of Birth Defects and the Effective Treatment of Disease. *Toxicol. Sci.* **2011,** *122* (1), 1–6.

    b. Stephens, T. D.; Brynner, R. *Dark Remedy: The Impact of Thalidomide and Its Revival as a Vital Medicine;* Perseus Books: New York, 2001.

129. The Durham-Humphrey Amendment. *J. Am. Med. Assoc.* **1952,** *149* (4), 371.

130. Peltzman, S. An Evaluation of Consumer Protection Legislation: The 1962 Drug Amendments. *J. Polit. Econ.* **1973,** *81* (5), 1049–1091.

131. Sokal, A. M.; Gerstenblith, B. A. The Hatch–Waxman Act: Encouraging Innovation and Generic Drug Competition. *Curr. Top. Med. Chem.* **2010,** *10* (18), 1950–1959.

132. Nick, C. The US Biosimilars Act: Challenges Facing Regulatory Approval. *Pharmaceut. Med. New Zealand* **2012,** *26* (3), 145–152.

133. Fleck, L. The Costs of Caring: Who Pays? Who Profits? Who Panders? *Hastings Cent. Rep.* **2006,** *36* (3), 13–17.

Classical Targets in Drug Discovery

As the drug discovery process has evolved, a great deal of effort has been focused on developing an understanding of the macromolecular targets. In the first half of the twentieth century, structural information and a true understanding of the mechanistic aspects of drug–protein interactions were limited by the technologies of the time. As technology and the drug discovery process evolved, however, tools such as X-ray crystallography, molecular modeling, PCR, and recombinant DNA technologies provided a sharper and sharper picture of the biological targets impacted by drugs. While the true number of potentially "druggable" targets (macromolecules that can be effectively manipulated with therapeutic entities) remains a subject of debate,[1a–1c] genomic sequencing of pathogenic organisms and humans has shed some light on the subject.

The Human Genome Project, which was completed in 2003, demonstrated that the 23 chromosomes necessary to support human existence are comprised of just over three billion DNA base pairs that encode approximately 20,000–25,000 protein coding genes.[2] Microbial genomes[3] are much smaller, of course, but still represent a significant number of protein-encoding genes. While not all of these genes and their gene-products are directly, or even indirectly, involved in disease progression, pathogenesis, or the therapeutic action of drugs, the sheer number of possible macromolecular targets for drug discovery is quite large. Recent estimates suggest that there are approximately 5000 potential "druggable" macromolecular targets suitable for small molecule therapeutics and an additional 3200 targets that may be suitable for biological therapeutics (Figure 3.1).[1]

Fortunately, nature is fond of recycling its methods, structures, and techniques, creating sets of distinct classes of macromolecules that can be studied together. In fact, an analysis of the full spectrum of marketed drugs demonstrates quite clearly that the vast majority of drugs target only four classes of macromolecules: enzymes, G-protein-coupled receptor (GPCR), ion channels, and transporters. While there are additional

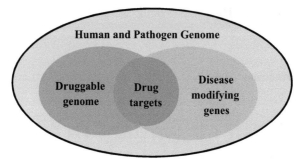

FIGURE 3.1 Completion of the human and pathogenic genomes provided a wealth of understanding of potential drug targets, but not all of the genes uncovered are useful as drug targets. Useful targets for therapeutic intervention sit within the juxtaposition of the druggable genome and disease-modifying genes. Druggable genes that do not modify diseases are not useful targets just as disease-modifying genes that cannot be successfully modulated to alter disease progression are unlikely to lead to novel therapeutics.

therapeutic targets, such as protein–protein interactions and DNA–protein interactions, an understanding of the four major classes of drug targets is essential for success in the field of drug discovery. These macromolecules are not just the targets of drug discovery programs. They are often the tools used to assess biological activity of potential therapeutics via *in vitro* screening. This topic will be discussed in Chapter 4. It is also worth noting that the pharmaceutical industry has just barely scratched the surface of drug targets. Currently, there are over 21,000 marketed drugs, but these products contain fewer than 1400 unique molecules that create a positive impact through interaction with just 324 drug targets (Figure 3.2).[1]

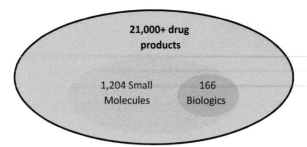

FIGURE 3.2 While the total number of marketed drug products exceeds 21,000 individual products, further analysis shows that the number of therapeutically useful compounds is actually far smaller. Elimination of supplements, imaging agents, vitamins, duplicate salt forms, and other redundancies reveals that there are fewer than 1400 unique drug molecules. Biologic, macromolecular therapeutic entities, represent only 12.2% of the total, but are growing in importance as technologies evolve to support their use. It has been estimated that the collection of marketed drugs exerts their influence through only 324 of the known drug targets.

PROTEIN STRUCTURE

Prior to considering the overall structural and functional aspects of the various drug target classes, it is necessary to have an understanding of the basic principles of protein structure. While DNA carries the genetic code that defines an organism, proteins are the tools that give rise to most biological function. Their remarkable scope of activity includes reaction catalysis, transport and storage, mechanical support, control of growth and cell differentiation, coordination of movement, and nerve impulse transmission. Proteins are involved in virtually all of the biological activities that support life, and while there are substantial differences between the various types of proteins, there are also a number of similarities that should be considered.

First and foremost, the structure of all proteins, irrespective of their origin, is built upon a small set of α-amino acids linked together through a series of amide bonds. In principle, the number of α-amino acids available is virtually unlimited but nature relies primarily on a set of 20 amino acids (Figure 3.3). There are some additional naturally occurring amino acids, such as ornithine and 4-hydroxyproline, that make rare appearances in protein sequences, but in general, the vast majority of protein structures (and therefore, drug targets) are built from this small set of building blocks. All but one, glycine, contain chiral centers, but nature only employs the enantiomers shown in Figure 3.3. The α-amino acids are linked together through a series of amide bonds to form linear polypeptide chains that can contain from a few dozen α-amino acids to thousands. The largest known polypeptides, the tintins, which contribute to muscle elasticity, contains between 27,000 and 33,000 α-amino acids.[4] The series of α-amino acids is referred to as the protein's linear sequence or primary structure. Proteins begin with an N-terminus, the amino side of the first amino acid in the sequence, and ends with the C-terminus, the carboxylic acids portion of the final α-amino acid in the chain.

Of course, it is well known that proteins are not simply long strings of α-amino acids. They have distinct three-dimensional shapes that are dictated by their primary structure. The linear sequence gives rise to the three-dimensional shape of proteins through a variety of different types of physicochemical interactions that occur within the framework of the protein itself. These interactions lead to folding, twisting, and bending of the protein chains to create three-dimensional structures that give rise to their biological function and activity. The spatial arrangements created by amino acids that are close to one another in a linear sequence are referred to as the secondary structure of the protein. In many cases, the secondary structures are highly ordered, presenting well-defined shapes such as α-helices and β-sheets. The combination of the secondary structures and interaction between amino acids that are linearly far apart, but brought

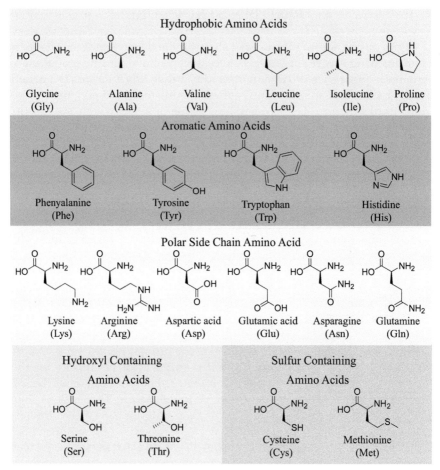

FIGURE 3.3   The 20 fundamental α-amino acids that are the building blocks for proteins.

together by physicochemical interactions within the protein is referred to as the tertiary structure of a protein.

It is often the case that the structure formed by a single protein is enough to provide the necessary biological function, but there are many cases where a single protein chain alone is not sufficient. In these instances, multiple protein chains are brought together as individual subunits of a greater overall molecular complex that can then perform a biological function. The organization of the subunits into the functional units and the nature of their contacts are referred to as the quaternary structure. The same physicochemical interactions that give rise to secondary and tertiary structure also support the formation of quaternary protein assemblies.

The physicochemical interactions that occur within a protein frame-work can be classified into one of several groups: covalent bonding, elec-trostatic interactions, and non-covalent interactions. The first category, covalent bonds, is predominately the result of the formation of a cova-lent bond between two cysteine residues in distal location of the protein

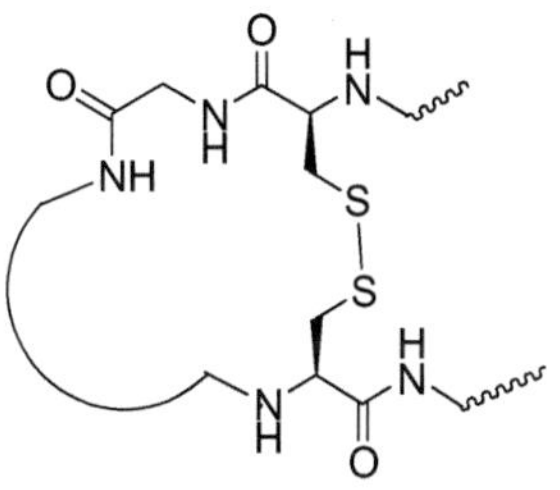

FIGURE 3.4 Disulfide bridge formed between two cys-teine residues.

(Figure 3.4). Their bond dissociation energy of 60 kcal/mol (251 kJ/mol) can provide significant stabilization to the structural features of a protein. While this interaction is the strongest interaction providing support for the secondary and tertiary structure of proteins, it is also the least com-mon. They can, however, form interwoven loops when multiple disulfide bridges are present in the same protein. These "cystine knots" were first identified in the X-ray crystal structure of nerve growth factor proteins, but have since been identified in a variety of other proteins.[5]

Electrostatic interactions, often referred to as salt bridges, also play an important role in protein structure. The majority of salt bridges in proteins are formed between deprotonated carboxylate side chains of aspartic or glutamic acid and protonated side chains of lysine or arginine residues (Figure 3.5). Other amino acids with ionizable side chains such

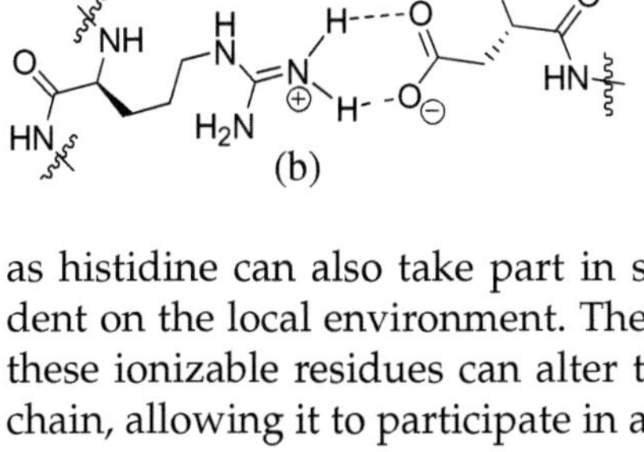

FIGURE 3.5 (a) Salt bridge formed between a lysine residue and an aspar-tic acid residue. (b) Salt bridge formed between a protonated arginine residue and an aspartic acid residue.

as histidine can also take part in salt bridges, but this is highly depen-dent on the local environment. The protein structure surrounding one of these ionizable residues can alter the apparent pKa of an ionizable side chain, allowing it to participate in an ionic bond that would not normally

occur in a fully aqueous environment. The strength of this class of bonding interaction, and therefore, its overall contribution to proteins stabilization, is a function of the distance between the two oppositely charged
residues. The further apart the residues are located, the lower the bonding
strength becomes, and charged residues that are greater than 4 Å apart are
generally considered too far apart for consideration as a salt bridge in a
protein.[6] While this kind of interaction is more common than the covalent
bonds formed between two cysteine residues, it is not the predominant
force that provides the overall support necessary for maintaining a protein's physical framework in place.

Non-covalent interactions, such as hydrophobic interactions, π-stacking,
π-cation interactions, and hydrogen bonding, play the predominant role
in the stabilization of the three-dimensional structure of proteins. While
none of these interactions are nearly as strong as cysteine–cysteine bonds
(disulfide bond) or salt bridges, the sheer number of interactions cumu-
latively available more than makes up for the limited bond strength provided by any single interaction. Each individual amino acid residue in
the linear sequence, for example, can participate in two separate hydrogen-bonding interactions, one as a hydrogen bond donor, the other as a
hydrogen bond acceptor. Multiply this by the number of amino acids in
the linear sequence of a protein, and the resulting energy of stabilization
grows quickly. The nature of each type of non-covalent interaction is different, however, and should be reviewed independently.

Although hydrophobic interactions are relatively weak in nature, they
are believed to be one of the major drivers of protein folding. In simplest
terms, the stabilization energy provided by this type of non-covalent
interaction is the result of clustering of non-polar side chains within
a protein (Figure 3.6(a)). Amino acids such as phenylalanine, alanine,

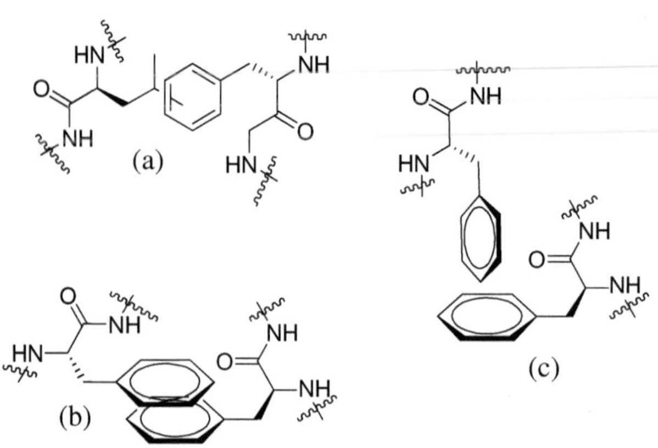

FIGURE 3.6   (a) Hydrophobic interaction between leucine and phenylalanine residues.
(b) Face to face π-stacking interaction between two phenylalanines. (c) t-Type π-stacking
interaction between two phenylalanines.

valine, leucine, and isoleucine fold toward each other in order to remove themselves from the surrounding water molecules, creating hydrophobic pockets within the protein structure. Since the non-polar side chains are incapable of forming hydrogen bonds, ionic interactions, or covalent bonds, the attractive forces between different side chains are limited to Van der Waals interactions. In isolation, these interactions are weak, but in summation across an entire protein sequence, they provide substantial stabilization to the secondary and tertiary structure of proteins.

Amino acids with aromatic side chains such as phenylalanine and tyrosine are capable of forming two additional energetically favorable interactions that support protein folding. The first is π-stacking, which refers to the attractive interaction formed between two aromatic systems. These interactions can occur in either a face-to-face sandwich arrangement (Figure 3.6(b)) or in a T-shaped face-to-edge arrangement (Figure 3.6(c)), but in either case the strength of the interaction is highly dependent on the distance between the two aromatic systems. The energetics of this interaction can also be impacted by the degree of overlap of the π-systems, a concept more easily understood in the face-to-face sandwich arrangement. The interactions are most favorable when the centers of the two rings are overlapping, and the bonding energy decreases as the ring centers move apart (parallel displacement). The addition of substituents to an aromatic system can also alter the strength of this non-covalent interaction. While this is not a significant issue for protein folding, modification of the substituents of an aromatic ring system is a useful technique for altering the binding properties of potential therapeutic agents.[7a–7c]

Protein folding can also be impacted by the interaction of an aromatic amino acid side chain with a positively charged amino acid side chain, an arrangement referred to as a π-cation bond (Figure 3.7). In this situation, the π-cloud of the electron-rich aromatic ring, which has a partial negative

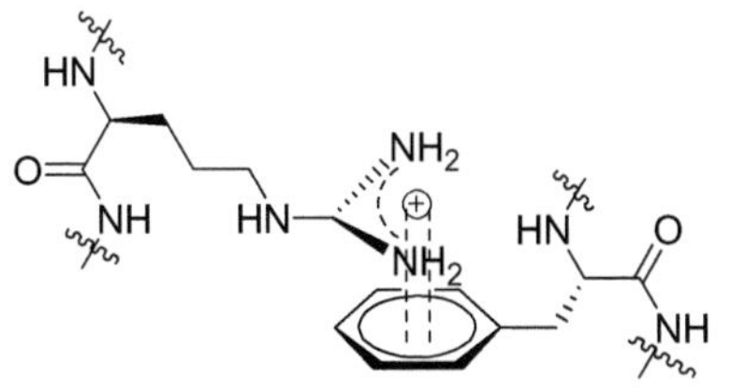

FIGURE 3.7   π-Cation interaction between a protonated arginine and phenylalanine.

charge both above and below the plane of the ring, interacts favorably with a positively charged amino acid side chain, such as a protonated lysine or arginine. Bonding energies produced can approach the same order of magnitude as those seen in hydrogen bonds and the strength of the interaction is dependent upon a number of factors, including distance between the two bonding groups, angle of the interaction, and the electrostatic potential surfaces of the aromatic ring system. This last factor has been employed to

modulate binding affinities of drug candidates through the use of substituent effects on aromatic rings. If all other factors remain equal, the additions of
electron-donating substituents to an aromatic ring will increase the strength
of a π-cation bond by increasing the electrostatic potential of the center of
the ring. Conversely, electron-withdrawing substituents decrease the electrostatic potential of the center of an aromatic system, which decreases
the strength of the π-cation bond. The relationship between π-cation bond
strength and aromatic substituents has been effectively employed to modulate the binding properties of potential therapeutic agents.[8a,8b]

Hydrogen bonding, of course, also plays a major role in determining
the three-dimensional shape of macromolecules. These dipole–dipole
interactions are formed between polarized hydrogen atoms of a hydrogen bond donor and a lone pair of electrons from a hydrogen bond
acceptor. Examples of hydrogen bond donors include alcohols, amines,
guanidines, and the NH of an amide bond, while typical hydrogen bond
acceptors include alcohols, amines, guanidines, and the oxygen atom of
an amide bonds (Figure 3.8). The strength of these interactions is highly

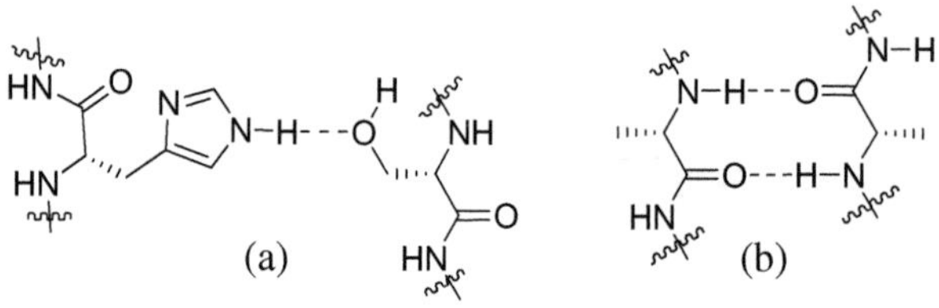

FIGURE 3.8   (a) Hydrogen-bonding interaction between histidine and serine side chains.
(b) Hydrogen-bonding interaction between backbone amide segments of a peptide chain.

distance and direction dependent, ranging in strength between 1 and
7 kcal/mol, but typically are between 3 and 5 kcal/mol. As was the case
with hydrophobic interactions, the strength of any single hydrogen bond
is relatively small, but collectively across an entire protein chain, the
stabilization energy provided by hydrogen bonding is significant. Each
amide bond within a protein's linear sequence contains both a hydrogen
bond donor (NH) and a hydrogen bond acceptor (CO) that can act to
stabilize a protein's folding pattern. Structural motifs such as the α-helix
(Figure 3.9(a)), β-sheet (Figure 3.9(b)), and β-turn (Figure 3.9(c)), for
example, are products of the hydrogen-bonding interaction provided by
the peptide backbone of a protein. In addition, many of the amino acid
side chains are also capable of acting as hydrogen bond donors, hydrogen bond acceptors, or both, providing additional opportunities for a
protein to generate stabilization energy through hydrogen bonding.[9a,9b]

The combination of stabilization energy provided by the various covalent and non-covalent interactions available within a protein's linear chain

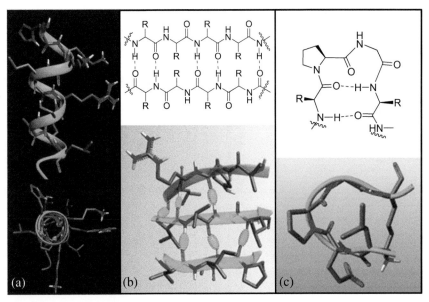

FIGURE 3.9 (a) The α-helix is a common structural motif in proteins that is formed by hydrogen-bonding interactions between the amide units in the backbone of a peptide chain. There are 3.6 amino acid residues per turn, and each complete turn is 5.4 Å in length. The side chains protrude from the barrel shape formed by the rotation of the peptide chain as seen in the two views provided. (b) β-Sheets are formed from extended peptide chains connected by a series of hydrogen bonds between the backbone amide moieties of each chain. The peptide chains can be either parallel or antiparallel to each other. (c) β-Turns occur when four to five amino acids in a peptide chain form a 180° turn. They are stabilized by hydrogen bonding and are often found between β-sheets.

defines its three-dimensional shape, and therefore its function. The major drug targets, enzymes, GPCRs, ion channels, and membrane transport proteins (transporters) are all protein structures created by the aforementioned forces. While there are certainly additional available drug targets, such as protein/protein interactions, DNA, and RNA, the pharmaceutical industry has focused most of its energy, sometimes unknowingly, on these classes of targets. As such, an exploration of each class is warranted.

ENZYMES

In simplest terms, enzymes are protein catalysts employed by nature to facilitate the chemical transformations required to sustain life. The concept of enzymes was first proposed by Wilhelm Kühne in 1877, but it was not until many years later that it would be recognized that enzymes are a kind of protein. Jack bean urease, the enzyme responsible for converting urea into ammonia and carbon dioxide, was crystallized by James B. Sumner in 1926 and was the first enzyme to be recognized as a protein.[10] Since that time, thousands of enzymes have been identified as key mediators of a wide range of biological functions such as signal transduction, muscle contraction, cell size regulation, viral infection, and fluorescence. Six classes of enzymes have been identified to date (Figure 3.10). Given their wide range of function and importance, it

Classification	Function	Examples
Oxidoreductases	Catalyzes redox chemistry, transfering electrons from one molecule to another, often with use of a cofactor	HMG-CoA Reductase, Cyclooxygenase, Monoamine Oxidase, Alcohol Dehydrogenase
Transferase	Catalyzes the transfer of a functional group from one compound to another.	Tyrosine kinase, Reverse Transcriptase, DNA Methyltransferase, Glycosyltransferases
Hydrolase	Catalyzes the hydrolysis of a chemical bond.	HIV Protease, Tyrosine Phosphatase, Carboxypeptidases, Influenza Neuraminidase
Lyase	Catalyzes the cleavage of a chemical bond in a manner other than hydrolysis or oxidation, often forming a double bond or ring.	Adenylate Cyclase, Pyruvate Decarboxylase, Maleate Hydratase, Isocitrate Lyase
Isomerase	Catalyzes structural rearrangement to form isomers of the substrate.	Topoisomerase, Retinol Isomerase, Mannose Isomerase, Isocitrate Epimerase
Ligase	Catalyzes the joining of large molecules with a chemical bond.	DNA Ligase, RNA Ligase, E3 Ubiquitin Ligase, Tyrosine—tRNA Ligase

FIGURE 3.10   Representative examples of six classes of enzymes.

should be no surprise that enzymes are often the target of drug discovery programs.

Structurally, enzymes are composed of a series of amino acids that folds and twists to form a specific three-dimensional shape based on the chemical interactions described earlier. The number of amino acids required for enzymatic activity is highly variable. One of the smallest enzymes, 4-oxalocrotonate tautomerase, which converts 2-hydroxymuconate into 2-oxo-3-hexenedioate, is comprised of just 62 amino acids.[11] On the other hand, fatty acid synthase, a key enzyme in the synthesis of fatty acids, is one of the largest enzymes with over 2500 amino acid residues.[12]

Despite their potentially immense size, enzymes are highly specific in nature, typically catalyzing only a single reaction on a very narrow range of substrates, and the business section of the protein, the active site, is only a small portion of the full length enzyme (Figure 3.11).[13] The remainder of the enzyme is essentially scaffolding required to create the active site, much in the same manner as framework of a building provides structural support that creates rooms within a building. The active site itself can be viewed as a cleft or crevice within the framework of an enzyme created by the residues surrounding the site. The amino acids that form the walls of the active site dictate an enzyme's specificity using the same kinds of interactions that drive the shape of the overall protein. The walls of the active site provide steric limitation on what will physically fit within the active site, and the various amino acid side chains can form positive interactions with a substrate. Aromatic side chains (e.g., phenylalanine) provide opportunities for π-stacking and π-cation interactions

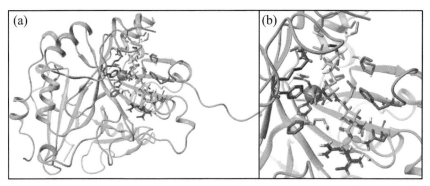

FIGURE 3.11 (a) Human glyoxalase 1, an enzyme responsible for the production of S-D-lactoylglutathione from the hemithioacetal of methylglyoxal and reduced glutathione, occurs as a dimer of two 183 amino acid residue proteins. An X-ray crystal structure of a monomeric unit bound to S-hexylglutathione is shown. (b) A magnified image of the active site of human glyoxalase 1 bound to S-hexylglutathione is shown. The catalytic site contains a zinc atom that is critical to enzymatic activity (RCSB 1BH5).

between a substrate and the enzyme, while hydrophobic interaction can occur between nonpolar amino acid side chains and nonpolar regions of a substrate molecule. Hydrogen bonds can also be formed between the substrate and the active site amino acids, either through their side chains (e.g., arginine, asparagine, etc.) or with the amide back bone of the protein. Compounds that cannot meet the strict criteria required for binding at the active site of an enzyme cannot be substrates for the enzyme.

The question of how enzymes accelerate chemical reactions has been considered by scientists for over 100 years. Perhaps the earliest hypothesis on enzymatic mechanisms was present by Emil Fischer in 1894. His "lock and key" model (Figure 3.12) proposed that enzymes and substrates must have complementary geometric shapes that exactly fit each other in order for an enzyme to function on a given molecule.[14] Fischer's theory was modified independently by Brown[15] and Henri,[16] each of whom suggested that enzymatic reactions occurred through an enzyme–substrate complex (Figure 3.12(a)). The theory was further modified in 1958 by Daniel Koshland with the addition of the concept of "induced fit." This additional aspect of enzyme mechanistic theory suggested that the binding of a substrate to the active site could induce conformational changes to the enzyme itself. These conformational changes could increase the catalytic activity of the enzyme by moving key residues into the proper orientation to catalyze the desired reaction (Figure 3.12(b)).[17]

Although they are far more complex when compared to simple acid catalysts that might be used to prepare an ester from an alcohol and a carboxylic acid, the principal features that define a catalyst are still the same.

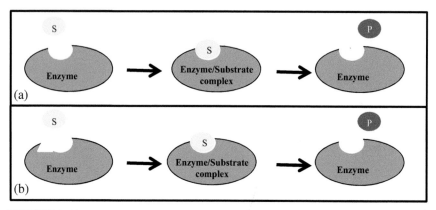

FIGURE 3.12   (a) Emil Fischer's 1894 theory that enzymes and substrates (S) must have complimentary shapes in order for catalysis of a reaction to form the product (P) forms the basis of modern understanding of enzymatic reactions. The concept of the formation of a transient enzyme/substrate complex as a reaction intermediate was later suggested by Brown and Henri. (b) "Induced fit," the theory that binding of a substrate to an enzyme could cause changes in the overall configuration of the enzyme to produce a structure capable of supporting reaction catalysis, was introduced by Koshland in 1958.

Enzymes act to increase the rate at which a preexisting reaction reaches its equilibrium by lowering the activation energy required for the reaction to proceed (Figure 3.13). Human carbonic anhydrase, for example, catalyzes

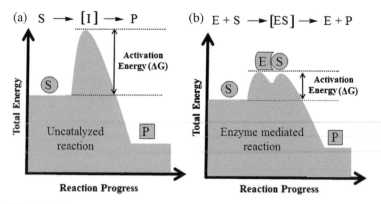

FIGURE 3.13   (a) In the absence of a catalyst, a given reaction will proceed from starting material (S) to product (P) by way of a transition state intermediate (I) that is higher in energy than the starting material. The energy required to reach the transition state is referred to as the activation energy (ΔG). In general, lowering the activation energy of a reaction leads to higher reaction rate. (b) An enzyme acts as a catalyst for a reaction, lowering the activation energy of a reaction by forming lower energy intermediates (an enzyme/substrate complex, ES). Binding interactions between the substrate and the enzyme decrease the energy requirements, in some cases forming multiple lower energy intermediates that lower the overall activation energy of a reaction.

the conversion of CO_2 to H_2CO_3 at nearly 10^8 times the rate of the uncatalyzed reaction, while orotidine-5′-phosphate decarboxylase increases the rate of conversion of orotidine-5′-phosphate into uridine-5′-phosphate by 10^{17} over the nonenzymatic reaction.[18] Also, like their non-biological counterparts, enzymes are unchanged by the catalytic process. Once the conversion from substrate to product is complete, the enzyme is free to encounter another substrate molecule and repeat the cycle until chemical equilibrium is reached.

An examination of the reaction mechanism of the cleavage of peptide bonds by serine protease (also known as serine endopeptidase) demonstrates the principles of enzyme reaction mechanisms (Figure 3.14). This enzyme employs a variety of hydrogen-bonding interactions and a key serine side chain to hydrolyze amide bonds in the peptide backbone of a protein. In the absence of a catalyst, hydrolysis of amide bonds requires extreme conditions, but serine protease can accomplish this reaction at an extremely high rate. Initial entry of a substrate into the active site is followed by the reaction of serine-195 with the substrate (a) to provide an enzyme–substrate complex intermediate (b). The formation of this intermediate is supported by the hydrogen-bonding interaction of other amino acids in the active site. Aspartic acid-102 and histidine-57 facilitate the deprotonation of serine-195 via hydrogen bonding, allowing it to react with the amide bond. The carbonyl of the amide substrate, on the other hand, is made more reactive through its interactions with serine-195 and glycine-193. The transient intermediate formed in this first step then reorganizes with loss of the amino-portion of the amide bond, leaving the former amide carbonyl bound to serine-195 as an ester (c). Although esters are less stable than amides, cleavage of esters is also slow in the absence of a catalyst. The serine-195 ester (c) reacts with a molecule of water with the assistance of the same active site amino acid side chains that supported the cleavage of the amide bond to form a second transient intermediate (d) that can rearrange to eject the C-terminal portion of the peptide substrate, leaving the enzyme ready for a new substrate (e).[19]

In some cases, enzymes require the presence of additional material, cofactors, in order to function properly (Figure 3.15). These cofactors or coenzymes can be a wide range of atoms and molecules. The matrix metalloproteinases,[20] which degrade collagen and gelatin, for example, require the presence of a zinc atom. In the absence of zinc, these enzymes will not function. Iron, magnesium, manganese, molybdenum, selenium, and copper have also been identified as required cofactors in a variety of enzymatic systems.[21]

Organic compounds also play a major role as coenzymes. Cytochrome P450 17A1,[22] also referred to as 17-α-hydroxylase/$C_{17,20}$-lyase, for example, has a key role in the production of a number of biologically important

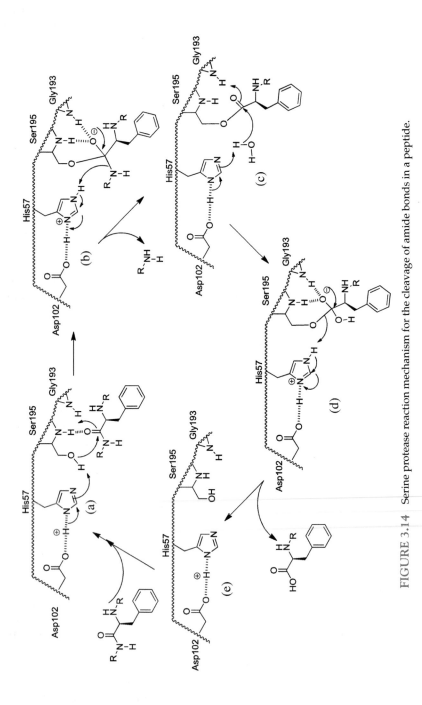

FIGURE 3.14   Serine protease reaction mechanism for the cleavage of amide bonds in a peptide.

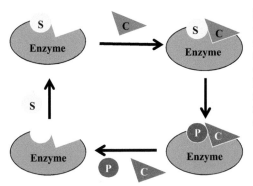

FIGURE 3.15 An enzyme binds with a substrate (S), which then binds to a coenzyme that supports the enzymatic process. After the reaction is complete, the product is released, and both the enzyme and coenzyme are recycled. In some cases, the coenzyme must be regenerated before the next reaction cycle.

compounds such as progestins, mineralocorticoids, glucocorticoids, androgens, and estrogens, but it requires the presence of an iron-based heme unit. Redox active compounds such as nicotinamide adenine dinucleotide phosphate (NADP) and flavin adenine dinucleotide (FAD) are critical cofactors in various metabolic processes. Additional important coenzymes include coenzyme A, adenosine-5′-triphosphate (ATP), coenzyme Q, and heme B (Figure 3.16). Recycling of the coenzymes is often achieved through an independent enzymatic pathway.[23]

INHIBITION OF ENZYMES

Although there are a large number of enzyme inhibitors, most of them can be categorized in to a relatively small number of distinct classes based on their general mode of actions: competitive inhibitors, irreversible inhibitors, and allosteric inhibitors (Figure 3.17). Compounds that act as competitive inhibitors of an enzyme are capable of occupying the active site of an enzyme (or a portion thereof), thus preventing entry of the natural substrate into the active site. In this case, inhibition is reversible, as no covalent bonds are formed between the enzyme and the inhibitor. The same forces that hold proteins in their native state (e.g., hydrogen bonding, hydrophobic interactions, etc.) allow the inhibitor to associate with the active site. The influenza drug Tamiflu® (Oseltamivir), for example, competitively inhibits influenza neuraminidase, an enzyme that catalyzes cleavage of sialic acid from glycoproteins, by producing transition state mimic (GS-4701) upon metabolism by patients (Figure 3.18). In this case, a series of energetically favorable interactions allow the drug to reversibly occupy the active site of the enzyme, blocking the natural ligand.[24a,24b]

Selectivity, however, can be an issue with competitive inhibitors. Consider the kinase family of enzymes, for example, which phosphorylate

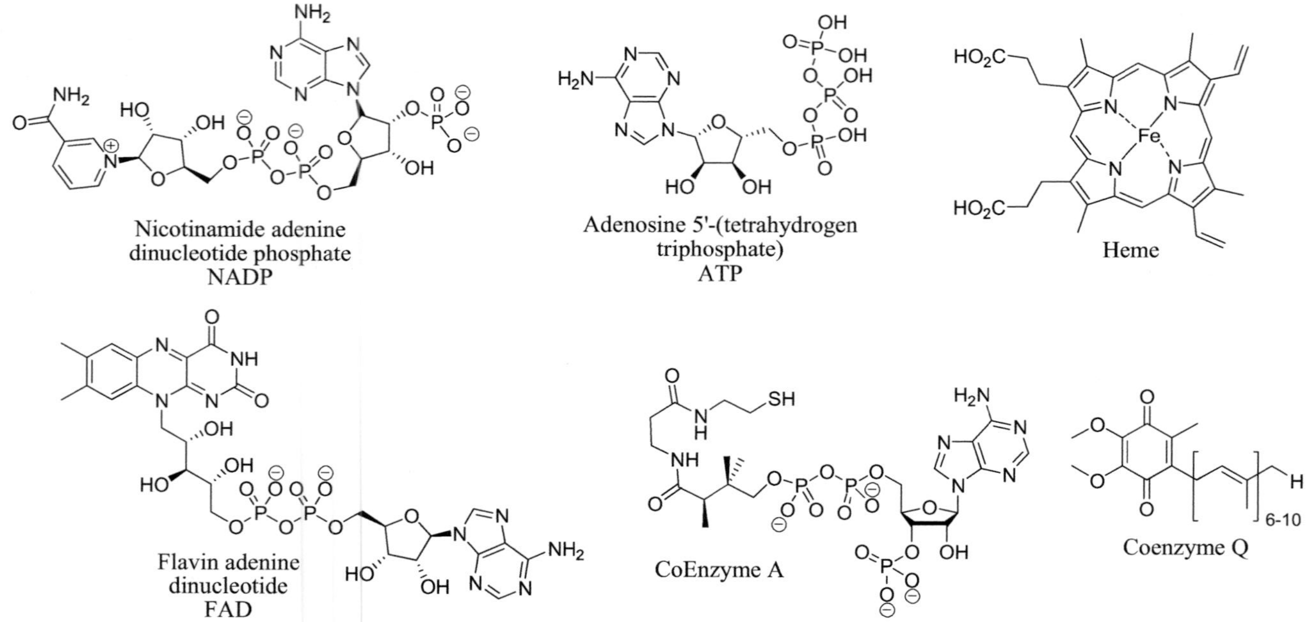

FIGURE 3.16 Coenzymes are required components of many enzymatic reactions. Nicotinamide adenine dinucleotide phosphate (NADP) and Flavin adenine dinucleotide (FAD) are employed in enzymatic redox chemistry, Adenosine-5'-triphosphate (ATP) is the most common energy transfer agent in cellular systems, Coenzyme A is an acyl transfer agent, and Coenzyme Q is part of electron transport chains that produce cellular energy. Heme B is the most abundant heme present in humans and is the carrier for oxygen in hemoglobin.

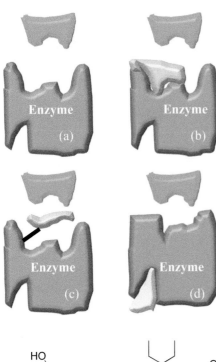

FIGURE 3.17 (a) In the normal enzyme process, substrates (green) interacts with active sites. (b) Competitive inhibitors (orange) reversibly blocks the active site. (c) Irreversible inhibitors (orange) covalently bind to the active site. (d) An allosteric inhibitor (yellow) binds to an allosteric binding site, altering the active site, preventing substrate binding.

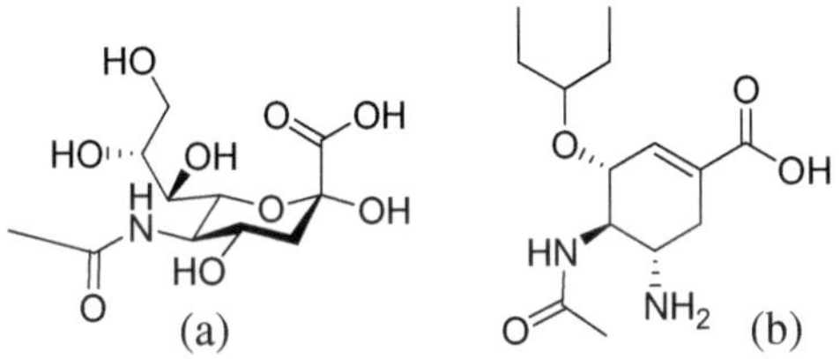

FIGURE 3.18 (a) Sialic Acid (b) GS-4071.

their substrates via an ATP-mediated process. With over 500 known examples, this class of enzymes has been the subject of intense study. They are critical to the majority of cellular functions, especially those related to signal transduction in both normal and disease states.[25] While the substrate for any particular kinase may be different, and therefore require different amino acid sequences for binding, ATP does not change its structure from one kinase to another. Nature has taken advantage of this by recycling the ATP-binding motif in the vast majority of kinase enzymes, leading to a significant level of homology within this family. While this may be efficient for nature, it can create a significant selectivity issue for competitive inhibitors of kinases that target the ATP-binding domain of the active site. In other words, kinases with very different phosphorylation substrates may have similar, perhaps identical ATP-binding domains, making the task of designing a compound that selectively inhibits one kinase over the others very challenging.

Allosteric inhibitors, by definition exert their influence at sites other than the active site of the enzyme in question just as their name implies. The active site of the targeted enzyme remains unoccupied, but is made inaccessible to the natural substrate as a consequence of the presence of an allosteric inhibitor. When the allosteric inhibitor binds to an allosteric binding site (a binding site on the enzyme that is different from the active site), it induces changes in the enzyme's overall configuration such that the binding site is no longer capable of interacting with the natural ligand (Figure 3.17(d)). MEK1, a member of the kinase class of enzymes that plays an important role in the progression of cancer, for example, can be allosterically inhibited by compounds such as CI-1040. Although this compound is a potent inhibitor of MEK1 and MEK2, it does not bind to the ATP-binding domain, which is the active site of MEK1. Rather, it binds to an adjacent binding site and produce inhibition of MEK1 via conformational changes induced by its presence in the allosteric site (Figure 3.19).[26]

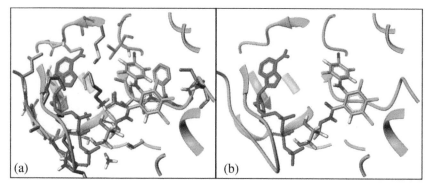

FIGURE 3.19   (a) CI-1040 bound adjacent to the active site of MEK1 occupies an allosteric binding site. ATP (red) and key side chain residues are displayed. (b) CI-1040 and ATP (red) bond to Mek1 with side chains hidden. *RCSB file 1S9J.*

Although the kinase family of enzymes shares a high degree of homology at the ATP-binding site, the same is not necessarily true of allosteric binding sites. This provides an opportunity to design compounds with a higher degree of selectivity within the kinase family.

Unlike competitive and allosteric inhibitors, irreversible inhibitors covalently attach to the active site of the target enzyme, blocking entry of the natural substrate and inactivating the enzyme. β-Lactams, such as benzylpenicillin (penicillin G), for example, react with a serine residue in the active site of penicillin-binding proteins (PBPs), deactivating the enzyme (Figure 3.20). PBPs are essential for the final stages of peptidoglycan synthesis, a major component of bacterial cell walls. Irreversible inhibition by β-lactams leads to decreased cell wall strength and cell death of the targeted microorganism.[27] Since there is no human counterpart to PBP,

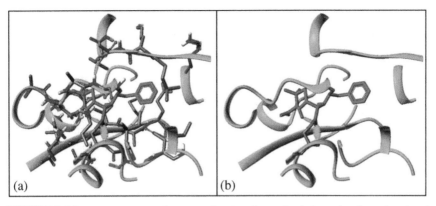

FIGURE 3.20   (a) Benzylpenicillin (Penicillin G, red) covalently bound to the active site of penicillin-binding protein A from *M. tuberculosis* with key side chains displayed. (b) Benzylpenicillin (Penicillin G, gray) covalently bound to the active site of penicillin-binding protein A from *M. tuberculosis* with key side chains hidden. *RCSB file 3UPO.*

the prospect of undesired side effect mediated by β-lactam antibiotics is generally low. The application of irreversible inhibitors for the treatment of conditions in which there is a human variant of the target enzyme can be an issue, however, as restoring enzymatic activity generally requires the synthesis of additional quantities of the target enzyme. Suppose, for example, an irreversible kinase inhibitor were developed targeting the ATP-binding site of MEK1. The irreversible inhibitor would bind to MEK1, suppressing its activity, but given the high degree of homology of the ATP-binding site within the kinase family, it is likely that many other kinases would also be irreversibly inhibited. Irreversible inhibition of enzymes that are involved in drug metabolism can also lead to significant negative consequences by altering the rate at which drugs are cleared from the body. This will be discussed in greater detail in Chapter 6. In general, the pharmaceutical industry favors the development of competitive and allosteric inhibitors over that of irreversible inhibitors.

G-PROTEIN-COUPLED RECEPTORS (GPCRs)

The flow of information across plasma membranes is absolutely essential to proper cellular function and coordination of bodily functions. Human cells must be in routine communication with neighboring cells, aware of their environment, and, in many cases, provide information to cells in distal locations of the body. Rapid responses are often required. The sensation of pain felt upon touching a hot surface, for example, must be relayed nearly instantaneously to the brain upon contact with the hot surface. Cellular functions, such as glucose uptake, proliferation, and even responses to photons in the eye, depend upon an intricately connected and responsive cellular communication system. Although enzymes are remarkable in their ability to facilitate chemical conversions, they are not suited to this task.

GPCRs play a significant role in signal transduction. This family of membrane-bound proteins is a major part of the communication network that provides cells with information necessary for proper bodily function. They are the largest class of human membrane-bound proteins,[28] and given their critical role in maintaining normal physiological activity, it should come as no surprise that they have been and continue to be the subject of intense drug discovery research.

Historically, the concept of membrane-bound receptors playing a role in cellular function was first suggested at the beginning of the twentieth century when the concept of "receptive substances" was introduced.[29a,29b] Although the mechanistic details were not understood at the time, Langley (1901) and Hale (1906) suggested that there were "receptive substances" present on the outer surface of cells that could respond to the surrounding environment. This hypothesis proved to be foundational in the modern understanding of GPCRs. The nature of the "receptive substance" remained largely unknown and the subject of intense debate for nearly 70 years, until 1970 when Robert Lefkowitz and his coworkers took the first steps to unravel the structural nature of GPCRs. Their pioneering work with radiolabeled adrenocorticotropic hormone (ACTH, corticotropin) allowed them to detect and visualize the binding of this agonist to its receptor in adrenal membrane preparations.[30] The structural and functional details of GPCRs are intimately related, and as such, once methods became available to detect and visualize them, the science advanced rapidly. Crystallographic evidence of GPCR structural motifs became available in 2000 with the publication of the first crystal structure of a mammalian GPCR, bovine rhodopsin (Figure 3.21(a)).[31] In 2007, the X-ray crystal structures of a human GPCR, β_2-adrenergic receptor, was reported (Figure 3.21(b)).[32]

Structurally, there are a number of features that all GPCRs have in common (Figure 3.22). First and foremost, all GPCRs are membrane-bound proteins whose interaction with a second class of proteins referred to as G-proteins, are controlled by the presence or absence of a ligand. The transmembrane portion of GPCRs forms a series of seven transmembrane segments designated TM-1 to TM-7, whose secondary structures are primarily α-helices. The segments themselves are interconnected by a series of three intracellular loops (designated IL-1 through IL-3) and three extracellular loops (designated EL-1 through EL-3). While the transmembrane regions are not identical from one GPCR to the next, there is a high degree of homology within the overall GPCR family. The presence of this common structural feature is the basis for an alternative name for the GPCR family, as they are also referred to as seven-transmembrane (7TM) receptors.

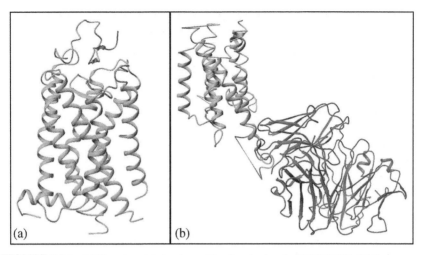

FIGURE 3.21   (a) X-ray crystal structure of bovine rhodopsin (RCSB 1F88). (b) X-ray crystal structure of human β2-adrenergic receptor (Green transmembrane domain, red cellular domain (RCSB 2R4R)).

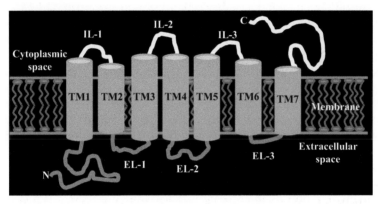

FIGURE 3.22   The typical G-protein-coupled receptor (GPCR) has seven transmembrane regions (TM-1 through TM-7, gray), three intracellular loops (IL-1 to IL-3, yellow), and three extracellular loops (EL-1 to EL-3, red).

The extracellular ends of GPCRs are oriented such that the carboxy terminus is on the cytoplasmic side of the cell membrane, whereas the amino terminus is on the extracellular side of the membrane. These regions, along with the TM5/6 intracellular loop, have a higher degree of variability than the transmembrane portions of GPCRs, and are primarily responsible for the ability of GPCRs to distinguish between a wide variety of ligands such that individual GPCRs are activated by only the specific molecular framework of its target ligand. In the absence of this ability to distinguish between different types of ligands, all GPCRs would be activated by the same ligand sets, substantially decreasing their ability to send and receive distinct chemical signals and maintain cellular function. The amine terminus, for example, tends to be smaller for monoamine and peptide receptors, containing 10–50 amino acid residues, while glycoprotein and glutamate receptors are significantly larger with 350–600 amino acid residues.[28]

While the exact number of distinct GPCRs has as yet to be determined, it is estimated that the structural variations described above has given rise to over 800 unique GPCRs in five families,[33] based on an analysis of the human genome. The largest subset in this group is the rhodopsin family with over 700 members divided into 19 separate classes (subfamilies A1 through A19). Members of this family are structurally related through their homology with rhodopsin, a GPCR that responds to photons striking the surface of the retina, but they have a broad range of function.[34] Serotonin receptors,[35] dopamine receptors,[36] angiotensin II receptors,[37] and prostaglandin receptors,[38] for example, are all members of this family, giving the rhodopsin family roles in central nervous system function, cardiovascular regulation, and pain perception. Despite their structural homology, each of these GPCRs respond to a different set of non-overlapping ligands.

There are four additional, smaller classes of the GPCRs based on their relative homology and structure. The secretin family[39] (15 known examples) contains the classic seven transmembrane regions that define the GPCR family, but all of the members of this family are homologous with the secretin receptor. The parathyroid hormone receptor[40] and glucagon receptors are also members of this family. Frizzled/taste GPCRs represent a third class of GPCRs, which have roles in functions ranging from cell differentiation and proliferation through the Wnt signaling pathway to the sensation of the taste as is implied by the name of the class. To date, there are 24 examples in this family.[41]

The glutamate receptors[42] comprise the fourth major class of GPCRs (15 members), and as their name implies, they are related through their response to glutamate. The response provided by ligand binding, however, is dependent upon the location of the GPCR. One of the critical functions for this family of GPCRs is the modulation of the excitability of synaptic cells, giving them an important role in nerve transmission.

Lastly, the adhesion class of GPCRs is typified by their chimeric structures. As is the case of all GPCRs, the adhesion GPCRs are based upon a seven-transmembrane domain, but they also possess an extracellular domain that is far larger than any of the other classes. In fact, adhesion class GPCRs are the largest (structurally) of the GPCRs, as the N-terminal extracellular domain can be the size of independent proteins. As the name implies, cellular adhesion is an important aspect of their function, and it is theorized that cellular adhesion of expressing cells with target cells (e.g., immune cells with foreign cells) is followed by a signal transduction event that elicits a cellular response. There are at least 24 adhesion GPCRs, and they have been identified in immune cells, the central nervous system, and reproductive tissue.[43]

G-Protein-Dependent Signaling Pathways

Signal transduction requires the presence of both machinery to produce the signal and a carrier to propagate the signal itself. The telephone, for example, is well suited to produce a signal that allows for communication over long distances, but cannot function in the absence of an electrical current, which acts as the signal to create sound at each end of the communication grid. In the same sense, GPCRs are the machinery that generate the signal (i.e., the telephone), but they are of little use in the absence of a signaling molecule of some type (i.e., the electricity) that transmits the signal. The signaling molecules that transmit signals from the GPCRs to the rest of the cellular machinery are often referred to as "second messengers" and can significantly amplify the signal strength by activating downstream cellular machinery. There are two major second messenger systems, the cAMP system (Figure 3.23), which uses cyclic adenosine monophosphate (cAMP, 3′-5′-cyclic adenosine monophosphate) system as its second messenger, and the phosphatidylinositol signaling system

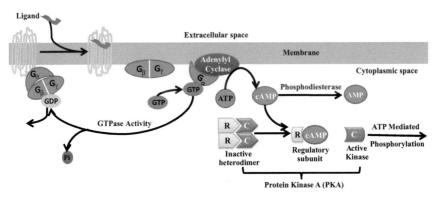

FIGURE 3.23   cAMP signaling begins with binding of a ligand to the GPCR. Conformational changes in the GPCR causes the G-protein complex to disassociate from the GPCR, the G_α protein and GDP are released, and GTP binds to the G_α protein. The GTP/G_α protein complex binds to adenylate cyclase, activating the enzyme, which produces cAMP. Binding of cAMP to the regulatory protein ("R") suppressing protein kinase A (RC) releases active protein kinase A (C), allowing it to phosphorylate molecular targets. The system is regulated by the GTPase activity of the G_α protein and cAMP phosphodiesterase.

(Figure 3.24), which generates inositol-1,4,5-trisphosphate (IP$_3$) and diacylglycerol (DAG) from phosphatidylinositol 4,5-bisphosphate to relay cellular signaling.[44]

cAMP Signaling

A review of the cAMP signaling system (Figure 3.23) begins with the resting state (non-signaling or basal level) of the pathway. In the resting state, a cAMP-dependent GPCR is associated with a guanyl nucleotide-binding protein, also referred to as a G-protein, on a the cytoplasmic side of the cell membrane. The G-protein is comprised of three subunits, the G_α, G_β, and G_γ subunits. In the absence of a ligand, the G_α subunit is bound to a guanosine diphosphate (GDP) molecule. When the natural ligand enters the extracellular binding site, conformational changes within the GPCR decreases its affinity for the G-protein assembly, and the G-protein assembly is released. This, in turn, initiates conformational changes in the G-protein assembly, causing it to release the GDP. At the same time, the G_α subunit separates from the G_β/G_γ complex. The G_α subunit then binds to guanosine triphosphate (GTP), and conformational changes within the G_α/GTP allow it to bind to the enzyme adenylyl cyclase. This enzyme is responsible for the conversion of adenosine-5'-triphosphate (ATP) into cAMP, the second messenger for this system. Binding of the G_α/GTP complex activates adenylyl cyclase, causing an increase in production of cAMP. Increased cAMP cellular concentrations then lead to changes in a targeted protein system. In the case of protein kinase A, cAMP binds to a regulator protein that is part of an enzyme/regulator complex that suppresses kinase activity. Binding of cAMP to the regulatory subunit, however, causes conformational changes in the regulatory protein, allowing it to disassociate from protein kinase A. Once it is released from the regulatory protein, protein kinase A becomes catalytically active, phosphorylating its substrate via an ATP-mediated pathway. Thus, ligand binding on the extracellular side of a GPCR is translated into kinase activity through a cascade of cellular signaling events.

Of course, once a signal is turned on, there needs to be a way to turn the signal off so that cellular activity can revert back to the way it was before the signal was initiated. Disassociation of the natural ligand from the binding site allows the GPCR to return to its inactive state and bind to a G-protein/GDP complex, but this will not release cAMP from the regulatory protein allowing it to suppress protein kinase A in the above example, nor will it separate the G_α/GTP complex from adenylyl cyclase, thereby halting the formation of additional cAMP. Fortunately, additional regulatory pathways exist to terminate the GPCR signal. cAMP levels, for example, are regulated by cAMP phosphodiesterase, which converts cAMP to AMP (adenosine monophosphate). Removal of cAMP by the action of cAMP phosphodiesterase, allows the regulatory protein to revert back to the configuration that suppresses protein kinase A activity, terminating the GPCR signal. Separately, the G_α protein is also a GTPase that slowly converts GTP to GDP. Once this occurs, the G_α protein is no longer able to bind to adenylyl cyclase, deactivating cAMP production and terminating the GPCR signal.

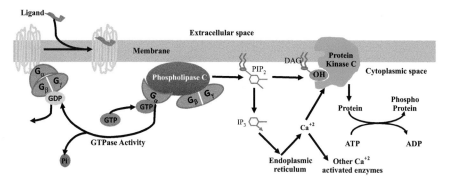

FIGURE 3.24 IP_3 signaling is initiated by ligand binding to the GPCR. Conformational changes in the GPCR cause the G-protein complex to disassociate from the GPCR, the G_α protein and GDP are released, and GTP binds to the G_α protein. The GTP/G_α protein complex binds to phospholipase C, which hydrolyzes PIP_2, releasing DAG and IP_3. Cytoplasmic IP_3 causes the release of cellular calcium stores, while membrane-bound DAG activates protein kinase C which phosphorylates molecular targets via ATP. Protein kinase C activity is augmented by the presence of calcium. The system is regulated by enzymatic degradation of IP_3 and DAG, GTPase activity of the G_α protein, and removal of cytoplasmic calcium.

IP_3 Signaling

The phosphatidylinositol signaling pathway (IP_3 signaling pathway, Figure 3.24) is similar to the cAMP pathway. The receptor is bound to a G-protein receptor complex composed of G_α, G_β, and G_γ subunits, which is, in turn, bound to a GDP. As in the cAMP pathway, ligand binding on the extracellular surface induces conformational changes in the GPCR that decrease its affinity for the G-protein/GDP complex, which disassociates and undergoes its own conformational changes. These changes lead to the separation of GDP from the G_α subunit and release of the G_α subunit from the G-protein complex. The G_α subunit then binds to GTP, forming a G_α/GTP complex. At this point, the IP_3 signaling pathway diverges from the cAMP pathway. In the IP_3 signaling pathway, the G_α/GTP complex interacts with phospholipase C, activating this enzyme, which then cleaves membrane-bound phosphatidyl-4,5-inositol bisphosphate (PIP_2) into two second messenger, inositol-1,4,5-triphosphate (IP_3) and diacylglycerol (DAG). The IP_3 is released into the cytoplasm of the cell and activates the release of calcium ion from cellular storage in the endoplasmic reticulum. DAG, on the other hand, remains membrane bound, and interacts with membrane-bound protein kinase C (PKC), which contains both a regulatory domain and a catalytic domain. The presence of DAG causes conformational changes that eliminate the inhibition caused by the regulatory domain, activating the enzyme, and increasing its affinity for calcium. Since PKC requires calcium for enzymatic activity, the release of calcium induced by IP_3 has a synergistic effect on PKC activity. The active PKC phosphorylates its targets in an

ATP-dependent system, thus propagating the signal induced at the cell surface by the ligand.

Signal termination is just as important in the IP_3 signaling pathway as it is in the cAMP pathway, and there are several ways in which the cascade events can be interrupted. The G_α protein's GTPase activity that slowly converts GTP to GDP causes G_α to be released from phospholipase C, deactivating it. This halts the production of IP_3 and DAG, terminating the signal. Also, removal of calcium from the cytosol by calcium ATPase pumps will dampen the signal (PKC requires calcium for enzymatic activity), while conversion of IP_3 to inositol by a series of phosphatase removes its influence from the signaling cascade. DAG, on the other hand, can be either converted to glycerol or phosphorylated by the appropriate enzymatic systems, but in either case, once it is removed from the signaling pathway, PKC reconfigures itself such that the regulatory domain suppresses its catalytic activity, halting signal propagation.

Modulating GPCR Activity

As complex as these systems are, it is important to understand that GPCRs do not exist in isolation, and the flow of information across a cell membrane is often not a linear event. In general, cells express multiple GPCRs that respond to the environment in different manners, have overlapping effects, can impact each other's activity, and form an integrated mosaic of information flow. Proteins such as β-arrestin[45] and G-protein-coupled receptor kinases[46] can also alter the signal pathways, giving rise to differential activities for ligand and GPCRs depending on their cellular context. The recent introduction of the concept of biased ligand, which suggests that there are multiple active conformations of a single GPCR that can create different downstream events depending on the chemical structure of the ligand, further complicates the picture (Figure 3.25). It should be clear that targeting a GPCR for new drug development is a complex endeavor.

Despite these complexities, the majority of drugs that interact with GPCRs can be described as falling into one of three categories, agonist, antagonists (also referred to as neutral antagonists), and inverse agonists (Figure 3.26). In simple terms, an agonist mimics the natural ligand of a given GPCR and produces the same cellular response as the natural ligand. The activity of an agonist is generally measured as a function of its binding affinity for the GPCR in question and its efficacy relative to the natural ligand. The cellular response prompted by saturation levels of the endogenous ligand is considered 100% efficacy, and potential drug candidates that provide this level of response are considered full agonists. Compounds that elicit a cellular response that is below that of the endogenous

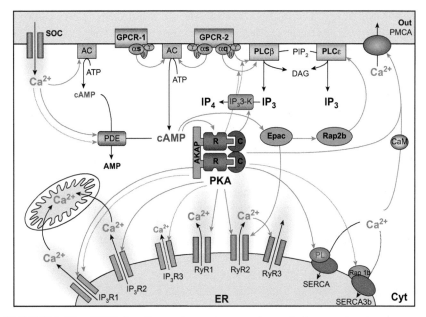

FIGURE 3.25 GPCR signaling pathways are complex and often overlapping systems. The biological impact of activation or deactivation of a GPCR depends on its physical/cellular location and downstream proteins that are impacted by changes in GPCR activity. Abbreviations: AC = adenylyl cyclase; PLC = phospholipase C; SOC = store-operated Ca^{2+} channel; IP_3-K = $InsP_3$ 3-kinase; PDE = phosphodiesterases; R = regulatory of PKA; C = catalytic subunit of PKA; AKAP = A-kinase anchoring protein. *Source: Reprinted from Bruce, J. I. E.; Straub, S, V.; Yule, D. I. Crosstalk between cAMP and Ca2+ signaling in non-excitable cells. Cell Calcium, 34 (6), 431–444, copyright 2003 with permission from Elsevier.*

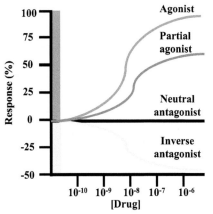

FIGURE 3.26 Full agonists (green) induce GPCR signaling equal to that of the endogenous ligand, while partial agonists (blue) activate GPCR signaling to a lesser extent. Neutral antagonists (black) do not induce GPCR activity, but will block agonist activity. Inverse agonists suppress basal activity of a constitutively active GPCR.

ligand are referred to as partial agonists. In some cases, partial agonists can effectively compete with the natural ligand, effectively lowering the response of the GPCR to signal induction.

Antagonists, on the other hand, bind to the GPCR, but do not elicit a cellular response. In the absence of an agonist or the endogenous ligand, an antagonist has no effect on cellular activity. However, the presence of an antagonist can prevent GPCR-mediated cellular responses, as binding of the antagonist prevents binding of the natural ligand or agonist, preventing conformational changes necessary for the initiation signaling events. Antagonists can be binding site mimics that directly block the natural ligand, or they can bind to an allosteric site that causes conformational changes that prevent the formation of the GPCR/ligand complex that starts the signaling cascade.

Inverse agonists can also block the activity of an endogenous ligand or agonists, thereby presenting themselves as antagonists, but they are also capable of producing a pharmacological response opposite to that mediated by the endogenous ligand in some situations. This inversion of action is only possible, however, if the GPCR target is constitutively active which provides an intrinsic or basal level of signaling that is present in the absence of the endogenous ligand. Basal signaling is the result of a GPCR spontaneously adopting the active conformation in the absence of the ligand. While GPCRs are often viewed as on/off switches, it is important to realize that GPCRs exist as an equilibrium mixture of conformations. Conformations that do not support signaling predominate in the absence of an activating ligand, but the equilibrium is often not 100% in favor of the deactivated GPCR signaling mechanism. GPCRs that display basal activity exist at least in some small level in the active signaling conformation irrespective of the presence of the natural ligand. The natural ligand tips the equilibrium heavily in favor of the active signaling conformation, boosting the signal beyond its basal level. An inverse agonist, on the other hand, binds to the same site as the endogenous ligand, but stabilizes the inactive, non-signaling conformation of the GPCR in question. This suppresses basal activity of the GPCR, blocking the signal that is normally present even when the endogenous ligand is not present.

It is interesting to note that despite their complexity, GPCR activity is often modulated by surprisingly simple molecules (Figure 3.27). Compounds such as serotonin,[35] histamine,[47] and dopamine[36] play critical roles in GPCR signaling, and yet they are tiny by comparison to the GPCRs that they stimulate. Fentanyl, a μ-opioid agonist, acts at the same GPCR as β-endorphin, which has a molecular weight of over 3400, despite the fact that it is only one-tenth the size of β-endorphin. Size and complexity are not the driving forces mediating GPCR function. Just as the GPCRs must obtain a specific conformation in order to propagate a signal, compounds that modulate GPCR activity must be able to adopt the specific conformation required by the binding site that they are targeting. Those interested in more detailed information on particular GPCRs are encouraged to consult the modern literature for a more detailed analysis.

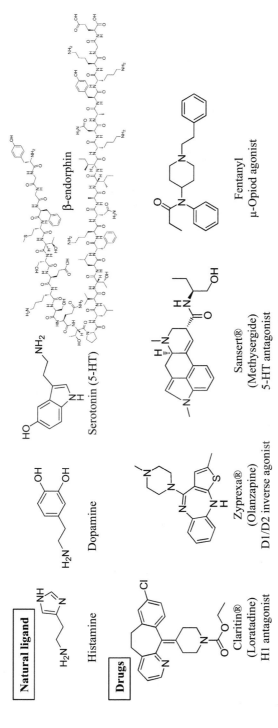

FIGURE 3.27 Functional activity at a given GPCR is not necessarily mediated by size. Claritin® (loratadine), Zyprexa® (olanzapine), and Sansert® (methysergide) suppress GPCR signaling and are significantly larger than the corresponding GPCR, but the same cannot be said of fentanyl and β-endorphin. In this case, the endogenous agonist is substantially larger than the synthetic agonist fentanyl, indicating that the proper agonist binding configuration can be achieved by molecules of diverse size and composition.

ION CHANNELS

Although living organisms are often viewed as complex machines run by chemical reactions, there are many critical functions in living organisms that cannot be accomplished solely by chemical means. In many cases, the generation of an electrical impulse or voltage gradient is required. Ion channels, transmembrane protein assemblies that regulate the flow of ions across biological barriers, play a major role in this process. Nerve impulse transmission,[48] muscle contraction,[49] and cardiovascular function, especially heart rate and rhythm,[50] all depend on the exquisitely balanced flow of ions created by a network of ion channels opening and closing in a coordinated fashion. T-cell activation in an immune response,[51] hormonal secretions (e.g., insulin), cellular proliferation (e.g., lymphocytes, cancer cells[52]), and even cell volume regulation[53] are all impacted by various ion channels. These proteins also play a major role in preventing cellular depolarization by counter balancing the impact of Na^+-coupled transporters (e.g., glucose transporter, amino acid transporters) and Ca^+-signaling events.[54] Modulation of ion channel activity has provided a number of important drugs and lethal toxins (Figure 3.28). Improper ion channel function has

FIGURE 3.28   (a) Norvasc® (Amlodipine), an antihypertensive agent that blocks calcium channels. (b) Amiodarone, an antiarrhythmic agent that blocks potassium channels. (c) Novocaine, a local anesthetic that blocks sodium channels. (d) Glipizide, an antidiabetic that blocks potassium channels in the pancreatic β-cells. (e) Phenytoin, an antiepileptic that blocks sodium channels. (f) Tetrodotoxin, a pufferfish toxin that blocks sodium channels, that is 100 times more lethal than cyanide.

been implicated in a number of important disease states (channelopathies, Table 3.1) such as cystic fibrosis,[55] epilepsy,[56] and long QT syndrome.

Modern efforts to understand the function of ion channels and the role of electrical currents in biological processes predate the age of modern drug discovery. The concept of "bioelectricity" was, in fact, explored as

TABLE 3.1 Disease States Associated with Ion Channels

Disease	Channel	Gene
Arrhythmia	Nav1.5	SCN5A
Arrhythmia	Kv1.5	KCNA5
Cystic fibrosis	CFTR	CFTR
Diabetes mellitus	Kir6.2	KCNJ11
Epilepsy	KCNQ2	KCNQ2
Epilepsy	Nav1.2	SCN2A
Episodic Ataxia	Kv1.1	KCNA1
Erythromelalgia	Nav1.7	SCN9A
Migraine	Cav2.1	CACNA1A
Fibromyalgia	Nav1.7	SCN9A
Long QT syndrome	hERG	KCNH2
Malignant hyperthermia	Cav1.1	CACNA1S
Neuropathic pain	TrpV1	TRPV1
Osteoporosis	ClC-7	CLCN7
Timothy syndrome	Cav1.2	CACNA2

early the mid-1840s by Carlo Matteucci[57] and Emil du Bois-Reymond,[58] who separately studied the impact of electrical currents on nerve and muscle tissue. Their experiments provided support for the role of electricity in biological processes, but did not provide an understanding of how living tissue could support electrical conductance. Hermann von Helmholtz provided additional insight into the role of electricity in living organisms in 1850 by demonstrating that the electrical signals traveled far slower in living tissue than they did in metal wires. His experiments indicated that simple conductance was not a viable explanation and suggested that an underlying chemical process might be involved.[59]

Nearly 50 years later, in 1902, Julius Bernstein introduced the "membrane theory" as a general explanation for bioelectrical events in living organism. The concept of semipermeable membranes was relatively new at the time, and Bernstein hypothesized that nerves and muscle were surrounded by semipermeable membranes. He further suggested that the electrical gradients across cellular barriers were the result of differences in ion concentrations on the inside and outside of cells created by the selective movement of ions across the cellular barrier. Bernstein referred to this effect as the formation of an "electrical double layer," but it is more often referred to as a "membrane potential" or "membrane voltage."[60] While his basic premise proved to be correct, a full understanding of the true nature of ion channel would remain a mystery for most of the twentieth century. In fact, at the dawn of the biotechnology revolution, Armstrong and his colleagues summed up the situation by writing "The ionic channels of nerve membrane and the gates that control ion movement through them are widely supposed to be composed of protein, but there is surprisingly little evidence on the question."[61]

The biotechnological tools and computer technology that became available in the last 30 years of the twentieth century provided the tools necessary to finally unravel the mystery of ion channels. Recombinant technologies, transfection methods, and advances in electronics allowed Erwin Neher and Bert Sakmann to develop the "patch clamp" method for studying ion channels directly. Prior to the advent of these technologies, ion channels experiments were limited to an analysis of macroscopic currents of a naturally occurring cell. Biotechnology provided the tools necessary to create cell lines that expressed a single type of ion channel. Neher and Sakmann's "patch clamp" technique placed a salt water containing micropipette against the surface of a single cell and measured the electrical currents generated by the flow of ions through an ion channel in much the same manner as electrical current are measured across a wire (Figure 3.29).

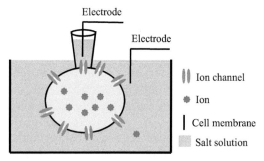

FIGURE 3.29 A basic patch clamp system consists of a micropipette with an opening on the order of 1 μm pressed against the surface of a cell. The inside of the micropipette covers a limited number of ion channels, and a seal with high electrical resistance ("gigaohm seal") is created by suction on the surface of the cell. An electrode, salt solution inside the micropipette, and the appropriate electrical amplification and monitoring systems can then be employed to either maintain a constant voltage while monitoring current or maintain a constant current while monitoring changes in membrane potential in the presence of test compounds.

Electrical signal amplification technologies made it possible to study the action of a single ion channel in a cell, providing for the first time a direct measurement of ion channel activity.[62] Neher and Sakmann were awarded the Nobel Prize in physiology or medicine in 1991 in recognition of the importance of their work.[63] Modernized version of this method[64] remains the gold standard for the study of ion channels and is one of only a few technologies available for the direct study of individual proteins.

Structural details of ion channels also began to emerge as a result of the technological advances that occurred at the end of the twentieth century. Protein sequences that encoded various ion channels were determined, and advances in molecular modeling were employed to predict the structural feature that would allow a transmembrane protein to transport a charged species through a lipophilic barrier. While various structural models were proposed, direct crystallographic evidence of the structural details of ion channels did not become available until 1998. Roderick Mackinnon's X-ray crystal structure of potassium channel designated KcsA from the soil bacteria *Streptomyces lividans* (Figure 3.30)

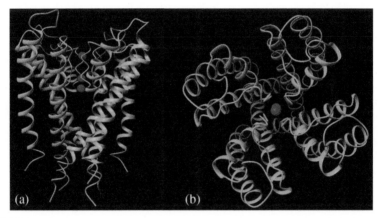

FIGURE 3.30   (a) Side view of X-ray crystal structure of *Streptomyces lividans* KcsA potassium channel. (b) Top view of X-ray crystal structure of *Streptomyces lividans* KcsA potassium channel. *RCSB 1BL8.*

provided the first complete view of an intact ion channel, an effort which won him the Nobel Prize in chemistry in 2003.[65a,65b] As of 2013, the RCSB Protein Data Bank contains over 3100 crystal structures categorized as ion channels.

Over 300 ion channels have been identified to date, and while their individual structures are designed to meet the need of their specific function, there are a number of common features that can be described. Much like the GPCRs, ion channels are integral membrane proteins comprised of a series of transmembrane domains that are linked by extracellular and intracellular loops. The majority of ion channels are multiunit assemblies, and functional channels can only be formed when multiple, compatible protein structure come together to form the active channel. Subunit homogeneity is not required, however, and this can lead to subtle differences in function. In many cases, the pore section of the channel is only wide enough for passage of a single ion at a time and passage is restricted to a single ion type. The selectivity of channels for a specific ion type is driven by the structural features of the proteins that makes up the channel. There are a number of sodium channels that cannot facilitate the movement of potassium ions across a membrane. Conversely, even though sodium ions are significantly smaller than potassium ions, there are potassium channels that do not accommodate sodium ions. To date, ion channels have been identified that support the flow of sodium, calcium, potassium, chloride, and hydrogen ions.

Gating Mechanisms

Another key feature of ion channels is the mechanism through which they are activated, also referred to as a "gating mechanism." In general, ion channels remain closed in the absence of an external stimulus. The action of a stimulating event or agent causes conformational changes in the proteins, opening the "gate" and allowing the flow of ions across a biological barrier. When the stimulus is removed, the channel reverts back to its closed state, stopping the flow of ions. The gating of a channel can be dependent on the presence of a ligand, environmental pH, temperature, or membrane voltage differences. Ion channels that are mechanosensitive (altered by mechanical deformations of a membrane such as tension and curvature changes) and light sensitive have also been identified. Ligand-gated and voltage-gated channels are the most extensively studied types of channels, and an examination of their modes of action can serve as a basis for understanding the nature of the remaining types of gating mechanisms.

Ligand-Gated Channels

Ligand-gated channels are activated when a ligand interacts with a specific binding site on the channel. When the ligand is removed or displaced, the channel closes, terminating the flow of ions (Figure 3.31). The nicotinic acetylcholine receptor (nAChR), a key player in neurotransmission, is a prototypical example. It is comprised of five 290 kDa subunits that arrange symmetrically to form a central pore, and each subunit contains four transmembrane domains that contribute to the overall structure of the channel. In the absence of a ligand such as acetylcholine, the pore is closed to traffic. When acetylcholine interacts with the binding site on the extracellular surface of the cell, however, the proteins undergo conformational changes that cause the channel to open, allowing the flow of ions through the membrane, creating an electrical signal. Removal of the agonist leads to a rearrangement of the proteins back to their closed state, terminating the signal.

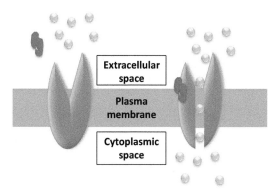

FIGURE 3.31 Ligand-gated channels are closed in the absence of a ligand (red). Binding of the ligand to channel leads to conformational changes that cause the channel to open, allowing migration of suitable ions through the channel. Removal of the ligand causes the channel to close, stopping ion flow. Ligand-gated channels can be activated by synthetic ligands or blocked with antagonists (compounds that bind to the ligand-binding site, but do not lead to channel opening). Direct blockade of the channel is also possible.

Modulation of ligand-gated channel activity can be accomplished in a number of ways. Activation of the channel can be accomplished with compounds that mimic the natural ligand. Nicotine, for example, is an agonist of nAChR, and its activity at this ligand-gated channel is at least partially responsible for activation of reward system of the brain by tobacco products.[66] The smoking-cessation medication Chantix® (Varenicline) is a partial agonist of nAChR, and provides a lower level of channel activity upon binding than nicotine.[67] It has been successfully employed to decrease the cravings and the pleasurable effects of nicotine, as it competes with nicotine for the same binding site on nAChR (Figure 3.32).[68]

Blocking activity of a ligand-gated channel is also possible. Compounds that compete for the natural ligands binding site, but do not cause the conformational changes associated with ligand binding will prevent opening of the channel, acting as functional antagonists. Similarly, compounds that bind to an allosteric site and either stabilize the closed form of the channel or cause conformational changes that prevent binding of the natural ligand also act as functional antagonists. The α-neurotoxins, for example, are a family of peptides from snake venom that are antagonists of postsynaptic nAChR located in neuromuscular synapses (Figure 3.33). These relatively small proteins (60–75 amino acid residues) tightly bind to nAChR in skeletal muscle, preventing acetylcholine-mediated neurotransmission through the opening of nAChR, causing paralysis in snake bite victims.[69] Of course, it is also possible to block the channel itself. In this case, the presence of an ligand opens the

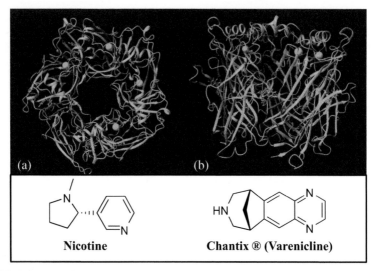

Nicotine   **Chantix ® (Varenicline)**

FIGURE 3.32   The acetylcholine-binding protein (AChBP) has been employed as a model system for the nicotinic acetylcholine receptor (nAChR). The crystal structure shows a ligand bound to the acetylcholine-binding sites. (a) Top view. (b) Side view. Nicotine and Chantix® (Varenicline) both bind to nAChR, but differential receptor responses provide an opportunity for therapeutic intervention in nicotine addiction. *RCSB 2XNT.*

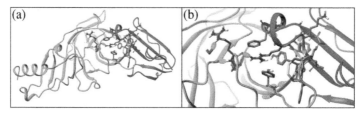

FIGURE 3.33   (a) Crystal structure of the extracellular domain of the nicotinic acetylcholine receptor 1 subunit (green) bound to α-bungarotoxin (red), from the venom of the snake Bungarus multicinctus, at 1.9 Å resolution. (b) Close-up of the binding-site interactions. *RCSB 2QC1.*

channel, but ion flow is prevented and the associated cellular response does not occur.

Allosteric activation of ligand-gated ion channels is also possible. The γ-aminobutyric acid type A receptor (GABA$_A$R), for example, is a ligand-gated chloride channel that plays a critical role in the central nervous system. Activation of GABA$_A$R by the endogenous ligand γ-aminobutyric acid (GABA, Figure 3.34(a)), an inhibitory neurotransmitter, opens the chloride channel of GABA$_A$R, which leads to hyperpolarization of neurons, inhibiting neurotransmission.[70] The presence of benzodiazepines and barbiturates increases the activity of GABA$_A$R. When compounds such as phenobarbital[71] and lorazepam[72] (Figure 3.34) bind to their respective allosteric sites on GABA$_A$R, they cause a conformational change in its

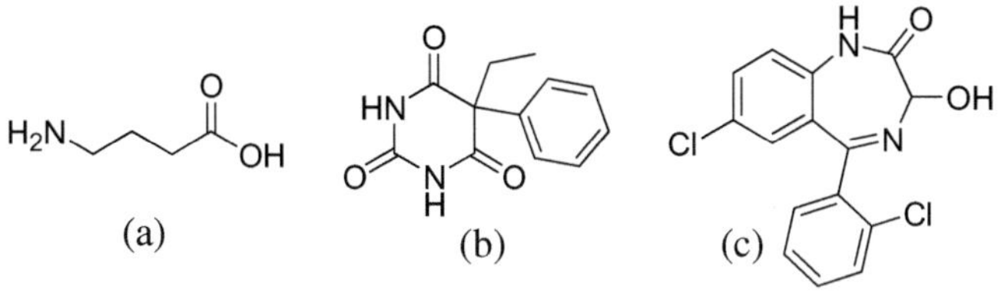

FIGURE 3.34 (a) γ-aminobutyric acid (GABA). (b) Phenobarbital, a barbiturate. (c) Lorazepam, a benzodiazepine.

structure generating a configuration with significantly higher affinity for GABA. This, in turn, increases the frequency of the opening of the associated chloride channel, increasing chloride transfer across the membrane, hyperpolarizing the associated neuron.

Voltage-Gated Channels

The voltage-gated channels represent another major class of ion channels. Unlike ligand-gated channels, voltage-gated channels have no natural ligand. They open and close as a result of changes in membrane potential produced as electrical currents move through biological systems. Voltage-sensing domains allow these channels to be exquisitely sensitive to changes in membrane potential, making them ideally suited for the propagation of nerve impulse through axons, muscle contraction, and cardiac function. The opening and subsequent closing of ion channels give rise to action potentials (Figures 3.35 and 3.36), the rapid rise and fall of the cellular membrane potential, which gives rise to the aforementioned functions. Mechanistically, voltage-gated channels are closed when the membrane electrical potential is at its resting potential. It is worth noting at this point that different types of voltage-gated channels will have different resting potentials, and thus will activate at different membrane potentials. If an electrical impulse (or other stimulus) causes the membrane potential to rise above the membrane threshold for activation, the channel will open via a series of conformational changes, causing a rapid change in membrane polarization via ion flow across the membrane. This leads to hyperpolarization of the cellular membrane, triggering inactivation of the voltage-gated channel through another series of conformational changes, stopping ion flow. Once the voltage-gated channel is inactivated by membrane hyperpolarization, it will not respond to another stimulus

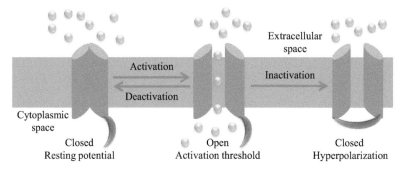

FIGURE 3.35   In the resting state, voltage-gated channels are closed. When the membrane potential reaches the proper level, conformational changes cause the channel to open, allowing the flow of ions across the membrane. This quickly leads to a hyperpolarized state, which induces another set of conformational changes that inactivate the channel. The channel cannot reopen until the resting potential is restored and its conformation shifts back to the closed resting potential state.

FIGURE 3.36   An electrical diagram of voltage-gated ion channel action over time, also referred to as an action potential, provides another view of channel activity. A stimulus must rise above the gating threshold in order to induce channel opening. Rapid depolarization caused by ion flow through the channel leads to hyperpolarization and closing of the inactivation gate. The inactivation gate remains closed until the membrane potential is reset by the action of opposing forces. Stimulation of the channel will not evoke a response until this "refractory period" has ended and the resting potential is restored.

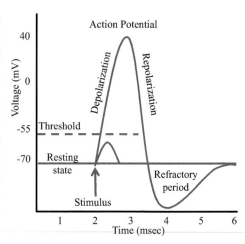

until the membrane potential has been "reset" to the resting potential, typically by the action of another voltage-gated channel with a different set of activation and deactivation parameters. This time period is referred to as a refractory period. Once the membrane potential has been reset, the voltage-gated channel reverts to its original conformation, ready for the next stimulus.[73]

Since voltage-gated channels do not have a natural ligand, modulation of their activity by replacing a natural ligand with either an agonist or antagonist is not an option. It is, however, possible to manipulate their activity in other ways. Blocking the open channel directly (Figure 3.37(a)) is perhaps the most obvious route to suppressing channel activity. Flecainide

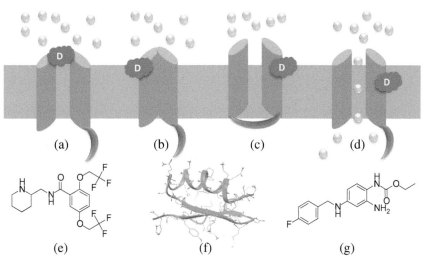

FIGURE 3.37 (a) Direct blockade of the open configuration of the channel by a drug (red) prevents ion flow through the pore. (b) Stabilization of the closed form of the channel by a compound (red) effectively increases the activation threshold, decreasing channel activity. (c) The hyperpolarized state of a voltage-gated channel can be stabilized by a drug (red), maintaining the position of the inactivation gate, slowing conformational changes required to reach the closed resting state. (d) Interaction of a drug (red) with the open channel can stabilize the open configuration, leading to increased ion flow across a cellular barrier. (e) Flecainide, a Nav1.5 blocker and antiarrhythmic agent. (f) Margatoxin, a 39-amino acid peptide found in the venom of *Centruroides margaritatus* (the Central American Bark Scorpion) and Kv1.3 channel blocker. (g) Retigabine, a Kv7.2 and Kv7.3 channel opener and antiseizure agent.

(Figure 3.37(e)), for example, blocks Nav1.5, a voltage-gated sodium channel that plays a major role in cardiac function, and is useful for the treatment of arrhythmia and the prevention of tachycardia.[74] Similarly, the scorpion venom Margatoxin (Figure 3.37(f)) blocks the Kv1.3 channel, a voltage-gated potassium channel found in a variety of cell types, including neuronal cells. Kv1.3 channel blockade alters the membrane potential of neuronal cells, leading to changes in the time required for action potential conduction and nerve transmission. The same channel is also present in T-lymphocytes, and Kv1.3 blockade can induce immunosuppression by decreasing T-cell proliferation.[75]

Alternative methods of modulating voltage-gated ion channel activity depend on the interaction of compounds with the protein at sites other than the pore region. This could be considered a form of allosteric modulation, as the "active site" of an ion channel is the pore through which ions move. As previously mentioned, the opening, hyperpolarization, and subsequent resetting of voltage-gated channel are accompanied by conformational changes, so compounds that interfere with these changes will have an impact on functional activity of the channel. Compounds that stabilize the closed resting potential conformation (Figure 3.37(b)), for example, will prevent channel activation, blocking activity. In a similar manner, compounds that stabilize the inactivated hyperpolarized state of the channel will prevent resetting of the channel after hyperpolarization (Figure 3.37(c)), preventing further channel activity. Conversely, compounds that stabilize the open conformation of a voltage-gated channel will increase ion channel activity (Figure 3.37(d)). Retigabine (Figure 3.37(g)), an anticonvulsant useful for the treatment of epilepsy and seizures, for example, stabilizes the open forms of voltage-gated potassium channels Kv7.2 and Kv7.3, leading to increased potassium flow and seizure suppression.[76]

Other Gating Mechanisms

Ligand gating and voltage gating are perhaps the most well-studied gating mechanisms, but there are other gating mechanisms that play important roles in both normal and disease states. Temperature-gated channels open and close based on distinct thermal thresholds and form the basis for the sensation of hot and cold.[77a,77b] Similarly, mechanosensitive ion channels are activated by mechanical deformations of membranes such as increased tension or changes in curvature and play a part in the sensation of touch.[78] pH gating has also been observed,[79a,79b] and in fact the KcsA potassium channel from *Streptomyces lividans* crystallized by Mackinnon in 1998 is pH-gated. Irrespective of the gating mechanism, however, the opening and closing of channels is intimately connected to conformational changes in the protein, and methods for modulating activity of these gating mechanisms are similar to those described for ligand- and voltage-gated channels.

MEMBRANE TRANSPORT PROTEINS (TRANSPORTERS)

It is well established that cellular survival requires that numerous compounds travel across cellular membranes. Nutrients must be brought into the cell, metabolic end-products may be excreted, and deleterious substances must be removed. In addition, intercellular communication requires that signaling molecules, such as serotonin, norepinephrine, dopamine, and glutamate, are either exported from or imported into cellular structures, depending on the environment and purpose of the cells in question. Even within cells, barriers between different cellular compartments must be traversed by a variety of molecules, both large and small, in order to maintain life. At the same time, however, biological membranes have evolved to tightly control access to the interior of a cell, protecting it from the surrounding environment. While there are a limited number of small molecules that are capable of freely diffusing across the lipid bilayers, simple diffusion is not sufficient to support cellular function. Endocytosis and exocytosis are possible in some cases, but the majority of transmembrane traffic of small molecules across biological barriers is accomplished by membrane transport proteins, also known as transporters. These critical integral membrane proteins play important roles in a variety of biological functions including nerve impulse transmission (serotonin,[35] norepinephrine,[80] and dopamine transporters[81]), metabolism (glucose transporter[82]), and muscle contraction (e.g., glucose transporter[83]). Improper transporter activity can have negative consequences and modulation of transporter activity is a major clinical tool in the treatment of psychiatric disorders.

As was the case with ion channels, the concept of cellular machinery operating to move material across membranes predates the modern era of drug discovery. It was recognized as early as the 1870s that cells were surrounded by a semipermeable "protoplasmic skin,"[84a,84b] and while this information was employed in the development of the "membrane theory" introduced in 1902 by Julius Bernstein to explain ion transport,[85] structural details of transporter proteins did not become available until nearly 100 years later. A true understanding of the structural intricacies of transporter proteins was not available until after the biotechnological revolution at the end of the twentieth century. As with ion channels and GPCRs, recombinant technologies and transfection methods that were developed beginning in the 1970s provided a solution to "supply issues" related to the low natural abundance of transporters. Advances in molecular modeling techniques led to structural models suggesting multiple membrane spanning regions and extracellular loops similar to those identified in GPCRs and ion channels. Validation of the models has been demonstrated by a number of X-ray crystal structures, the first of which was reported in 1996 (Figure 3.38).[86] As of 2013, the RCSB Protein Data Bank contains over 800 crystal structures categorized as transporters.

The majority of transporters contain 12 transmembrane regions and do not require multisubunit assemblies for activity.[87] Members of the major facilitator superfamily (MFS)[88] of transporters such as the norepinephrine transporter (NET),[89] glucose transporters (GLUT1, Figure 3.39),[90a,90b] and the ATP-binding cassette transporters (ABC-transporters)[91] share this structural feature. The P-glycoprotein efflux pumps (Pgps) are a particularly important subclass of the ABC-transporters family, as they are the most common "molecular pumps" that protect cells from toxic materials and xenobiotics. Unlike the majority of transporters, the Pgps are capable of interacting with a broad array of compounds, and are a major issue in drug discovery programs.[92] Potential drug candidates that are Pgp substrates

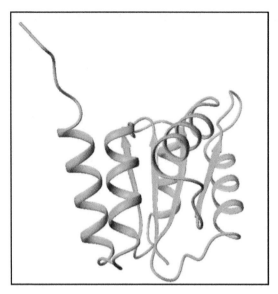

FIGURE 3.38   The mannose transporter, a member of the phosphoenolpyruvate-dependent phosphotransferase system, was the first transporter to be structurally verified by X-ray crystallography (1.7 Å resolution). The monomer from *Escherichia coli* forms a dimer that is biologically active (RCSB 1PDO).

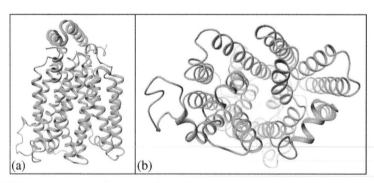

FIGURE 3.39   (a) The proton/xylose symporter of *E. coli* has a high degree of homology with human glucose transporter proteins 1, 2, 3, and 4 (GLUT1-4). The X-ray crystal structure of this member of the major facilitator superfamily contains the typical 12 transmembrane segments and has been the basis of computer models of the GLUT-4 system. (a) Side view. (b) Through membrane view (RCSB 4GBZ).

may not be able to reach their intended target, as Pgps rapidly export them. Also, Pgps are a major source of multidrug resistance in cancer,[93] while similar transporters impart drug resistance to bacteria.[94] In both cases, upregulation of Pgp expression enhances the cells ability to eject normally effective drugs, leading to decreased efficacy. Additional details of the impact of Pgps on drug discovery will be described in Chapter 6.

Transporters with fewer than 12 transmembrane domains are less common, but still account for a significant portion of this class of proteins. Members of the small multidrug transporter (SMR) family such as EmrE, a multidrug transporter from *E. coli*, for example, have only four transmembrane domains and require oligomerization for functional activity.[95a,95b] The mitochondrial carrier family of transporters, which are responsible for facilitating the transfer of material into and out of the mitochondria, on the other hand typically contain six transmembrane units and do not generally require oligomerization for activity (Figure 3.40(a)).[96] At the other end of the spectrum, NADH: ubiquinone oxidoreductase (also known as respiratory complex I), one of the largest known membrane-protein complexes, contains multiple transporter proteins that contain as many as 14 transmembrane domains (Figure 3.40(b)).[97]

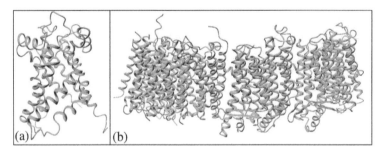

FIGURE 3.40   (a) The mitochondrial ADP/ATP carrier was one of the first mitochondrial carrier family (MCF) transporter to be characterized. It is the most abundant MCF transporter and is responsible for shuttling nucleotides across the inner mitochondrial membrane (RCSB 2C3E). (b) Respiratory complex I is the first protein system in respiratory chain that couples electron transfer between NADH and ubiquinone. It is comprised of six subunits, NuoL, NuoM, NuoN, NuoA, NuoJ, and NuoK, with a total of 55 transmembrane helices (RCSB 3RKO).

Irrespective of their size and shape, membrane transporters share the same basic function, the movement of material across cellular boundaries that would otherwise be impenetrable to the substrate. In a sense, their function is similar to that of ion channels, but there are some key differences. While ion channels form a tunnel for passage of material, membrane transporters employ a binding site that is only available on one side of a cellular membrane at a time. Conformational changes induced by the binding of a solute molecule lead to the transfer of the solute molecule from one side of the membrane to another. There are three basic types of membrane transporters, uniporter, symporters, and antiporters (Figure 3.41). As the name implies, uniporters move a single compound across a barrier. Symporters and antiporters, on the other hand, coordinate the transfer of two or more molecules. Symporters, also referred to as cotransporters, move two or more molecules across a cellular barrier in the same direction, whereas antiporters, also referred to as exchangers or counter-transporters, move two or more molecules in opposite directions.

FIGURE 3.41 Uniporters move a single molecule in one direction down a concentration gradient, while symporters and antiporters move multiple molecules. Symporters move molecules in the same direction, while antiporters move molecules in opposite directions.

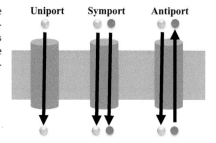

Membrane transporters can also be categorized according to their use of passive or active transport systems. In passive transport systems, facilitated diffusion occurs. Binding of a solute compound leads to conformational changes that transfer the solute from an area of high solute concentration to an area of low solute concentration (down the concentration

FIGURE 3.42 Facilitated diffusion moves a solute down a concentration gradient without expending cellular energy. Binding of the solute molecule induces conformational changes that move the solute across the membrane at a rate substantially greater than possible by simple diffusion.

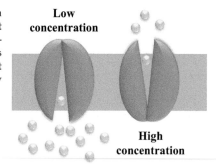

gradient, Figure 3.42). Since this process is entropy driven, it does not require the expenditure of cellular energy (e.g., cleavage of ATP to ADP). The transport of glucose into erythrocytes, for example, employs a uniporter that employs passive diffusion. Erythrocyte glucose transporters, also referred to as GLUT-1, 2, 3, 4, and 5, increase the rate of passage of glucose across the cellular membrane by a factor of 50,000 relative to simple diffusion. No energy input is required, as the flow of glucose into erythrocytes moves down the concentration gradient.[98]

Erythrocytes also contain an antiporter system that operates through a facilitated diffusion event, the chloride–bicarbonate exchanger, more commonly referred to as anion exchanger 1 (AE1) or the band 3 protein. This membrane transporter is critical to the elimination of carbon dioxide generated in tissues (e.g., muscles) through respiration. Carbon dioxide is converted to bicarbonate ions in erythrocytes through the action of carbonic anhydrase, and the bicarbonate is then released from erythrocytes by AE1. The process requires that bicarbonate and chloride ions move across the cell membrane in opposite directions in order to ensure that there is no net change in the electrical potential of the cells. Both ions are moving with their respective concentration gradients, so no net energy expenditure is required, but the actions of AE1 enhance the permeability of erythrocytes to bicarbonate by six orders of magnitude (1,000,000 times the normal diffusion rate).[98]

Membrane transporters that employ active transport systems, on the other hand, move solutes from areas of low concentration to areas of high concentration, leading to the accumulation of a solute on one side of a membrane. The movement of material is often against an electrochemical gradient and requires the input of cellular energy, such as cleavage of ATP to ADP, in order for the process to occur. There are two types of active transport, primary active transport and secondary active transport. Primary active transporters employ cellular energy directly, whereas the secondary active transport is not directly coupled to energy utilization. A prototypical example of a primary active transporter is the sodium potassium ATPase (Na^+/K^+ ATPase) pump, an antiporter system that maintains cellular sodium and potassium levels by importing two potassium ions (K^+) into a cell while moving three sodium ions (Na^+) out of the cell. The resulting charge separation makes the cells electrochemically negative relative to the outside of the cell and results in a membrane potential of -50 to -70 mV. Initial binding of three Na^+ to the outward facing binding site of the transporter and an ATP is followed by phosphorylation of the protein via its ATPase activity. This induces a conformational change that moves the sodium ions into the cell. Release of the sodium ions is followed by uptake of two potassium ions, which induces dephosphorylation of the transporter. This, in turn, causes the transporter to reset itself to its original conformation while carrying the two K^+ out of the cell (Figure 3.43). Na^+/K^+ ATPase activity is crucial to the conduction of action potential in neurons and muscles, both smooth and striated, and it is estimated that 25% of energy producing metabolism in a resting individual is dedicated to support of Na^+/K^+ ATPase activity.[99a,99b]

In contrast to primary active transport system, in which solute movement is directly coupled to energy utilization through metabolic changes (e.g., ATP/ADP conversions), secondary active systems employ energy stored in the form of electrochemical gradients. The sodium–glucose cotransporters (also referred to as sodium–glucose linked transporters), for example, utilize the sodium gradient created by Na^+/K^+ ATPase pumps to drive the transport of glucose into cells. This symporter transfers sodium and glucose simultaneously across

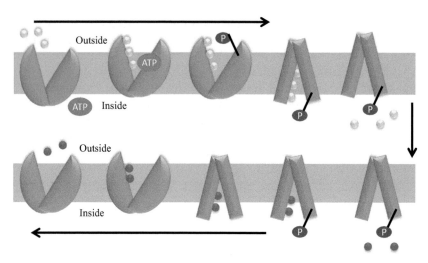

FIGURE 3.43   Na$^+$/K$^+$ ATPase pumps bind three Na$^+$ (yellow) and one ATP (red) to the transporter. Phosphorylation (dark purple) leads to conformational changes that transfer the Na$^+$ to the interior of the cell. The Na$^+$ is released and two K$^+$ (light purple) associate with the protein. This induces dephosphorylation, which causes the transporter to revert to its original structure and transfer the K$^+$ to the outside of the cell at the same time.

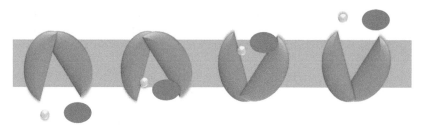

FIGURE 3.44   Glucose (red) and Na$^+$ (yellow) bind to the sodium–glucose cotransporters (SGLT) which moves both across the cell membrane. Na$^+$/K$^+$ ATPase pumps (not shown) create a Na$^+$ gradient that drives this process.

a membrane in the same direction, and the rate of glucose transport is dependent on the Na$^+$ concentration gradient across the membrane (Figure 3.44).[100]

Given their important function, the transporters have been and are likely to remain an important target for drug discovery. Inhibition of transporter function, for example, has been an effective tool in altering the outcome of various biological processes. Transporter inhibition can be accomplished by occupying the substrate binding site with a substrate mimic that is not transported through the membrane, thus "clogging the pipe." Alternatively, a substrate mimic that is preferentially transported as compared to the normal substrate would also decrease the rate of transport of the normal substrate. It is also possible to prevent transporter activity by binding to an allosteric site on the transporter in a manner that prevents the transporter from undergoing the conformational changes required for transport of the natural substrate.

Irrespective of whether an allosteric or orthosteric mechanism is employed, the end result is the same, disruption of the transmembrane movement of substrate molecules. Whether this has a positive or negative impact on any given biological system, however, depends on the downstream impact of altering the flow of substrate molecules through the targeted transporter. Therapeutics that target the serotonin transporter (SERT), for example, have been very successful in treating a variety of psychiatric disorders such as attention deficit hyperactivity disorder, obsessive compulsive disorder, schizophrenia, and depression. Selective serotonin reuptake inhibitors (SSRIs) such as fluoxetine (Prozac®),[101a, 101b] citalopram (Celexa®),[102] and sertraline (Zoloft®)[103a, 103b] inhibit SERTs mediated transport of serotonin by presynaptic cells of neural junctions. This, in turn, increases the serotonin concentration in the synaptic cleft available to act upon the serotonin receptors of postsynaptic cells, and eventually, down regulation of serotonin receptor expression occurs.[104] X-ray crystal structures and homology modeling studies using the bacterial leucine transporter (LeuT), a transporter protein with a high degree of homology with SERT, have demonstrated that both sertraline and fluoxetine bind to the inner end of the extracellular cavity of SERT (Figure 3.45).[105] The binding site of these compounds in LeuT is distinct from that of substrate, leucine, suggesting that while the SSRIs bind in the transport cavity, they do not occupy the same space as the natural ligand. Interestingly, the tricyclic antidepressant Desipramine binds to LeuT in the same location as the SSRIs.[106]

While it is clear that the SSRIs elicit a positive pharmacological response, interfering with the operation of transporter can also have negative consequences. As mentioned earlier, the downstream impact of transporter inhibition plays a major role in determining the nature of the pharmacological response elicited by a compound. Cocaine is an excellent example of the potential negative consequences of transport inhibition. One of the most widely used addictive substances, cocaine, exerts its influence by blocking dopamine transporters (DAT) in presynaptic cells. Blocking dopamine reuptake results in a rapid increase in extracellular dopamine levels that produce the euphoric feeling associated with cocaine exposure.[107a, 107b] Chronic cocaine exposure leads to an upregulation of DAT, and this increase in receptor density may contribute to the increasing level of cocaine required to deliver the same effects as chronic use continues.[108] At the molecular level, the binding site for cocaine overlaps with that of the natural substrate, dopamine, blocking its access to the transporter.[109]

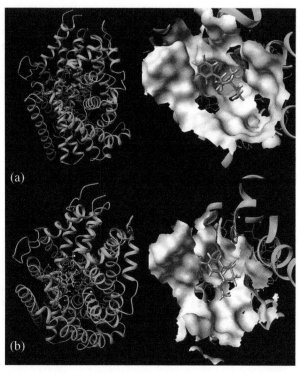

FIGURE 3.45   (a) Leucine transporter with sertraline, full enzyme (left) and binding-site close-up (right) (RCSB 3GWU). (b) Leucine transporter with R-Fluoxetine, full enzyme (left) and binding-site close-up (right) (RCSB 3GWV).

EMERGING TARGETS

Although historically the majority of drugs research has focused on the aforementioned target classes, some new approaches to drug discovery have emerged over the last few decades. Protein/protein, protein/DNA, and protein/RNA interactions have emerged as potential avenues for therapeutic intervention. The importance of protein/protein interactions can be seen in the GPCR pathways, which require the interaction of multiple protein types in order for signal transduction to occur, and interrupting any one of these interactions would alter the physiological outcome. A compound that interfered with binding of the Gα protein to protein kinase C, for example, would block signal transduction through the cAMP pathway, while a compound that stabilized this interaction would augment the signal. In a similar manner, a compound that disrupts the interaction of a regulatory protein with its enzymatic target would increase the amount of free enzyme available to facilitate chemical reactions. In theory, any physiological process that requires the formation of protein quaternary structures could be disrupted by compounds that interfere with the ability of the specified proteins to interact with each other. Since the expression and utilization of both DNA and RNA depend on interactions with a variety of proteins, blocking the interaction of key proteins with DNA and RNA could provide significant therapeutic benefit.

Designing compounds capable of disrupting macromolecular interactions has, however, proved challenging. Unlike small molecules that generally bind in small, specific areas of a target protein, the interaction between two (or more) macromolecules can occur over wide surface areas of the macromolecules in question. The ability to understand the nature of these interactions continues to advance as additional structural data are acquired and computer models advance. Also, binding interactions that lead to the formation of macromolecular complexes (e.g., hydrogen bonds, hydrophobic interactions, etc.) are the same as those considered for designing compounds that interact with historically important targets described earlier in this chapter. As such, information gleaned from the design of small molecule can be reapplied to the identification of compounds that interfere with or augment the formation of superstructures that support various biological functions. Protein therapeutics may also have a significant role to play in this emerging area. The importance of these targets is likely to grow in the future as our understanding of macromolecular interactions improves.

QUESTIONS

1. Provide the general function of the six major classes of enzymes.
    a. Oxidoreductases
    b. Transferases
    c. Hydrolases
    d. Lyases
    e. Isomerases
    f. Ligases
2. Define the following non-covalent interactions.
    a. Hydrophobic interactions
    b. Electrostatic/salt bridge
    c. Hydrogen bond
    d. π-stacking
    e. π-cation interactions
3. Name three methods of enzyme inhibition and define each.
4. What are the three key structural features of G protein-coupled receptors (GPCRs)?
5. What are the two major signaling pathways for GPCRs?

6. Define the following with respect to ion channels:
  a. Ligand gated ion channel
  b. Voltage gated ion channel
  c. Temperature gated ion channel
  d. Mechanosensitive gated ion channel
7. What is a Channelopathy?
8. What is a passive transport system?
9. What is an active transport system?

References

1. a. Imming, P.; Sinning, C.; Meyer, S. Drugs, Their Targets and the Nature and Number of Drug Targets. *Nat. Rev. Drug Discov.* **2006**, *5*, 821–834.
   b. Overington, J. P.; Al-Lazikani, B.; Hopkins, A. L. How Many Drug Targets Are There? *Nat. Rev. Drug Discov.* **2006**, *5*, 993–996.
   c. Hopkins, A. L.; Groom, C. R. The Druggable Genome. *Nat. Rev. Drug Discov.* **2002**, *1*, 727–730.
2. International Human Genome Sequencing Consortium. Finishing the Euchromatic Sequence of the Human Genome. *Nature* **2001**, *409*, 861–921.
3. Pallen, M. J.; Wren, B. W. Bacterial Pathogenomics. *Nature* **2007**, *449*, 835–842.
4. Labeit, S.; Kolmerer, B. Titins: Giant Proteins in Charge of Muscle Ultrastructure and Elasticity. *Science* **1995**, *270* (5234), 293–296.
5. McDonald, N. Q.; Lapatto, R.; Rust, J. M.; Gunning, J.; Wlodawer, A.; Blundell, T. L. New Protein Fold Revealed by a 2.3-Å Resolution Crystal Structure of Nerve Growth Factor. *Nature* **1991**, *354*, 411–414.
6. Kumar, S.; Nussinov, R. Close-range Electrostatic Interactions in Proteins. *ChemBioChem* **2002**, *3* (7), 604–617.
7. a. McGaughey, G. B.; Marc Gagné, M.; Rappé, A. K. Pi-stacking Interactions. Alive and Well in Proteins. *J. Biol. Chem.* **1998**, *273* (25), 15458–15463.
   b. Ringer, A. L.; Sinnokrot, M. O.; Lively, R. P.; Sherrill, C. D. The Effect of Multiple Substituents on Sandwich and T-shaped Pi-pi interactions." *Chem. Eur. J.* **2006**, *12* (14), 3821–3828.
   c. Hunter, C. A.; Sanders, J. K. M. The Nature of pi.-.Pi. Interactions. *J. Am. Chem. Soc.* **1990**, *112* (14), 5525–5534.
8. a. Dougherty, D. A.; Ma, J. C. The Cation-π Interaction. *Chem. Rev.* **1997**, *97* (5), 1303–1324.
   b. Burley, S. K.; Petsko, G. A. Amino-Aromatic Interactions in Proteins. *FEBS Lett.* **1986**, *203* (2), 139–143.
9. a. Lehninger, A.; Nelson, D. L.; Cox, M. M. *Lehninger Principles of Biochemistry*, 5th ed.; W. H. Freeman: New York, New York, 2008. 113–122.
   b. Kabsch, W.; Sander, C. Dictionary of Protein Secondary Structure: Pattern Recognition of Hydrogen-Bonded and Geometrical Features. *Biopolymers* **1983**, *22* (12), 2577–2637.
10. Sumner, J. B. The Isolation and Crystallization of the Enzyme Urease: Preliminary Paper. *J. Biol. Chem.* **1926**, *69*, 435–441.
11. Chen, L. H.; Kenyon, G. L.; Curtin, F.; Harayama, S.; Bembenek, M. E.; Hajipour, G.; Whitman, C. P. 4-Oxalocrotonate Tautomerase, an Enzyme Composed of 62 Amino Acid Residues per Monomer. *J. Biol. Chem.* **1992**, *267* (25), 17716–17721.
12. Smith, S. The Animal Fatty Acid Synthase: One Gene, One Polypeptide, Seven Enzymes. *FASEB J.* **1994**, *8* (15), 1248–1259.

13. Ridderstrom, M.; Cameron, A. D.; Jones, T. A.; Mannervik, B. Involvement of an Active-site Zn^{2+} Ligand in the Catalytic Mechanism of Human Glyoxalase I. *J. Biol. Chem.* **1998**, *273*, 21623–21628.

14. Fischer, E. Einfluss der Configuration auf die Wirkung der Enzyme. *Berichte Dtsch. Chem. Ges.* **1894**, *27* (3), 2985–2993.

15. Brown, A. Enzyme Action. *J. Chem. Soc., Trans.* **1902**, *81*, 373–388.

16. Henri, V. Theorie generale de l'action de quelques diastases. *C. R. Acad. Sci. Paris* **1902**, *135*, 916–919.

17. Koshland, D. E., Jr Application of a Theory of Enzyme Specificity to Protein Synthesis. *Proc. Natl. Acad. Sci.* **1958**, *44* (2), 98–104.

18. Radzicka, A.; Wolfenden, R. A Proficient Enzyme. *Science* **1995**, *267* (5194), 90–93.

19. Hedstrom, L. Serine Protease Mechanism and Specificity. *Chem. Rev.* **2002**, *102* (12), 4501–4524.

20. Page-McCaw, A.; Ewald, A. J.; Werb, Z. Matrix Metalloproteinases and the Regulation of Tissue Remodeling. *Nat. Rev. Mol. Cell Biol.* **2007**, *8*, 221–233.

21. Aggett, P. J. Physiology and Metabolism of Essential Trace Elements: An Outline. *Clin. Endocrinol. Metab.* **1985**, *14* (3), 513–543.

22. Gilep, A. A.; Sushko, T. A.; Usanov, S. A. At the Crossroads of Steroid Hormone Biosynthesis: the Role, Substrate Specificity and Evolutionary Development of CYP17. *Biochim. Biophys. Acta* **2011**, *1814* (1), 200–209.

23. Lehninger, A.; Nelson, D. L.; Cox, M. M. *Lehninger Principles of Biochemistry*, 5th ed.; W. H. Freeman: New York, 2008; pp 183–234.

24. a. Dong-Qing Wei, D. Q.; Qi-Shi Du, Q. S.; Sun, H.; Chou, K. C. Insights from Modeling the 3D Structure of H5N1 Influenza Virus Neuraminidase and its Binding Interactions with Ligands. *Biochem. Biophys. Res. Commun.* **2006**, *344*, 1048–1055.
    b. Kim, C. U.; Lew, W.; Williams, M. A.; Wu, H.; Zhang, L.; Chen, X.; Escarpe, P. A.; Mendel, D. B.; Laver, W. G.; Stevens, R. C. Structure-activity Relationship Studies of Novel Carbocyclic Influenza Neuraminidase Inhibitors. *J. Med. Chem.* **1998**, *41*, 2451–2460.

25. Manning, G.; Whyte, D. B.; Martinez, R.; Hunter, T. .; Sudarsanam, S. The Protein Kinase Complement of the Human Genome. *Science* **2002**, *298*, 1912–1934.

26. Ohren, J. F.; Chen, H.; Pavlovsky, A.; Whitehead, C.; Zhang, E.; Kuffa, P.; Yan, C.; McConnell, P.; Spessard, C.; Banotai, C.; Mueller, W. T.; Delaney, A.; Omer, C.; Sebolt-Leopold, J.; Dudley, D. T.; Leung, I. K.; Flamme, C.; Warmus, J.; Kaufman, M.; Barrett, S.; Tecle, H.; Hasemann, C. A. Structures of Human MAP Kinase Kinase 1 (MEK1) and MEK2 Describe Novel Noncompetitive Kinase Inhibition. *Nat. Struct. Mol. Biol.* **2004**, *12*, 1192–1197.

27. Fedarovich, A.; Nicholas, R. A.; Davies, C. The Role of the β5–α11 Loop in the Active-site Dynamics of Acylated Penicillin-binding Protein a from *Mycobacterium tuberculosis*. *J. Mol. Biol.* **2012**, *418*, 316–330.

28. Kobilka, B. K. G Protein Coupled Receptor Structure and Activation. *Biochim. Biophys. Acta* **2007**, *1768* (4), 794–807.

29. a. Langley, J. N. Observation on the Physiological Action of Extracts of the Supra-renal Bodies. *J. Physiol.* **1901**, *1901* (17), 231–256.
    b. Dale, H. H. On Some Physiological Actions of Ergot. *J. Physiol.* **1906**, *34*, 163–206.

30. a. Lefkowitz, R. J.; Roth, J.; Pricer, W.; Pastan, I. ACTH Receptors in the Adrenal: Specific Binding of ACTH-125I and its Relation to Adenylyl Cyclase. *Proc. Natl. Acad. Sci.* **1970**, *65*, 745–752.
    b. Lefkowitz, R. J.; Roth, J.; Pastan, I. Radioreceptor Assay for Adrenocorticotropic Hormone: New Approach to Assay of Polypeptide Hormones in Plasma. *Science* **1970**, *170*, 633–635.

31. Palczewski, K.; Kumasaka, T.; Hori, T.; Behnke, C. A.; Motoshima, H.; Fox, B. A.; Trong, I. L.; Teller, D. C.; Okada, T.; Stenkamp, R. E.; Yamamoto, M.; Miyano, M. Crystal Structure of Rhodopsin: A G Protein-Coupled Receptor. *Science* **2000**, *289* (5480), 739–745.

32. Rasmussen, S. G.; Choi, H. J.; Rosenbaum, D. M.; Kobilka, T. S.; Thian, F. S.; Edwards, P. C.; Burghammer, M.; Ratnala, V. R.; Sanishvili, R.; Fischetti, R. F.; Schertler, G. F.; Weis, W. I.; Kobilka, B. K. Crystal Structure of the Human β2-adrenergic G-Protein-Coupled Receptor. *Nature* **2007**, *450* (7168), 383–387.

33. Fredriksson, R.; Lagerström, M. C.; Lundin, L. G.; Schiöth, H. B. The G-Protein-Coupled Receptors in the Human Genome Form Five Main Families. Phylogenetic Analysis, Paralogon Groups, and Fingerprints. *Mol. Pharmacol.* **2003**, *63* (6), 1256–1272.

34. Joost, P.; Methner, A. Phylogenetic Analysis of 277 Human G-protein-coupled Receptors as a Tool for the Prediction of Orphan Receptor Ligands. *Genome Biol.* **2002**, *3* (11), 1–16.

35. Nichols, D. E.; Nichols, C. D. Serotonin Receptors. *Chem. Rev.* **2008**, *108* (5), 1614–1641.

36. Girault, J. A.; Greengard, P. The Neurobiology of Dopamine Signaling. *Arch. Neurol.* **2004**, *61* (5), 641–644.

37. de Gasparo, M.; Catt, K. J.; Inagami, T.; Wright, J. W.; Unger, T. International Union of Pharmacology. XXIII. The Angiotensin II Receptors. *Pharmacol. Rev.* **2000**, *52* (3), 415–472.

38. Thierauch, K. H.; Dinter, H.; Stock, G. Prostaglandins and Their Receptors: I. Pharmacologic Receptor Description, Metabolism and Drug Use. *J. Hypertens.* **1993**, *11* (12), 1315–1318.

39. Siu, F. K.; Lam, I. P.; Chu, J. Y.; Chow, B. K. Signaling Mechanisms of Secretin Receptor. *Regul. Pept.* **2006**, *137* (1–2), 95–104.

40. Mannstadt, M.; Jüppner, H.; Gardella, T. J. Receptors for PTH and PTHrP: Their Biological Importance and Functional Properties. *Am. J. Physiol.* **1999**, *277*, F665–F675. 5, 2.

41. Huang, H. C.; Klein, P. S. The Frizzled Family: Receptors for Multiple Signal Transduction Pathways. *Genome Biol.* **2004**, *5* (7), 234. 1–7.

42. Rousseaux, C. G. A Review of Glutamate Receptors I: Current Understanding of Their Biology. *J. Toxicol. Pathol.* **2008**, *21* (1), 25–51.

43. Yona, S.; Lin, H. H.; Siu, W. O.; Gordon, S.; Stacey, M. Adhesion-GPCRs: Emerging Roles for Novel Receptors. *Trends Biochem. Sci.* **2008**, *33* (10), 491–500.

44. Berg, J. M.; Tymoczko, J. L.; Stryer, L. *Biochemistry*, 6th ed.; W. H. Freeman and Company: New York, 2007.

45. Reiter, E.; Ahn, S.; Shukla, A. K.; Lefkowitz, R. J. Molecular Mechanism of β-Arrestin-Biased Agonism at Seven-transmembrane Receptors. *Annu. Rev. Pharmacol. Toxicol.* **2012**, *52*, 179–197.

46. Premont, R. T.; Gainetdinov, R. R. Physiological Roles of G Protein–Coupled Receptor Kinases and Arrestins. *Annu. Rev. Physiol.* **2007**, *69*, 511–534.

47. Khardori, N.; Rahat Ali Khan, R. A.; Tripathi, T. *Biomedical Aspects of Histamine: Current Perspectives;* Springer-Verlag: New York, 2010.

48. Moran, M. M.; Xu, H.; Clapham, D. E. TRP Ion Channels in the Nervous System. *Curr. Opin. Neurobiol.* **2004**, *14* (3), 362–369.

49. Thorneloe, K. S.; Nelson, M. T. Ion Channels in Smooth Muscle: Regulators of Intracellular Calcium and Contractility. *Can. J. Physiol. Pharmacol.* **2005**, *83* (3), 215–242.

50. Blass, B. E.; Fensome, A.; Trybulski, E.; Magolda, R.; Gardell, S.; Liu, K.; Samuel, M.; Feingold, I.; Huselton, C.; Jackson, C.; Djandjighian, L.; Ho, D.; Hennan, J.; Janusz, J. Selective Kv1.5 Blockers: Development of KVI-020/WYE-160020 as a Potential Treatment for Atrial Arrhythmia. *J. Med. Chem.* **2009**, *52* (21), 6531–6534.

51. Cahalan, M. D.; Chandy, K. G. Ion Channels in the Immune System as Targets for Immunosuppression. *Curr. Opin. Biotechnol.* **1997**, *8*, 749–756.

52. Pardo, L. A.; del Camino, D.; Sanchez, A.; Alves, F.; Bruggemann, A.; Beckh, S.; Stuhmer, W. Oncogenic Potential of EAG K⁺ Channels. *EMBO J.* **1999,** *18* (20), 5540–5547.

53. Lang, F.; Föller, M.; Lang, K.; Lang, P.; Ritter, M.; Vereninov, A.; Szabo, I.; Huber, S. M.; Gulbins, E. Cell Volume Regulatory Ion Channels in Cell Proliferation and Cell Death. *Methods Enzym.* **2007,** *428,* 209–225.

54. Berridge, M. J.; Lipp, P.; Bootman, M. D. The Versatility and Universality of Calcium Signaling. *Nat. Rev. Mol. Cell Biol.* **2000,** *1* (1), 11–21.

55. Welsh, M. J. Abnormal Regulation of Ion Channels in Cystic Fibrosis Epithelia. *FASEB J.* **1990,** *4* (10), 2718–2725.

56. Lerche, H.; Jurkat-Rott, K.; Lehmann-Horn, F. Ion Channels and Epilepsy. *Am. J. Med. Genet.* **2001,** *106* (2), 146–159.

57. Matteucci, C. *Essai Sur Les Pheń Omeǹ Es Électriques Des Animaux;* Carilian-Goeury et Vr. Dalmont: Paris, 1840.

58. du Bois-Reymond, E. *Untersuchungen U¨ber Thierische Elektricita¨t;* Reimer: Berlin, 1848.

59. von Helmholtz, H. L. F. Vorläufiger Bericht über die Fortpflanzungs-Geschwindigkeit der Nervenreizung. In *Archiv für Anatomie, Physiologie und wissenschaftliche Medicin;* Jg. Veit & Comp: Berlin, 1850; pp S. 71–73.

60. Bernstein, J. Untersuchungen zur Thermodynamik der bioelektrischen Ströme. *Pflügers Archiv: Eur. J. Physiol.* **1902,** *92* (10–12), 521–562.

61. Armstrong, C. M.; Bezanilla, F.; Roja, E. *J. Gen. Physiol.* **1973,** *62,* 375–391.

62. Neher, E.; Sakmann, B. Single-channel Currents Recorded from Membrane of Denervated Frog Muscle Fibres. *Nature* **1976,** *260,* 799–802.

63. *The Nobel Prize in Physiology or Medicine;* 1991. http://www.nobelprize.org/nobel_prizes/medicine/laureates/1991/. Nobelprize.org.

64. Hamill, O. P.; Marty, A.; Neher, E.; Sakmann, B.; Sigworth, F. J. Improved Patch-clamp Techniques for High-resolution Current Recording from Cells and Cell-free Membrane Patches. *Pflügers Archiv: Eur. J. Physiol.* **1981,** *391* (2), 85–100.

65. a. Doyle, D. A.; Cabral, J. M.; Pfuetzer, R. A.; Kuo, A.; Gulbis, J. M.; Cohen, S. L.; Chait, B. T.; MacKinnon, R. The Structure of the Potassium Channel: Molecular Basis of K⁺ Conduction and Selectivity. *Science* **1998,** *280,* 69–77.
    b. Mackinnon, R. Potassium Channels and the Atomic Basis of Selective Ion Conduction (Nobel Lecture). *Angew. Chem. Int. Ed.* **2004,** *43,* 4265–4277.

66. Kenny, P. J.; Markou, A. Nicotine Self-administration Acutely Activates Brain Reward Systems and Induces a Long-lasting Increase in Reward Sensitivity. *Neuropsychopharmacol.* **2006,** *31* (6), 1203–1211.

67. Mihalak, K. B.; Carroll, F. I.; Luetje, C. W. Varenicline Is a Partial Agonist at Alpha4beta2 and a Full Agonist at Alpha7 Neuronal Nicotinic Receptors. *Mol. Pharmacol.* **2006,** *70* (3), 801–805.

68. Akdemir, A.; Rucktooa, P.; Jongejan, A.; Elk, R. V.; Bertrand, S.; Sixma, T. K.; Bertrand, D.; Smit, A. B.; Leurs, R.; De Graaf, C.; De Esch, I. J. Acetylcholine Binding Protein (AChBP) as Template for Hierarchical *in Silico* Screening Procedures to Identify Structurally Novel Ligands for the Nicotinic Receptors. *Bioorg. Med. Chem.* **2011,** *19* (20), 6107–6119.

69. Hodgson, W. C.; Wickramaratna, J. C. *In Vitro* Neuromuscular Activity of Snake Venoms. *Clin. Exp. Pharmacol. Physiol.* **2002,** *29* (9), 807–814.

70. Sivilotti, L.; Nistri, A. GABA Receptor Mechanisms in the Central Nervous System. *Prog. Neurobiol.* **1991,** *36* (1), 35–92.

71. Rho, J. M.; Donevan, S. D.; Rogawski, M. A. Direct Activation of GABAA Receptors by Barbiturates in Cultured Rat Hippocampal Neurons. *J. Physiol.* **1996,** *497* (2), 509–522.

72. Riss, J.; Cloyd, J.; Gates, J.; Collins, S. Benzodiazepines in Epilepsy: Pharmacology and Pharmacokinetics. *Acta Neurol. Scand.* **2008,** *118* (2), 69–86.

73. Yellen, G. The Moving Parts of Voltage-gated Ion Channels. *Q. Rev. Biophys.* **1998,** *31* (3), 239–295.

74. Ramos, E.; O'leary, M. State-dependent Trapping of Flecainide in the Cardiac Sodium Channel. *J. Physiol.* **2004,** *560* (1), 37–49.

75. Garcia-Calvo, M.; Leonard, R. J.; Novick, J.; Stevens, S. P.; Schmalhofer, W.; Kaczorowski, G. J.; Garcia, M. L. Purification, Characterization, and Biosynthesis of Margatoxin, a Component of Centruroides Margaritatus Venom that Selectively Inhibits Voltage-dependent Potassium Channels. *J. Biol. Chem.* **1993,** *268* (25), 18866–18874.

76. Main, M. J.; Cryan, J. E.; Dupere, J. R.; Cox, B.; Clare, J. J.; Burbidge, S. A. Modulation of KCNQ2/3 Potassium Channels by the Novel Anticonvulsant Retigabine. *Mol. Pharmacol.* **2000,** *58* (2), 253–262.

77. a. Reubish, D.; Emerling, D.; Defalco, J.; Steiger, D.; Victoria, C.; Vincent, F. Functional Assessment of Temperature-gated Ion-channel Activity Using a Real-time PCR Machine. *Biotechniques* **2009,** *47* (3), 3–9.

    b. Dhaka, A.; Viswanath, V.; Patapoutian, A. TRP Ion Channels and Temperature Sensation. *Annu. Rev. Neurosci.* **2006,** *29*, 135–161.

78. Sachs, F. Stretch-activated Ion Channels: what Are They? *Physiology* **2010,** *25* (1), 50–56.

79. a. Gu, Q.; Lee, L. Y. Acid-sensing Ion Channels and Pain. *Pharmaceuticals* **2010,** *3*, 1411–1425.

    b. Wemmie1, J. A.; Price, M. P.; Welsh, M. J. Acid-sensing Ion Channels: Advances, Questions and Therapeutic Opportunities. *Trends Neurosci.* **2006,** *29* (10), 578–586.

80. Zhou, J. Norepinephrine Transporter Inhibitors and Their Therapeutic Potential. *Drugs Future* **2004,** *29* (12), 1235–1244.

81. Mohamed Jaber, M.; Jones, S.; Giros, B.; Caron, M. G. The Dopamine Transporter: A Crucial Component Regulating Dopamine Transmission. *Mov. Disord.* **1997,** *12* (5), 629–633.

82. Thorens, B.; Mueckler, M. Glucose Transporters in the 21st Century. *Am. J. Physiol. Endocrinol. Metab.* **2010,** *298*, E141–E145.

83. Jessen, N.; Goodyear, L. J. Contraction Signaling to Glucose Transport in Skeletal Muscle. *J. Appl. Physiol.* **2005,** *99* (1), 330–337.

84. a. de Vries, H. Arch. Neé Rl. *Physiology* **1871,** *6*, 117.

    b. Pfeffer, W. *Osmotische Untersuchungen: Studien zur Zellmechanik;* W. Engelmann: Leipzig, 1877.

85. Armstrong, C. M.; Bezanilla, F.; Roja, E. Destruction of Sodium Conductance Inactivation in Squid Axons Perfused with Pronase. *J. Gen. Physiol.* **1973,** *62*, 375–391.

86. Nunn, R. S.; Housley, Z. M.; Genovesio-Taverne, J. C.; Flukiger, K.; Rizkallah, P. J.; Jansonius, J. N.; Schirmer, T.; Erni, B. Structure of the IIA Domain of the Mannose Transporter from *Escherichia coli* at 1.7 Å Resolution. *J. Mol. Biol.* **1996,** *259*, 502–511.

87. Veenhoff, L. M.; Heuberger, E. H. M.L.; Poolman, B. Quaternary Structure and Function of Transport Proteins. *Trends Biochem. Sci.* **2002,** *27* (5), 242–249.

88. Pao, S. S.; Paulsen, I. T.; Saier, M. H., JR. Major Facilitator Superfamily. *Microbiol. Mol. Biol. Rev.* **1998,** *62* (1), 1–34.

89. Torres, G. E.; Gainetdinov, R. R.; Caron, M. G. Plasma Membrane Monoamine Transporters: Structure, Regulation and Function. *Nat. Rev. Neurosci.* **2003,** *4*, 13–25.

90. a. Henderson, P. J. F.; Baldwin, S. A. Bundles of Insights into Sugar Transporters. *Nature* **2012,** *490*, 348–350.

    b. Sun, L.; Zeng, X.; Yan, C.; Sun, X.; Gong, X.; Rao, Y.; Yan, N. Crystal structure of a Bacterial Homologue of Glucose Transporters GLUT1-4. *Nature* **2012,** *490* (7420), 361–366.

91. Schmitt, L. The First View of an ABC Transporter: The X-ray Crystal Structure of MsbA from *E. coli. ChemBioChem* **2002,** *3*, 161–165.

92. Goodsell, D. "P-Glycoprotein" Molecule of the Month, March 2010, RCSB, http://www.rcsb.org/pdb/101/motm.do?momID=123.

93. Demant, E. J. F.; Sehested, M.; Jensen, P. B. A Model for Computer Simulation of P-glycoprotein and Transmembrane ΔpH-mediated Anthracycline Transport in Multi-drug-resistant Tumor Cells. *Biochim. Biophys. Acta* **1990,** *1055,* 117–125.

94. van Veen, H. W.; Callaghan, R.; Soceneantu, L.; Sardini, A.; Konings, W. N.; Higgins, C. F. A Bacterial Antibiotic-resistance Gene that Complements the Human Multidrug-resistance P-glycoprotein Gene. *Nature* **1998,** *391* (6664), 291–295.

95. a. Schuldiner, S.; Granot, D.; Mordoch, S. S.; Ninio, S.; Rotem, D.; Soskin, M.; Tate, C. G.; Yerushalmi, H. Small Is Mighty: EmrE, a Multidrug Transporter as an Experimental Paradigm. *Physiology* **2001,** *16,* 130–134.

    b. Chen, Y. J.; Pornillos, O.; Lieu, S.; Ma, C.; Chen, A. P.; Chang, G. X-ray Structure of EmrE Supports Dual Topology Model. *Proc. Natl. Acad. Sci.* **2007,** *104* (48), 18999–19004.

96. Kunji, E. R. S. The Role and Structure of Mitochondrial Carriers. *FEBS Lett.* **2004,** *564,* 239–244.

97. Rouslan, G. E.; Sazanov, L. A. Structure of the Membrane Domain of Respiratory Complex I. *Nature* **2011,** *476,* 414–422.

98. Amidon, G. L., Sadee, W., Eds. *Membrane Transporters as Drug Targets;* Kluwer Academic Press: New York, 2002; pp 9–10.

99. a. In *Membrane Transporters as Drug Targets;* Amidon, G. L., Sadee, W., Eds. Kluwer Academic Press: New York, 2002; pp 11–12.

    b. Berg, J. M.; Tymoczko, J. L.; Stryer, L. *Biochemistry;* W. H. Freeman: New York, 2010. 374–375.

100. Wright, E. M.; Turk, E. The Sodium/glucose Cotransport Family SLC5. *Pflugers Arch: Eur. J. Physiol.* **2004,** *447* (5), 510–518.

101. a. Wong, D. T.; Horng, J. S.; Bymaster, F. P.; Hauser, K. L.; Molloy, B. B. Selective Inhibitor of Serotonin Uptake. Lilly 110140, 3-(*p*-trifluoromethylphenoxy)-*n*-methyl-3-phenylpropylamine. *Life Sci.* **1974,** *15* (3), 471–479.

    b. Lemberger, L.; Rowe, H.; Carmichael, R.; Crabtree, R.; Horng, J. S.; Bymaster, F.; Wong, D. Fluoxetine, a Selective Serotonin Uptake Inhibitor. *Clin. Pharmacol. Ther.* **1978,** *23* (4), 421–429.

102. Pawlowski, L.; Ruczynska, J.; Gorka, Z. Citalopram: a New Potent Inhibitor of Serotonin (5-HT) Uptake with Central 5-ht-mimetic Properties. *Psychopharmacology* **1981,** *74* (2), 161–165.

103. a. Koe, B. K.; Weissman, A.; Welch, W. M.; Browne, R. G. Sertraline, 1S,4S-methyl-4-(3, 4-dichlorophenyl)-1,2,3,4-tetrahydro-1-naphthylamine, a New Uptake Inhibitor with Selectivity for Serotonin. *J. Pharmacol. Exp. Ther.* **1983,** *226* (3), 686–700.

    b. Welch, W. M.; Kraska, A. R.; Sarges, R.; Koe, B. K. Nontricyclic Antidepressant Agents Derived from Cis- and Trans-1-amino-4-aryltetralins. *J. Med. Chem.* **1984,** *27* (11), 1508–1515.

104. Rothman, R. B.; Baumann, M. H. Therapeutic and Adverse Actions of Serotonin Transporter Substrates. *Pharmacol. Ther.* **2002,** *95,* 73–88.

105. Zhou, Z.; Zhen, J.; Karpowich, N. K.; Law, C. J.; Reith, M. E. A.; Wang, D. N. Antidepressant Specificity of Serotonin Transporter Suggested by Three LeuT-ssri Structures. *Nat. Struct. Mol. Biol.* **2009,** *16* (6), 652–657.

106. Zhou, Z.; Zhen, J.; Karpowich, N. K.; Goetz, R. M.; Law, C. J.; Reith, M. E. A.; Wang, D. N. LeuT-desipramine Structure Reveals How Antidepressants Block Neurotransmitter Reuptake. *Science* **2007,** *317,* 1390–1393.

107. a. Volkow, N. D.; Wang, G. J.; Fischman, M. W.; Foltin, R. W.; Fowler, J. S.; Abumrad, N. N.; Vitkun, S.; Logan, J.; Gatley, S. J.; Pappas, N.; Hitzemannt, R.; Shea, C. E. Relationship between Subjective Effects of Cocaine and Dopamine Transporter Occupancy. *Nature* **1997,** *386,* 827–830.

    b. Ritz, M. C.; Lamb, R. J.; Goldberg, S. R.; Kuhar, M. J. Cocaine Receptors on Dopamine Transporters Are Related to Self-administration of Cocaine. *Science* **1987,** *237* (4819), 1219–1223.

108. Letchworth, S. R.; Nader, M. A.; Smith, H. R.; Friedman, D. P.; Porrino, L. J. Progression of Changes in Dopamine Transporter Binding Site Density as a Result of Cocaine Self-administration in Rhesus Monkeys. *J. Neurosci.* **2001,** *21* (8), 2799–2807.

109. Beuming, T.; Kniazeff, J.; Bergmann, M. L.; Shi, L.; Gracia, L.; Raniszewska, K.; Newman, A. H.; Javitch, J. A.; Weinstein, H.; Gether, U.; Loland, C. J. The Binding Sites for Cocaine and Dopamine in the Dopamine Transporter Overlap. *Nat. Neurosci.* **2008,** *11* (7), 780–789.

In vitro Screening Systems

The identification of new and useful drugs depends on the ability of scientists to discriminate between compounds that have the desired physiological effect and those that do not. How this is accomplished has evolved as our understanding of the underlying science has improved. As mentioned in Chapter 2, up until the beginning of the twentieth century, new drugs were identified primarily through serendipitous events. It was not until the dawn of the modern age of drug discovery that truly systematic screening efforts began to be employed in the process. Paul Ehrlich's efforts to identify a new treatment for syphilis represent the first such attempt. His recognition that differences in "chemoreceptors" in different species could be exploited to develop therapeutically useful compounds was a revolutionary concept when he first suggested it at the turn of the twentieth century, but has since formed the cornerstone of modern drug discovery efforts. The discovery of Salvarsan through the screening of over 600 diaminodioxyarsenobenzene analogs provided clear and convincing evidence that systematic screening of compounds could be used to identify important, new medications.[1]

The basic premise of systematic screening of compounds that Paul Ehrlich employed remains intact, but the strategies and technologies employed have changed dramatically over time. Early drug screening efforts were dominated by phenotypic screens designed to identify compounds with a desired physiological effect (such as antimicrobial activity), without necessarily providing a true understanding of the underlying mechanism of action of the candidate compound. While there can be no doubt that numerous important drugs were developed in this manner, the modern drug discovery process now focuses primarily on distinct molecular targets and the biochemical readouts that they provide. This shift is partly the result of scientific innovation, but it is also a consequence of the need to increase the efficiency of drug discovery process. As indicated in Chapter 1, the drug discovery and development process is a costly endeavor.

From a scientific standpoint, several technical hurdles had to be overcome in order to facilitate the transition from primarily phenotypic screening to the modern *in vitro* screening systems that dominate the modern drug discovery process. First, supply-side issues had to be addressed. The use of biochemical and cellular screens was relatively common place in the drug discovery process by the 1960s and 1970s, but the number of screens that could be run was limited by the ability of conventional methods to isolate protein targets and the natural abundance of cells displaying a target of interest. These issues were largely eliminated with by the biotechnological revolution that began in the 1970s. Technological breakthroughs, such as recombinant DNA,[2] transfection science,[3] polymerase chain reaction (PCR),[4] and related technologies made it possible for scientists to generate cell lines that overexpressed targeted biomolecules, essentially eliminating the supply limitations of the past. Recombinant proteins could be harvested from cellular factories, providing ample quantities of target proteins. Custom cell lines could be designed to incorporate biomolecular targets to support cellular screening assays. Further, miniaturization of *in vitro* assays through automation, the use of 96, 384, and 1536 well plates, robotics, and increasingly sophisticated detection system substantially decreased reagent supply requirements. Finally, the rise of parallel organic synthesis, along with existing compound collections (synthetic and natural products), provided drug discovery operations with access to increasingly diverse compound collections for *in vitro* screening purposes.

The following sections explore the key *in vitro* screening methods that have developed as the process of modern drug discovery and development has evolved, along with key language required to understand and interpret the resulting data sets. As discussed in Chapter 1, the multidisciplinary nature of drug discovery requires the successful interaction of individuals with diverse background and skills set. In order for this to occur, it is necessary for all of the players in the process to have a basic understanding of the technology required for *in vitro* screening. Those interested in a more in-depth analysis and detailed information on screening methods are encouraged to consult texts such as Ramakrishna Seethala and Litao Zhang's *Handbook of Drug Screening*.[5]

THE LANGUAGE OF SCREENING: BASIC TERMS

Understanding and interpreting the information developed through screening methods, high throughput or otherwise, requires an understanding of the "language of screening." Like most scientific fields, there are many terms that have been developed over time that have become embedded within the field. While it would be beyond the scope of this text to provide a comprehensive list of terms and their associated meanings, there are a number of key terms that a drug discovery scientist should understand, irrespective of their primary field of expertise.

A medicinal chemist, for example, would be hard-pressed to utilize the data provided by a screening scientist's assay in the absence of an understanding of the terms used to describe the data. It would be equally difficult for an *in vivo* pharmacologist to determine if screening results for a biochemical or cellular assay had any significant correlation with their animal models. An understanding of the following set of terms will provide a basis for communication between scientists from multiple disciplines and support a better understanding of the data and screening systems employed as part of a drug discovery program.

Concentration Response Curves and IC_{50}s

The examination of a compound's biological activity generally begins by determining a compound's ability to interact with a selected biological target. In many cases, an initial assay is performed at a fixed concentration to "weed out" compounds with activity below a desired threshold, such as 50% inhibition of a targeted enzyme at $1.0\,\mu M$ concentration. In order to further characterize the "hit" compounds (those with the desired minimum potency in the assay), the assay is repeated with the test compounds, but in the second run the *in vitro* assay is performed in the presence of a test compound at varying concentrations, usually eight to twelve concentrations separated by a ½ log concentration difference. Ideally, the biological activity of the test compound increases with increasing concentration of the compound, providing an indication of the concentration dependency of the compound's ability to interact with a given target. Plotting this information on an X–Y graph of signal output (assay results) versus concentration of the test compound provides a dose response curve (Figure 4.1). With this information in hand, it is possible to compare and rank order different test compounds in order to determine which compounds provide stronger (or weaker) responses in a given assay.

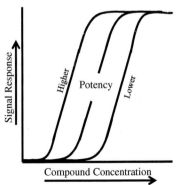

FIGURE 4.1   The relative potency of sets of compounds can be compared by determining the change in signal intensity with an increasing concentration of each compound. The curve created by plotting the concentration against the signal intensity will shift to the left as the potency increases and to the right as potency decreases. The concentration that provides 50% of the maximal response is the IC_{50} of a compound under the assay conditions employed.

Comparing the full concentration response curve can be readily accomplished if the number of compounds is relatively small, but comparing concentration response curves for dozens, perhaps hundreds or even thousands of compounds would be challenging at best. In order to simplify this process, the half maximal inhibitory concentration (the concentration at which 50% of the assay signal is blocked), or IC_{50}, of test compounds are compared and rank ordered. The meaning of this value, however, is dependent on the nature of the assay. In the case of an enzyme, the IC_{50} generally refers to a compound's ability to inhibit enzymatic action. On the other hand, an IC_{50} derived from a binding assay targeting a GPCR and employing a radioligand provides an IC_{50} for ligand displacement, but does not generally provide an assessment of functional activity (i.e., determining if the compound is an agonist or antagonist the targeted GPCR). In addition, assay conditions such as substrate/ligand concentration or cofactor concentrations can impact IC_{50} values for test compounds. This can lead to different, but no less accurate IC_{50} values for the same test compounds targeting the same macromolecular target, but using different assay conditions. Consider, for example, an enzyme inhibition assay in which the enzyme requires a cofactor for enzymatic activity. A 5 fold difference in the concentration of the cofactor employed in two otherwise identical assays could provide different results for a series of test compounds. Interpretation of an IC_{50} value must be done within the context of the assay used to develop the data.[6]

Dissociation Constants (K_d) and Inhibition Constants (K_i)

While the IC_{50} of a test compound in a particular assay is an important tool used for comparison purposes, it is important to understand that IC_{50}s are not binding constants. They do not directly represent a compound's affinity for a macromolecular target, but rather its ability to reduce the maximum signal in a given assay by 50%. Binding constants, on the other hand, are a measure of the strength of the interaction between molecules. Comparison of compound binding across multiple assay conditions is best accomplished through the use of binding constants for the natural ligand, substrate, or inhibitor. The affinity of a ligand for a biomolecule is its dissociation constant, K_d. In the simplest scenario in which there is only one ligand binding site and no cofactors, a ligand and a macromolecule can form a complex in a reversible fashion according to reaction Scheme 4.1.

$$P + L \; \underset{}{\overset{K_d}{\rightleftharpoons}} \; CPL$$

SCHEME 4.1   The interaction of a protein (P) with a ligand (L) to form a protein/ligand complex (CPL) is defined by its dissociation constant (K_d).

The value of K_d is defined by the concentration of the ligand (L), protein (P), and the protein/ligand complex (CPL) in Eqn (4.1). From a practical standpoint K_d represents the concentration at which half of the binding site on the target macromolecule are occupied. Smaller concentration values for K_d represent stronger binding interactions, as lower concentrations of the ligand are required to occupy half of the binding sites available at a set concentration of target molecules. Thus, compounds with a K_d in the nanomolar (nM) range bind more strongly with a target molecule than compounds whose K_d is in the micromolar range (μM).

$$K_d = \frac{[P]\,[L]}{[CPL]}$$

(4.1)

It is worth noting at this point, that the interaction of macromolecules and their associated ligands can be far more complicated. Requirements for cofactors, multiple binding sites, and cooperativity in binding can have a significant impact on the situation. If, for example, a protein contains multiple binding sites for a ligand, it is entirely possible that binding of the first ligand will alter the conformation of the macromolecule such that the second binding site is more readily occupied by another ligand. The mathematical derivations of K_d become more complex, but ultimately, the assertion that lower concentration values equates to stronger binding affinity remains true.

$$K_i = \frac{[E]\,[I]}{[EI]}$$

(4.2)

In a similar fashion, inhibition constants, K_i's, for a given inhibitor of a macromolecular target can be used to compare compounds. As was the case with the natural ligand, in the simplest scenario with only one binding site and no cofactors, the interaction of an inhibitor with a target biomolecule is described in Scheme 4.2 in which the inhibitor (I) and biomolecule (E) interact to form a complex (EI). The mathematical representation of K_i (Eqn (4.2)) is similar to that used to describe K_d, as K_i is defined by the concentration of the inhibitor, biomolecule, and the biomolecule/inhibitor complex. In a practical sense, K_i represents the concentration at which half of the biomolecule concentration is bound to the inhibitor, and is expressed in units of concentration in the same manner as K_d values (lower values indicate stronger binding energy for a given

$$E + I \xrightleftharpoons{K_i} EI$$

SCHEME 4.2   The interaction of an enzyme (E) with an inhibitor (I) resulting in the formation of an enzyme/inhibitor complex (EI) is defined by its inhibition constant (K_i).

set of compounds). Since K_i values are independent of assay conditions, they are a true measure of a compound's ability to inhibit a particular biomolecule and can be effectively used to compare test compounds across multiple assay conditions measuring the same biological function. They can also be used to compare a compound's affinity for multiple biological targets.[6]

As was the case with K_d values, K_i determinations become significantly more complex if additional factors come into play. Issues such as requirements for cofactors, multiple binding sites, and cooperative binding effects will have an impact on the mathematical derivation of K_i values, but once again, it is still generally true that lower concentration values equate to stronger binding affinity. The full mathematical derivation of these terms under more complex scenarios is beyond the scope of this text. Those interested in the detailed mathematical determination of binding constants and associated reaction kinetics of biomolecules, natural ligands, and test compounds are encouraged to consult biochemistry textbooks.[7]

Efficacy versus Binding: $EC_{50}s$

In many cases, the determination of an IC_{50} for a compound at a given target is not enough to provide a full assessment of the compound's potential importance to a program. As previously discussed, binding assays for GPCRs provide insight into how well a test compound displaces a radioligand from it binding site, but provides little, if any, indication as to the functional effect of the compound. Determining the functional impact of a test compound is the ultimate goal of a drug discovery program, and this is accomplished in a similar manner as an IC_{50}. Test compounds are assessed in a functional assay at varying concentrations, usually eight to twelve concentrations separated by a ½ log concentration difference, and the functional activity of the compound is determined. In the case of a GPCR, the readout from a functional assay could be the amount of second messenger produced by the test compound. The concentration that produces 50% of the maximum effect produced by the *test compound* is defined as its EC_{50}. It is important to understand, however, that the maximum effect of the natural ligand and a given test compound may not be the same. The natural ligand's maximal effect is used to establish an *in vitro* assay's "100%" efficacy level for a given GPCR, and some test compounds may never reach the same signal intensity as that observed with the natural ligand. In other words, compounds with the same EC_{50} values may provide substantially different levels of activity, as the maximum

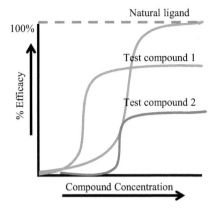

FIGURE 4.2 In the assay data portrayed, the natural ligand (blue) establishes the "100%" efficacy level for a given compound. Test compound 1 (green) has a significantly improved EC_{50}, but it does not reach 100% efficacy. Test compound 2 (red) has the same EC_{50} as the natural ligand, but is substantially less efficacious, as the maximum response is well below that of the natural ligand. Both EC_{50} and percent efficacy must be considered in determining the relative importance of candidate compounds.

response provided by each in a given assay may be different (Figure 4.2). Therefore, a true understanding of a test compound's performance in functional assay requires knowledge of both its EC_{50} and the maximum response provides by the test compound. It should be noted that the term EC_{50} is also employed outside of the GPCR field as an indicator of the concentration of a test compound that provides 50% of the maximal response of the assay in question.[8]

Agonist, Partial Agonist, Antagonist, Allosteric Modulators, and Inverse Agonists

As discussed earlier, in many cases, a full understanding of a test compound's impact on a biological system or assay requires information beyond a simple measurement of the test compound's ability to interact with the biological target of interest. Determining the functional activity of a test compound is often an important consideration. For enzymatic systems, test compounds are generally inhibitors of enzyme function, so the determination of function is relatively straightforward. Signal transduction systems such as GPCRs are, however, more complex, as binding of a compound to the macromolecule is not the end of the story. While it is possible for non-natural ligands to induce the same response as the natural ligand, this is not the only possible outcome. Ligand binding could also block the action of the natural ligand, preventing downstream events that would have been caused by the presence of the natural ligand. It is also worth noting that there are many examples of compounds that provoke changes in the outcomes of biological pathways by interacting with macromolecular targets in areas that are distal to the natural ligand binding site. In general, compounds that impact functional activity of a biological system can be described as agonists, partial agonists, antagonists, positive allosteric modulators, negative allosteric modulators, and inverse agonists.

Agonists and Partial Agonists

Compounds that elicit the same functional biological response as the natural ligand are referred to as agonists. The biological activity of an agonist is defined by its intrinsic activity, which is the ability of a test compound to activate a macromolecular target as a function of receptor binding and produce efficacy. Efficacy, in turn, is the ability of a population of macromolecular target molecules in a given biological system to elicit a maximum response when occupied by an adequate number of agonist molecules. For the purposes of comparing sets of test compounds, agonist activity is generally defined with multiple terms that described its ability to bind to the biological target (IC_{50}), induce a functional response (EC_{50}), and the level of response relative to the natural ligand. (Percent efficacy: The natural ligand's functional response is defined as 100% efficacy.) It is important to understand that a given agonist may not provide the same level of biological response as that observed with the natural ligand, in which case the test compound is described as a partial agonist.[9] This difference in functional activation of a biological system is independent of a compound's IC_{50} and EC_{50}. Thus, it is possible for two compounds to have nearly identical IC_{50}s and EC_{50}s, but different functional impact, as their percent efficacy relative to the natural ligand is different (Figure 4.3). In some cases, partial agonists can effectively compete with the natural ligand, effectively lowering the functional response observed in a biological setting.

FIGURE 4.3 Full agonists (green) induce GPCR signaling equal to that of the endogenous ligand, while partial agonists (blue) activate GPCR signaling to a lesser extent. Neutral antagonists (black) do not induce GPCR activity, but will block agonist activity. Inverse agonists suppress basal activity of a constitutively active GPCR.

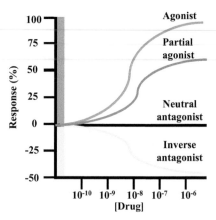

Antagonists

In some cases, test compounds exhibit strong binding interactions with a macromolecular target, but do not elicit a functional response. In other words, the compound is capable of effectively competing with the natural ligand for the binding site, but binding of the test compound produces no measurable response. Compounds of this type have no impact in the absence of the natural ligand, but can effectively block the action of the natural ligand when both are present. In practice, screening for antagonist activity must be done in the presence of the natural ligand (or a ligand that elicits the same response), and signal reduction (such as lowering of cAMP production) is measured to determine antagonist activity.

Basal Activity and Inverse Agonists

While biological systems such as GPCRs are often viewed as on/off switches, it is important to realize that they actually exists as an equilibrium mixture of conformations. Conformations that do not support functional activity predominate in the absence of an activating ligand, but the equilibrium is often not 100% in favor of the inactive conformation. The presence of low levels of conformations that support functional activity provides an intrinsic level of biological activity that occurs in the absence of a ligand or agonist. Such systems are described as constitutively active, and basal activity is the result of a macromolecule spontaneously adopting the active conformation in the absence of the ligand. Compounds that bind to the ligand binding site and stabilize the inactive conformation decrease the basal activity in systems that are constitutively active. Although they bind to the same site that would be occupied by an agonist, compounds of this type produce the opposite pharmacological response (decreased signaling activity), and are thus referred to as inverse agonists. Compounds of this type can also block the action of the endogenous ligand or other agonists, making them functional antagonists in certain circumstances.[10]

Allosteric Modulation

The most direct route toward impacting the biological activity of a given macromolecule is interaction at the binding site for its natural substrate/ligand, often referred to as the orthosteric binding site. There are, however, other alternatives. As discussed in Chapter 3, it is often possible to alter the course of biochemical events by employing molecules that interact with binding site that are remote from the active site of the targeted macromolecule. Such compounds bind at an allosteric site and induce conformational changes that alter the ability of the targeted biomolecules to perform its designated function. Molecules that bind to an allosteric site and enhance the ability of the target macromolecule to perform its function are referred to as positive allosteric modulators (PAMs), while those that diminish function are referred to as negative allosteric modulators (NAMs).[12] In many cases, the presence of a PAM or NAM has no impact on biological function in the absence of the substrate or ligand for the targeted biological system. Within the context of a GPCR signaling pathway, for example, the presence of a PAM could induce conformational changes that increase its affinity for its natural ligand, but if no ligand is present, the PAM will have no impact on biological function. The importance of PAMs and NAMs is evident by the success of the benzodiazepine drug class. Although it was not known when they were originally discovered, it has since been demonstrated benzodiazepines (Figure 4.4) function as

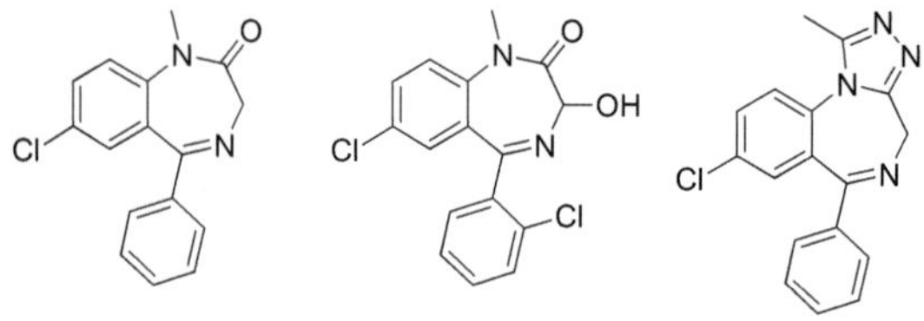

Valium®	Ativan®	Xanax®
(Diazepam)	(Lorazepam)	(Alprazolam)

FIGURE 4.4   Valium® (Diazepam), Ativan® (Lorazepam), and Xanax® (Alprazolam) are representative examples of the highly successful benzodiazepine class of psychoactive drugs.

positive allosteric modulators of the γ-aminobutyric acid$_A$ receptor (GABA$_A$R).[13] There are currently over 30 different benzodiazepines available for the treatment of a variety of conditions such as anxiety, panic disorder, and insomnia.

Receptor Reserve

It is often assumed that there is a one to one relationship between a receptor and the underlying signaling pathway, but this is not always the case. In many cases, the number of receptor sites available on the surface of a cell exceeds the signaling capacity of the cell. Functionally, this means that it is not necessary to occupy 100% of the receptor sites on the surface of a cell in order to generate the maximum signaling response for a given receptor. If, for example, there is a 10 to 1 ratio of receptor sites to signaling machinery, then occupying 10% of the available receptor sites will produce a maximum response. The remaining receptor density, or receptor reserve, is incapable of producing increased signal intensity even if occupied by a ligand or agonist, as the cellular mechanisms of the signaling process has already been fully activated.[11] It is also worth noting that if two different tissues express the same receptor, but with different receptor reserve levels, then the impact of a set concentration of an agonist may have different physiological outcomes. Consider, for example, the case in which one cell type has a no receptor reserve (the number of receptors is equal to the level of signal transduction capacity) and a second cell type has a 100/1 ratio of receptor to signal transduction capacity. If there are no other differences, then full response requires 100% occupancy for the first cell type, but only 1% occupancy for the second cell type. In other words, an agonist will appear to be more potent in cell with higher receptor reserve. This can be an advantage if target cells have high receptor reserve as compared to other cells containing the same receptor, or a disadvantage if the opposite is true. Also, it is important to consider that receptor overexpression in artificial cell lines can lead to higher receptor reserve than seen in physiologically relevant systems, potentially leading to overstatement of the potency of test compounds. If, for example, an overexpressing cell line used for an assay has a 100/1 ratio of receptor to signal transduction capacity, while the naturally occurring cell is closer to 10/1, then compounds screened could appear to be nearly 10 fold more potent in the artificial cell line.

STREPTAVIDIN AND BIOTIN

In many cases, it is necessary to create a linkage point between materials that are useful in the elucidation of biochemical process without significant chemical modifications. Binding proteins to the surface of a 96 plate, for example, can simplify high throughput screening procedures, but chemical reaction required for attachment of the protein to the surface of a plate are not always conducive to retaining activity of the protein. The exceptionally strong interaction between biotin and the protein streptavidin (Figure 4.5[14]) is often employed in biotechnology in order to create stable, strong, non-covalent interactions.[15] The streptavidin/biotin binding

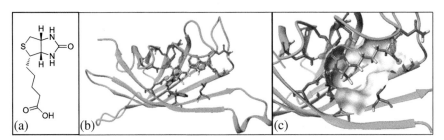

FIGURE 4.5 (a) Biotin (b) Crystal structure of Streptavidin monomer bound to biotin. (c) Close up image of biotin binding site. Hydrogen bonds between biotin and the protein backbone are indicated by blue ellipses and the surface of the binding site is designated by the gray surface (RCSB file 3RY2).

interaction is one of the strongest non-covalent interactions known, with a dissociation constant in the femtomolar range, making this bonding interaction virtually inert under conditions generally used for biological screening. Binding of a protein to the surface of a 96-well plate can be accomplished by taking advantage of the strength of this bond and the presence of four biotin binding sites on streptavidin (Figure 4.6). In this instance, the plastic surface

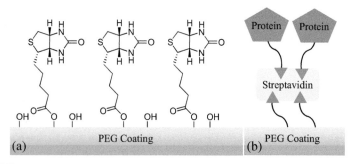

FIGURE 4.6   (a) Polyethylene glycol coating of the surface of a microtiter plate provides open hydroxyl groups that can be capped with biotin. (b) Streptavidin binds to the bicyclic portion of biotin (blue triangle) linked to the surface of the microtiter plate. A second biotin-tagged molecule will also bind to the streptavidin, creating a labeled surface suitable for use in various assay platforms.

of a 96-well plate could be covalently linked to a biotin with an ester or amide bond. Addition of streptavidin to the 96-well plate would lead to binding of streptavidin to biotin on the surface of the 96-well plate. A second protein suitably functionalized with a biotin molecule on the external surface of the protein, away from the active site, could be added to the 96-well plate and would bind to the streptavidin coating. The strength of the streptavidin/biotin interaction creates a surface on the 96-well plate that is effectively coated with the protein of interest that will stay attached to the surface of the plate under the vast majority of biological screening conditions. Alternatively, a streptavidin-labeled protein in solution could be brought together with a biotin-tagged molecule also in solution in order to generate an assay signal that can be used as a biological readout to determine the biological activity of test compounds (e.g., FRET, TRFRET, SPA assays systems).

BIOCHEMICAL VERSUS CELLULAR ASSAYS

The vast majority of modern screening assays can be divided into two basic categories, biochemical and cellular assays. Each has advantages and disadvantages and the choice of which is appropriate for a given study is largely dependent on the goal of the program. Typically, drug discovery programs employ multiple, sequential screening systems and both types of assays. Biochemical assays employ cell-free conditions designed to study a small subset of cellular activity and are capable of providing direct information on the nature of the interaction of candidate compounds with a molecular target of interest. This type of assay requires that sufficient quantities of the necessary biological target can be produced for prosecution of the screen, and that the biological target will maintain its functional activity in the absence of other cellular components. Enzyme inhibition assays are a typical example of a biochemical assay. These assays are often designed to monitor either the formation of a labeled product or the loss of a labeled starting material through a variety of different detection methods that will be discussed later in this chapter.

It is also possible to employ biochemical assay systems to interrogate non-enzymatic events such as protein/protein interactions and receptor binding. It is important to realize, however, that biochemical assays by their nature are conducted in the absence of a cellular context. As such, there are some important limitations. The cellular consequences of enzyme inhibition, for example, cannot be determined in a biochemical assay. Similarly, while it may be possible to measure a compound's ability to bind to a given GPCR using a biochemical preparation such as a membrane fragment, the absence of cellular constructs responsible for signaling cascades prevents assessment of the functional activity of a candidate compound. In other words, biochemical binding assays are not suitable for determining if a compound is an agonist, antagonist, or inverse agonist at a given GPCR. Events downstream from a GPCR signaling event also cannot be determined in simple biochemical assays. It is also important to realize that membrane permeability generally does not play a role in biochemical assays. Thus, the absence of cellular context may lead to the identification of compounds that are active in a biochemical assay, but incapable of providing the desired impact on intact cells due to poor membrane permeability. Despite these limitations, biochemical assays are widely used in the drug discovery industry as a first step in identifying potential chemical leads for drug discovery.

Of course, cellular assays play a major role in drug discovery, as there are many targets and biological processes that are not suitable for biochemical screening assays. Ion channel activity, membrane transport, signal transduction pathways, antibacterial activity, and proliferation studies all require the use of cellular systems. Unlike biochemical assays, cell-based assays provide a full complement of cellular functionality, and thus more closely mimic the *in vivo* setting. This provides an opportunity to determine the functional properties of a compound. With regard to signal transduction pathways, for example, cellular assays can be used to determine if a compound is an agonist or antagonist in a particular pathway. Events downstream from candidate compound binding can also be monitored in ways that are not possible in biochemical assays.

In some cases, it is not possible to isolate and purify the functional protein responsible for biological activity due to scientific limitations or competitive patent coverage, leaving cellular assessment as the only course of action. Irrespective of the rationale for choosing to pursue a cellular assay, however, there are some disadvantages that should be considered. First, off-target activity must be considered when results are interpreted. Unlike biochemical assay which are designed to interrogate subcomponents of cellular activity, cell-based assays have no such restrictions. Off-target effects (those associated with cellular systems other than the ones that are the subject of the screen) can produce both false positive and false negative results in a cell-based screening assay. For intracellular targets, membrane permeability effects may decrease the observed signal response. If a compound is unable to pass through a membrane, either by virtue of its physical properties or as the result of the action of efflux pump systems (proteins designed to pump xenobiotics out of the cell), the amount of the test compound that reaches the target of interest inside the cell will be lower than expected based on the conditions of the assay, lowering the apparent activity in a given assay system. This can be viewed in both a positive and negative manner. Compounds with lower permeability are less likely to lead to drug candidates, so eliminating them early could be a positive outcome. However, negative screening results from a cellular assay that are the results of poor cellular permeability may mask important structure activity data that might have been available in a cell-free system.

Understanding the inherent benefits and limitations of biochemical and cellular assays is critically important to the successful interpretation of the data set provided by the selected screening method. In the absence of such an understanding, it will be difficult to recognize false positive and false negatives when they occur, making data analysis more difficult and time-consuming. Careful consideration must be given to the choice of assay type, as the ability to identify biologically relevant molecule and new therapeutic entities is heavily dependent on proper assay selection. Additional information on the benefits and risks associated with each class of assay is available in articles such as K. Moore's and S. Rees's review on the subject.[16]

ASSAY SYSTEMS AND METHODS OF DETECTION

The ability to identify biologically relevant molecules is heavily dependent on the choice of detection systems and equipment employed for a given assay. Unfortunately, there is no universal system that can be employed to determine the relative activity of compounds of interest at a biological target. Each system has advantages and disadvantages that must be considered in order to ensure that an assay system meets the needs of the research project in question. It is also worth noting that drug discovery programs are likely to employ multiple *in vitro* assays, as it is often the case that a single assay will only give a partial answer to the questions at hand. In a GPCR program, for example, a preliminary binding assay may be followed by a functional assay, each of which employs entirely different detection methods.

There are a wide range of tools and platforms available to meet the needs of an *in vitro* screening scientist, but they can be divided into several different categories. The following sections provide a basic understanding of the most common detection system available in modern drug discovery laboratories. In principle, thousands of assays could be designed using the tools described below, but the basic premise of each would be the same. Variations in design and even the merging of different assay systems are limited only by the creativity of scientists.

RADIOLIGAND ASSAY SYSTEMS

In general, modern radioligand binding assays employ scintillation principles as a means of quantifying interactions between macromolecular targets and candidate compounds. In simple terms, materials that undergo scintillation (a scintillant) emit light when they are excited by ionizing radiation, such as a β-particle emitted by a ^3H, ^{14}C, ^{33}P, and ^{35}S, or an Auger electron emitted by ^{125}I. Compounds containing these isotopes are, therefore capable of inducing luminescence in scintillants, and the amount of light generated is proportional to the intensity of ionizing radiation.[17] Typical scintillants include crystals of polyvinyltoluene, polystyrene, yttrium silicate (YSi), and yttrium oxide (YOx).[18] In a standard radioligand assay, membranes or cells containing a target of interest are incubated with a mixture of a candidate compound and a radiolabeled ligand for a set period of time. After incubation, filtration and washing separates the membranes or cells from any unbound ligands, and then a scintillant is added to induce light emission from the resulting dry material. Compounds that compete with the radioligand for a binding site will displace it from the molecular target, and free radioligand will be washed away in the filtration step. This will decrease scintillation in a quantifiable manner, and the change in signal intensity provides direct insight into the strength of binding interaction between the candidate compound and the macromolecular target.

There are, of course, some complicating factors that must be considered in this type of assay. First, non-specific binding (NSB) must be considered. It is often the case that a radioligand is capable of binding to material in the assay other than the target of interest. Materials such as cellular membranes, plastic, filter material, and even other receptors can bind to a radioligand essentially creating a "background" signal that is independent of the target of interest. The level of NSB can be determined by measuring the amount of scintillation that occurs in the presence of an unlabeled ligand that completely occupies the binding site (displaces all of the radioligand of interest). Any signal that remains after filtration and washing is the result of binding of the radioligand to sites other than the binding site of the target molecule and represents the NSB of the assay in question. The specific binding (SB) of a radioligand to its target of interest can be determined as the difference between the total binding (TB) in an assay (the signal observed in the assay with the radioligand and no unlabeled material) and the NSB (SB = TB − NSB). The SB for a radiolabeled compound in a particular assay is important as it sets the maximum signal that can be observed in an assay and the signal to noise ratio is determined by the ratio of specific binding versus total binding (%SB = (SB/TB) × 100). In an ideal assay, a radioligand will have high affinity for the target (dissociation constant (K_D) 0.1–30 nM), high %SB (70–100%), and low NSB.

Another important aspect that must be considered is the nature of the radioligand itself. As mentioned earlier, the amount of light generated via scintillation is proportional to the amount of ionizing radiation produced. In the case of a radioligand, the amount of ionizing radiation produced is dictated by the radioactive output of the labeled atoms, referred to as their specific activity, and their decay rate (half-life). Tritium (^3H) and ^{125}I are the most commonly used isotopes. There are several advantages to using ^3H. Incorporation of ^3H into a protein binding site or ligand of interest has no direct impact on binding interactions, as it is a simple isotopic exchange. Also, radioactive decay of ^3H produces energy that is incapable of penetrating tissue, making it significantly safer to employ in a laboratory setting than more energetic species. The downside of the low energy emission of ^3H is that it can be difficult to detect. Its specific activity is only 30 Ci/mmol, but since there are often multiple positions in a compound or macromolecule that can be modified to incorporate a ^3H, it is often possible to augment the energy output. Tritium labeling of a methyl substituent, for example, will triple the specific activity of the labeled molecule (3×30 Ci/mmol = 90 Ci/mmol), increasing the signal intensity created in the presence of a scintillant. Finally, the over 12-year half-life ($T_{1/2}$) of ^3H provides ample time for application in screening assays.

Radioactive iodine is also routinely employed, but there are a number of important differences in the application of ^{125}I. Unlike ^3H, incorporation of an iodine atom may have a significant impact on binding interactions between the ligand and the target macromolecule, as its presence is often a result of adding an additional atom to the labeled material rather than a simple isotopic exchange. Energetically speaking, the specific activity is significantly higher (2000 Ci/mmol), so substantially less radioactive material is required for an assay. Also, the rate of radioactive decay is much shorter than ^{3}H ($T_{1/2} = 60$ days). It is important to realize that safety precautions required with ^{125}I are considerably different from those associated with tritium. Radioactive decay of ^{125}I produces energy (γ, ε, and X-ray decay) capable of penetrating thick plastics, making dense physical shielding a requirement in assays incorporating this isotope. Accidental exposure to this material is also a significant issue, as it is known to accumulate in the thyroid, where it can do significant and permanent damage. Great care should be taken in working with this material.

Other radioisotopes, such as ^{14}C, ^{33}P, and ^{35}S, have also been successfully employed in scintillation assays. ^{33}P has been employed in the study of kinases and other systems that require the relocation of a phosphate group, while ^{35}S radioactivity can be exploited by incorporating it into sulfur containing compounds such as Acetyl-CoA and cysteine residues of proteins. Of course, ^{14}C can be employed as an isotopic substitution in any position in which tractable chemistry can be designed to incorporate the radiolabel, but its radioactive half-life of 5730 years makes it less desirable.[18]

Simple scintillation assays remain an important aspect of drug discovery, but there are some key limitations that have limited their utility. The required filtration steps necessitated by the need to remove unbound radiolabeled material make these assays more labor- and time-intensive than other modern assays. In addition, disposal of radiolabeled material is often a factor in determining whether or not to pursue a radiolabeled platform. Each filtration step generates additional radioactive waste material that must be properly handled. Efforts to substantially reduce the amount of hazardous waste material generated as a consequence of radiolabeled screening techniques lead to the development of the Scintillation Proximity Assay platform.

Scintillation Proximity Assay (SPA)

The majority of modern radioactive assays employed in modern drug discovery programs are homogeneous in nature and require no filtration to remove unbound radiolabeled material, but are still dependent on scintillation as a means of detection. Eliminating filtration steps has greatly facilitated the use of radiolabeled assays on high throughput screening platforms, but required important changes in how the assays were performed. The key to Scintillation Proximity Assay (SPA) technology is the distance limitation created by the presence of the assay media itself and the ability to immobilize scintillants in specially designed microspheres. While it is true that a scintillant will emit light in the presence of a source of ionizing radiation, if the distance between the source material (radiolabeled material) and the scintillant is too large, the energy of the β-particle (or Auger electron in the case of ^{125}I) will be dispersed into the environment and scintillation will not occur. Within the context of a biological assay, the path length limitation for induction of scintillation is determined by the average path length available to a corresponding β-particle (or Auger electron in the case of ^{125}I) in an aqueous environment. In the case of ^{3}H and ^{125}I, ^{3}H emits a single β-particle with a path length of 1.5 μm, while ^{125}I emits two Auger electrons with average path lengths of 1 and 17.5 μm respectively.[17] Microspheres and microtiter plates containing scintillants and an immobilized biological target can be prepared and employed for compound screening utilizing the aforementioned physical properties. Thus, in the presence of a radiolabeled ligand, the immobilized biological target will ensure that β-particle emission occurs within the path length limitation, inducing scintillation, which can be quantified with a scintillation counter. Compounds that compete with the radioligand for the same binding site will displace the radioligand, and while the unbound radioligand continues to emit β-particles, in the unbound state it is too far away from the scintillant for excitation to occur. This leads to a decrease in luminescence that can be quantified to determine the relative binding strength of the test compounds as compared to the radioligand (Figure 4.7).

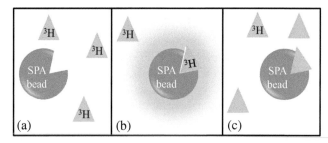

FIGURE 4.7   (a) An SPA bead containing a scintillant and a target protein will emit no light when radiolabeled material (orange ^{3}H) is unbound. (b) Binding of the radiolabeled material brings the radiation source within the path limitation, causing scintillation. (c) Unlabeled material (orange) can compete with labeled material and quench scintillation.

Alternatively, SPA beads or plates can be tagged with enzyme substrates as a means of measuring enzymatic activity. If, for instance, a SPA bead or plate is coated with a radiolabeled substrate for a protease and cleavage of the substrate by the protease releases the radiolabel, the signal intensity (scintillation/light emission) would decrease with the action of the enzyme of interest. Inhibitors of this enzyme would block cleavage of the radiolabel portion of the substrate, maintaining the signal in quantifiable manner (Figure 4.8(a)).

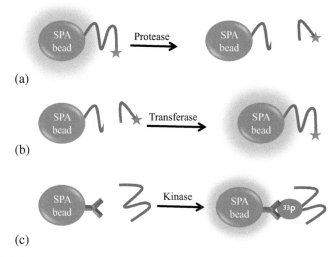

FIGURE 4.8 (a) Protease activity cleaves a peptide (blue line), releasing the radiolabel, quenching the signal. (b) Transferase activity attaches the radiolabeled tag to the peptide (blue line) on the APS bead, inducing scintillation. (c) Phosphorylation of the kinase substrate with ^{33}P-labeled phosphate leads to binding of the radiolabeled product to the antibody-tagged SPA bead, producing scintillation.

Enzymes such as transferase, on the other hand, could be examined with SPA technology using an unlabeled SPA surface and a labeled reagent. Enzymatic action of the transferase would attach the radiolabeled reagent to the SPA surface, inducing scintillation. Candidate compounds that block transferase activity would prevent scintillation, suppressing the increased signal that would be observed in the absence of the inhibitor in a quantifiable manner (Figure 4.8(b)).

Antibody-coated SPA beads or plates can also be employed to assess biological activity of candidate compounds. If the antibody is designed to interact with the product of an enzymatic reaction, such as the phosphorylated product of a kinase, then the rate of formation of the product can be monitored by employing a radiolabeled phosphorylating agent, such as radiolabeled cyclic AMP. In the absence of an inhibitor, kinase activity will generate radiolabeled phosphorylated products that will bind to the antibody on the SPA beads or plates, inducing luminescence in a quantifiable manner. Introduction of a kinase inhibitor will block this process, decreasing the signal intensity, providing a method of assessing the biological activity of test compounds (Figure 4.8(c)).[19]

While SPA assay technology has certainly simplified the process involved in the prosecution of radiolabel-dependent screening systems, it should be kept in mind that SPA technology does not address a key drawback of radiolabeled screening techniques. SPA screening system still requires the use of radioactive material, and as such, necessitates isolated equipment, staff skilled and certified in the use of radiolabeled material, as well as the generation and disposal of radioactive waste material. These drawbacks must be considered before committing to the prosecution of an assay incorporating radiolabeled material, conventional or SPA, as other technologies may be able to provide similar data output at a lower cost.

ENZYME-LINKED IMMUNOSORBENT ASSAY (ELISA)

The binding interaction between an antibody and its target antigen is both highly stable and exceptionally selective. These features have led to the development of a number of assay systems designed to take advantage of the highly specific nature and strength of the antigen/antibody complex. Radioimmunoassay designed to monitor biological processes, particularly the concentration of antigens in blood samples, were described in by Rosalyn Yalow and Solomon Berson as part of their efforts to quantify blood insulin levels.[20] Their methods, along with many modernized variations, are still employed today, as radioimmunoassays are both extremely sensitive and inexpensive. Efforts to move away from the use of radioactive material, however, led to the development of a similar assay system that eliminates the need for radiolabeled material.

The enzyme-linked immunosorbent assay, more commonly referred to as ELISA, takes advantages of the positive aspects of radioimmunoassay techniques, but quantification of biological signals is accomplished by monitoring enzymatic activity instead of changes in radioactivity. Originally developed independently by Anton Schuurs and Bauke van Weemen in the Netherlands[21] and Peter Perlmann and Eva Engvall at Stockholm University in Sweden,[22] ELISA methods employ antibodies that have been linked to an enzyme capable of generating a signal when in the presence of a specific substrate. In many cases, the signal change is a colorimetric change that can be observed with a simple spectrophotometer. In the direct ELISA method, the surface of a microtiter plate is coated with an antigen and an enzyme-linked antibody is added. Removal of unbound antibodies using standard plate washing methods is followed by addition of a substrate that can be converted to a colored material by the enzyme. Monitoring the rate of appearance of color can be used to quantitate biological material or biological events. If the amount of antigen applied to the plate is *unknown*, then the rate of color change will be directly related to the amount of antigen applied to each well, and the amount of antigen in solution can be determined based on colorimetric changes. Alternatively, if the amount of antigen applied to each well is a *known constant*, then the rate of color change can be used to identify enzyme inhibitors. The rate of color change will decrease in the presence of compounds that block enzyme activity (Figure 4.9).

FIGURE 4.9   A surface coated with an antigen (black) is treated with a compatible antibody that has been covalently linked to an enzyme (orange). Addition of a suitable substrate (open blue circle) will produce a color change (yellow) whose intensity is dependent upon the amount of substrate and antigen present. The presence of an enzyme inhibitor (red) can also be detected based on changes the intensity of color produced.

Another important variation of the ELISA technique, the sandwich ELISA, requires multiple antibodies, the last of which is linked to an enzyme that can be used to provide a quantitative signal. In one scenario, a microtiter plate (or other surface) is coated with a capture antibody. Application of a solution containing an antigen that interacts with the antibody forms a stable complex, and standard plate washing techniques are employed to remove any unbound material. The addition of a second antibody linked to an enzyme capable of producing a signal and binding to the antigen produces a three-membered complex that sandwiches an antigen between two antibodies. Addition of an appropriate enzyme substrate produces a signal that can be monitored and quantified (Figure 4.10). While sandwich ELISAs are employed in drug discovery, this technique has enjoyed broad

FIGURE 4.10   A surface coated with an antibody (blue) is treated with a solution containing a compatible antigen (black). The stable complex that is formed is then treated with a second antibody (green) linked to an enzyme (orange), creating a "sandwich" complex. Addition of a suitable substrate (open blue circle) will produce a color change (yellow) whose intensity is dependent upon the amount of substrate and antigen present.

application by consumers in need of information on circulating levels of human chorionic gonadotropin. Increasing concentrations of this hormone can be detected via sandwich ELISA using properly designed urine test strips, and this forms the basis of the common in-home pregnancy test kits.[23]

There are many variations of ELISA techniques and their application has extended well beyond the field of drug discovery. *In vitro* diagnostic kits are common place in medical laboratories,[24] and the food industry employs ELISA methods to detect potential food allergens such as peanuts in food products.[25] Some types of illicit drugs can also be rapidly detected using ELISA systems.[26] It is also worth noting that the basic principles of ELISA technology (antibody/antigen complexes designed for detection) have been successfully employed in non-ELISA assays, making the original development of ELISA much more important to the drug discovery process.

FLUORESCENCE-BASED ASSAY SYSTEMS

Although the detection of signals created by the interaction of radiolabeled compounds and an appropriate scintillator can be an effective means of measuring biological function, the inherent limitations and the desire to avoid the generation of radioactive waste material led to the development of alternative detection systems. A number of modern assay systems take advantage of the fluorescent properties of selected classes of compounds. In general, fluorescence occurs when a substance absorbs light energy at one wavelength and then undergoes a near-immediate release of the energy at a lower wavelength (lower energy) in the visible spectrum due to non-radiative energy loss to the lowest vibrational energy level of the excited state. The difference in energy between the absorbed and emitted light is referred to as the Stokes shift (Figure 4.11).[27] The emission of

FIGURE 4.11   When a photon of energy from a light source is absorbed by a fluorophore (a compound that undergoes fluorescence, orange), an excited energy state is created. Relaxation of the excited state back to the ground state occurs with the emission of a photon at lower energy.

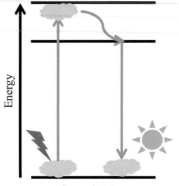

Ground state

light from the substance in question stops almost immediately after the incident radiation is terminated, and the level of fluorescence can be measured and quantified over time.

Fluorescence intensity assays are the simplest systems designed to take advantage of fluorescent properties for the purposes of monitoring biological systems and examining test compounds. There are two basic types. Assay systems can be designed to measure either the increasing or decreasing levels of fluorescence over time to monitor biological activity. Consider, for example, a fluorogenic assays in which an enzyme converts a non-fluorescent substrate into a fluorescent product. Compounds that inhibit the enzyme will slow the rate of formation of the product in a quantifiable manner, and this information can be used to determine the IC_{50} of test compounds. Alternatively, an enzyme could convert a fluorescent substrate into a non-fluorescent substrate. In this case, the biological activity of test compounds would be determined by monitoring the loss of fluorescence, as enzyme inhibitors would slow the conversion of the fluorescent substrate into the non-fluorescent material. In either case, it is important to be aware that readouts from a fluorescence intensity assay are potentially subject to interference created by natural fluorescence of test compounds, colored compounds, and test compounds that are also fluorescence quenchers on their own. These issues can create both false positive and false negative readouts in fluorescent intensity assays and should be considered when assay data are reviewed.

Fluorescence Polarization (FP)

Simple fluorescence intensity assays have proven to be very valuable tool for scientists in a number of fields, and it should come as no surprise that more advanced fluorescence assays have been developed in order to take advantage of the properties of light. Fluorescence polarization technology relies on differences in polarization of light created by changes in molecular size. Polarized light was first described by Eitenne-Louise Malus in 1808,[28] well before the appearance of organized drug discovery programs. Over 100 years later, Perrin[29] and Weigert[30] separately described the relationship between molecular size, polarization of light, and fluorescence, and how these properties could be used to study molecular interactions. Instrumentation and homogeneous assay were later developed by Weber[31] and Dandliker[32] respectively. Their work forms the basis of screening technologies that are widely used in modern high throughput screening laboratories.

As previously described, excitation of a fluorophore leads to emission of light at a lower wavelength in a predictable and quantifiable manner.

If excitation is achieved through the use of polarized light, the nature of the ensuing fluorescent emission will depend upon the molecular motion available to the fluorophore. If the fluorophore is immobilized, then the fluorescent emission will remain polarized and occur in the same plane as the original excitation source. On the other hand, if the fluorophores are in solution, the level of polarization of the emitted light will decrease in a predictable manner based on the size of the fluorophore. When the fluorophore is small, such as a potential drug molecule, molecular motion in solution is fast relative to the time interval between excitation and emission, leading to randomization/depolarization of the fluorescent signal. If, however, the fluorophore is part of a larger complex, such as a ligand/protein complex, molecular motion is substantially slower than the ligand alone. In this scenario, there is less opportunity for randomization/depolarization of the emission signal, leading to an increase in the amount of polarized emission relative to the free ligand emission. Thus, quantification of the ratio of polarized versus depolarized emission can be used to measure binding interaction and monitor biological processes (Figure 4.12).[33]

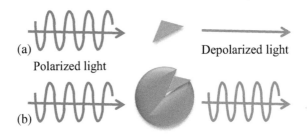

FIGURE 4.12   (a) Irradiation of an unbound small molecule fluorophore (green) in solution with polarized light produces depolarized emission. (b) Interaction of the same small molecule with a biomolecule (blue) will preserve polarization upon emission.

In practice, fluorescence polarization assay techniques have been applied to a wide range of biological interactions. Monitoring the enzymatic reactions, for example, can be accomplished by monitoring either increases or decreases in polarization depending on the nature of the enzyme and assay design. Screening for protease inhibitor can be accomplished with a fluorescently tagged protease substrate. In the absence of the protease, a large protein tagged with a fluorophore will rotate relatively slowly in solution. Upon excitation with polarized light, polarization will be maintained upon emission, leading to a higher signal. Protease activity will cleave the protein into smaller fragments that rotate more quickly in solution. Emission induced by excitation with polarized light

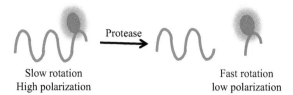

Slow rotation
High polarization

Fast rotation
low polarization

FIGURE 4.13 Polarized irradiation of a fluorescently tagged protein in solution will produce a fluorescent signal that maintains the polarization to a greater extent than fragments of the protein produced by the action of a suitable protease. The degree of depolarization in the presence of candidate compounds can be used to identify enzyme inhibitors.

will produce fluorescence, but the level of polarization will be decreased (Figure 4.13). Compounds that inhibit protease activity will decrease the rate of loss of polarization, and monitoring these changes allows quantification of the inhibitory activity of test compounds.[34]

Combinations of FP methods and antibody technology have provided assays capable of monitoring phosphorylation of substrates by various kinases (Figure 4.14). Substrate peptides labeled with a suitable

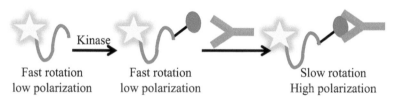

Fast rotation
low polarization

Fast rotation
low polarization

Slow rotation
High polarization

FIGURE 4.14 A fluorescently tagged peptide (blue) will rotate quickly in solution, leading to depolarization of emitted light as a result of irradiation with polarized light. Phosphorylation by a kinase in the presence of a suitable antibody (green), however, decreases rotation al speed. This preserves polarization upon emission.

fluorophore will emit a depolarized fluorescent signal upon irradiation with a polarized source. Kinase activity to produce the corresponding phosphorylated product in the presence of an antibody designed to bind with the phosphorylated product would produce an antibody/antigen complex that is significantly larger. The increased size decreases the rotational speed of the fluorophore in solution, leading to increased retention of polarization upon emission. In the presence of a kinase inhibitor, the rate of formation of the antibody/antigen complex is slowed, decreasing the rate of increase in the polarized fluorescent signal, providing insight into the potential utility of test compounds as kinase inhibitors.[35]

It is also possible to design FP assays capable of detecting changes in protein/protein interactions and DNA/protein interactions (Figure 4.15). Labeling one of the two partners in an interaction of macromolecules with

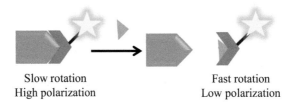

Slow rotation
High polarization

Fast rotation
Low polarization

FIGURE 4.15   Compounds that disrupt the interaction between two macromolecules, one of which is fluorescently labeled, will cause a loss of polarized emission in a fluorescence polarization assay. Disruption of the macromolecular complex decreases the size of the tagged material, increasing its rotational speed and loss of polarization upon emission.

a fluorescent tag provides the necessary signal source, while the inherent change in molecular weight associated with formation and disassociation of the macromolecular complex of interest provides the change in size required to measure a difference in polarization of a fluorescent signal. Compounds that interfere with the formation of the macromolecular complex will produce changes in the observed FP in a quantifiable manner that can be used to determine binding constants of test compounds.[36]

Fluorescence Resonance Energy Transfer (FRET)

A number of biological assay have been designed to take advantage of the Theodor Förster's discovery of fluorescence resonance energy transfer.[37] In this process, a fluorescent donor molecule absorbs electromagnetic energy, and this energy is transferred to a nearby fluorescent acceptor molecule, which then fluoresces at a lower wavelength than the donor molecule. The exchange of energy between a donor and an acceptor pair occurs without the generation of thermal energy, requires no molecular collision, but is distance-dependent. Much like scintillation proximity assays, the amount of energy transferred, and thus the level of fluorescents, decreases as the distance between the donor and acceptor molecules increases. The distance requirements are larger than interatomic distance, usually on the order of 10–100 Å, depending on the nature of the donor/acceptor pair. In addition, there must be an overlap between the donor molecule's fluorescent spectrum and the acceptor molecule's absorbance spectrum in order for resonance energy transfer to occur. In effect, the fluorescent signal that would normally be emitted by the donor upon irradiation with a lamp or laser is decreased at the expense of the acceptor's fluorescent signal.[38]

Since the energy input wavelength, donor fluorescence, and acceptor fluorescence all occur at different wavelengths, it is possible to measure each and monitor the course of events in a properly arranged biological system. Early efforts to identify HIV protease inhibitors, for example, benefited from the application of FRET technology. Scientists at Abbot Laboratories identified a small peptide substrate of HIV protease, Ser–Gln–Asn–Tyr–Pro–Ile–Val–Gln, that is cleaved at the Tyr–Pro bond. The peptide chain was converted to a FRET donor/acceptor system by attaching 5-(2′-aminoethyl)aminonaphthalene sulfonic acid (EDANS) to the carboxy terminus (Gln) of the peptide and 4′-(Dimethylamino)-4-azobenzenecarboxylic acid (DABCYL) to the amino-terminus (Ser) of the peptide. In isolation, excitation of EDANS by irradiation at 340 nM results in a fluorescent response at 490 nM. However, within the context of the aforementioned modified peptide, the normal fluorescent response by EDANS is quenched by a resonance energy transfer to DABCYL. In the presence of HIV protease, however, the modified peptide is cleaved, separating the donor/acceptor complex, allowing EDANS to fluoresce at 490 nM. Thus, the enzymatic activity of HIV protease can be quantified by monitoring the intensity of EDANS fluorescence over time. HIV protease inhibitors will prevent cleavage of the donor/acceptor complex, preventing EDANS fluorescence in a quantifiable manner (Figure 4.16).[39]

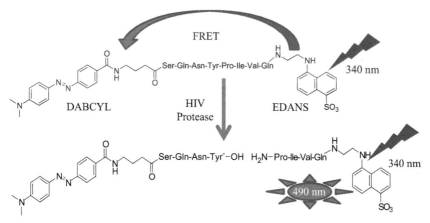

FIGURE 4.16   The seven amino acid HIV protease substrate maintains the EDANS/DABCYL donor/acceptor pair within the minimum distance required for energy transfer, which quenches the fluorescent emission of EDANS (490 nm). Peptide cleavage by HIV protease separates the pair, allowing EDANS fluorescence. HIV protease inhibitors can be identified and compared by monitoring changes in EDANS fluorescence intensity.

FRET technology has also been employed in the study of membrane potentials and ion channels. Labeling of the outside of a cell membrane with a fluorescently tagged protein provides half of the requisite donor/acceptor pair. If the second half of the FRET pair were a lipid soluble compound that could move freely through the bilayer, then the presence or absence of a FRET interaction could be controlled by the position of the lipid-soluble compound relative to the tagged protein. If the lipid-soluble compound is close to the outside surface of the cell, then FRET occurs. On the other hand, if the lipid-soluble compound is close to the inside surface of the cellular membrane, then no FRET interaction occurs as the distance between the donor and acceptor is too great. Thus, controlling the location of the lipid-soluble acceptor molecule is the key to successfully using FRET technology in this type of system. When the lipid-soluble acceptor is charged, alteration of the cell membrane potential will control the position of the acceptor molecule (Figure 4.17). Since membrane potentials are controlled by the action

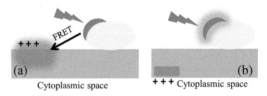

FIGURE 4.17   A cell surface protein (yellow) tagged with a FRET donor (red) can be coupled with a charged, lipid-soluble FRET acceptor (blue) that is sensitive to electrical charge. (a) When the exterior of the cell is positively charged, a negatively charged FRET acceptor will move to the exterior of the lipid layer, inducing FRET emission. (b) On the other hand, reversal of the membrane potential will cause the FRET acceptor to move toward the cytoplasmic space, preventing FRET emission.

of ion channels, this assay system can be used to monitor ion channel activity and screen for compounds that modulate ion channel function.[40]

The combination of FRET technology and fluorescent proteins provides a method for studying GPCR activity. In this case, the donor/acceptor pair is derived from genetically altered versions of the fluorescent proteins originally isolated from the jellyfish *Aequorea victoria*.[41] While the first fluorescent protein isolated produced a green color and was cleverly named green fluorescent protein (GFP), mutations in the protein sequence have provided access to fluorescent protein that provide a signal in different parts of the light spectrum, and pairs of overlapping absorbance/fluorescence spectra can be employed as FRET pairs. Importantly, in most cases, tagging of functional proteins with the jellyfish-derived fluorescent proteins has minimal impact on protein function, and this provides useful tools for monitoring protein function and location.[42] In the case of GPCRs, activation of the signaling pathway could be monitored by tagging the various subunits of the G-protein. For instance, tagging the Gα subunit with yellow fluorescent protein (YFP) and the Gγ subunit with a cyan fluorescent protein (CFP) would provide a donor/acceptor pair capable of undergoing a FRET interaction, provided they are

Pharmacol Rev 64:299–336, 2012

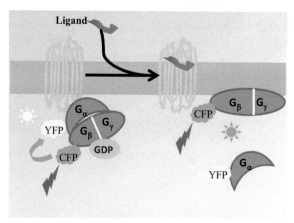

FIGURE 4.18 Functional activity of test compounds can be determined with double-labeled GPCR assemblies. Attaching two FRET compatible fluorescent proteins to different subunits of the G-protein complex enables a FRET interaction in the absence of an activating ligand. Activation of the GPCR by a functional agonist will separate the donor/acceptor pair, suppressing FRET emission. Functional antagonists will block separation of the FRET pair, preserving FRET emission.

in close enough proximity (Figure 4.18). As discussed in Chapter 3, in the absence of the natural ligand or agonist, the G-protein complex associates with the GPCR and GDP. Irradiation of CFP in the inactive complex produces fluorescence in the YFP through a FRET-mediated pathway. An activating ligand, however, will cause the G-protein complex to disassociate, disrupting the FRET interaction, and irradiation of CFP will produce CFP fluorescence rather than YFP fluorescence. This assay system can be used to identify both agonists and antagonists. Agonists will activate the system in the absence of the natural ligand, leading to an increase in CFP fluorescence at the expense of YFP fluorescence. On the other hand, the presence of an antagonist in the presence of the natural ligand will result in decreased activation of the GPCR signaling pathway, maintaining YFP fluorescence at the expense of CFP fluorescence. Variations on this general strategy have been employed to successfully study multiple GPCRs and their related signaling pathways.[43]

FRET technology has been successfully employed to study many other biochemical processes such as conformational changes in proteins,[44] assembly and dissociation of protein complexes,[45] and the distribution and transport of lipids.[46] Biosensors have also been designed to take advantage of FRET interactions. Proteins designed with molecular recognition sites that either bring together or separate a donor/acceptor FRET pair can be used to detect and quantitate analytes in solution (i.e., small molecules, proteins) by monitoring fluorescence at the proper wavelength.[47] This technology has been widely adopted since its initial development, and a full description of the vast array of FRET assays is well beyond the scope of this text. In general, if it is possible to create a donor/acceptor pair that can assemble and disassemble as the result of a biological process, then a FRET-based assay can be designed to monitor the process.

There are, however, some important limitations and issues that should be kept in mind when reviewing data from a FRET assay. Interference with the signal can occur in a variety of forms, leading to false positives and false negatives. Test compounds that fluoresce, quench fluorescence, or are colored can interfere with detection methods. In addition, background fluorescence of the assay matrix, proteins, cellular material, and even the plastic of the assay plate can create interference that may limit the sensitivity of a FRET-based assay.[48]

Time-Resolved Fluorescence Resonance Energy Transfer (TRFRET)

As mentioned above, a significant limitation of FRET technology is the presence of background fluorescence that, irrespective of the source, can create misleading data and decrease the sensitivity of a FRET-based assay. In the vast majority of cases, background fluorescence emits energy in the same short-lived time frame as the donor/acceptor pairs employed in FRET experiments. The key to eliminating background fluorescence is the application of lanthanide metals such as Europium (Eu), Terbium (Tb), and other lanthanide (rare earth) elements (Figure 4.19).[49] On their own, the lanthanides are poor

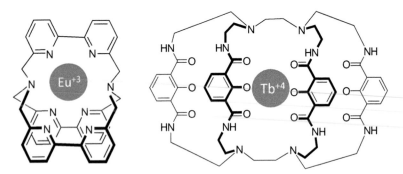

FIGURE 4.19 Organic scaffolds employed in TRFRET systems provide a suitable microenvironment that supports increased fluorescent lifetime of lanthanide-based fluorophores. Fluorescent signals from caged lanthanides outlast background fluorescence, providing a measurement window free of interference.

fluorophores, and cannot be used for FRET experiments. When captured in an appropriate organic scaffold, however, the lanthanides provide fluorescent species that can not only be employed as part of a donor/acceptor FRET pair, but also have fluorescent emission profiles that last substantially longer than typical background fluorescent sources. Background emission sources such as cellular material, plastics, and small organic compounds typically fluoresce in time windows measured in microseconds, while lanthanide-based organometallic systems fluoresce with decay rates that extend into millisecond time scales. Thus, a small delay between initial excitation and fluorescent detection, 50–150 μs, creates a measurement window during which background fluorescence has occurred and extinguished itself, but the lanthanide acceptor is still fluorescent, eliminating a significant amount of background fluorescence (Figure 4.20). Commercially

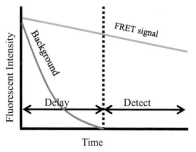

FIGURE 4.20 Irradiation of the components of a TRFRET assays system produce both background fluorescence and the FRET signal associated with the donor/acceptor pair. The application of lanthanide-based organofluorophores, however, extends the duration of the FRET signal beyond the decay limit of background fluorescence. This provides a time window for measurement of FRET signal with improved signal to noise ratios.

available TRFRET systems include Perkin Elmer's Lance® technology that utilizes europium chelates,[50] Cisbio's HTRF® assay systems that employ europium and terbium cryptates,[51] and Invitrogen's LanthaScreen® platforms that take advantage of terbium and europium chelates.[52]

Much like FRET technology, as long as it is possible to label a biological system with a donor/acceptor pair that will associate or disassociate as a result of a biological process, it is possible to apply TRFRET technology to quantify the biological activity of test compounds. Kinase activity, for example, can be measured through the use of antibodies tagged with a TRFRET fluorophore (Figure 4.21). In this case, a biotinylated kinase substrate could undergo phosphorylation, and the resulting phosphorylated species would

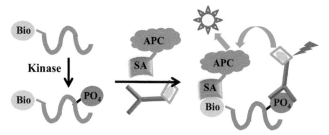

FIGURE 4.21   In this TRFRET kinase assay, phosphorylation of the biotinylated (Bio) target by a kinase produces an antigen capable of interacting with an antibody tagged with a lanthanide-based fluorophore. Addition of a streptavidin (SA)-linked fluorescent protein such as Allophycocyanin (APC) establishes a donor/acceptor pair that will produce light via a FRET pathway upon irradiation.

then bind to a TRFRET-labeled antibody that targets the phosphorylated species. Addition of a streptavidin-tagged acceptor molecule, such as the protein Allophycocyanin, a fluorescent protein, provides the necessary donor/acceptor pairing, which will undergo a long-lived FRET interaction upon irradiation. Test compounds that inhibit kinase activity will block phosphorylation, leading to a decrease in the formation of the FRET donor/acceptor complex, decreasing the intensity of the assay signal in a quantifiable manner.[53]

Downstream cellular activity can also be monitored with TRFRET technology (Figure 4.22). As mentioned in Chapter 3, activation of a GPCR signaling pathway can lead to a number of downstream cellular events

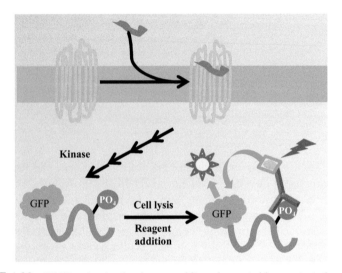

FIGURE 4.22   GPCR activation by the natural ligand or suitable agonist induces downstream events culminating in the phosphorylation of a GFP-labeled substrate. Cell lysis followed by the addition of an antibody tagged with a TRFRET fluorophore leads to close association of the donor/acceptor pair and FRET signaling. Antagonist will block induction of the downstream events leading to FRET emission by preventing GPCR activation by the natural ligand.

including the activation of various kinases. If a substrate is known to be phosphorylated by a kinase that is activated as a downstream event as a consequence of GPCR activation, then tagging the substrate with a fluorescent label, such as green fluorescent protein (GFP) provides an opportunity to establish a TRFRET assay to measure GPCR activation in cellular system. This requires stable cell lines that produce the substrate with a GFP tag in a way that does not interfere with normal kinase activity. Activation of the GPCR signaling pathway with the natural ligand or a synthetic agonist will produce the phosphorylated substrate. Cell lysis and addition of a lanthanide-tagged antibody TRFRET reagent and irradiation with the proper wavelength of light will produce a measurable fluorescent signal through a FRET interaction. Quantification of the signal intensity provides a means for quantifying agonist activity of test compounds. In a similar manner, the presence of an antagonist will block activation of the GPCR signaling pathway, decreasing phosphorylation of the GFP-tagged substrate. This will decrease the intensity of the FRET signal produced upon irradiation with light, thus providing a method for measuring antagonist activity of test compounds.[52]

Amplified Luminescent Proximity Homogeneous Assay (AlphaScreen™)

The need to develop assay capable of avoiding background fluorescence from assay components, reagents, plates, and other materials also lead to the exploration of methods that could produce fluorescent signals in the absence of a FRET interaction. The prospect of inducing fluorescent signals for the purposes of monitoring biological processes through the generation of singlet oxygen was first introduced in by Ullman in 1994.[54] Originally referred to as luminescent oxygen channeling assay (LOCI), reagents and assay kits that employ this technology are now readily available for drug discovery purposes under the name AlphaScreen™ and are marketed by Perkin Elmer. In some ways, the AlphaScreen™ is similar to FRET and TRFRET assay systems, as all three systems depend on the interaction of a donor and acceptor for the production of a measurable signal. Unlike FRET and TRFET, however, irradiation of the donor does not lead to an energy transfer to the acceptor followed by light emission. AlphaScreen™ technology employs the coupling of the photosensitizer phthalocyanine, which produces singlet oxygen (an excited state oxygen) upon irradiation at 680 nm, and thioxene derivatives, which emit light (520–620 nm) in the presence of singlet oxygen (Figure 4.23).[55]

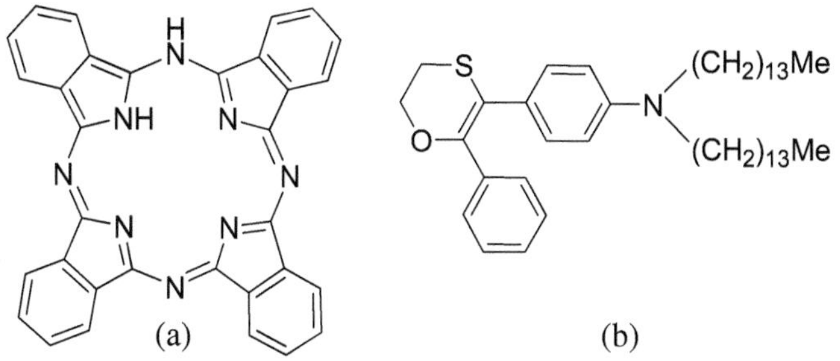

FIGURE 4.23 AlphaScreen™ assays depend on phthalocyanine derivatives (a) ability to generate singlet oxygen upon absorption of light and the reaction of thioxene derivatives (b) with singlet oxygen coupled with the emission of light.

Of course, in order to be useful, it must be possible to control the interaction of the donor and acceptor in order for the assay to provide useful information, and in this case, the two parties are separated through the use of specially designed beads, one that contains the donor and the other that contains the acceptor. Both the acceptor and donor beads are coated with a hydrogel that limits nonspecific interaction between donor and acceptor beads and also provides chemical linkage points for assay reagents (i.e., proteins, ligands, etc.). The beads are also small enough that they are suspended in biological media (they do not sediment out) and can be manipulated as if they were in solution (standard pipetting techniques and robotic platforms). In the absence of an interaction capable of bringing the acceptor and donor beads together, singlet oxygen produced upon irradiation of the donor beads simply reverts back to the ground state. If, however, the two beads contain complementary binding groups, such as a protein and a binding ligand, the interaction between the two species will bring the beads into close proximity. This, in turn, allows the singlet oxygen generated by the donor beads to interact with the thioxene derivatives incorporated into the acceptor beads and produce a fluorescent signal (Figure 4.24).

FIGURE 4.24 When the donor and acceptor beads in an AlphaScreen are separated by more than the path length of singlet oxygen in water, fluorescent signaling will not occur. Interactions that bring the donor and acceptor beads together allow singlet oxygen produced by phthalocyanine derivatives in the donor bead to interact with thioxene derivatives in the acceptor beads, generating a fluorescent signal.

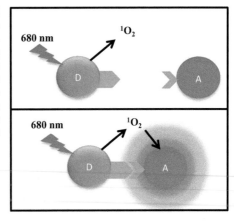

The development of assays and methods designed to take advantage of AlphaScreen™ technology has followed a similar path as that described for both FRET and TRFRET. For the most part, as long as it is possible to design an assay in which the interaction between donor and acceptor beads can be manipulated, then it is possible to create an effective AlphaScreen assay™. Kinase inhibitors, for example, have been identified using a biotinylated substrate, streptavidin-coated donor beads, and acceptor beads tagged with an antibody that will bind to the phosphorylated substrate. Phosphorylation of the substrate by the kinase produces a substrate that binds to the donor beads through the streptavidin/biotin interaction and to the acceptor bead via an antibody/antigen interaction. This positions the donor and acceptor beads in close proximity and singlet oxygen produced upon irradiation at 680 nm produces a fluorescent response in the acceptor beads. Compounds that interfere with kinase activity will block phosphorylation of the substrate, preventing the formation of bidentate interaction that brings the donor and acceptor beads together, thus decreasing the intensity of the fluorescent signal. Quantification of the changes in signal intensity produced in the presence of various concentrations of potential kinase inhibitors allows IC_{50} determinations for test compounds (Figure 4.25).[56]

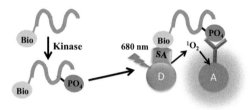

FIGURE 4.25 In the presence of an appropriate kinase, a biotinylated peptide is phosphorylated. When streptavidin-labeled donor beads and acceptor beads tagged with an antibody to the phosphorylated peptide are added, the donor/acceptor pair is brought together. Irradiation with light induces a fluorescent response in the acceptor bead. Compounds that block kinase activity will decrease fluorescent intensity in a concentration-dependent manner.

AlphaScreen technology has found widespread acceptance in the pharmaceutical industry. Applications for this assay system have been developed to monitor enzyme activity,[57] identify compounds that are functionally active in signal transduction cascades,[58] and to study the formation and disruption of macromolecular complexes.[59] As with FRET and TRFRET technology, as long as one can design a system in which a biological event mediates the relative position of the donor and acceptor beads, an AlphaScreen assay is possible.

Fluorescent Detection of Calcium Flux

The ability to detect changes in cellular calcium utilization and mobilization has lead to the development of numerous assay systems designed to generate fluorescent signals based on calcium concentration. Some of these systems depend on calcium-sensitive dyes such as Fluo-3 and Fluo-4

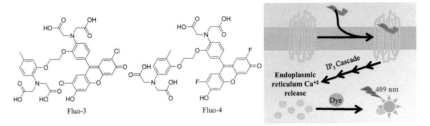

Fluo-3   Fluo-4

FIGURE 4.26 The presence of Ca^{+2} increases the fluorescent intensity produced upon irradiation of Flou-3 or Fluo-4. When a cell is preloaded with calcium-sensitive dyes, activation of a GPCR-IP_3 -mediated signaling event leads to downstream release of Ca^{+2} (orange circles) from the endoplasmic reticulum. The interaction between the Ca^{+2} and the dye produces a fluorescent signal upon irradiation with light. Changes in fluorescent signal intensity can be used as a means of identifying compounds that are functionally active at a target GPCR.

(Figure 4.26).[60] Both of these dyes fluoresce upon irradiation with a light source, such as an Argon laser (488 nm), but more importantly, the intensity of the fluorescent signal increases in the presence of increasing Ca^{+2} concentration. This phenomenon can be harnessed to monitor biological events that lead to changes in calcium concentration. GPCR signaling that proceeds through the IP_3 signaling cascade, for example, can be monitored, as activation of this signaling cascade by an agonist leads to the release of Ca^{+2} from cellular storage. In the presence of a fixed concentration of the aforementioned dyes, this increase in Ca^{+2} concentration leads to increased fluorescent signal intensity that can be monitored to determine the ability of test compounds to acts as functional agonists for the GPCR in question. In a similar manner, antagonists will block the ability of the GPCR's natural ligand to activate the signaling cascade, preventing calcium release from cellular stores. This also prevents the increase in fluorescent signal intensity that would have been observed with the natural ligand, providing a means for assessing a test compound's ability to behave as a functional antagonist for a particular GPCR.[61]

Of course, in order for these assays to function properly, the fluorescent dye must be capable of entering the cell and have little to no impact on other cellular functions. Also, like any other fluorescent system, interference from background fluorescence can be an issue. An alternative system, the aequorin-based assays, needs neither dyes nor outside irradiation in order to monitor calcium concentration changes related to cellular activity. These assays depend on the bioluminescence induced by the presence of Ca^{2+}, coelenterazine, and apoaequorin, the apo-enzyme of aequorin. This 22 kDa photoprotein was originally isolated from bioluminescent jellyfish, specifically the *Aequorea victoria* jellyfish, and is responsible for the blue light emitted by these creatures (Figure 4.27).[62] In the absence of Ca^{2+},

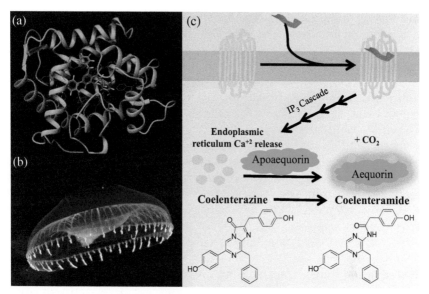

FIGURE 4.27 (a) The X-ray crystal structure of aequorin and the ligand coelentera-zine-2-hydroperoxide (RCSB file 1EJ3). (b) The *Aequorea victoria* jellyfish, also referred to as the crystal jellyfish. Image provided by Sierra Blakelym Wikipedia web page http://en.wikipedia.org/wiki/File:Aequorea4.jpg. (c) Activation of a GPCR by its natural ligand (red) causes conformational changes that leads to a downstream release of Ca^{+2} stores (orange) from the endoplasmic reticulum. Calcium binding by aequorin converts it to an active conformation that is capable of converting coelenterazine into coelenteramide. This process releases carbon dioxide and produces a fluorescent emission that can be monitored for the purposes of identifying compounds that are functionally active at a targeted GPCR.

apoaequorin is inactive. However, when Ca^{2+} binds to the three calcium binding sites present in this protein, a conformational change occurs that converts the protein into its active form, aequorin, which oxidizes coelenterazine into coelenteramide, producing carbon dioxide and blue light (469 nm).[63]

The gene for this protein has been isolated and successfully transfect into non-native environments, providing a means of harnessing changes in aequorin/Ca^{2+}-induced bioluminescence for the study of cellular processes. If, for example, a cell line were developed that expressed both a GPCR that utilizes the IP_3 pathway and aequorin, then calcium release induced by GPCR activation would produce a blue emission, provided that the cells are treated with the cofactor coelenterazine. Screening for functional agonists of an IP_3 associated GPCR could thus be accomplished by monitoring the intensity of blue light emission created in the presence of test compounds. On the other hand, the identification of functional antagonists is also possible. Addition of the natural ligand to an assay system that has been pretreated with functional antagonists will produce less blue light emission as a result of decreased IP_3 activation caused by the presence of test compounds that are functional antagonists.[64]

Since the generation of an assay signal does not depend upon irradiation with an outside light source, background fluorescence is generally not an issue. There are, however, other important limitations of aequorin assays. First and foremost, it is critical that the production of aequorin not interfere with cellular function. As with any assay system that depends on the addition of foreign genes, if insertion of the signaling genes significantly alters cellular function, then correlation to normal conditions (native cells) becomes problematic. In addition, compounds that interfere with the conversion of coelenterazine into coelenteramide will provide erroneous data, as this will block light production. Despite these limitations, aequorin-based assay systems are routinely employed in modern drug discovery research programs. Assay for a variety of GPCRs, ion channels, and transporters have been reported.[64]

REPORTER GENE ASSAYS

Differential gene expression plays a key role in cellular functions such as communication between cells, cellular development, growth regulation, and proliferation. The process is tightly regulated and dysregulation can lead to a number of pathological conditions. Monitoring changes in gene expression has become an important aspect of drug discovery, and reporter gene assays have become an important tool for this purpose. Essentially, the expression of a native gene (one normally present in a cell) is coupled to the expression of a non-native gene, a reporter gene, whose protein product can be used to monitor changes in expression of the native gene. Every time the native gene is expressed, the non-native gene is also expressed (Figure 4.28).

FIGURE 4.28   In a reporter gene assay system, expression of target gene is tied to expression of a non-native gene that will produce a measurable signal upon expression.

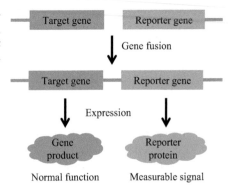

Compounds that increase expression of the gene of interest will enhance expression of the gene product of the reporter gene. Conversely, compounds that suppress expression of the gene of interest will decrease production of the reporter gene's end product. The development of this assay platform was enabled by the scientific discoveries of the biotechnological revolution discussed in Chapter 2, along with the identification of proteins such as chloramphenicol acetyltransferase (CAT),[65] β-lactamase,[66] and luciferase.[67] A full review of reporter gene assays is well beyond the scope of this text, but an examination of three aforementioned systems will provide a solid basis of understanding for the general assay method.

Chloramphenicol Acetyltransferase (CAT)

The first practical application of reporter gene technology took advantage of the bacterial enzyme responsible for the metabolism of chloramphenicol, a broad spectrum bacteriostatic antimicrobial.[68] Chloramphenicol acetyltransferase catalyzes the transfers of an acetyl group from acetyl-CoA to chloramphenicol (Figure 4.29),[69] eliminating

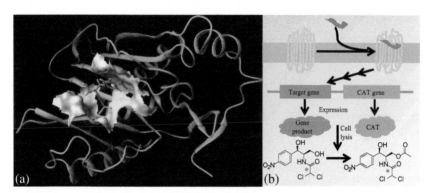

FIGURE 4.29   (a) An X-ray crystal structure of chloramphenicol acetyltransferase (RCSB file 3CLA). The binding pocket surface is highlighted in gray. (b) Activation of a GPCR by its natural ligand or suitable agonist triggers expression of a gene of interest. Fusion of the gene of interest to the CAT gene forces co-expression of CAT. Cell lysis followed by addition of radiolabeled chloramphenicol provides a means of quantifying gene expression by monitoring the acylation of chloramphenicol by CAT via acetyl-CoA.

its ability to bind to ribosomes and its antibacterial activity.[70] There is no mammalian counterpart exhibiting similar properties, so coupling the expression of the CAT gene to a target gene provides an opportunity to monitor gene expression by monitoring the rate of acetylation of chloramphenicol. Compounds that increase expression of the target gene will also increase CAT levels, increasing the rate of acetylation of chloramphenicol, while compounds that suppress the expression of the target gene will have the opposite effect. If, for example, activation of a GPCR pathway by the natural ligand was known to activate expression of a particular gene, then coupling of the expression of the target gene to expression of the CAT gene could be used to identify functional agonist and antagonists of the GPCR in question. Agonist activation of the GPCR would lead to increased expression of CAT, while antagonists would block the activity of the normal ligand, suppressing its ability to induce expression of the target gene and CAT.[71] The most significant limitation of CAT reporter gene systems, however, is their dependence on radiolabeled materials. Specifically, the CAT assays monitors the conversion of ^{14}C-labeled chloramphenicol to ^{14}C-labeled acetyl chloramphenicol. Despite this limitation, this reporter system has been widely employed to study biological systems relevant to a wide range of therapeutic endpoints.

β-Lactamase Reporter Assays

Although CAT reporter gene assays remain a viable platform for the monitoring of biological processes, the desire to avoid generating radioactive waste material led to the development of alternative reporter gene systems. Of course, if radioactive signals or radiolabeled material were to be eliminated from the process, then an alternate signaling mechanism would be required. In the β-lactamase reporter assay system, the overall concept of coupling the expression of a target gene to a reporter gene remains the same, but radiolabeling is replaced with FRET signaling. As its name implies, β-lactamase cleaves β-lactam antibiotics, such as penicillins and cephalosporins, and is encoded by the ampicillin resistance gene (amp[r]). There is no mammalian counterpart to this enzyme, and coupling to target genes is readily accomplished using modern biotechnological tools. In principle, once a target gene is coupled to amp[r], compounds that stimulate expression of the target gene would also stimulate expression of β-lactamase in a quantifiable manner. The key to the success of this assay, however, was the identification of suitable β-lactam substrates that contains a FRET donor/acceptor pair.[66]

Zlokarnik et al.[72] demonstrated the first such system in which 7-hydroxycoumarin-3-carboxamide, a FRET donor, and fluorescein, a FRET acceptor, are linked through a cephalosporin bridge. In the absence of β-lactamase, irradiation at 409 nm produces a FRET signal at 518 nm. On the other hand, the presence of β-lactamase eliminates the FRET. Cleavage of the β-lactam ring by β-lactamase causes the molecule to eject the fluorescein component, terminating the FRET signal (Figure 4.30). It is worth noting at this point that compounds that inhibit β-lactamase activity can significantly impact assay results.

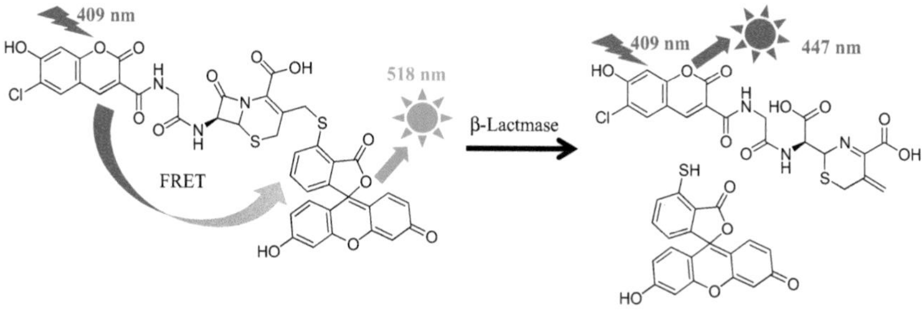

FIGURE 4.30 In the β-lactamase reporter gene reporter system, expression of the target gene of interest is linked to the expression of β-lactamase. In order to provide a quantitative signal for the β-lactamase activity, disruption of a FRET signal can be monitored. Compounds that induce greater expression of the target gene will also increase expression of β-lactamase, leading to decreased FRET signaling.

Luciferase Reporter Assays

Bioluminescence, the production of light by living organisms through the use of enzymatic systems, has also been adopted as a tool for monitoring a variety of biological processes. Specifically, luciferase enzymes from fireflies (*Lampyridae*),[73] click beetles (*Elateridae*),[74] glowworms (*Phengodidae*),[75] and sea pansies (*Renilla reniformis*)[76] have all been cloned, isolated, and expressed for the purpose of developing assay systems. Firefly (*Photinus pyralis*) luciferase is the most widely used of this class. This monomeric 61 kDa protein catalyzes a two-step oxidation of luciferin to oxyluciferin in the presence of ATP, oxygen, and magnesium that produces a burst of green/yellow light (550–570 nm).[77] Renilla luciferase, a 36 kDa monomeric enzyme, is also routinely employed, but its substrates are quite different. This enzyme produces light through the conversion of coelenterazine to coelenteramide, and produces blue light (480 nm) [78] (Figure 4.31).

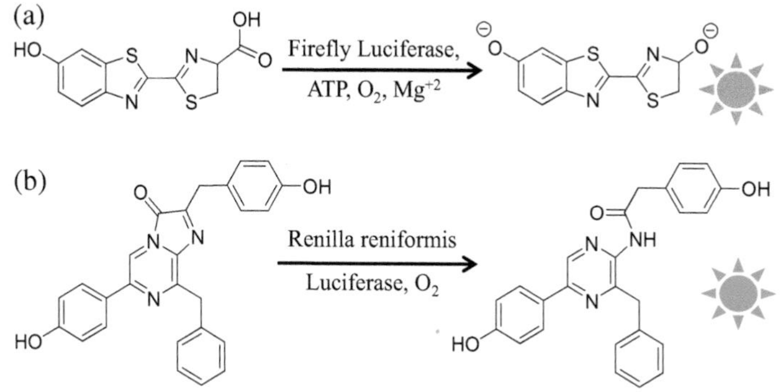

FIGURE 4.31 (a) Firefly luciferase produces light upon conversion of luciferin into oxyluciferin in the presence ATP, oxygen, and magnesium. (b) Renilla luciferase converts coelenterazine to coelenteramide in the presence of oxygen and emits blue light.

Within the context of reporter assays, the application of luciferase technology is essentially the same as CAT and β-lactamase reporter assays systems. Coupling of the expression of a target gene with a luciferase gene creates a situation in which compounds that promote the expression of the target gene also promote the expression of luciferase. In the presence of the proper substrate, increased gene expression leads to increased light emission that can be quantified as means of monitoring gene transcription. Conversely, compounds that block expression of the target gene will also decrease the production of luciferase, providing a means of identifying compounds that prevent gene expression.[67]

Both the β-lactamase and luciferase reporter assays systems avoid the use of radiolabeled materials, but luciferase reporter assay systems have an additional advantage. As discussed earlier, β-lactamase reporter systems depend upon the presence or absence of a FRET signal, which requires light irradiation at the acceptor compound's absorbance frequency and detection of the donor compound's emission frequency. Potential fluorescent interference from the assay media, cellular material, assay plates, etc. is an issue in FRET-based reporter assay in the same manner as it is in any other type of FRET-based assay. Thus compounds that fluoresce or act as fluorescence quenchers can produce misleading results in FRET reporter systems.

Luciferase reporter assays, on the other hand, generate light emission as a consequence of the chemical reaction mediate by the enzyme and do not require irradiation with an external light source. As such, the risk of interference via fluorescence of other assay components (assay media, cellular material, assay plates, etc.) is significantly lower. Of course, compounds that inhibit luciferase activity will impact screening results. Counterscreening of compounds for luciferase inhibition in the absence of the reporter gene system is an important aspect of data analysis in luciferase reporter gene assay.

KINETIC FLUORESCENT MEASUREMENT SYSTEMS

While there are many biological processes that can be measured with simple fluorescent plate readers, there are some cellular events that are transient in nature. Changes in membrane potential, calcium mobilization events, and GPCR activation can occur on an extremely short timescale and are often transient in nature. Membrane potential in cardiomyocytes, also shifts rapidly and in opposing directions with each heartbeat. Monitoring the kinetic time courses of these events can provide a wealth of information on the impact of test compounds, provided it is possible to measure changes in signal emission within the event window. The Fluorescent Imaging Plate Reader (FLIPR) developed by Molecular Devices[79] and the Functional Drug Screening System (FDSS) [80] produced by Hamamatsu Photonics were both designed to address this need. Although there are some differences in capabilities between the two instruments, essentially, each instrument is equipped with an energy source for irradiation of sample (an argon laser, light emitting diodes, etc.) in a microtiter plate (96, 384 well) and a charge-coupled device camera capable of collecting data from all plate wells simultaneously. Multiple images of the assay plate are recorded through the course of a single experiment so that changes in fluorescent intensity can be monitored and interpreted by software that drives the instruments. Plate images can be captured in rapid succession at intervals of less than 1 s per image, making these systems well suited toward the examination of transient cellular events that would be otherwise impossible to monitor (Figure 4.32).

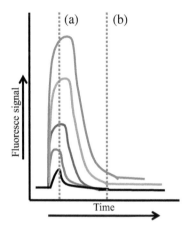

FIGURE 4.32 Kinetic assay systems are capable of monitoring changes in fluorescent intensity over time. Each color represents the impact of increasing concentrations of a test compound on the fluorescent intensity. The importance of this technology becomes apparent when one considers the difference in observed potency at time points (a) and (b). If only data from time point (b) were available, this test compound would appear to lack potency, even though there is a potent effect at an early time point.

Both of these systems are capable of supporting a wide range of fluorescent assay systems (FRET, TRFRET, AlphaScreen, etc.) and have been successfully employed to study a wide range of transient cellular events.

LABEL-FREE ASSAY SYSTEMS

The overwhelming majority of *in vitro* assay screening systems employed in modern drug discovery research rely on fluorescent labeling or radiolabeling of small molecules, proteins, DNA, antibodies, or related materials. While labeling techniques have clearly been successfully used to develop sophisticated high throughput screening assays capable of driving drug discovery programs, it is important to understand that the labels themselves have the potential to influence screening results. Attaching a label to a biologically relevant compound can alter its conformation, its overall molecular properties, and its ability to interact with biological systems necessary for functional activity. Fluorescent labeling of an enzyme's natural substrate, for example, can have a direct impact upon the rate at which the enzyme converts the now-altered substrate to its end product. In order to circumvent this type of issue, assays system that measure changes in physical or chemical properties of cellular systems, particularly cell monolayers, have been developed. Changes in cell volume, pH, refractive index, membrane potential, electrical impedance, refractive index, and optical properties have been employed as means of determining the impact of test compounds on cellular function. Although it is not possible to review all of the available label-free technologies, some key examples include the CellKey™ assay platform (Molecular Devices, LLC, impedance measurements), Corning's Epic™ assay system, surface plasmon resonance assays (Biacore™ platform, GE Healthcare), and electrophysiological patch clamp (membrane potential).

Cellular Dielectric Spectroscopy

Monitoring the ability of electrical currents to flow through a cell monolayer forms the basis of cellular dielectric spectroscopy (CDS), the label-free assay that forms the basis of the CellKey™ assay platform (Molecular Devices, LLC). In principle, a monolayer of cells grown on the surface of an electrode will create a barrier to the passage of an electrical current. If a high enough voltage (V) is applied, however, the electrical current (I) will flow through the monolayer, and the ratio of the voltage applied to the current is defined as the electrical impedance (Z) of the system ($Z = V/I$). Of course, cell monolayers often have gaps between the cells where current flow is not impeded by cellular material (Figure 4.33). The total current

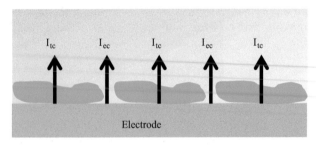

FIGURE 4.33 An electrical current passing through a monolayer of cells (blue) is the sum of the current that flows between the cells (I_{ec}) and the current that passes through the cells (I_{tc}).

that passes through a monolayer of cell is comprised of the current that flows through the cell (transcellular current, I_{tc}) and the current that moves through gaps in the monolayer (extracellular current, I_{ec}).

Measuring the electrical impedance of a cell monolayer in the absence of an outside stimulus, such as an agonist for a GPCR of interest, provides a baseline impedance value. Activation of a biological pathway in the cellular monolayer can produce changes in electrical impedance through changes in adherence of cells to the electrode, changes in cell–cell interactions, and changes in cell shape and volume (Figure 4.34).

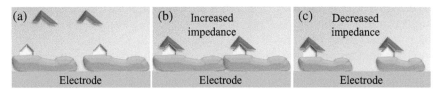

FIGURE 4.34   (a) A baseline measurement of cellular impedance is established for a cell monolayer (blue) that expresses a GPCR of interest (yellow). (b) Binding of a ligand (red) could lead to changes in cellular shape that decrease the intracellular space and increase cellular impedance. (c) Alternatively, activation of the target GPCR could increase the space between cells, decreasing impedance.

The change in electrical impedance (ΔZ) in the presence of increasing concentrations of test compounds can then be quantified as a means of identifying biologically active compounds. Alternatively, blocking a natural ligand's ability to cause changes in impedance can also be employed as a strategy to characterize potential therapeutic agents. Kinetic measurement of cellular events can also be monitored by applying voltages at specific time intervals, making it possible to monitor transient cellular events via cellular dielectric spectroscopy. Specially designed 96 and 384 well plates with interdigitated electrodes at the bottom of each well are commercially available to facilitate the application of this technology to high throughput screen in conjunction with the CellKey platform.[81]

Optical Biosensors

The identification of biologically active compounds by monitoring for changes in the refractive index of light by a surface, coated with either a macromolecular target or a cell monolayer, is another effective screening method. The Epic™ System employs specially designed plates that contain an optical biosensor referred to as a resonant waveguide grating (RWG) biosensors. An RWG biosensor consists of a transparent substrate (glass), a thin film layer containing an embedded waveguide grating structure, and a biological substrate layer. Macromolecular targets of interest can be chemically attached to the biological substrate layers in order to facilitate biochemical assays. Alternatively, a cell monolayer can be employed to study cellular response to changes in the surrounding environment (i.e., the addition of test compounds). In the absence of external influences, illumination of an RWG biosensor with broadband light will produce reflected light only at the wavelength that is resonant with the combination of the embedded waveguide grating structure and the biological surface layer. The wavelength of the reflected light is sensitive to the refractive index of the RWG biosensor, which changes as the nature of the biological substrate layer changes.[82]

The application of this technology to the identification of enzyme inhibitors could be accomplished by functionalizing the biological substrate layer with a substrate for an enzyme of interest. Protease inhibitors, for example, could be identified by attaching a peptide substrate to the biological substrate layer and monitoring the changes in signal achieved in the presence of the enzyme and potential inhibitors (Figure 4.35).

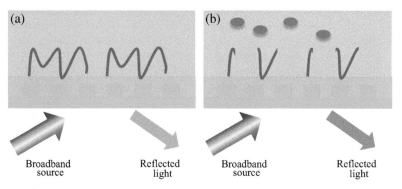

| (a) | (b) |
| Broadband source | Reflected light | Broadband source | Reflected light |

FIGURE 4.35   (a) An RWG biosensor with a peptide attached to its surface (blue) will produce a specific signal upon irradiation with broadband light. (b) Addition of a protease (red) will alter the nature of the peptide attached to the RWG biosensor, causing change in the signal produce upon irradiation with broadband light that can be quantified for the purpose of identifying compounds that block protease activity.

In the absence of the protease, irradiation of the RWG biosensor produces reflected light based on the refractive index of the intact substrate. Cleavage of the peptide substrate by addition of a protease will change the nature of the RWG biosensor, altering the refractive index and reflected light upon irradiation of the RWG biosensor. Compounds that inhibit the protease will prevent this change in a quantifiable manner, allowing identification of protease inhibitors.[83]

In a similar manner, if the biological substrate layer is tagged with a receptor, binding of the natural ligand to the receptor will produce

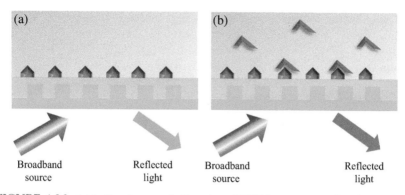

FIGURE 4.36   (a) In the absence of a ligand, an RWG biosensor tagged with a receptor (blue) produces a specific signal output. (b) Addition of a ligand (red) will alter the signal output by changing the physical nature of the biological substrate layer surface, providing a means of identifying compounds that bind to the receptor.

a measurable change in both the refractive index and reflected wavelength proportionate to the amount of natural ligand bound to the receptor (Figure 4.36). Test compounds that bind to the receptor will produce changes to the refractive index and reflected wavelength proportionate to the amount of binding that occurs. This information can be used to determine binding constants (IC_{50}s) for the natural ligand and test compounds.

The situation is somewhat more complex when the biological substrate layer is a cell monolayer. Once again, in the absence of external influence, illumination of an RWG biosensor with broadband light will produce reflected light only at the wavelength that is resonant with the combination of the embedded waveguide grating structure and the cell monolayer. Changes in cellular activity induced by changes in the environment of the cell monolayer, however, will produce changes in the refractive index and reflected wavelength observed that is proportional to the changes in the environment. Activation of a GPCR signaling pathway by an agonist, for example, will alter cellular functions such as trafficking of molecules and molecular assemblies to different cellular compartments. This dynamic translocation of cellular material is referred to as dynamic mass redistribution (DMR), and is proportional to the changes in refractive index and reflected wavelength. These changes are in turn proportional to the concentration of the agonist, providing a means of determining the biological activity of potential therapeutic entities. Antagonists can also be identified in a similar manner, as these compounds will block the ability of the natural ligand to induce changes in the refractive index and reflected wavelength in a concentration-dependent manner (Figure 4.37).[82]

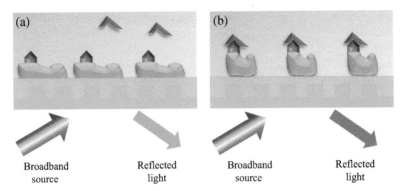

Broadband source   Reflected light   Broadband source   Reflected light

FIGURE 4.37   (a) Cell expressing a GPCR of interest will present in a specific manner when aligned as a monolayer on an RWG biosensor. (b) Activation of the GPCR will cause dynamic mass redistribution (DMR) that is proportional to both agonist concentration and the signal change induced by the agonist.

Surface Plasmon Resonance Technology

An alternative method of monitoring changes in the refractive index of light for the purposes of identifying biologically active compounds depends upon changes in surface plasmon resonance (SPR). In this system, one side of an electrically conducting thin metal film, most often gold, is chemically modified so that a biologically relevant molecule, such as an antibody or receptor protein can be bound to the surface. The opposite side is placed against a prism. At the interface of the prism and the metal film, the surface electrons will undergo a coherent oscillation referred to as a surface plasmon. Illumination of the prism/metal interface will create a surface plasmon resonance at a specific angle of incidence and at the frequency of light that matches the oscillation frequency of the electrons at the prism/metal interface. The angle of reflection of the light associated with the surface plasmon resonance can be determined with an appropriate detector, as the intensity of the reflected light at this angle will be modified by this effect, providing a baseline measurement for the prism/metal interface in the absence of an outside influence.[84]

Surface plasmon resonance is extremely sensitive to changes in the prism/metal interface, particularly changes in the refractive index close to the surface. The refractive index of the surface is, in turn influenced by changes in the mass at the prism/metal interface. When the surface of the thin metal film is tagged with an antibody, protein, or one half of a binding pair, the presence of a compound capable of interacting with the material bound to the surface will increase the mass at the prism/metal interface. These changes cause a detectable change in the angle of reflection of light associated with the surface plasmon resonance that can be quantified to measure binding affinities. In practice, this is accomplished by exposing the tagged-side of the thin metal film to a continuous flow of a solution containing potential binding partners in a flow cell. The application microfluidics technology provides a means for miniaturization of this technology to create "labs on a chip." The most common drug discovery application of surface plasmon resonance is the Biacore assay system. In this particular variation, a gold surface is coated with a layer of carboxymethylated dextran, creating a hydrophilic environment that preserves the non-denatured state of attached biomolecules. Although the throughput of these systems is not as high as other methods, the high degree of sensitivity has led to wide adoption of Biacore assays in the drug discovery process (Figure 4.38).[85]

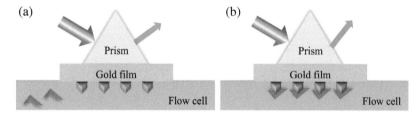

FIGURE 4.38 (a) In the absence of a binding partner, a tagged gold biosensor will undergo a surface plasmon resonance at a specific angle. (b) Addition of a binding partner via a flow cell will change the angle of reflected light associated with the surface plasmon resonance in a concentration-dependent manner.

ELECTROPHYSIOLOGICAL PATCH CLAMP

As discussed in Chapter 3, the flow of electrical currents and creation of voltage gradients across cellular membranes are critical to a wide range of cellular functions. Ion channels are an important aspect of these processes, and as such, monitoring their activity can provide a great deal of insight into the biochemical processes for both normal and pathological conditions. Earlier sections of this chapter have described fluorescent-based systems that are capable of providing indirect evidence regarding ion channel activity (e.g., calcium-sensitive dyes, charged lipid-soluble FRET acceptors), but none of the methods discussed thus far are capable of providing a direct measurement of channel activity, as each uses a surrogate marker (a fluorescent signal). To date, electrophysiological patch clamp remains the gold standard for assessing ion channel activity, as it is the only available method capable of providing direct measurement of ion flow through cellular barriers. Ultimately, indirect methods must be validated by defining a correlation with patch clamp results, as patch clamp results are the definitive evaluation of a compound's ability to influence ion channel activity.

A basic patch clamp system consists of a micropipette with an opening on the order of 1 μm pressed against the surface of a cell. The inside of the micropipette covers a limited number of ion channels, and a seal with high electrical resistance ("gigaohm seal") is created by suction on the surface of the cell. An electrode, salt solution inside the micropipette, and the appropriate electrical amplification and monitoring systems can then be employed to either maintain a constant voltage while monitoring current or maintain a constant current while monitoring changes in membrane potential in the presence of test compounds.[86] The process itself is time-consuming, labor-intensive, and even well-trained electrophysiologist can characterize only a few compounds per day. Ultimately, this has limited the use of patch clamp methods in high throughput screening programs, but efforts to increase screening throughput have begun to emerge (Figure 4.39).

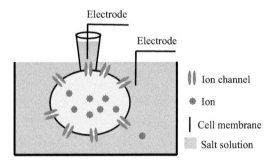

FIGURE 4.39   A basic patch clamp system consists of a micropipette with an opening on the order of 1 μm pressed against the surface of a cell. The inside of the micropipette covers a limited number of ion channels, and a seal with high electrical resistance ("gigaohm seal") is created by suction on the surface of the cell. An electrode, salt solution inside the micropipette, and the appropriate electrical amplification and monitoring systems can then be employed to either maintain a constant voltage while monitoring current or maintain a constant current while monitoring changes in membrane potential in the presence of test compounds.

Plate-based perforated patch clamp technology increased the patch clamp screening capacity when it was introduced along with the Ion-Works™ HT platform. In this assay system, a specially designed 384-well plate (referred to as a PatchPlate) configured with a 1–2 μm hole at the bottom of each well allows for the suspension of a single cell over the

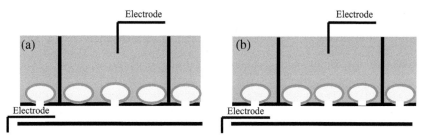

FIGURE 4.40 The development of 96 plates suitable for electrophysiological experiments has facilitated the identification of compounds capable of modulating ion channel activity. In the IonWorks™ PatchPlate system (a), each well of the plate has a single hole at the bottom that can be used to access the inside of a single cell for electrophysiological experiments on one cell per well. Population patch clamp (PPC) plates (b) have 64 holes at the bottom of each well and provide information on electrical changes across a small population of cells per plate well.

hole in each plate (Figure 4.40(a)). Application of a slight vacuum to the underside of the plate keeps the single cell in place, while the addition of the pore forming antibiotic amphotericin provides electrical access to the cell, thus permitting patch clamp experiments in a plate-based system. Automated liquid handlers built into the IonWorks™ HT platform further streamline the process.[87]

In theory, moving from single compound screening to a 384 well plate-based system should provide a 384-fold increase in screening capacity, but in practice the increase in throughput is much lower. The electrical seals established in perforated patch clamp method have substantially lower electrical resistance than manual methods. This leads to well to well variability, which is further complicated by a relatively high failure of cells to be properly placed over the holes at the bottom of the well. These limitations necessitate multiple wells per assay condition in order to provide useful data.

The introduction of population patch clamp (PPC) mode of operation and the IonWorks™ Quattro system have overcome some of these limitations by not relying on a single cell recording per well. The PatchPlates employed in PPC mode have 64 holes per well and measure the electrical changes as a population average measurement (Figure 4.40(b)). Although this does not address the issue of lower electrical resistance seals than that available via manual methods, it does decrease the failure rate by decreasing the likelihood that cells are not properly positioned within the microtiter plate. Irrespective of this limitation, however, the ability to use plate-based technology for patch clamp determination represents a significant advance in capacity over manual methods.[88]

GENERAL CONSIDERATION FOR ALL SCREENING METHODS

Irrespective of the *in vitro* screening methods selected for a particular assays system, there are some key factors that must be considered before a drug discovery program moves into a full capacity mode. Cost of the assay per data point is, of course, an important issue, especially as the number of compounds screened increases, but is not necessarily the driving force leading to the choice of one assay platform over another. Data quality has a major impact on assay selection and platform choice. An assay that is cheap to run, but produces uninterruptible data is a waste of time and resources, while an exceptionally robust assay with high costs may be too expensive in both time and money for the examination of the number of compounds necessary to identify a therapeutically relevant compound. This begs the question how does an organization determine the cost and quality of data required in order to effectively pursue an *in vitro* assay?

The question of tolerable cost for any assay is highly dependent on the budget available to an organization, with large pharmaceutical companies able to pursue more costly screening campaigns than academic labs or small biotech operations. In the competitive environment of the modern pharmaceuticals industry, however, it is important for all organization to get as much "bang for the buck" as possible. Understanding the true cost of an assay is of critical importance, and calculating this amount should include an analysis of the number of assay plates, pipette tips, assay reagents, and assay solvents that will be consumed in the course of the assay. Equally important, though occasionally overlooked, are the costs associated with replenishing screening samples for future assays and waste disposal costs. Finally, while modern robotics are capable of operating unattended to some degree, human intervention is often necessary at some point in the process, instruments require routine maintenance to operate smoothly, and someone must be available at the end of the process to evaluate the data generated so that it can be put to practical use. Of course, some of these numbers are difficult to define, so the math as a whole can become a little fuzzy. In the absence of at least some level of analysis of the total cost of an assay, however, it is very difficult to judge whether or not an organization can truly afford to run the assay in question, or if a cheaper method needs to be found (a different assay method, or perhaps a smaller set of compounds screened).

Once the decision has been made to run an assay, especially a high throughput assay, the question of data quality comes into play. In order to compare assays run on different microtiter plates, on different days, or weeks apart, it is necessary to have an understanding of the variability of the data produced by the assay. Benchmark compounds are most often used to measure assay variability and to define the nature of a useful compound. If, for example, an assay designed to identify enzyme inhibitor uses a known benchmark with a reported IC_{50} of 15 nM, then this compound could be included in each 96 (or 384) well plate run of the assay. If each assay run produces a value of close to 15 nM (statistical parameters are set to determine how close is close enough), then the assay is consider to have a low level of variability. If, however, the assay routinely provides data that is all over the map (15 nM in one run, 1060 nM, in another, 75 nM in a third, etc.), then the assay is considered to have a high level of variability and will produce data of low quality. Statistical methods designed to understand the average value (mean), standard deviation (SD), and coefficient of variation ($CV = SD/mean \times 100\%$) are all employed in order to determine the reliability of a data set produced by an *in vitro* screen. These values are used to determine if the signal to noise ratio (S/N) is high enough to make the data from a given assay rise above the variability of the assay itself and thus provide useful data. The term Z' is often used to quantify this information in a more readily understandable fashion. The Z' value for any given assay is defined in the Eqn (4.3) where SD_{max} is the standard deviation of the maximum signal of a positive control, SD_{min} is the standard deviation of the minimum signal of a negative control, $mean_{max}$ is the average of the maximum signal of a positive control, and $mean_{min}$ is the average value of the minimum signal of a negative control.

$$Z' = 1 - 3x \left[(SD_{max} + SD_{min}) / |(Mean_{max} - Mean_{min})| \right] \qquad (4.3)$$

In an ideal assay, the Z' is 1.0, but life is rarely ideal. From a practical standpoint, assay with Z' values between 1.0 and 0.5 are considered acceptable, while assays with a Z' value below 0.5 are marginal and should be expected to produce data with significant variability. Secondary screen to confirm activity or improvements to the assay should be considered when Z' is below the 0.5 threshold.[89]

Another important aspect that must be considered is correlation between assays. It is often the case that multiple assays are employed within a single program in order to determine if a candidate compound will have the desired disease modifying effect. Each assay moves sequentially from the biochemical level, through cellular and tissue levels assessment, and finally to whole animal models. If two assays designed to measure the same overall effect do not point in the same direction, then there is a lack of correlation between assays that must be understood in order to move forward.

Consider, for example, an assay designed to measure changes in membrane potential using a voltage-sensitive dye and a cell line overexpressing the channel of interest. The assay is highly reproducible and has a high degree of precision, but the results provided do not match up with (i.e., correlate with) the results of manual patch clamp methods, the gold standard used for determining the impact of candidate compounds on ion channels. In this case, the primary screen (the voltage-sensitive dye assay) does not correlate with the manual patch clamping methods and cannot be used to select compounds for advancement, even though the primary assay reliably reproduce its results. In this situation, the primary assay is not answering the same question as the secondary assay. A good analogy would be asking a child what is the sum of two plus two. If the child always answers fish, no matter what the circumstance, the child is reliably reproducing the same answer, but not the answer to the question at hand, just as the primary assay described above. An assay that reliably reproduces answers that do not correlate with higher-tiered assay, *in vivo* models, or human clinical data is of little use in drug discovery.

In summary, there is an exceptionally large range of assays that can be explored in order to understand the nature, function, and utility of potential therapeutic compounds. A thorough review of program needs, desired outcomes, costs constraints, and a multitude of other factors should be considered before selecting an assay or set of assay for a drug discovery program. In the final analysis, the ability to design screens for compound activity is limited only by the ingenuity of the scientist, as the tools available for modern drug discovery provide a means to study an extremely wide range of biochemical and cellular events.

QUESTIONS

1. Define the term IC_{50}.
2. What is an EC_{50}?
3. Define the following with respect to GPCRs:
   a. Agonist
   b. Antagonist
   c. Inverse agonist
4. How can high receptor reserve impact *in vitro* screening results?
5. How are streptavidin and biotin used to facilitate screening?
6. Describe how a scintillation proximity assay works.
7. Describe the basic principles of an ELISA assay.
8. What is the difference between a standard FRET assay and a time-resolved FRET assay?
9. What is a reporter gene assay?
10. What is a label free assay system?

References

1. a. Bosch, F.; Rosich, L. The Contributions of Paul Ehrlich to Pharmacology: A Tribute on the Occasion of the Centenary of His Nobel Prize. *Pharmacology* **2008**, *82*, 171–179.
   b. Drews, J. Drug Discovery: A Historical Perspective. *Science* **2000**, *287*, 1960–1964.
   c. Brownstein, M. J. "A Brief History of Opiate, Opioid Peptides, and Opioid Receptors." *Proceeding of the National Academy of Sciences of the United States of America*, 1993, 90, 5391–5393. .
2. 4, 237,224.
3. a. Gordon, J.; Ruddle, F. Integration and Stable Germ Line Transmission of Genes Injected into Mouse Pronuclei. *Science* **1981**, *214* (4526), 1244–1246.
   b. Costantini, F.; Lacy, E. Introduction of a Rabbit β-globin Gene into the Mouse Germ Line. *Nature* **1981**, *294* (5836), 92–94.
4. Lawyer, F.; Stoffel, S.; Saiki, R.; Chang, S.; Landre, P.; Abramson, R.; Gelfand, D. High-level Expression, Purification, and Enzymatic Characterization of Full-length Thermus Aquaticus DNA Polymerase and a Truncated Form Deficient in 5′ to 3′ Exonuclease Activity. *PCR Methods Appl.* **1993**, *2* (4), 275–287.
5. Seethala, R., Zhang, L., Eds. *Handbook of Drug Screening*, 2nd ed.; Informa Healthcare Inc: New York, New York, 2009.
6. Yung-Chi, C.; Prusoff, W. H. Relationship between the inhibition constant (K_i) and the concentration of inhibitor which causes 50 per cent inhibition (IC_{50}) of an enzymatic reaction. *Biochem. Pharmacol.* **1973**, *22*, 3099–3108.
7. Nelson, D. L., Cox, M. M., Eds. *Lehninger Principles of Biochemistry*, W. H. Freema: New York, New York, 2008.
8. Neubig, R.; R., R.; Spedding, M.; Kenakin, T.; Christopoulos, A. International Union of Pharmacology Committee on Receptor Nomenclature and Drug Classification. XXXVIII. Update on Terms and Symbols in Quantitative Pharmacology. *Pharmacol. Rev.* **2003**, *55* (4), 597–606.
9. Zhu, B. T. Mechanistic Explanation for the Unique Pharmacologic Properties of Receptor Partial Agonists. *Biomed. Pharmacother.* **2005**, *59* (3), 76–89.
10. Milligan, G. Constitutive Activity and Inverse Agonists of G Protein-coupled Receptors: a Current Perspective. *Mol. Pharmacol.* **2003**, *64* (6), 1271–1276.
11. Offermanns, S., Rosenthal, W., Eds. *Encyclopedia of Molecular Pharmacology*, 2nd ed.; Springer: Berlin Heidelberg, 2008.
12. a. Groebe, D. R. Screening for Positive Allosteric Modulators of Biological Targets. *Drug Discov. Today* **2006**, *11* (13–14), 632–639.
    b. Epping-Jordan, M.; Le Poul, M.; Rocher, J. P. Allosteric Modulation: a Novel Approach to Drug Discovery. *Innovations Pharm. Technol.* **2007**, *24*, 24–26.
13. Melancon, B. J.; Hopkins, C. R.; Wood, M. R.; Emmitte, K. A.; Niswender, C. M.; Christopoulos, A.; Conn, P. J.; Lindsley, C. W. Allosteric Modulation of Seven Transmembrane Spanning Receptors: Theory, Practice, and Opportunities for Central Nervous System Drug Discovery. *J. Med. Chem.* **2012**, *55*, 1445–1464.
14. Le Trong, I.; Wang, Z.; Hyre, D. E.; Lybrand, T. P.; Stayton, P. S.; Stenkamp, R. E. Streptavidin and its Biotin Complex at Atomic Resolution. *Acta Crystallogr. D* **2011**, *67*, 813–821.
15. Dechancie, J.; Houk, K. N. The Origins of Femtomolar Protein–ligand Binding: Hydrogen Bond Cooperativity and Desolvation Energetics in the Biotin–(Strept) Avidin Binding Site. *J. Am. Chem. Soc.* **2007**, *129* (17), 5419–5429.
16. Moore, K.; Rees, S. Cell-based versus Isolated Target Screening: How Lucky Do You Feel? *J. Biomol. Screen.* **2001**, *6* (2), 69–74.
17. Wu, S.; Liu, B. Application of Scintillation Proximity Assay in Drug Discovery. *Biodrugs* **2006**, *19* (6), 383–392.
18. Glickman, J. F.; Schmid, A.; Ferrand, S. Scintillation Proximity Assays in High-throughput Screening. *Assay Drug Dev. Technol.* **2008**, *6* (3), 433–455.

19. Cook, N. D. Scintillation Proximity Assay: a Versatile High-throughput Screening Technology. *Drug Discov. Today* **1996,** *1* (7), 287–294.
20. a. Yalow, R. S.; Berson, S. A. Immunoassay of Endogenous Plasma Insulin in Man. *J. Clin. Investig.* **1960,** *39* (7), 1157–1175.
    b. Glick, S. Rosalyn Sussman Yalow (1921–2011). *Nature* **2011,** *474* (7353), 580.
21. Van Weemen, B. K.; Schuurs, A. H. Immunoassay Using Antigen-enzyme Conjugates. *FEBS Lett.* **1971,** *15* (3), 232–236.
22. Engvall, E.; Perlman, P. Enzyme-linked Immunosorbent Assay (ELISA). Quantitative Assay of Immunoglobulin G. *Immunochemistry* **1971,** *8* (9), 871–874.
23. Bandi, Z. L.; Schoen, I.; DeLara, M. Enzyme-linked Immunosorbent Urine Pregnancy Tests. Clinical Specificity Studies. *Am. J. Clin. Pathol.* **1987,** *87* (2), 236–242.
24. a. Gnann, J. W., Jr; Michael Oldstone, M. *Hiv-1-Related Polypeptides, Diagnostic Systems Assay Methods;* 1993. EP 0329761B1.
    b. Yamamoto, N. *Diagnostic and Prognostic ELISA Assay of Serum A-N-Acetylgalactosaminidase for Influenza;* 1998. WO 1998030906 A1.
25. Hurst, W. J.; Krout, E. R.; Burks, W. R. A Comparison of Commercially Available Peanut ELISA Test Kits on the Analysis of Samples of Dark and Milk Chocolate. *J. Immunoassay Immunochem.* **2002,** *23* (4), 451–459.
26. Pujol, M. L.; Cirimele, V.; Tritsch, P. J.; Villain, M.; Kintz, P. Evaluation of the IDS One-step ELISA Kits for the Detection of Illicit Drugs in Hair. *Forensic Sci. Int.* **2007,** *170* (2), 189–192.
27. Abbyad, P.; Childs, W.; Shi, X.; Boxer, S. G. Dynamic Stokes Shift in Green Fluorescent Protein Variants. *Proc. Natl. Acad. Sci. U.S.A.* **2007,** *104* (51), 20189–20194.
28. Malus, E. L. *Nouveau Bull de la Societé Philomatique, vol. 1;* 1809. p. 266.
29. Perrin, F. Polarization of Light of Fluorescence, Average Life of Molecules. *J. Phys. Radium* **1926,** *7,* 390–401.
30. Weigert, F. *Verh. d.D. Phys. Ges.* **1920,** *1,* 100.
31. a. Weber, G. Polarization of the Fluorescence of Macromolecules. Theory and Experimental Method. *Biochem. J.* **1952,** *51,* 145–155.
    b. Weber, G. Polarization of the Fluorescence of Macromolecules. 2. Fluorescent Conjugates of Ovalbumin and Bovine Serum Albumin. *Biochem. J.* **1952,** *51,* 155–167.
32. a. Dandliker, W. B.; Feigen, G. A. Quantification of the Antigen-antibody Reaction by the Polarization of Fluorescence. *Biochem. Biophys. Res. Commun.* **1961,** *5,* 299.
    b. Dandliker, W. B.; Halbert, S. P.; Florin, M. C.; Alonso, R.; Schapiro, H. C. Study of Penicillin Antibodies by Fluorescence Polarization and Immunodiffusion. *J. Exp. Med.* **1965,** *122,* 1029.
33. Lea, W. A.; Simeonov, A. Fluorescence Polarization Assays in Small Molecule Screening. *Exp. Opin. Drug Discov.* **2011,** *6* (1), 17–32.
34. Bolger, R.; Checovich, W. A New Protease Activity Assay Using Fluorescence Polarization. *Biotechniques* **1994,** *17* (3), 585–589.
35. Seethala, R.; Menzel, R. A Homogeneous, Fluorescence Polarization Assay for Src-family Tyrosine Kinases. *Anal. Biochem.* **1997,** *253* (2), 210–218.
36. Zhang, M.; Huang, Z.; Yu, B.; Ji, H. New Homogeneous High-throughput Assays for Inhibitors of B-catenin/Tcf Protein–protein Interactions. *Anal. Biochem.* **2012,** *424,* 57–63.
37. Förster, T. Intermolecular Energy Migration and Fluorescence. *Ann. Phys. (Leipzig)* **1948,** *2,* 55–75.
38. Corry, B.; Jayatilaka, D.; Rigby, P. A Flexible Approach to the Calculation of Resonance Energy Transfer Efficiency between Multiple Donors and Acceptors in Complex Geometries. *Biophys. J.* **2005,** *89,* 3822–3836.
39. a. Wang, G. T.; Matayoshi, E.; Huffaker, H. J.; Krafft, G. A. Design and Synthesis of New Fluorogenic HIV Protease Substrates Based on Resonance Energy Transfer. *Tetrahedron Lett.* **1990,** *31* (45), 6493–6496.

b. Matayoshi, E. D.; Wang, G. T.; Krafft, G. A.; Erickson, J. Novel Fluorogenic Substrates for Assaying Retroviral Proteases by Resonance Energy Transfer. *Science* **1990,** *247* (4945), 945–958.

40. Gonzalez, J. E.; Tsien, R. Y. Voltage Sensing by Fluorescence Resonance Energy Transfer in Single Cells. *Biophys. J.* **1995,** *69,* 1272–1280.

41. Ormö, M.; Cubitt, A.; Kallio, K.; Gross, L.; Tsien, R.; Remington, S. Crystal Structure of the Aequorea Victoria Green Fluorescent Protein. *Science* **1996,** *273* (5280), 1392–1395.

42. Shaner, N. C.; Steinbach, P. A.; Tsien, R. Y. A Guide to Choosing Fluorescent Proteins. *Nat. Methods* **2005,** *2,* 905–909.

43. Lohse, M. J.; Nuber, S.; Hoffmann, C. Fluorescence/Bioluminescence Resonance Energy Transfer Techniques to Study G-protein-coupled Receptor Activation and Signaling. *Pharmacol. Rev.* **2012,** *64,* 299–336.

44. Kajihara, D.; Abe, R.; Iijima, I.; Komiyama, C.; Sisido, M.; Hohsaka, T. FRET Analysis of Protein Conformational Change through Position-specific Incorporation of Fluorescent Amino Acids. *Nat. Methods* **2006,** *3,* 923–929.

45. a. Fernández-Dueñas, V.; Llorente, J.; Gandía, J.; Borroto-Escuela, D. O.; Agnati, L. F.; Tasca, C. I.; Fuxe, K.; Ciruela, F. Fluorescence Resonance Energy Transfer-Based Technologies in the Study of Protein–protein Interactions at the Cell Surface. *Methods* **2012,** *57,* 467–472.

b. Song, Y.; Madahar, V.; Liao, J. Development of FRET Assay into Quantitative and High-throughput Screening Technology Platforms for Protein–protein Interactions. *Ann. Biomed. Eng.* **2011,** *39* (4), 1224–1234.

46. Heberle, F. A.; Buboltz, J. T.; Stringer, D.; Feigenson, G. W. Fluorescence Methods to Detect Phase Boundaries in Lipid Bilayer Mixtures. *Biochim. Biophys. Acta* **2005,** *1746,* 186–192.

47. Ibraheem, A.; Campbell, R. E. Designs and Applications of Fluorescent Protein-based Biosensors. *Curr. Opin. Chem. Biol.* **2010,** *14,* 30–36.

48. Hemmila, I.; Webb, S. Time-resolved Fluorometry: An Overview of the Labels and Core Technologies for Drug Screening Applications. *Drug Discov. Today* **1997,** *2* (9), 373–381.

49. a. Alpha, B.; Lehn, J. M.; Mathis, G. Energy Transfer Luminescence of Europium(III) and Terbium(III) Cryptates of Macrobicyclic Polypyridine Ligands. *Angew. Chem. Int. Ed. Engl.* **1987,** *26* (3), 266–267.

b. Petoud, S.; Cohen, S. M.; Bunzli, J. C. G.; Raymond, K. N. Stable Lanthanide Luminescence Agents Highly Emissive in Aqueous Solution: Multidentate 2-Hydroxyisophthalamide Complexes of Sm^{3+}, Eu^{3+}, Tb^{3+}, Dy^{3+}. *J. Am. Chem. Soc.* **2003,** *125,* 13324–13325.

50. Hemmila, I. LANCE™: Homogeneous Assay Platform for HTS. *J. Biomol. Screen.* **1999,** *4* (6), 303–307.

51. Degorce, F.; Card, A.; Soh, S.; Trinquet, E.; Knapik, G. P.; Xie, B. HTRF: A Technology Tailored for Drug Discovery - a Review of Theoretical Aspects and Recent Applications. *Curr. Chem. Genom.* **2009,** *3,* 22–32.

52. a. Carlson, C. B.; Robers, M. B.; Vogel, K. W.; Machleidt, T. Development of Lantha-Screen™ Cellular Assays for Key Components within the PI3K/AKT/mTOR Pathway. *J. Biomol. Screen.* **2009,** *14* (2), 121–132.

b. Robers, M. B.; Machleidt, T.; Carlson, C. B.; Bi, K. Cellular LanthaScreen and B-Lactamase Reporter Assays for High-throughput Screening of JAK2 Inhibitors. *Assay Drug Dev. Technol.* **2008,** *6* (4), 519–529.

53. Legault, M.; Roby, P.; Beaudet, L.; Rouleau, N. *Comparison of LANCE Ultra TR-fret to PerkinElmer's Classical LANCE TR-fret Platform for Kinase Applications;* PerkinElmer Life and Analytical Sciences Application Note: Shelton, CT, 2006.

54. a. Ullman, E. F.; Kirakossian, H.; Switchenko, A. C.; Ishkanian, J.; Ericson, M.; Wartchow, C. A.; Pirio, M.; Pease, J.; Irvin, B. R.; Singh, S.; Singh, R.; Patel, R.; Dafforn, A.; Davalian, D.; Skold, C.; Kurn, N.; Wagner, D. B. Luminescent Oxygen Channeling Assay (LOCI): Sensitive, Broadly Applicable Homogeneous Immunoassay Method. *Clin. Chem.* **1996,** *42,* 1518–1526.

b. Ullman, E. F.; Kirakossian, H.; Singh, S.; Wu, Z. P.; Irvin, B. R.; Pease, J. S.; Switchenko, A. C.; Irvine, J. D.; Dafforn, A.; Skold, C. N. Luminescent Oxygen Channeling Immunoassay: Measurement of Particle Binding Kinetics by Chemiluminescence. *Proc. Natl. Acad. Sci. U.S.A.* **1994**, *91*, 5426–5430.

55. Eglen, R. M.; Reisine, T.; Roby, P.; Rouleau, N.; Illy, C.; Bossé, R.; Bielefeld, M. The Use of AlphaScreen Technology in HTS: Current Status. *Curr. Chem. Genom.* **2008**, *1*, 2–10.

56. Guenat, S.; Rouleau, N.; Bielmann, C.; Bedard, J.; Maurer, F.; Allaman-Pillet, N.; Nicod, P.; Bielefeld-Sévigny, M.; Beckmann, J. S.; Bonny, C.; Bossé, R.; Roduit, R. Homogeneous and Nonradioactive High-throughput Screening Platform for the Characterization of Kinase Inhibitors in Cell Lysates. *J. Biomol. Screen.* **2006**, *11* (8), 1015–1026.

57. a. Hou, Y.; Mcguinness, D. E.; Prongay, A. J.; Feld, B.; Ingravallo, P.; Ogert, R. A.; Lunn, C. A.; Howe, J. A. Screening for Antiviral Inhibitors of the HIV Integrase–ledgf/p75 Interaction Using the AlphaScreen™ Luminescent Proximity Assay. *J. Biomol. Screen.* **2008**, *13* (5), 406–415.

    b. Von Leoprechting, A.; Kumpf, R.; Menzel, S.; Reulle, D.; Griebel, R.; Valler, M. J.; Büttner, F. H. Miniaturization and Validation of a High-throughput Serine Kinase Assay Using the AlphaScreen Platform. *J. Biomol. Screen.* **2004**, *9* (8), 719–725.

58. Taouji, S.; Dahan, S.; Bossé, R.; Chevet, E. Current Screens Based on the AlphaScreen™ Technology for Deciphering Cell Signalling Pathways. *Curr. Genom.* **2009**, *10*, 93–101.

59. Mills, N. L.; Shelat, A. A.; Guy, R. K. Assay Optimization and Screening of RNS-protein Interactions by AlphaScreen. *J. Biomol. Screen.* **2007**, *12*, 946–956.

60. Li, N.; Sul, J. Y.; Haydon, P. G.A Calcium-induced Calcium Influx Factor, Nitric Oxide, Modulates the Refilling of Calcium Stores in Astrocytes. *J. Neurosci.* **2003**, *23* (32), 10302–10310.

61. Zima, A. V.; Blatterm, L. A. Inositol-1,4,5-trisphosphate-dependent Ca^{2+} Signaling in Cat Atrial Excitation–contraction Coupling and Arrhythmias. *J. Physiol.* **2004**, *555* (3), 607–615.

62. Head, J. F.; Inouye, S.; Teranishi, K.; Shimomura, O. The Crystal Structure of the Photoprotein Aequorin at 2.3 a Resolution. *Nature* **2000**, *405*, 372–376.

63. a. Brini, M.; Marsault, R.; Bastianutto, C.; Alvarez, J.; Pozzan, T.; Rizzuto, R. Transfected Aequorin in the Measurement of Cytosolic Ca^{2+} Concentration ($[Ca^{2+}]c$): A Critical Evaluation. *J. Biol. Chem.* **1995**, *270*, 9896–9903.

    b. Prasher, D.; McCann, R. O.; Cormier, M. J. Cloning and Expression of the cDNA Coding for Aequorin, a Bioluminescent Calcium-binding Protein. *Biochem. Biophys. Res. Commun.* **1985**, *126*, 1259–1268.

    c. S Inouye, S.; Noguchi, M.; Sakaki, Y.; Takagi, Y.; Miyata, T.; Iwanaga, S.; Miyata, T.; Tsuji, F. I. Cloning and Sequence Analysis of cDNA for the Luminescent Protein Aequorin. *Proc. Natl. Acad. Sci. U.S.A.* **1985**, *82* (10), 3154–3158.

64. a. Brough, S. J.; Shah, P. Use of Aequorin for G Protein-coupled Receptor Hit Identification and Compound Profiling. *Methods Mol. Biol.* **2009**, *552*, 181–198.

    b. Dupriez, V. J.; Maes, K.; Le Poul, E.; Burgeon, E.; Detheux, M. Aequorin-based Functional Assays for G-protein-coupled Receptors, Ion Channels, and Tyrosine Kinase Receptors. *Recept. Channels* **2002**, *8* (5–6), 319–330.

    c. George, S. E.; Schaeffer, M. T.; Cully, D.; Beer, M. S.; McAllister, G. A High-throughput Glow-type Aequorin Assay for Measuring Receptor-mediated Changes in Intracellular Calcium Levels. *Anal. Biochem.* **2000**, *286* (2), 231–237.

65. Shaw, W. V.; Leslie, A. G. W. Chloramphenicol Acetyltransferase. *Annu. Rev. Biophys. Biophys. Chem.* **1991**, *20*, 363–386.

66. Qureshi, S. A. β-Lactamase: an Ideal Reporter System for Monitoring Gene Expression in Live Eukaryotic Cells. *BioTechniques* **2007**, *42*, 91–96.

67. Thorne, N.; Inglese, J.; Auld, D. S. Illuminating Insights into Firefly Luciferase and Other Bioluminescent Reporters Used in Chemical Biology. *Chem. Biol. Rev.* **2010**, *17*, 646–657.

68. Gorman, C. M.; Moffat, L. F.; Howard, B. H. Recombinant Genomes Which Express Chloramphenicol Acetyltransferase in Mammalian Cells. *Mol. Cell. Biol.* **1982,** 1044–1051.

69. Leslie, A. G. Refined Crystal Structure of Type III Chloramphenicol Acetyltransferase at 1.75 a Resolution. *J. Mol. Biol.* **1990,** *213,* 167–186.

70. Shaw, W. V.; Packman, L. C.; Burleigh, B. D.; Dell, A.; Morris, H. R.; Hartley, B. S. Primary Structure of a Chloramphenicol Acetyltransferase Specified by R. Plasmids. *Nature* **1979,** *282,* 870–872.

71. a. Thomas, R. F.; Holt, B. D.; Schwinn, D. A.; Liggett, S. B. Long-term Agonist Exposure Induces Upregulation of F33-Adrenergic Receptor Expression via Multiple cAMP Response Elements. *Proc. Natl. Acad. Sci. U.S.A.* **1992,** *89,* 4490–4494.

    b. Collins, S.; Bouvier, M.; Bolanowski, M. A.; Caron, M. G.; Lefkowitz, R. J. cAMP Stimulates Transcription of the F82-Adrenergic Receptor Gene in Response to Short-term Agonist Exposure. *Proc. Natl. Acad. Sci. U.S.A.* **1989,** *86,* 4853–4857.

72. Zlokarnik, G.; Negulescu, P. A.; Knapp, T. E.; Mere, L.; Burres, N.; Feng, L.; Whitney, M.; Roemer, K.; Tsien, R. Y. Quantitation of Transcription and Clonal Selection of Single Living Cells with B-Lactamase as Reporter. *Science* **1998,** *279,* 84–88.

73. de Wet, J. R.; Wood, K. V.; DeLuca, M.; Helinski, D. R.; Subramani, S. Firefly Luciferase Gene: Structure and Expression in Mammalian Cells. *Mol. Cell. Biol.* **1987,** *7* (2), 725–737.

74. Vázquez, M. E.; Cebolla, A.; Palomares, A. J. Controlled Expression of Click Beetle Luciferase Using a Bacterial Operator-repressor System. *FEMS Microbiol. Lett.* **1994,** *121* (1), 11–18.

75. Viviani, V. R.; Arnoldi, F. G.; Ogawa, F. T.; Brochetto-Braga, M. Few Substitutions Affect the Bioluminescence Spectra of Phrixotrix (Coleoptera: Phengodidae) Luciferases: a Site-directed Mutagenesis Survey. *Luminescence* **2007,** *22* (4), 362–369.

76. Srikantha, T.; Klapach, A.; Lorenz, W. W.; Tsai, L. K.; Laughlin, L. A.; Gorman, J. A.; Soll, D. R. The Sea Pansy Renilla Reniformis Luciferase Serves as a Sensitive Bioluminescent Reporter for Differential Gene Expression in Candida Albicans. *J. Bacteriol.* **1996,** *178* (1), 121–129.

77. a. Gould, S. J.; Subramani, S. Firefly Luciferase as a Tool in Molecular and Cell Biology. *Anal. Biochem.* **1988,** *175,* 5–13.

    b. Vieites, J. M.; Navarro-García, F.; Pérez-Diaz, R.; Pla, J.; Nombela, C. Expression and *in Vivo* Determination of Firefly Luciferase as Gene Reporter in *Saccharomyces cerevisiae.* *Yeast* **1994,** *10,* 1321–1327.

    c. Gailey, P. C.; Miller, E. J.; Griffin, G. D. Low-cost System for Real-time Monitoring of Luciferase Gene Expression. *BioTechniques* **1997,** *22,* 528–534.

78. Hori, K.; Charbonneau, H.; Hart, R. C.; Cormier, M. J. Structure of Native Renilla Reinformis Luciferin. *Proc. Natl. Acad. Sci. U.S.A.* **1977,** *74* (10), 4285–4287.

79. a. Schroeder, K. S.; Neagle, B. D. FLIPR: A New Instrument for Accurate, High Throughput Optical Screening. *J. Biomol. Screen.* **1996,** *1* (2), 75–80.

    b. Benjamin, E. R.; Skelton, J.; Hanway, D.; Olanrewaju, S.; Pruthi, F.; Ilyin, V. I.; Lavery, D.; Victory, S. F.; Valenzano, K. J. Validation of a Fluorescent Imaging Plate Reader Membrane Potential Assay for High-throughput Screening of Glycine Transporter Modulators. *J. Biomol. Screen.* **2005,** *10* (4), 365–373.

80. a. Menon, V.; Ranganathn, A.; Jorgensen, V. H.; Sabio, M.; Christoffersen, C. T.; Uberti, M. A.; Jones, K. A.; Babu, P. S. Development of an Aequorin Luminescence Calcium Assay for High-throughput Screening Using a Plate Reader, the LumiLux. *Assay Drug Dev. Technol.* **2008,** *6* (6), 787–793.

    b. Choi, Y.; Baek, D. J.; Seo, S. H.; Lee, J. K.; Pae, A. N.; Cho, Y. S.; Min, S. J. Facile Synthesis and Biological Evaluation of 3,3-diphenylpropanoyl Piperazines as T-type Calcium Channel Blockers. *Bioorg. Med. Chem. Lett.* **2011,** *21,* 215–219.

    c. Mori, T.; Itami, S.; Yanagi, T.; Tatara, Y.; Takamiya, M.; Uchida, T. Use of a Real-time Fluorescence Monitoring System for High-throughput Screening for Prolyl Isomerase Inhibitors. *J. Biomol. Screen.* **2009,** *14* (4), 419–425.

81. a. Verdonk, E.; Johnson, K.; McGuinness, R.; Leung, G.; Chen, Y. W.; Tang, H. R.; Michelotti, J. M.; Liu, V. F. Cellular Dielectric Spectroscopy: A Label-free Comprehensive Platform for Functional Evaluation of Endogenous Receptors. *Assay Drug Dev. Technol.* **2006,** *4* (5), 609–620.

    b. Leung, G.; Tang, H. R.; McGuinness, R.; Verdonk, E.; Michelotti, J. M.; Liu, V. F. Cellular Dielectric Spectroscopy: A Label-free Technology for Drug Discovery. *J. Lab. Autom.* **2005,** *10,* 258–269.

82. a. Lee, P. H.; Gao, A.; van Staden, C.; Ly, J.; Salon, J.; Xu, A.; Fang, Y.; Verkleeren, R. Evaluation of Dynamic Mass Redistribution Technology for Pharmacological Studies of Recombinant and Endogenously Expressed G Protein-coupled Receptors. *Assay Drug Dev. Technol.* **2008,** *6* (1), 83–94.

    b. Fang, Y.; Frutos, A. G.; Verkleeren, R. Label-free Cell-based Assays for GPCR Screening. *Comb. Chem. High Throughput Screen.* **2008,** *11,* 357–369.

83. O'Malley, S. M.; Xie, X.; Frutos, A. G. Label-free High-throughput Functional Lytic Assays. *J. Biomol. Screen.* **2007,** *12* (1), 117–126.

84. Fan, X.; White, I. M.; Shopova, S. I.; Zhu, H.; Suter, J. D.; Sun, Y. Sensitive Optical Biosensors for Unlabeled Targets: A Review. *Anal. Chim. Acta* **2008,** *620,* 8–26.

85. Zeng, S.; Yong, K. T.; Roy, I.; Dinh, X. Q.; Yu, X.; Luan, F. A Review on Functionalized Gold Nanoparticles for Biosensing Applications. *Plasmonics* **2011,** *6* (3), 491–506.

86. a. Neher, E.; Sakmann, B. Single-channel Currents Recorded from Membrane of Denervated Frog Muscle Fibres. *Nature* **1976,** *260,* 799–802.

    b. Hamill, O. P.; Marty, A.; Neher, E.; Sakmann, B.; Sigworth, F. J. Improved Patch-clamp Techniques for High-resolution Current Recording from Cells and Cell-free Membrane Patches. *Pflügers Archiv Eur. J. Physiol.* **1981,** *391* (2), 85–100.

87. Schroeder, K.; Neagle, B.; Trezise, D. J.; Worley, J. IonWorks™ HT: A New High-throughput Electrophysiology Measurement Platform. *J. Biomol. Screen.* **2003,** *8* (1), 50–64.

88. a. Dale, T. J.; Townsend, C.; Hollands, E. C.; Trezise, D. J. Population Patch Clamp Electrophysiology: a Breakthrough Technology for Ion Channel Screening. *Mol. Biosyst.* **2007,** *3* (10), 714–722.

    b. John, V. H.; Dale, T. J.; Hollands, E. C.; Chen, M. X.; Partington, L.; Downie, D. L.; Meadows, H. J.; Trezise, D. J. Novel 384-Well Population Patch Clamp Electrophysiology Assays for Ca^{2+} Activated K^+ Channels. *J. Biomol. Screen.* **2007,** *12* (1), 50–61.

89. Zhang, J. H.; Chung, T. D. Y.; Oldenburg, K. R. A Simple Statistical Parameter for Use in Evaluation and Validation of High Throughput Screening Assays. *J. Biomol. Screen.* **1999,** *4* (2), 67–73.

Medicinal Chemistry

The ultimate goal of the vast majority of drug discovery and development programs is the identification of a single, novel compound capable of altering the outcome of a biological process. Whether the biological process is pathological, such as cancer, or a lifestyle issue, such as male pattern baldness, the end goal is essentially the same, a marketable therapeutic entity. In the previous chapters, common biochemical targets and methods of monitoring biological activity were described. While these topics are important to the process, knowledge in these areas alone is insufficient to drive a drug discovery program. If these were the only tools available, the identification of compounds with interesting biological properties would be possible, but moving beyond the initial starting set of biologically active compounds would be a difficult task. In order to move beyond initial starting points, such as literature leads, competitor therapeutics, or *in vitro* screening hits, an additional set of skills and knowledge is required. Program scientists must understand the relationship between the structures of biologically interesting compounds and their properties, both biological and physicochemical. In addition, they must have an understanding of the impact that subtle changes in chemical structure can have on these properties. In other words, knowledge of medicinal chemistry is a key component of drug discovery and development.

The American Chemical Society's division of medicinal chemistry defines medicinal chemistry as "the application of chemical research techniques to the synthesis of pharmaceuticals." This rather restrictive definition implies that the field of medicinal chemistry is the sole purview of those capable of synthesizing compounds, i.e., synthetic chemists. In reality, however, this is not the case. A more accurate description of medicinal chemistry is a field focused on understanding the chemical basis of the biological effects of compounds by integrating fundamental concepts from different fields such as synthetic chemistry, biochemistry, pharmacology, physiology, and molecular biology (Figure 5.1). Certainly, some of the team members in a drug discovery program must be well versed

FIGURE 5.1 Medicinal chemistry applies knowledge derived from synthetic chemistry, biochemistry, pharmacology, physiology, and molecular biology in an effort to understand the biological impact of candidate compounds. It is an interdisciplinary science whose primary goal is the identification of therapeutically useful compounds.

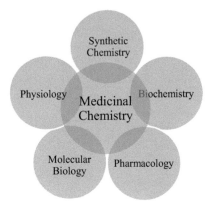

in synthetic chemistry so that new molecules can be prepared, but this is not a requirement for contribution to medicinal chemistry efforts. To be clear, an in-depth knowledge of how to make compounds for screening is not a prerequisite for being able to understand and contribute to the design aspects of medicinal chemistry, and it is the design elements of this process that enable a program to move forward toward a new therapeutic entity. This, of course, begs the question what are the design aspects of medicinal chemistry and how are they applied as a program moves forward?

STRUCTURE–ACTIVITY RELATIONSHIPS AND STRUCTURE–PROPERTY RELATIONSHIPS

A deceptively simple answer to this question begins with the premise that a compound's biological properties are a function of its chemical structure. In addition, in a given series of related compounds, structural changes across the series will lead to predictable changes in biological properties. This relationship is referred to as a structure–activity relationship (SAR). The impact of structural changes on biological properties such as binding potency, functional activity, and selectivity can be monitored in order to establish structure–activity relationships for each property. In principle, program scientists simply need to understand relationships between chemical structure and biological activity in a series of related compounds in order to identify the compound with the most promising biological properties and design compounds with improved properties.

In a similar manner, the physicochemical properties of a series of related compounds are tied to the structure of the compounds. Changes in structure will lead to changes in physicochemical properties such as solubility, membrane permeability, lipophilicity, and total polar surface area.

The relationship between a series of compounds and their physicochemical properties is referred to as structure–property relationships. In theory, one needs to only understand the relationship between chemical structure and physical properties in order to identify a compound with optimal properties.

Pharmacokinetic (PK) properties such as absorption, distribution, metabolism, and excretion, all of which are useful in describing how the body deals with a compound *in vivo*, are also all a function of chemical structure. These properties will be discussed in greater detail in Chapter 6, but at this juncture, it is worth noting that changes in chemical structure will lead to changes in these properties that can be predicted using the same principles noted for structure–activity relationship and structure–property relationships. Knowledge of the impact of structural changes on these properties should provide the ability to quickly identify compounds with optimal pharmacokinetic properties.

In practice, none of these principles are as simple as they appear on paper. As discussed in earlier chapters, biologically relevant macromolecular targets are complex, three-dimensional structures whose configurations are controlled by a multitude of molecular interactions (e.g., hydrogen bonds, salt bridges, hydrophobic interactions, etc. See Chapter 3). These interactions give rise to their function and, more importantly within the context of drug discovery, afford-binding sites that are critical to their function. Binding of a ligand, natural or otherwise, requires that the ligand interact with a macromolecule at a preferred binding site, employing the same forces that control the configuration of the macromolecule as a whole. As such, in order to have a *complete* understanding of the structure–activity relationship between a target macromolecule and a series of related compounds, one must understand how these interactions change as subtle structural changes are applied across a series of potential ligands. Understanding these interactions is further complicated by the fact that macromolecules are not static entities. Binding of a ligand to a macromolecule can induce changes in the binding site that create additional favorable interactions and increase binding strength. It is also important to note that in any given chemical series, it may be possible to access hundreds, if not thousands, of possible candidate compounds. It is neither economically feasible nor practical to prepare and screen more than a small fraction of chemical space within a given series.

Fortunately, it is not necessary to have a *complete* picture of all aspects of the structure–activity relationships for a series of compounds against a given target in order to understand how one can manipulate biological properties by making subtle changes in structure across the series. A structure–activity relationship can be viewed as a jigsaw puzzle. Long before all of the pieces are put into place, it is possible to anticipate the full image that will be provided upon completion of a jigsaw puzzle. In the same sense, a small series of compounds with appropriately applied structural changes can provide a wealth of information. These data can then be used to predict biological properties of potential candidate compounds and direct program choices toward desirable properties and away from potential dead ends. These concepts are best explained through simplified examples in which one assumes that the structure of the macromolecular target is static and that the overall structure of the series of compounds in question are identical in all respects, except in the areas indicated in each example.

Consider, for example, a class of compounds that contains a substituted aromatic ring in the same position, but are otherwise identical in structure (Figure 5.2). If the aromatic ring in question is directed into solvent space

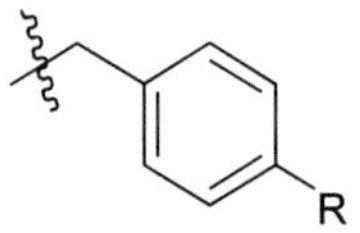

Entry	R	IC_{50} (nM)	Entry	R	IC_{50} (nM)
1	Methoxy	10000	4	Fluoro	50
2	Methyl	1000	5	CN	20
3	Chloro	100	6	NO_2	2

FIGURE 5.2   Altering the substituents on an aromatic ring can have a significant impact on biological activity. In this hypothetical set of compounds, potency at the biological target increases as the benzene ring becomes increasingly electron poor. These data suggest that additional electron-donating R-groups such as an NH_2 would have lower potency than R-groups that are electron withdrawing, such as a CF_3.

(that is, outside of the binding pocket), then changes in the substitution pattern of the aromatic ring would be expected to have little impact on the potency of the compounds. On the other hand, if the aromatic ring is positioned toward the binding pocket, differences in substitution patterns on the aromatic ring could have a significant impact on potency. In this hypothetical example, potency at the target is strongly influenced by the electronic nature of the R-group. Electron-donating substituents, such as methyl and methoxy, decrease potency, while electron-withdrawing groups, such as chlorine, bromine, or trifluoromethyl, increase potency. Using this structure activity information, one would conclude that other electron-donating groups in this position will not provide potent compounds, while alternative electron-withdrawing groups would be likely to provide compounds with a high degree of potency. For example, an analog wherein the R-group is a trifluoromethyl group ($-CF_3$), an electron-withdrawing substituent, would be predicted to be relatively potent in this assay. Importantly, if one were interested in identifying additional analogs in this series of compounds by changing other areas of the molecule, the above structure–activity relationship analysis would suggest that new compounds should contain aromatic rings with electron-withdrawing groups in the position shown.

In a similar manner, physical restrictions of a ligand-binding site can be elucidated by examining binding data. As in the previous case, consider a series of compounds that are identical, except for the presence of a variable R-group on an aromatic ring. In this case, however, rather than altering the electronic nature of the R-group, a series of increasingly sterically demanding substituents are employed (Figure 5.3). The blue line represents the outer

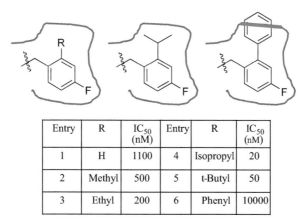

Entry	R	IC_{50} (nM)	Entry	R	IC_{50} (nM)
1	H	1100	4	Isopropyl	20
2	Methyl	500	5	t-Butyl	50
3	Ethyl	200	6	Phenyl	10000

FIGURE 5.3   In any given ligand-binding site, the "walls" of the binding site are comprised of the backbone and side chain features of the macromolecular target. In this example, the outer perimeter of the biding cavity is designated by the blue line. Binding of the hypothetical ligand becomes more efficient as the R-group fills the available space and hydrophobic interaction increases (entries 2–5). Exceeding the size of the cavity (entry 6), however, will cause a dramatic loss in binding efficiency, as the candidate compound no longer fits within the allowed space (as indicated by the red line).

boundary of a hydrophobic pocket of a binding site. Increasing the size of the lipophilic substituent increases hydrophobic interactions between the ligand and the binding pocket, increasing binding strength as long as the R-group does not exceed the size of the lipophilic space. Once the ligand exceeds the size limit of the hydrophobic pocket, binding potency falls off rapidly. The compound no longer fits within the binding site, as indicated by the red line in the hypothetical binding pocket in Figure 5.3. With this information in hand, one can reasonably predict that compounds in which the R-group is an aromatic ring will be significantly less potent than those where smaller substituents are employed, irrespective of the nature of the appended aromatic ring. There is not enough room in the binding site to accommodate larger groups in this position.

The placement and orientation of functionality capable of forming hydrogen bonds can also have a significant impact on ligand binding, and an analysis of binding data within the framework of a structure–activity relationship can provide a great deal of insight when considering the design

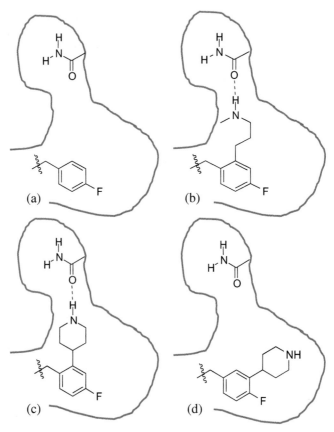

FIGURE 5.4   In this hypothetical scenario, the binding site of a protein is defined by the blue line. In (a), a 4-fluorobenyl substituent occupies a small portion of the binding site. Addition of an amine side chain meta to the fluorine atom (b) allows the hypothetical ligand to form a hydrogen bond with an amide side chain, strengthening the binding interaction. Replacing the propyl amine side chain with a piperidine ring (c) adds rigidity to the ligand while maintaining the newly accessible hydrogen bonding interaction. Relocating the piperidine ring so that it is ortho to the fluorine atom (d), however, prevents the ligand from accessing the amide hydrogen bond partner, as the orientation of the ligand in the binding pocket no longer favors this interaction.

of next-generation compounds. Once again, a hypothetical series of compounds that are identical with the exception of the substitution pattern of an aryl ring (Figure 5.4) is considered. In this case, an amide side chain of an amino acid in the binding site is available to form a hydrogen bond with a suitably functionalized ligand. The monofluoronated species (a) is incapable of forming a hydrogen bond, and thus derives no additional binding strength from the presence of the amide side chain. If, however, the ligand also contained a properly positioned propyl amine side chain (b), the resulting compound would be able to form the hydrogen bond indicated, strengthening the interaction between the ligand and the binding pocket. Knowledge of the increased binding energy, as indicated by IC_{50} screening data against the target of interest, would suggest that replacing the propyl amine side chain with a piperidine (c) would provide a compound of similar potency, as both compounds would provide the framework necessary for the formation of a critical hydrogen bond. In contrast, one could predict that relocating the amine hydrogen bond donor ortho to the fluorine atom (d) would position the potential hydrogen bond donor away from the hydrogen bond acceptor (amide). No significant increase in binding strength would be expected in this scenario.

THE ROLE OF CHIRALITY

All of the aforementioned examples describe the relationship between structural changes and biological activity in a two-dimensional framework. In reality, of course, biological activity, cellular function, and life in general occur in three-dimensional space. Macromolecular targets that are the focus of drug discovery and development efforts possess complex three-dimensional structures. This forms the basis of their ability to differentiate between potential ligands, even in cases where a pair of compounds are simply mirror images of each other. As such, a discussion of structure–activity relationships must include an examination of how changes in the three-dimensional structure of a series of compounds impact biological activity.

The concept of chirality is fundamental to understanding structure–activity relationships and was originally postulated by Lord Kelvin more than 100 years ago. He stated "I call any geometrical figure, or group of points, 'chiral', and say that it has chirality if its image in a plane mirror, ideally realized, cannot be brought to coincide with itself."[1] Although this statement applies to any arrangements of materials, within the context of drug discovery the most common occurrence of chirality is a tetrahedral carbon atom with four different substituents (Figure 5.5). In this scenario, the substituents can be

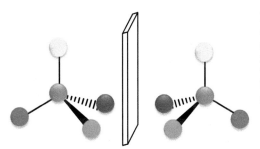

FIGURE 5.5 Chiral compounds cannot be superimposed on their mirror image. The two isomers are referred to as enantiomers (R and S). Chirality often plays a key role in biological activity, as nature is a chiral environment.

attached to the central carbon atom in one of two arrangements, which are non-superimposable mirror images of each other. The two separate isomers are referred to as enantiomers or optical isomers, and in the absence of a chiral environment, they are identical in all respects, except for the direction in which they rotate plane polarized light. Biological macromolecules, being chiral, offer an environment that predictably discriminates between enantiomers, leading to dramatic changes in the functional activity between pairs of enantiomers. Darvon[2] and Novrad[2a,3] (Figure 5.6), for example, are identical

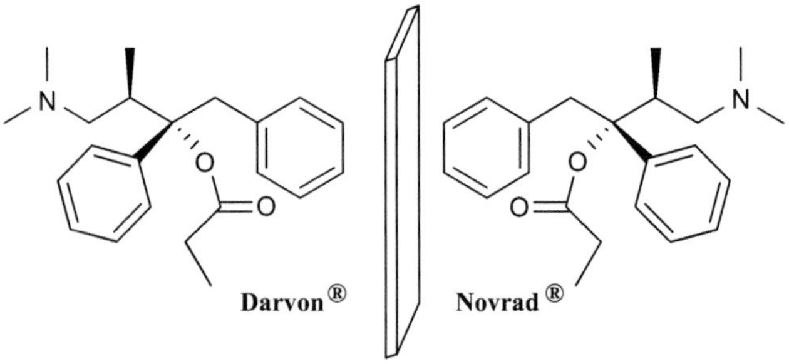

FIGURE 5.6   Darvon® and Novrad® are identical, save for the nature of the chiral center present in each. The enantiomers have very different properties in a biological setting. Darvon® is an analgesic that activates the μ-opioid receptor, while Novrad® is an antitussive agent with minimal efficacy as an analgesic. Unlike its enantiomer, Novrad® has little affinity for the μ-opioid receptor.

except for the difference in chirality, and this leads to substantially different biological activity. The former is a potent analgesic that acts as a μ-opioid receptor agonist,[4] while the latter is an antitussive with minimal potency at the μ-opioid receptor.[5] This underscores the importance of how a deceptively simple change in molecular structure, the inversion of a chiral center, can have a significant influence on therapeutic action. In fact, a single change at a chiral center is a drastic change in three-dimensional space and is certainly not viewed as a subtle change to a medicinal chemist. From the perspective of a drug discovery scientist, there is also the underlying question of how one could employ knowledge of the impact of chiral inversions in order to design better candidate compounds. In other words, what is the relationship between structural changes induced by chiral inversion and biological activity?

An understanding of the impact of chiral inversion on biological activity, and thus understanding the role of chirality in structure–activity relationships, is best accomplished through the examination of a practical example. Consider, for example, Tamiflu® (oseltamivir), an influenza neuraminidase inhibitor pro-drug useful for the treatment of the common flu.[6] X-ray crystallographic studies of this compound in the binding site of influenza

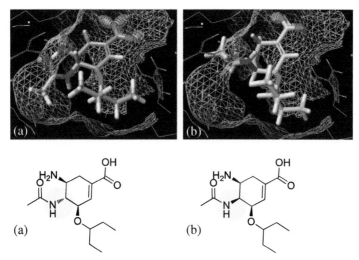

FIGURE 5.7   (a) The active agent derived from Tamiflu® (oseltamivir) occupies the binding site of influenza neuraminidase, the boundaries of which are delineated by the wire mesh. (b) Inversion of the amide chiral center causes the acetamide to protrude from the binding cavity in this model, demonstrating the importance of the orientation of this chiral center (RCSB 2QWK).

neuraminidase (Figure 5.7(a)) portray a tight-fitting interaction, which explains its potent impact on enzymatic activity. The translucent gray surface represents the outer edges of the binding site, marking the physical limits for potential inhibitors of this enzyme.[7] Hypothetically, one could invert a single chiral center of this compound and produce an alternative structure that, in all other respects, is structurally identical to the active component of Tamiflu®. As indicated in Figure 5.7(b), however, this modification has a significant impact on the interaction of the hypothetical ligand with the binding site. Rather than fitting within the confines of the binding site, the proposed ligand protrudes through the theoretical boundary defined by the atoms of the binding site. The negative interaction produced by the "bumping" of atoms of the ligand and the target have a negative impact on binding interactions, which in turn decreases biological potency.

In the case of Tamiflu®, knowledge of the X-ray structure and the availability of advanced protein modeling techniques can provide insight into how theoretical compounds might fit into the binding site of influenza neuraminidase, providing guidance for prioritization. If an X-ray structure is not available, however, one can still apply design principles in order to predict activity and prioritize compound preparation and screening. Comparison of *in vitro* screening data for a series of related compounds that are identical except for the orientation of a single chiral center, for example, would provide indirect knowledge of the limitation of the ligand-binding site. For the purposes of illustration, consider a series of compounds with an embedded amino acid

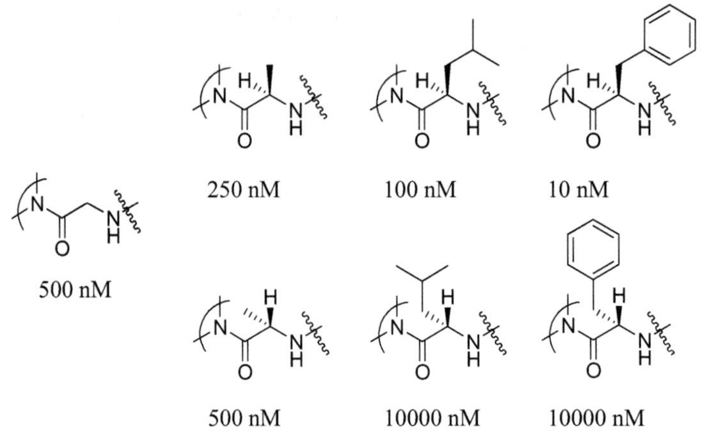

250 nM   100 nM   10 nM

500 nM

500 nM   10000 nM   10000 nM

FIGURE 5.8 In a series of otherwise identical compounds, differences in chirality can have a significant impact on biological activity. Although an X-ray crystal is a valuable tool in identifying areas where changes in chirality will have the largest impact, it is not required. In this hypothetical example, potency differences between the R-series (top) and the S-series (bottom) clearly demonstrate that the R-series is heavily favored.

(Figure 5.8). When the embedded amino acid is a glycine moiety, there is no chirality at the α-carbon, as both substituents are hydrogen. Incorporation of either naturally occurring alanine or its enantiomer, however, renders the molecule chiral, as would the incorporation of the enantiomers of leucine or phenylalanine. A comparison of the screening results from this set of seven compounds indicates that the ligand-binding site strongly favors one of the two possible enantiomers. *In vitro* potency improves with increased steric bulk in one enantiomeric series (the R-series, top), but the same changes in the opposite enantiomeric series (the S-series, bottom) leads to significant loss of potency. With these screening results in hand, one could reasonably predict that compounds with the S-configuration at this chiral center would be significantly less potent than the R-series and should be deprioritized when considering potential candidate compounds.

PUSH AND PULL IN STRUCTURE–ACTIVITY RELATIONSHIPS

The examples discussed so far are all based on the assumption that changes to the candidate compound are isolated to one location within the molecular framework. In practice, however, this is not always the case. As drug discovery programs evolve through the process of identifying a hit in an assay, developing a lead series, and finally choosing a clinical candidate, multiple changes to the molecular framework of the original hit will be examined. These changes often produce similar compound classes that may have similar SAR patterns. It is important to keep in mind, however, that "similar" is not "same." Given two related, but ultimately different compound classes, there may be significant overlap between the SAR patterns of the two classes of compounds, but there are also likely to be differences. If, for example, a series of imidazopyridines (Figure 5.9(a)) had

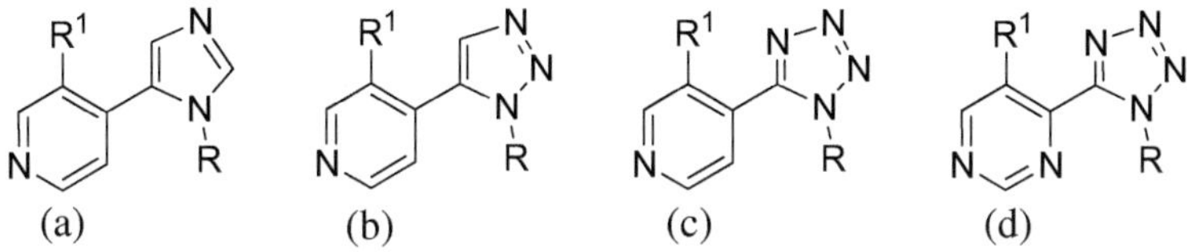

(a)   (b)   (c)   (d)

FIGURE 5.9 Compounds (a), (b), (c), and (d) have similarities that could lead to overlapping SAR, but the increasing number of nitrogen atoms present in each structure changes the character of the aromatic rings. If the aromatic rings' sole function is structural support for the functionality required for binding, then SAR patterns may be comparable. On the other hand, if the aromatic rings are critical to binding, then changes in nitrogen counts may lead to very different SAR patterns for each series of compounds.

a known SAR pattern of activity as kinase inhibitors, it would be reasonable to expect that the related triazolopyridines series (Figure 5.9(b)) might also be active against the same kinase. Similar SAR patterns would likely emerge in the triazolopyridines class, but there may be differences as a result of the presence of an additional nitrogen atom in the five-membered ring. The corresponding tetrazolopyridines (Figure 5.9(c)) are also similar to the original lead series, as are the tetrazolopyrimidines (Figure 5.9(d)), but each step away from the original compound class changes the chemical nature of the compound class and can result in unexpected changes is SAR patterns. Making the assumption that the SAR patterns of similar compound classes will be identical is a risky proposition. It is well worth the effort to test the limits of the overlap of SAR patterns between related series.

QUANTITATIVE STRUCTURE–ACTIVITY RELATIONSHIPS

Identifying structure activity trends is an integral part of the drug discovery process, but ultimately this aspect of the process is inherently limited by the number of compounds that can be prepared in the lifetime of a discovery campaign. Although it may be possible to generate several hundred, or perhaps even a few thousand analogs in any given series, the theoretical number of compounds that could be prepared is virtually unlimited. Efforts to merge the concepts of structure–activity relationships with computational methods and growing computer power led to the development of quantitative structure–activity relationship (QSAR) models. In simple terms, QSAR models represent an attempt to mathematically quantify molecular properties and correlate changes in these properties to changes in biological activity.

This information is then used to predict the impact of potential alterations in the molecular structure on the biological activity of potential candidate compounds. In theory, if one could develop a QSAR model capable of accurately predicting biological activity in a given series of compounds, then the computational power of modern computer systems could be employed to determine which compounds to prepare, greatly simplifying the selection process.

In practice, QSAR models require that molecular properties be distilled to mathematical equations, and this is not easily accomplished. Early efforts to develop mathematical relationships between structure and activity include the Hammett equation. Originally proposed in 1937 by Louis Plack Hammett,

$$\log (K/Ko) = \rho\sigma \qquad (5.1)$$

Equation (5.1): The Hammett equation: K and K_0 are equilibrium constants for the hydrolysis of benzoic acid derivatives, where K is the value for a substituted benzoic acid and K_0 is the value for the unsubstituted reference compound, ρ is a reaction constant, and σ is a substituent constant that is dependent only on the electronic effects of the substituent.

$$\log (K/Ko) = \rho * \sigma * + \delta E_s \qquad (5.2)$$

Equation (5.2): The Taft Equation: Ko and K are the equilibrium constants for a reference reaction and a reaction in which substituents are present, $\sigma*$ describes the electrical field and inductive effective of a substituent in the reaction, and $\rho*$ is a sensitivity constant that defines the impact of polar effects on the rate of the reaction. Steric effects in the reaction are defined by the term E_s, and δ is a sensitivity factor that considers the sensitivity of the reaction to steric effects.

$$\log (1/C) = a\log P + b(\log P)^2 + \rho\sigma + \delta E_s + d \qquad (5.3)$$

Equation (5.3): The Hansch equation: The terms ρ, σ, δ, and E_s are defined in the same manner as in Eqns (5.1) and (5.2); logP is the octanol/water partitioning coefficient; and a, b, and d are mathematical constants derived through linear regression analysis of experimental data. The concentration at which a biological response is being measured for data analysis is defined by the term "C."

this formula proposes that the rate of hydrolysis of a benzoic ester is a function of the electronic effects of the substituents of a benzene ring according to Eqn (5.1).[8] Although limited in its scope and not directly applicable to drug discovery, the concepts proposed by Hammett were considered revolutionary at the time. Robert W. Taft modified the Hammett equation in 1952 by incorporating the concept of steric effects in an attempt to broaden the applicability of the equation (Eqn (5.2)),[9] but many additional factors remained outside the realm of this model. Corwin Hansch, often referred to as the father of QSAR, was one of the early pioneers in developing mathematical representations of the interactions of drugs and their biological targets. He was the first scientist to demonstrate that biological activity could be quantitatively related to physicochemical characteristics of compounds (Eqn (5.3)). His equations built upon the work of Hammett and Taft, but he added concepts such as lipophilicity (logP) and "active concentration" (C) in order to improve the utility of the formula for predictive purposes.[10]

Of course, the complexity of the equations required to predict the molecular properties of candidate compounds increases with each parameter added, and there are many to consider. Additional molecular factors that could be considered as part of an effort to predict biological and physicochemical attributes of candidate compounds include molecular weight, total polar surface area, interatomic distance and angles, potential for hydrogen bonding (as a donor or acceptor), and three-dimensional shape just to name a few. In the absence of the appropriate computer technology and software, it would be nearly impossible to perform the necessary calculation on a sufficient number of compounds within the time frame of the typical drug discovery project. Up until the end of the twentieth century and the beginning of the twenty-first century, computer capacity and software technology limited the translation of computational data into drug discovery efforts and compound design.

Efforts to apply the principles of QSAR to drug design on a practical level began with the development of "decision trees" designed to guide compound design. Rather than preparing a large number of analogs, one could explore the SAR parameters of a particular compound class by following the guide points set in the decision tree. The Topliss Scheme, for example, was introduced in 1970[11] as a means of systematically examining the impact of changing aromatic substation on the biological activity of candidate compounds (Figure 5.10). This system takes into consideration factors such as inductive electronic effects and lipophilicity, and begins with the unsubstituted aromatic ring. Modification of the structure by incorporation of a chlorine atom in the 4-position followed by biological screening provides data for the first decision point. At this juncture, increased activity would suggest further modification of the aromatic ring by incorporation of a second chlorine atom in the 3-position, while decreased activity would suggest that incorporation of a 4-methoxy substituent might lead to an improved outcome. If biological activity remains the same (within a reasonable margin), then the Topliss tree path indicates that the next analog of interest is the 4-methyl analog. As each analog in the tree is prepared and screened, a new decision point is reached and new compounds are suggested. Decision trees of this type are based on empirical observations across multiple data sets, but are an imperfect guide for molecular design strategies. Subtle differences between compound classes can lead to deviations from the expected results. The advantages of not requiring computational analysis in order to decrease the potential

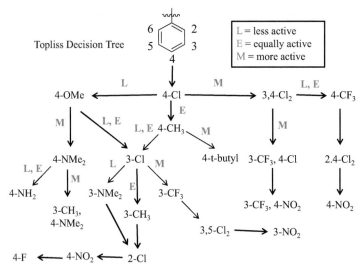

FIGURE 5.10 The Topliss Decision Tree is designed to facilitate the decision-making process involved in the identification of biologically active compounds containing functionalizable benzene rings. Systematic changes to an unsubstituted benzene ring (the top of the tree) based on changes in observed biological activity guide compound selection toward one set of substituents and away from others. At the first decision point, for example, if incorporation of a chlorine atom in the 4-position leads to decreased potency, the next analog to prepare should be a 4-methoxy analog. The next set of biological data would then trigger a choice at the next branch point in the tree, leading to either a 3-chloro analog or a 4-N,N-dimethyl analog. In theory, each step through the tree should lead to an increase in potency.

number of analogs needed in order to establish an SAR pattern, however, cannot be understated.

Over the last several decades, molecular modeling software and computing power has increased substantially. Modern QSAR methods incorporate numerous mathematical algorithms that are designed to incorporate a wide range of physicochemical properties. Software packages such as Tripos Sybyl-X have become substantially more user friendly, and for many tasks, it is no longer necessary to have an advanced understanding of computational chemistry in order to apply the tools necessary to predict changes in biological activity, provided enough structural data are available. For example, potential analogs of a known lead compound can be created *in silico*, energy minimized through the click of a mouse (and an automated algorithm designed for this purpose), and then compared to the lead compound. Physical overlaps can be observed, electrostatic surfaces can be compared, and positions of hydrogen bond donor and acceptors can be examined to determine the differences and similarities between compounds. Automated systems capable of comparing thousands, if not millions, of compounds with a single lead compound and ranking them for similarity to the lead compound are available. *In silico* screening of compounds can be accomplished in a relatively short time frame (hours).

The development of sophisticated computer modeling designed to provide a more comprehensive understanding of the relationship between molecular properties and biological activity also provided scientists with the ability to visualize compounds in the way that macromolecular targets "see" them. On paper, scientists use lines and juncture points to describe molecules, both large and small, but in reality, potential therapeutic entities and macromolecular targets "see" each other as something far more complicated than simply a collection of atoms joined together. They "see" overall three-dimensional shapes with contoured regions of high electron density and low electron density, areas of high and low lipophilicity, and hydrogen bond forming potential (hydrogen bond donors or acceptors). The visualization tools developed as a consequence of advances in QSAR modeling provide scientists with the ability to generate three-dimensional images of compounds and view them in the way that they are "seen" by their intended biological targets. Substantially different compounds that modulate the same target could be more readily compared to gain a better understanding of similarities between the apparently different classes of compounds that generate the same biological response. The selective serotonin reuptake inhibitors sertraline,[12] citalopram,[13] fluoxetine,[14] for example, have very different chemical structures, yet they interact with the same biological target (Figure 5.11). When viewed through the

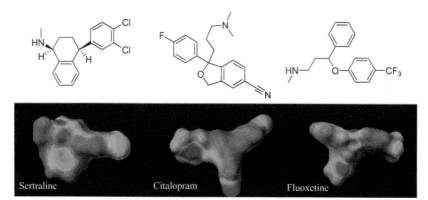

FIGURE 5.11 In order to facilitate communicating structural information, scientists typically draw compounds using line structures such as those portrayed for the three selective serotonin reuptake inhibitors (SSRI) sertraline, citalopram, and fluoxetine. It is important to remember, however, that this representation is only a shadow of the full structure of the molecule. A more accurate depiction can be created using advanced computer software and graphical display systems that can provide renderings of molecules that display features beyond the simple connectivity of the atoms. In this case, electrostatic potential surface maps are superimposed over the three-dimensional structures of the SSRIs in order to visualize the similarities and differences. Images of this type can help explain why compounds with apparently different structures can behave in a similar manner within the context of a biological system.

lens of QSAR modeling, however, similarities between the three compounds become apparent. Overlaps between regions of high and low electron density become apparent, lipophilic regions can be compared, and the orientation of hydrogen bond donors and acceptors provides a greater understanding of the relationship between these three compounds. Thus, QSAR modeling provides the scientific basis for understanding why compounds with dissimilar chemical structures interact with the same macromolecular targets by providing a more comprehensive view of candidate compounds. In simpler terms, compounds with very different chemical structures can modulate the same biological target if they "look" similar to each other through the QSAR lens.

THE PHARMACOPHORE

Another aspect of structure–activity relationships considers the relative importance of various sections of an overall molecular framework, rather than individual changes to a series of lead compounds. In many cases, a binding interaction between a lead compound and its macromolecular target requires only a subset of the atoms and functionalities available within the molecular framework of the lead compound. These sections of the molecule are referred to as the pharmacophore of the structure, and changes in these parts of the molecule can have a dramatic impact on the strength of binding interactions and biological activity. The official definition provided by the International Union of Pure and Applied Chemistry (IUPAC) states that a pharmacophore is "an ensemble of steric and electronic features that is necessary to ensure the optimal supramolecular interactions with a specific biological target and to trigger (or block) its biological response."[15]

The remainder of the compound, portions of which may provide structural support for the position of the functionality within the pharmacophore but play no direct role in binding events, is often referred to as the auxophore. In principle, since the auxophore is not directly involved in binding events, changes to this section of a potential lead compound should have minimal impact on binding interactions, provided that the changes do not impact the overall shape of the pharmacophore. These sections of the compound may be oriented away from the binding surface of the macromolecule, so changes in these regions, including excision from the lead compound, can often be accomplished while maintaining biological activity. It is also possible to take advantage of auxophore space for the purposes of altering physical properties of a compound. Consider, for example, a compound with poor solubility that contains a side chain that points away from the binding surface of the target of interest and into solvent space (i.e., water). Since this side chain is outside of the pharmacophore, changes to this side chain such as the addition of a basic amine should have little impact on the desired biological activity. It is, however, likely that the addition of a basic amine will have a positive impact on solubility, thereby improving a physical property of the compound without negatively impacting biological function.

Understanding the distinction between the pharmacophore and the auxophore can provide significant design opportunities. Consider, for example, the four compounds: morphine,[16] levorphanol,[17] metazocine,[18] and meperidine[19] (Figure 5.12). All of these compounds are analgesic

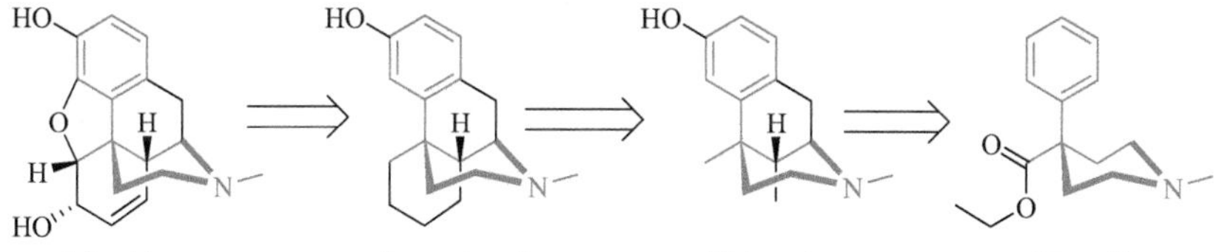

FIGURE 5.12 A pharmacophore can be described as the minimum structural requirements that must be present in order for a compound to have biological activity at a given target. The μ-opioid receptor family of analgesics, typified by morphine, levorphanol, metazocine, and meperidine, can be described by the minimum pharmacophore outlined in red in each structure. Features outlined in black can have a significant impact on biological activity, but can also be modified (levorphanol), or even removed entirely (metazocine and meperidine), while still maintaining the ability to bind to the μ-opioid receptor. Loss of the features in red, however, leads to a loss of μ-opioid receptor binding.

agents that bind to the μ-opioid receptor, but with varying degrees of potency. The common pharmacophore, outlined in red, represents the minimal structural feature required for biological activity. Morphine is, of course, a powerful analgesic, but levorphanol, a less-known structural analog in which one of the two alcohols and the oxygen-containing ring have been removed, is three to four times *more* potent than morphine. This indicates that these regions of the molecule are not part of the critical binding elements for the μ-opioid receptor pharmacophore and were actually impeding binding interactions.

Metazocine and meperidine are also part of the μ-opioid receptor family of analgesics, but they are both less potent than morphine and levorphanol. In both cases, the basic pharmacophore is intact, but molecular elements responsible for rigidity and shape have been removed. In the case of metazocine, the lower cyclohexane ring has been truncated, leading to a loss in activity relative to morphine, but the excision of both cyclohexane rings in meperidine leads to an even more dramatic loss of potency. Meperidine possesses only 10–12% of the analgesic potency of morphine, even though the overall pharmacophore remains intact. The ability to excise portions of the auxophore may be limited by their importance in maintaining the overall three-dimensional shape of candidate compounds.[20]

Understanding the nature of a pharmacophore is greatly facilitated when a set of related compounds such as the μ-opioid receptor family of analgesics is available for analysis, but it is often the case that apparently unrelated compounds elicit a biological response by binding to the same macromolecular target. Identifying a unifying pharmacophore within a structurally diverse set of compounds may require a more abstract viewpoint. Rather than defining a distinct molecular framework as in the previous example, it may be necessary to define required chemical functionalities and their relative positions in space. In the case of the voltage-gated chloride channels ClC-Ka, for example, Liantonio et al.[21] described a pharmacophore containing two aromatic rings and one carboxylic acid positioned within a specific distance from each other (Figure 5.13). The

FIGURE 5.13 A pharmacophore is not always described as a substructure of a larger class of compounds as seen in the μ-opioid receptor family of analgesics. In many cases, a pharmacophore for a macromolecular target can be defined by key features and their relative positions in space (distances, torsional angles, etc.). The pharmacophore model of the voltage-gated chloride channels ClC-Ka provides guidance for compound design and optimization even though it is not a complete structure.

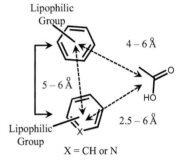

nature of the linker units between each region of the pharmacophore is defined only by the distance between the regions and their ability to hold the aromatic rings and acid moiety in the proper positions relative to each other. The ability to generate constructs that are capable of defining the physical and chemical requirements for biological activity at a given target has been greatly facilitated by the development of software capable of analyzing molecular interactions on a broad scale.

DEVELOPING AN SAR DATA SET

The drug discovery process depends on the ability of scientists to make informed decisions based on the structure–activity relationships observed through the course of their program. In the beginning of a program, however, it is often the case that very little biological data is available, as only a few compounds, usually benchmark compounds and literature leads, have been examined in biological assays. This, of course, begs the question of how does one begin the process of developing an SAR database that can be used as a basis for design strategy and project decisions?

It is often the case that these efforts begin with the high-throughput screening of large compound libraries in an effort to identify groups of related compounds that possess the desired biological activity. Ideally, the compound libraries should be composed of a diverse set of pharmacophores that cover a large portion of "druglike" chemical space and contain enough members of representative chemical classes such that an initial set of SAR data will be provided in the first round of screening. The number of compounds necessary for this kind of exercise to succeed is a subject that has been debated in the literature and no clear answer has emerged. Some institutions maintain compound collections of over 1 million discrete compounds.

In practice, libraries available for screening are usually biased based on their origin. Private compound collections available to pharmaceutical companies and research institutions tend to be populated with historical compounds from internal efforts. This increases the likelihood of identifying novel leads and groups of related compounds, but creates an institutional bias in the collection's coverage of chemical space. Large commercial compound collections are also readily available from a variety of vendors (Maybridge, Enamine, Life Chemicals, etc.) and can be effective tools as initial screening sets. However, they are composed of known compounds, making intellectual property protection potentially more difficult to obtain (see Chapter 12) and are often expensive, especially from the perspective of small biotech companies and academic institutions. Searchable databases of commercially available compounds are available, the largest of which is the Zinc database with over 21 million entries.[22] Custom compound collections can be designed and acquired in order to decrease the cost burden, but intellectual property issues will remain.

Natural product and biological extract collections have also been used as starting points for screening programs designed to generate initial SAR data, but it is important to realize that these libraries may contain complex mixtures. Identifying compounds with positive results may be challenging, and it is also possible that the mixtures contain previously identified active components with previously established intellectual property. Compound supply issues should also be considered, as natural products can be exceptionally complex to prepare in scales required for follow-up activities. That being said, a large portion of marketed pharmaceutical agents are natural products, or derivatives thereof, so this avenue of approach is worth considering as compound collections are built.

Irrespective of the source of the library, once the compound collection is chosen, the next step in building an SAR data set is screening the compounds for biological activity in an assay. This can be accomplished either via physical screening of the sample collection or by virtual screening

FIGURE 5.14 An initial data set for a drug discovery program can be derived from a physical screening of a compound library or a virtual screening of the same library if an appropriate computer model is available. Each method has advantages and disadvantages that should be considered, and neither method is exclusionary of the other. It is, however, very important to validate the *in silico* screening results with physical screening methods to ensure that the virtual screening is truly predictive of real-world results.

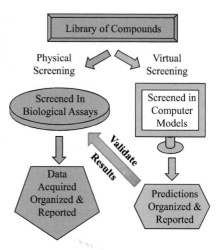

(Figure 5.14). As discussed in Chapter 4, there are a multitude of assay platforms and technologies that can be crafted to screen thousands of compounds for biological activity in 96-, 384-, or even higher throughput modes on robotic systems, providing access to potentially millions of data points. If, for example, a library of 100,000 compounds were examined using a 12-point dilution curve to determine IC_{50} values at a biological target, this would produce 2.4 million data points when run in duplicate, not including positive and negative controls used to ensure assay integrity. Although this may seem like a large amount of data, many large pharmaceutical companies maintain compound collections containing 500,000 to 1.5 million samples, which would produce a proportionately larger number of data points to analyze if screened in its entirety. Clearly, this is a staggering level of data output that cannot be processed in a reasonable length of time in the absence of the proper computer software. Fortunately, there are a number of commercial software systems capable of analyzing screening data, generating IC_{50} data, and associating that data (or any other data set) with a structure-searchable chemical database. Once the screening data have been matched with the chemical structures, the data can be sorted to identify active compounds and collections of related compounds in order identify any nascent SAR trends that may be discernible in the first round of screening.

While high throughput screening is an effective tool for identifying initial leads and early SAR trends, the previous discussion neglects to consider a major factor associated with high throughput screening. The cost of this process can be quite high. Even if a compound collection is already in place, the cost of consumable materials can be substantial. Consider again a compound collection with 100,000 members. A high throughput screening assay run in 96-well plates using three concentrations run in duplicate would require 7500 microtiter plates, thousands of disposable pipette tips for reagents and solvent transfer, as well as reagents and solvents necessary for the assay itself. A single high throughput screening cycle for a compound collection of this size can cost thousands of dollars, creating a significant cost barrier that must be considered.

As an alternative to physical screening of compounds, many organizations choose to employ a virtual screening approach. Rather than acquiring large compound collections, electronic databases such as the Zinc database mentioned earlier are examined *in silico* using computer models designed to identify potentially active compounds. In some cases, the model is based on an X-ray crystallographic structure of the target macromolecule bound to a known ligand. Binding interactions of the known ligand are defined in the model system, the ligand is then removed from the model-binding site, and database compounds are then "docked" in the binding site (Figure 5.15). The relative binding strength of each member

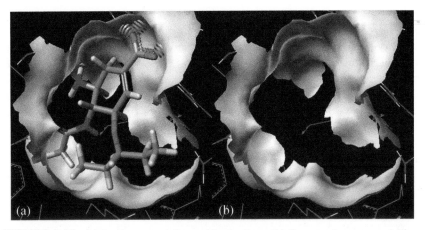

FIGURE 5.15   (a) An X-ray crystallographic structure of influenza virus neuraminidase with an inhibitor bound to the active site provides information that can be used in the design of additional inhibitors. The boundaries of the binding site (gray) and important interactions are displayed, including hydrogen bonds (barrels). (b) *In silico* excision of the inhibitor from the binding site using protein modeling software provides a template for virtual examination candidate compounds. RCSB 2QWK.

of the collection is calculated and compared with the calculated binding strength of the original ligand. The information derived from this process is used to identify compounds with binding "scores" equal to or better than the original ligand. In theory, the compounds identified in this process would have binding affinities for the target macromolecule that are similar to or better than the original ligand.

If an X-ray crystallographic structure for the biological target is not available, it may be possible to create a homology model *in silico* using structural data from a macromolecule with a high enough degree of similarity to the target of interest. As noted earlier, it has been demonstrated that proteins with similar primary structures also possess similar secondary and tertiary structures.[23] Given the relationship between primary structure and the overall three-dimensional structure of a protein, this should not be surprising. The process of building the model itself begins with an alignment of the linear peptide sequences (the primary structure) of the two macromolecules. Once the sequences are aligned, the known structural data (an X-ray crystal structure[24] or a structure built with protein NMR data[25]) are used as a template to establish a three-dimensional model of the macromolecule of interest. Of course, there will be some areas within the model structure that will require adjustment in order to account for differences in the primary sequences of the template and the model. Structural changes to the model are generally accomplished through energy minimization, the addition of water molecules that may have been present in the original X-ray structure, and the elimination of overlaps within the structure. The model is then tested by docking ligands in the theoretical binding site, scoring their relative binding affinities, and comparing the *in silico* results with real-world data. This process has been successfully employed to create model systems for a number of GPCRs based on X-ray crystal structures of rhodopsins, a family of light-sensing transmembrane proteins.[26] Structural data derived from the X-ray crystal structures of bacterial rhodopsin, for example, were successfully adapted to establish a homology model for the dopamine D2 receptor (Figure 5.16). Although

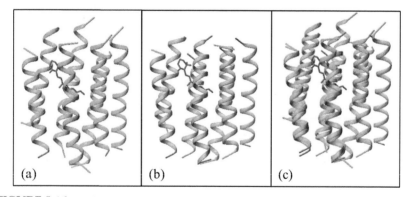

FIGURE 5.16   In the absence of an X-ray crystallographic structure, it is possible to build a model *in silico* to support virtual screening efforts. Structural data for bacterial rhodopsin ((a), RCSB 1BRD) were successfully mapped onto the dopamine D2 receptor to design a homology model ((b), RCSB 1I15) suitable for docking and evaluating candidate compounds. An overlay of the original structure with the model structure (c) demonstrates significant similarities between the two structures, but also a number of differences.

the overlap between the two structures is not perfect, the homology structure provided significant insight into the structural details of the dopamine D2 receptor that would not have been available otherwise.[27]

In the absence of structural data for the biological target of interest, it is still possible to perform a virtual screen of compound databases in order to focus screening efforts. Rather than docking compounds into a binding site, known ligands can be used as templates to identify potential lead compounds from within a structural database. This process is often referred to as ligand-based design. Modern molecular modeling software is applied to a starting compound, often a natural ligand or a known drug molecule, to determine its molecular properties, mapping out lipophilic potential surfaces, electrostatic potential surfaces, areas suitable for hydrogen bonding, lipophilic regions, and a wide array of molecular properties that define the compound in question (Figure 5.17). If multiple ligands

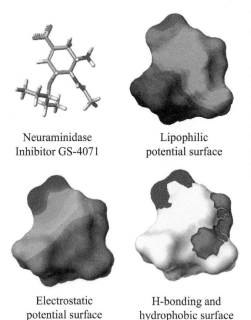

Neuraminidase
Inhibitor GS-4071

Lipophilic
potential surface

Electrostatic
potential surface

H-bonding and
hydrophobic surface

FIGURE 5.17 The neuraminidase inhibitor GS-4071 is structurally described by the classical stick figure (top left), but computer modeling can provide improved visualization of molecular properties such as lipophilic potential surfaces (top right), electrostatic potential surface (bottom left), and hydrogen bonding/hydrophobic surfaces (bottom right). These "maps" can be combined to provide a model system useful for the identification of compounds that are similar to GS-4071.

are available, it may be possible to identify common features from each ligand/drug to define a pharmacophore for the binding site of interest. The aforementioned property sets are determined for all of the members of the compound collection, the individual members of the compound collections are compared with the lead compound, and each is assigned a score based on its overall similarity to the lead compound. Compounds with higher scores (those that more closely resemble the lead compound) can be identified in this manner and similar compounds can be grouped together to select compounds for physical screening, once again significantly decreasing the overall cost of physical screening.

Irrespective of whether virtual screening results are based on an X-ray crystal structure, a homology model, or comparison of compound libraries with a known ligand, it is important to realize that virtual screening is a *model* of reality. As such, it must be validated with physical screening to ensure that there is a correlation between the results provided by the virtual screening and the physical screening. If, for example, compounds with high binding scores in a model of an enzyme were demonstrated to be potent inhibitors of enzymatic activity, and that as scoring values decreased, inhibitory action of the compounds decreased, then the *in silico* screening is validated for identifying compounds that are inhibitors of the enzyme in question (Figure 5.18(a)). On the other hand, if there

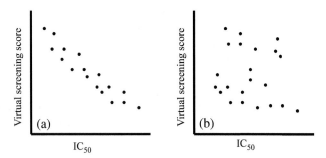

FIGURE 5.18   In order for a virtual screen to be useful, it must correlate with real-world results (a). In the absence of this correlation (b), data provided by the virtual screen cannot be used to predict biological activity of candidate compounds.

is no relationship between the scores generated through virtual screening and the inhibitory action of compounds physically screened against the enzyme, then the *in silico* model cannot be used as a predictive tool for virtual screening (Figure 5.18(b)). Of course, no model is a perfect representation of reality, so a perfect correlation should not be expected. Additional data from physical screening or new structural data for the target can be used to refine models to improve their ability to predict biological activity.

The strategic application of virtual screening can significantly reduce the number of compounds examined in a physical screen, decreasing the required monetary commitment. There are, however, costs associated with computer capacity and state-of-the-art molecular modeling software (including annual license fees) that should be considered. These costs can rival that of the purchase of a sizable physical library. Also, when virtual screens do not correlate to the physical screening system that they are designed to represent, there is a loss in productivity that is more difficult to quantify but important to consider. It is worth pointing out at this juncture that physical and virtual screening methods are not mutually exclusive strategies for drug discovery. They are often used in tandem, as they can be complimentary pathways to success.

Although screening of large compound collections can be an effective method of identifying biologically relevant compounds, the process is not without its issues. As previously mentioned, the costs associated with screening even a medium size collection of 100,000 compounds can be significant. In addition, the process itself is essentially a random selection process. Potentially millions of compounds are screened, whether physically or *in silico*, relying on the compound designs of previously defined compound collections. Compounds that are part of a historical collection may be unique to a given institution, but they were most likely prepared for a specific purpose which may not be useful for a current program (It could, in fact, represent a problem if off-target activity becomes an issue). It is also quite possible that these compounds have already been disclosed to the public as part of the patent application process for their project of origin, potentially creating intellectual property issues. A similar issue exists for commercially acquired samples, as they are part of the public domain, again creating a potential issue for patent protection, a necessary facet of drug discovery.

The sheer volume of data resulting from a high throughput screen can also be quite daunting. High-throughput screening as a means of identifying biologically active compounds has been compared to looking for a needle in a haystack, and with good reason. A high throughput screen of a collection of 500,000 compounds producing a positive result of only 0.2% will provide a set of 1000 compounds for examination and potential follow-up. While there are software packages capable of simplifying data handling, at the end of the day, scientists must review the data and determine how to move forward. There is also the issue of false positive and false negative results. The aforementioned high throughput screening run of 500,000 compounds would produce 100 erroneous results if the false positive/negative rate 0.02%. If they are equally split between false positives and false negatives, then there will be 50 compounds incorrectly identified as biologically active. The remaining 50 compounds that are false negatives are more difficult to track down, as they will blend in with the rest of the compounds collection that did not meet the desired potency range.

While the randomized approach discussed thus far has been successfully applied to numerous drug discovery programs across a wide array of disease states and biological targets, it is by no means the only method of identifying a starting point for structure–activity relationship studies. An alternative approach employed by many drug discovery scientists takes advantage of the significant advances in our understanding of the relationship between structure and function in a more direct fashion. In this approach, often referred to as rational drug design or structure-based drug design, lead series and eventually lead compounds are developed through a detailed analysis of structural information on the targeted macromolecule, known ligands (both natural and synthetic), or both.

When an X-ray crystal structure of a macromolecule bound to a small molecule is available, for example, modern molecular modeling software can be employed to determine points of positive interaction (hydrogen bonding, salt bridges, lipophilic interactions, etc.), clash points between the ligand and the biomolecule (negative interactions), and areas of empty space within the binding pocket that might be accessible if changes were made to the ligand. Compounds that improve upon positive interactions, perhaps by optimizing the distance between a hydrogen bond donor/acceptor pair, removing a side chain that hinders binding, or adding a new side chain to create an additional binding interaction, are then designed based on the available structural data. The new compounds can be "prepared" *in silico*, docked with the X-ray structure, and scored for binding energy relative to the original ligand to assess whether or not preparation is warranted. Biological screening of the actual compounds should be used to validate the *in silico* methods to determine the reliability of the model (i.e., docking of the compounds in the X-ray crystal structure), but if it is validated by real-world results, it can be used to prioritize medicinal chemistry efforts.

These techniques have been employed to identify inhibitors against a wide range of enzymes such as human protein tyrosine phosphatase-β (HPTP-β), a receptor-type phosphatase that is expressed in endothelial cells that form the protective lining of blood vessels.[28] X-ray crystallographic studies of the catalytic domain of HPTP-β bound to a vanadium tetraoxide ion indicated that this compound formed multiple hydrogen bonding interactions in the binding site (Figure 5.19(a)). It was then

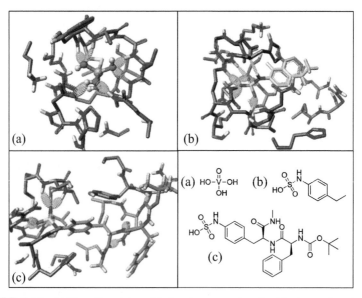

FIGURE 5.19 (a) Vanadium tetraoxide ion forms a network of hydrogen bonds (green barrels) and binds tightly to the active site of human protein tyrosine phosphatase-β (HPTP-β) (RCSB 2I4E) (b) 4-Ethylbenzene sulfamic acid forms a similar series of hydrogen bonds (RCSB 2I5X) and was used as a starting point for a drug discovery program that eventually produced a highly selective HPTP-β inhibitor ((c), RCSB 2I4H).

surmised that small molecules capable of replicating this intricate network of hydrogen bonds should be able to occupy the same space in the HPTP-β catalytic site, inhibiting enzyme action. Synthesis, biological evaluation, and additional X-ray crystallography studies verified this hypothesis, as 4-ethylbenzene sulfamic acid formed a stable complex with catalytic domain of HPTP-β, inhibiting enzyme action (Figure 5.19(b)). Molecular modeling studies using this information lead to the identification of more advanced inhibitors capable of forming additional hydrogen bonding interactions with nearby amino acid residues (Figure 5.19(c)), which lead to a significant increase in potency against the target.[29]

Fragment-based screening[30] has also been employed as a means of developing structure–activity relationship, but as the name implies, the starting points for these programs are typically smaller than the average drug molecule. In this approach, low-molecular weight compounds, typically 100–250 AMU, are screened to determine their binding affinity with a macromolecule. Compounds of this size typically have low binding affinities (>100 μM), so sensitive techniques such as X-ray crystallography,[31] protein NMR,[32] or surface plasmon resonance (SPR)[33] are required, and throughput is low compared to high-throughput screening methods. Compounds that are identified in these processes, however, generally create efficient binding interactions with the macromolecules when their low molecular weight is taken into consideration.[34] As such, they can be employed as starting points for further optimization. This is typically accomplished by either "growing" the fragment compound or "linking" multiple fragments molecules together. The "linking" strategy depends upon synergies that may be created by forming linkage points between fragments that bind to the target in neighboring regions. Just as the individually weak interaction in protein structure can collectively form very stable three-dimensional structures, joining together multiple fragment structures can have a synergistic effect, leading to tighter binding than is observed for the individual fragments, possibly greater than the sum of the individual components (Figure 5.20). The linker itself may also play a role in binding if additional

FIGURE 5.20   In one approach to fragment-based screening, low-molecular weight compounds are screened to determine binding affinities using sensitive screening techniques such as NMR, SPR, and X-ray crystallography. Low-molecular weight compounds that demonstrate binding are then stitched together with linker units designed to allow the individual components to access their individual binding sites. The tethered individual units act collectively to create a single compound with improved biological activity based on synergistic binding.

binding interactions are available. If, for example, the linker contained an amide bond, then it could form a hydrogen bond with an appropriate functional group in the targeted macromolecule, further enhancing binding strength.

As an alternative, one can "grow" a fragment molecule into unoccupied space in the binding site of target molecule. This process is greatly facilitated by the availability of either an X-ray structure of the fragment bound to the target or a homology model that can be used to visualize potential areas of interest in the binding site. Using this technique, indazole was identified as a fragment "hit" with very low binding potency (185 μM) for cyclin-dependent kinase 2 (CDK-2) and eventually transformed into AT7519, a significantly more potent CDK-2 inhibitor (47 nM, Figure 5.21).[35]

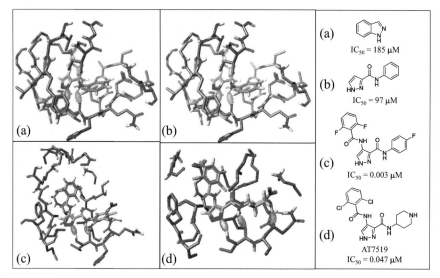

FIGURE 5.21   (a) Screening of a fragment library via X-ray crystallography identified indazole as a weak CD-2 binder (RCSB 2VTA) anchored into the binding site with a hydrogen bond (cyan barrel). (b) Removal of the benzene ring and growth of the fragment via addition of an N-phenyl-acetamide provided access to a neighboring binding site and an additional hydrogen bond interaction (RCSB 2VTL). (c) Growing the structure in the opposite direction by adding a benzamide created additional opportunities for hydrophobic interactions in a proximal section of the binding site, leading to a dramatic improvement in potency (RCSB 2VTP). (d) Incorporation of a piperidine to increase solubility and cellular activity led to the identification of the clinical candidate AT7519 (RCSB 2VU3).

Although each of the techniques described thus far has been discussed in isolation, it is important to realize that they are not mutually exclusive. In fact, it is often the case that multiple approaches are employed within the context of a given drug discovery program. Each approach has strengths and weaknesses, but none of them have proved to be superior to the others in identifying a clinical candidate, let alone a drug. Each of these methods is most appropriately viewed as a tool to be employed when appropriate.

THE STRUCTURE–ACTIVITY RELATIONSHIP CYCLE

Ideally, once the first round of screening is complete and the data have been analyzed, it will be possible to identify a set of compounds that have varying degrees of biological activity at the target of interest. Compounds with similar core structures can then be grouped together to determine how differences in their structures impact observed biological activity, providing an early indication of structure–activity relationships within the series. An iterative cycle of design, synthesis, screening, and analysis begins at this point. New molecules with structural changes are designed based on the available compound data set, synthesized, and screened for biological activity. The resulting data are analyzed to determine the impact of structural changes on biological activity. If the structural changes incorporated into the new set of compounds have a positive impact on screening results, then the changes are retained. On the other hand, if biological activity decreases, then the changes are discarded. In either case, the new information provided by the preparation and screening of the second set of compounds is incorporated into the SAR information available, and a new round of design, synthesis, screening, and analysis is initiated. This iterative process (Figure 5.22) is repeated over and over until a molecule with the desired biological properties is identified.

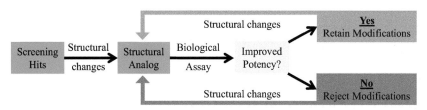

FIGURE 5.22   Identifying an initial set of hits in a drug discovery program, irrespective of their source and methods employed, is generally the beginning of an iterative process. Analogs containing specific structural changes are prepared and assessed for biological activity. Favorable changes are maintained, unfavorable changes are discarded, and a new round of compounds is prepared. Each successive round of synthesis and biological assessment builds on the previous round with the ultimate goal of optimizing biological activity.

BIOISOSTERISM

As discussed above, simple structural changes, such as increasing the length of a side chain (methyl, ethyl, propyl, etc. Figure 5.3), adding functionality to a benzene ring as described in the Topliss Tree (Figure 5.10), or the additions of hydrogen bond acceptors/donors (Figure 5.4) can lead to improvements in biological properties. There are, however, additional avenues that can be explored. The concept of isosterism, originally developed by Langmuir[36] and then broadened by Erlenmeyer,[37] can also be applied to drug discovery. Langmuir (1919) and Erlenmeyer (1932) hypothesized that compounds, functional groups, or atoms with the same arrangement and number of electrons in their outermost electron shells would have similar physical and chemical properties based on their isoelectronic features. The scope of Erlenmeyer's original premise was eventually expanded by Harris L. Friedman in 1951 when he defined compounds as bioisosteric "if they fit the broadest definition for isosteres and have the same type of biological activity."[38] In a similar manner, functional groups and atoms within a more complex molecule that can be interchanged while maintaining the same or similar biological properties are referred to as bioisosteres of each other. Modification of candidate compounds by applying the premise of bioisosterism is commonplace in drug discovery programs, as it is a useful tool for expanding the structural diversity of a lead series to identify novel chemical space and to alter physicochemical properties. These subjects will be addressed in greater detail in Chapters 12 and 6, respectively.

Bioisosteres can be divided into two broad classes: classic and nonclassic bioisosteres, as first proposed by Alfred Burger.[39] According to this system of classification, classic bioisosteres are broadly defined as atoms, molecular fragments, or functional groups that have the same valence and ring equivalents (Table 5.1). Typical examples include exchanging halogen atoms and replacing methylene units with an oxygen or sulfur atom. A practical application of this principle is seen in efforts to develop alternatives to the antihypertensive drug rilmenidine (Albarel). In this instance, replacing the oxygen atom of rilmenidine's 4,5-dihydro-oxazole ring system with a methylene unit provided the 4,5-dihydro-pyrrole analog

TABLE 5.1   Classic Bioisosteres

Monovalent	Divalent	Trivalent	Tetravalent
$-OH, -NH_2, -CH_3, -OR$	$-CH_2-$	$=CH-$	$=C=$
$-F, -Cl, -Br, -I, -SH, -PH_2$	$-O-$	$=N-$	$=Si=$
$-SiR_3, -Sr$	$-S-$	$=P-$	$=N^+=$

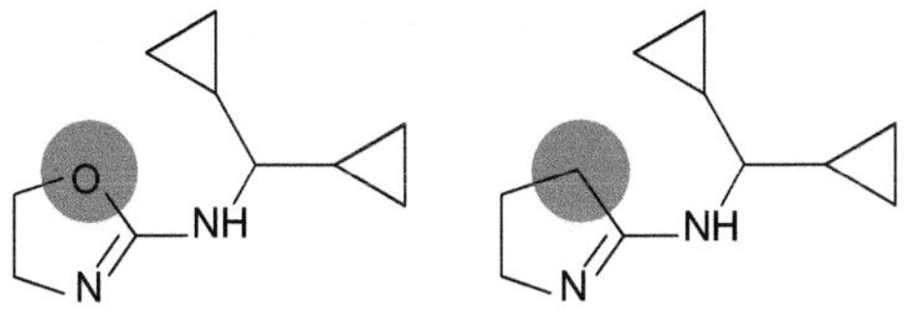

Rilmenidine   Methylene isostere

FIGURE 5.23 In this example of classic bioisosterism, exchanging a carbon atom for an oxygen atom (highlighted in red) eliminates an undesired off-target activity. While both compounds are potent modulators of the I_1 imidazoline receptor, rilmendine's potent α_2-adrenoceptor binding is eliminated by this simple change.

(Figure 5.23). Both compounds share similar potency at the I_1 imidazoline receptor, a key player in the regulation of blood pressure. On the other hand, the new series of compounds demonstrated improved target selectivity, as although rilmendine potently binds to the α_2-adrenoceptors, the methylene analogs showed no affinity for this receptor. This successful application of classic bioisosterism clearly demonstrates that simple changes in chemical structure can be employed to alter some properties (abolishing α_2-adrenoceptor binding), while successfully preserving others (potency at the I_1 imidazoline receptor).[40]

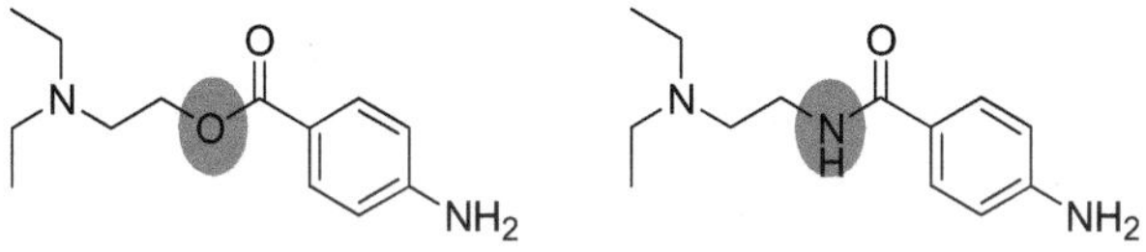

Procaine   Procainamide

FIGURE 5.24 While both procaine and procainamide block the same sodium channel, only procainamide can be used as a treatment for arrhythmia. In this instance, bioisosteric replacement of an oxygen atom with a nitrogen atom provides a compound with similar biological activity and substantially improved metabolic stability.

Procaine and procainamide (Figure 5.24) are also excellent examples of the impact that can be achieved with a classic bioisosteric replacement. Both of these compounds exert their pharmacological impact via sodium channel blockade. Procaine, more commonly referred to as Novocain, is only useful as a topical anesthetic, while procainamide is an orally administered antiarrhythmic agent. In this instance, a classic bioisostere exchange is employed to improve metabolic properties. The ester linkage of procaine is hydrolyzed quickly *in vivo*, limiting its utility to topical application, while the amide linkage in procainamide is significantly more stable *in vivo*. The enhanced metabolic stability conferred to procainamide versus procaine is critical to its utility as an oral antiar-rhythmic agent.[41]

Non-classic bioisosteres represent a much broader range of structural diversity, as this category is defined as anything that does not fit the definition of a classic bioisostere. It should come as no surprise that the vast majority of examples of bioisosterism fall into this category and that a full characterization of this area of drug discovery is beyond the scope of this chapter. The concepts involved, however, can be described in a few examples. Replacing carboxylic acids in candidate compounds, for example, has been widely studied, as this functional group can create undesirable physicochemical properties. Compounds containing car-boxylic acids tend to have low permeability through biological mem-branes, are poorly absorbed when dosed orally, and are rapidly cleared by the body. Their ability to cross the blood–brain barrier is also often limited at best. In other words, if a compound with a high degree of potency at a biological target of interest contains this functional group, moving the compound down the drug discovery and development path-way may prove challenging. Replacing this functionality with a suitable bioisostere may alleviate these issues, and some examples are shown

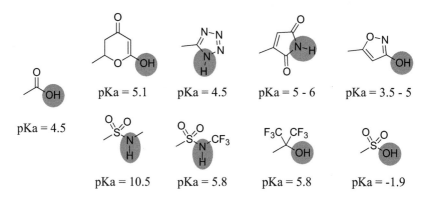

FIGURE 5.25 There are many potential non-classic bioisosteric replacements for a car-boxylic acid, but they are by no means equivalent. Structural diversity is readily apparent, but an examination of the pKa values of each (highlighted in blue) provides an alternative view of the differences between the various possible carboxylic acid mimetics.

in Figure 5.25. In each instance, the dissociable proton of the carboxylic acid is maintained, but the remainder of the functionality is replaced with a collection of atoms capable of mimicking the original structure in a biological setting. The successful application of any one of these car-boxylic acid bioisosteres is dependent upon the nature of the interaction between the candidate compound and the macromolecular target. In one successful example, one of the two carboxylic acids of Carbenicillin was

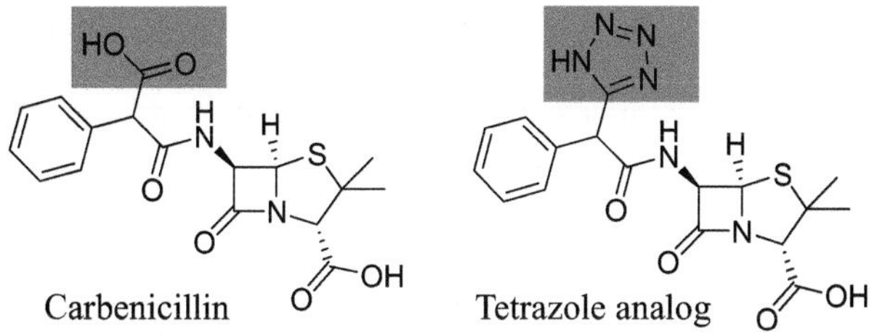

FIGURE 5.26 Although carbenicillin is a useful antibiotic, its use is hampered by spontaneous decarboxylation in acidic media. The tetrazole analog is similarly effective as an antibiotic, but does not suffer from instability in an acidic media as a result of the non-classic bioisosteric replacement (highlighted in red).

replaced with a tetrazole (Figure 5.26). In this instance, antibacterial activity was maintained, and an increase in chemical stability was imparted. Carbenicillin can spontaneously decarboxylate, while the tetrazole derivative is incapable of decomposing in this manner.[42]

In some cases, non-classic bioisosterism can be employed to replace one scaffold with another. Biological activity is maintained, but physicochemical properties and target selectivity may be very different. Further, synthetic accessibility may be simplified, and greater intellectual property space may also be available around the new scaffold. It has been demonstrated, for example, that the steroid framework of 17β-estradiol, a major female sex hormone, can be effectively replaced with a simple *trans*-stilbene (Figure 5.27).

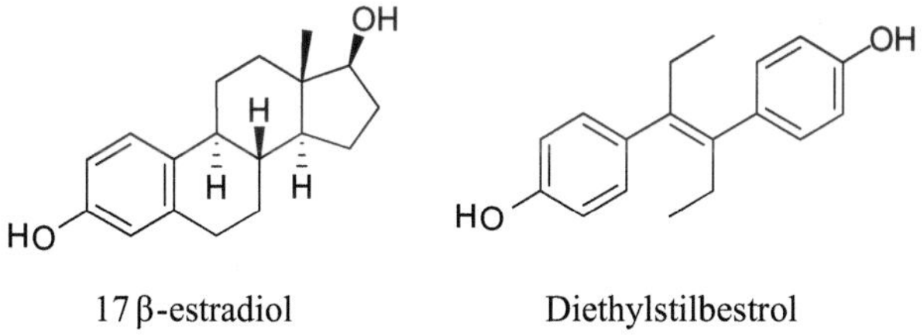

FIGURE 5.27 In this instance, three of the four rings in the steroidal framework are replaced by a stilbene framework. At the time of this discovery, this non-classical bioisosteric replacement provided novel intellectual property space and dramatically simplified the chemistry required to prepare novel estrogen receptor modulators.

Even though the two central rings of 17β-estradiol have been removed and a third ring has been replaced with a benzene ring, *trans*-diethylstilbestrol remains a potent estrogen receptor agonist.[43]

Bioisosteric replacement, whether classical or non-classical, is an exceptionally important concept in the field of drug discovery and development. As indicated by the examples provided, even small bioisosteric changes can have a dramatic impact on the physicochemical and biochemical properties of a compound. There are hundreds of examples of potentially useful bioisosteres that one could consider in a drug discovery program, and a full discussion of this concept is well beyond the scope of this chapter. Fortunately, numerous review articles are available on the subject for those interested in an in-depth analysis of the subject.[44]

STRUCTURE–ACTIVITY RELATIONSHIP, SELECTIVITY AND PHYSICOCHEMICAL PROPERTIES

Although the focus of the previous sections has been on the identification of increasingly more potent compounds through the analysis of structure–activity relationships, potency at a biological target is not the only consideration in a drug discovery program. As noted in Chapter 1, target selectivity is often critical to the identification of clinically relevant compounds, and physicochemical properties such as solubility and metabolic stability are major factors that must be considered. Fortunately, the same principles that have been discussed with respect to optimization of binding interactions at the intended molecular target can be employed to optimize any molecular property, provided there is a means of measuring the impact of structural changes on the property of interest. If, for example, one were targeting Kv1.5, a voltage-gated potassium channel that has clinical relevance in atrial arrhythmia,[45] optimizing this activity would be necessary. It would also be necessary, however, to establish that candidate compounds did not negatively impact the hERG channel, a voltage-gated potassium channel associated with Torsades de pointes and sudden cardiac death.[46] In this instance, a structure–activity relationship could be established for activity at the hERG channel, focusing on minimizing the potency at this channel. Thus, structural changes that *decreased* hERG activity would be maintained, while those that increased hERG activity would be dropped. Successful compounds would be identified by analyzing the SAR of *both* Kv1.5 and hERG in an effort to maximize one while minimizing the other.

In a similar manner, physicochemical properties can be optimized through structure–property relationships. Just as changes in molecular structure will have an impact of the ability of a compound to bind to a biomolecule, changes in molecular structure will also have an impact on physicochemical properties. Quantifying the impact of structural changes on a property of interest provides a means of optimizing that property through manipulation of the molecular structure. Some of these properties, for example, solubility and metabolic stability, require physical assessment. Fortunately, high-throughput screens have been developed to facilitate this process. Other properties, such as lipophilicity and polar surface area, can be calculated using the appropriate software tools. In either case, once knowledge of structural changes is paired with the appropriate data set, patterns can be established and properties can be optimized appropriately. The importance of physicochemical assessment of compounds will be discussed in greater detail in Chapter 6, but the recognition of the high level of importance of physicochemical properties dramatically changed the way new drugs are pursued.

"DRUGLIKE" GUIDELINES

Although modern science has provided researchers with the ability to rapidly synthesize and screen compounds for biological activity across a wide range of targets, the small molecule universe is simply too large to explore exhaustively. It is estimated that there are over 10^{60} synthetically accessible small molecules, and only an infinitesimally small portion of this chemical space is likely to ever be explored.[47] Even if one were to consider only compounds within a specific class, the number of possible compounds is still extremely large. There are, for example, over 60,000 1,4-benzodiazepines (Figure 5.28)[48] in the literature as of 2013, and tens of

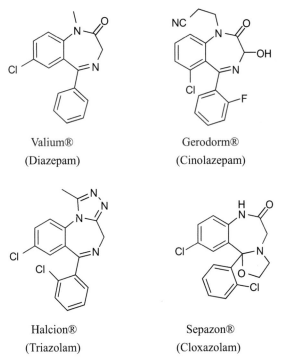

Valium®
(Diazepam)

Gerodorm®
(Cinolazepam)

Halcion®
(Triazolam)

Sepazon®
(Cloxazolam)

FIGURE 5.28 Valium® (Diazepam), Gerodorm® (Cinolazepam), Halcion® (Triazolam), and Sepazon® (Cloxazolam) are all members of the benzodiazepine family of compounds, a "privileged scaffold." Chemical frameworks that have been the source of large numbers of biologically important compounds are referred to as "privileged scaffolds."

thousands more could be prepared with commercially available building blocks. Given the enormity of chemical space within even a single series of compounds, the need to establish a general method of prioritization becomes evident. Structure–activity relationships are useful tools, but they are inherently limited by the need to synthesize and screen compounds in order to move a program forward.

In an effort to simplify what appears to be a near intractable problem, Christopher Lipinski and his colleagues examined the physicochemical properties of over 2000 drugs and drug candidates in clinical trials at the time (mid-1990s).[49] While biological activity is often the major focus of research programs, Lipinski et al. took heed of a 1991 Tufts University study on experimental drugs and why they fail. The Tufts study indicated that 30% of clinical failure could be attributed to a lack of efficacy, but that nearly 40% the programs that failed were unsuccessful due to poor pharmacokinetic profiles.[50] The candidate compounds may have been potent and selective, but they were unable to reach their intended target. Lipinski and his colleagues hypothesized that successful drugs and clinical candidates occupied specific portions of the larger chemical universe and that compounds that failed as a result of poor pharmacokinetic profiles existed outside of this "druglike" chemical space. They further proposed that if one could understand the limitation of this region of the chemical universe, then one could predict whether or not a new compound would be inside or outside of "druglike" chemical space.

The results of their studies had a profound and lasting influence on the pharmaceutical industry. They determined that the overwhelming majority of the compounds in their test set shared certain key characteristics referred to as the Lipinski Rule of Five (Figure 5.29). Successful compounds had a

Lipinski's Rule of 5 and the Veber Extension

1) Molecular weight <500 AMU
2) LogP below 5
3) Fewer than 5 hydrogen bond donors
4) Fewer than 10 hydrogen bond acceptors
5) Fewer than 10 rotatable bonds
6) Polar surface area less than 140 Å^2

FIGURE 5.29   Although there are exceptions, these guidelines have been developed to help identify compounds that likely to have physicochemical properties consistent with successful drugs.

molecular weight below 500 AMUs, a logP less than 5 (the logarithm of the partition coefficient between water and 1-octanol; a measure of lipophilicity), fewer than 5 hydrogen bond donors, and fewer than 10 hydrogen bond acceptors. These guidelines to "druglike" space were based on the

90th percentile values of the properties distribution, and assume that compounds are absorbed into cells via passive diffusion (Compounds that are actively transported across a cell membrane are exceptions to these limitations.). Additional observations by Daniel Veber et al. lead to an expansion of the Lipinski guidelines to include the number of rotatable bonds (less than 10) and the concept of polar surface area (less than 140Å^2).[51] There are, of course, exceptions to these guidelines, as many natural products fall outside of the scope of these rules, but the Lipinski Rule of Five has been broadly accepted by the pharmaceutical industry and are often examined via *in silico* prediction prior to preparation of proposed new compounds.

In an ideal world, a novel series of compounds could be optimized through medicinal chemistry and screening science to provide a perfect clinical candidate. Target selectivity would be absolute, eliminating the risk of off-target effects, and physicochemical properties would support a simple dosing regimen. In practice, however, it is rarely possible to identify a perfect compound. Clinical candidates, as well as the overwhelming majority of marketed drugs, have "warts" that must be considered as they are studied or used in a clinical setting. Medicinal chemistry in combination with screening science provides the tools to maximize the positive traits and minimize the negative characteristics of candidate compounds. It is up to program scientists to decide when medicinal chemistry efforts have provided candidate compounds that are suitable for further study (e.g., pharmacokinetic studies, *in vivo* animal models of efficacy, and safety studies) or that despite their best efforts, a series of compounds should be abandoned, as it is not possible to remove enough "warts" to justify further effort.

QUESTIONS

1. Define medicinal chemistry and briefly discuss what it entails.
2. Define the term structure activity relationship (SAR).
3. How does chirality impact binding of candidate compounds to their intended biological target?
4. Define the term quantitative structure activity relationship (QSAR)?
5. What is a pharmacophore?
6. What does the term auxophore refer to?
7. Describe the iterative process that is used to acquire knowledge of the SAR of a series of molecules.
8. Briefly describe fragment-based drug design.
9. What are two general approaches used to establish an?
10. What are two general approaches used in virtual screening?
11. What is a bioisostere and why are they employed?

References

1. Kelvin, W. T. *Baltimore Lectures on Molecular Dynamics and the Wave Theory of Light;* C.J. Clay and Sons, Cambridge University Press Warehouse: London, 1904.
2. a. Pohland, A. Esters of Substituted Aminobutanes US 2728779, 1955.
   b. Niesenbaum, L.; Deutsch, J.; Moss, N. H. Analgesic Efficacy of Dextro-propoxyphene Hydrochloride in the Postoperative Patient. *J. Albert Einstein Med. Cent. Phila.* **1962**, *10*, 188–192.
3. Galli, A. D. Oral Compositions Containing Antitussives and Benzydamine WO 9523602 A1, 1995.
4. a. Neil, A.; Terenius, L. D-Propoxyphene Acts Differently from Morphine on Opioid Receptor-Effector Mechanisms. *Eur. J. Pharmacol.* **1981**, *69* (1), 33–39.
   b. Neil, A. Affinities of Some Common Opioid Analgesics towards Four Binding Sites in Mouse Brain. *Naunyn-Schmiedeberg's Arch. Pharmacol.* **1984**, *328* (1), 24–29.
5. a. Carter, C. H. A Clinical Evaluation of the Effectiveness of Novrad and Acetylsalicylic Acid in Children with Cough. *Am. J. Med. Sci.* **1963**, *245*, 713–717.
   b. Strapkova, A.; Nosalova, G.; Korpas, J. Effects of Antitussive Drugs under Normal and Pathological Conditions. *Acta Physiol. Hung.* **1987**, *70* (2–3), 207–213.
6. Li, W.; Escarpe, P. A.; Eisenberg, E. J.; Cundy, K. C.; Sweet, C.; Jakeman, K. J.; Merson, J.; Lew, W.; Williams, M.; Zhang, L.; et al. Identification of GS 4104 as an Orally Bioavailable Prodrug of the Influenza Virus Neuraminidase Inhibitor GS 4071. *Antimicrob. Agents Chemother.* **1998**, *42* (3), 647–653.
7. Varghese, J. N.; Smith, P. W.; Sollis, S. L.; Blick, T. J.; Sahasrabudhe, A.; McKimm-Breschkin, J. L.; Colman, P. M. Drug Design against a Shifting Target: A Structural Basis for Resistance to Inhibitors in a Variant of Influenza Virus Neuraminidase. *Structure* **1998**, *6*, 735–746.
8. Hammett, Louis P. The Effect of Structure upon the Reactions of Organic Compounds. Benzene Derivatives. *J. Am. Chem. Soc.* **1937**, *59*, 96–103.
9. a. Taft, R. W. Linear Free Energy Relationships from Rates of Esterification and Hydrolysis of Aliphatic and Ortho-Substituted Benzoate Esters. *J. Am. Chem. Soc.* **1952**, *74*, 2729–2732.
   b. Taft, R. W. Polar and Steric Substituent Constants for Aliphatic and *o*-Benzoate Groups from Rates of Esterification and Hydrolysis of Esters. *J. Am. Chem. Soc.* **1952**, *74*, 3120.
   c. Taft, R. W. Linear Steric Energy Relationships. *J. Am. Chem. Soc.* **1953**, *75*, 4538–4539.
10. a. Debnath, A. K. Quantitative Structure-Activity Relationship (QSAR) Paradigm – Hansch Era to New Millennium. *Mini-Rev. Med. Chem.* **2001**, *1* (2), 187–195.
    b. Hansch, C. A Quantitative Approach to Biochemical Structure-Activity Relationships. *Acc. Chem. Res.* **1969**, *2*, 232–239.
    c. Hansch, C.; Leo, A.; Taft, R. W. A Survey of Hammett Substituent Constants and Resonance and Field Parameters. *Chem. Rev.* **1991**, *91*, 165–195.
11. Topliss, J. G. Utilization of Operational Schemes for Analog Synthesis in Drug Design. *J. Med. Chem.* **1972**, *15* (10), 1006–1011.
12. Hirschfeld, R. M. A. Sertraline in the Treatment of Anxiety Disorders. *Depression Anxiety* **2000**, *11* (4), 139–157.
13. Keller, M. B. Citalopram Therapy for Depression: A Review of 10 Years of European Experience and Data from U.S. Clinical Trials. *J. Clin. Psychiatry* **2000**, *61* (12), 896–908.
14. Wong, D. T.; Perry, K. W.; Bymaster, F. P. The Discovery of Fluoxetine Hydrochloride (Prozac). *Nat. Rev. Drug Discov.* **2005**, *4*, 764–774.
15. Wermuth, C. G.; Ganellin, C. R.; Lindberg, P.; Mitscher, L. A. Glossary of Terms Used in Medicinal Chemistry (IUPAC Recommendations 1998). *Pure Appl. Chem.* **1998**, *70* (5), 1129–1143.
16. Novak, B. H.; Hudlicky, T.; Reed, J. W.; Mulzer, J.; Trauner, D. Morphine Synthesis and Biosynthesis – An Update. *Curr. Org. Chem.* **2000**, *4*, 343–362.

17. Prommer, E. Levorphanol: The Forgotten Opioid. *Supportive Care Cancer* **2007,** *15* (3), 259–264.

18. a. Berzetei-Gurske, I.; Loew, G. H. The Novel Antagonist Profile of (-)metazocine. *Prog. Clin. Biol. Res.* **1990,** *328,* 33–36.

    b. Hori, M.; Ban, M.; Imai, E.; Iwata, N.; Suzuki, Y.; Baba, Y.; Morita, T.; Fujimura, H.; Nozaki, M.; Niwa, M. Novel Nonnarcotic Analgesics with an Improved Therapeutic Ratio. Structure-Activity Relationships of 8-(methylthio)- and 8-(acylthio)-1, 2,3,4,5,6-hexahydro-2,6-methano-3-benzazocines. *J. Med. Chem.* **1985,** *28* (11), 1656–1661.

19. Kaiko, R. F.; Foley, K. M.; Grabinski, P. Y.; Heidrich, G.; Rogers, A. G.; Inturrisi, C. E.; Reidenberg, M. M. Central Nervous System Excitatory Effects of Meperidine in Cancer Patients. *Ann. Neurol.* **1983,** *13* (2), 180–185.

20. Silverman, R. B. *The Organic Chemistry of Drug Design and Drug Action,* 2nd ed.; Elsevier Academic Press: Oxford, UK, 2004. 17–20.

21. Liantonio, L.; Picollo, A.; Carbonara, G.; Fracchiolla, G.; Tortorella, P.; Loiodice, F.; Laghezza, A.; Babini, E.; Zifarelli, G.; Pusch, M.; et al. Molecular Switch for CLC-K Cl-Channel Block/Activation: Optimal Pharmacophoric Requirements towards High-Affinity Ligands. *Proc. Natl. Acad. Sci. U.S.A.* **2008,** *105,* 41369–41373.

22. Irwin, J. J.; Sterling, T.; Mysinger, M. M.; Bolstad, E. S.; Coleman, R. G. ZINC: A Free Tool to Discover Chemistry for Biology. *J. Chem. Inf. Model.* **2012,** *52* (7), 1757–1768.

23. a. Chothia, C.; Lesk, A. M. The Relation between the Divergence of Sequence and Structure in Proteins. *EMBO J.* **1986,** *5* (4), 823–826.

    b. Kaczanowski, S.; Zielenkiewicz, P. Why Similar Protein Sequences Encode Similar Three-Dimensional Structures? *Theor. Chem. Acc.* **2010,** *125,* 543–550.

24. a. Sharma, H.; Cheng, X.; Buolamwini, J. K. Homology Model-Guided 3D-QSAR Studies of HIV-1 Integrase Inhibitors. *J. Chem. Inf. Model.* **2012,** *52* (2), 515–544.

    b. Joshi, U. J.; Shah, F. H.; Tikhele, S. H. Homology Model of the Human 5-HT1A Receptor Using the Crystal Structure of Bovine Rhodopsin. *Internet Electron. J. Mol. Des.* **2006,** *5* (7), 403–415.

25. a. Menon, V.; Vallat, B. K.; Dybas, J. M.; Fiser, A. Modeling Proteins Using a Super-Secondary Structure Library and NMR Chemical Shift Information. *Structure* **2013,** *21* (6), 891–899.

    b. Kitchen, D.; Hoffman, R. C.; Moy, F. J.; Powers, R. Homology Model for Oncostatin M Based on NMR Structural Data. *Biochemistry* **1998,** *37* (30), 10581–10588.

26. a. Henderson, R.; Schertler, G. F. X. The Structure of Bacteriorhodopsin and Its Relevance to the Visual Opsins and Other Seven Helix G-Protein Coupled Receptors. *Philos. Trans. R. Soc. Lond. Ser. B Biol. Sci.* **1990,** *326,* 379–389.

    b. Ovchinnikov, Y. A. Rhodopsin and Bacteriorhodopsin: Structure Function Relationships. *FEBS Lett.* **1982,** *148,* 179–191.

27. Teeter, M. T.; Froimowitz, M.; Stec, B.; DuRand, C. J. Homology Modeling of the Dopamine D2 Receptor and Its Testing by Docking of Agonists and Tricyclic Antagonists. *J. Med. Chem.* **1994,** *37,* 2874–2888.

28. a. Fachinger, G.; Deutsch, U.; Risau, W. *Oncogene* **1999,** *18,* 5948–5953.

    b. Krueger, N. X.; Streuli, M.; Saito, H. *EMBO J.* **1990,** *9,* 3241–3252.

    c. Wright, M. B.; Seifert, R. A.; Bowen-Pope, D. F. *Arterioscler. Thromb. Vasc. Biol.* **2000,** *20,* 1189–1198.

29. Evdokimov, A. G.; Pokross, M.; Walter, R.; Mekel, M.; Cox, B.; Li, C.; Bechard, R.; Genbauffe, F.; Andrews, R.; Diven, C.; et al. Engineering the Catalytic Domain of Human Protein Tyrosine Phosphatase Beta for Structure-Based Drug Discovery. *Acta Crystallogr. Sect. D Biol. Crystallogr.* **2006,** *D62,* 1435–1445.

30. Murray, C. W.; Rees, D. C. The Rise of Fragment-Based Drug Discovery. *Nat. Chem.* **2009,** *1,* 187–192.

31. Hartshorn, M. J.; Murray, C. W.; Cleasby, A.; Frederickson, M.; Tickle, I. J.; Jhoti, H. Fragment-Based Lead Discovery Using X-ray Crystallography. *J. Med. Chem.* **2005,** *48,* 403–413.

32. Lepre, C. A.; Moore, J. M.; Peng, J. W. Theory and Applications of NMR-Based Screening in Pharmaceutical Research. *Chem. Rev.* **2004,** *104,* 3641–3676.

33. Neumann, T.; Junker, H. D.; Schmidt, K.; Sekul, R. SPR-Based Fragment Screening: Advantages and Applications. *Curr. Top. Med. Chem.* **2007,** *7,* 1630–1642.

34. Hann, M. M.; Leach, A. R.; Harper, G. Molecular Complexity and Its Impact on the Probability of Finding Leads for Drug Discovery. *J. Chem. Inf. Comput. Sci.* **2001,** *41,* 856–864.

35. Wyatt, P. G.; Woodhead, A. J.; Berdini, V.; Boulstridge, J. A.; Carr, M. G.; Cross, D. M.; Davis, D. J.; Devine, L. A.; Early, T. R.; Feltell, R. E.; et al. Identification of N- (4-piperidinyl)-4-(2, 6-dichlorobenzoylamino)-1H-pyrazole-3-carboxamide (AT7519), a Novel Cyclin Dependent Kinase Inhibitor Using Fragment-Based X ray Crystallography and Structure Based Drug Design. *J. Med. Chem.* **2008,** *51,* 4986–4999.

36. Langmuir, I. Isomorphism, Isosterism and Covalence. *J. Am. Chem. Soc.* **1919,** *41,* 1543–1559.

37. Erlenmeyer, H.; Leo, M. Über Pseudoatome. *Helv. Chim. Acta* **1932,** *15,* 1171–1186.

38. Friedman, H. L. *Influence of Isosteric Replacements upon Biological Activity,* Vol. 206; National Academy of Sciences National Research Council Publication, 1951. 295.

39. Burger, A. *Medicinal Chemistry,* 3rd ed.; John Wiley & Sons: New York, 1970. p. 127.

40. Schann, S.; Bruban, V.; Pompermayer, K.; Feldman, J.; Pfeiffer, B.; Renard, P.; Scalbert, E.; Bousquet, P.; Ehrhardt, J. D. Synthesis and Biological Evaluation of Pyrrolidinic Isosteres of Rilmenidine. Discovery of Cis-/trans-Dicyclopropylmethyl-(4,5-dimethyl-4, 5-dihydro-3H-pyrrol-2-yl)-amine (LNP 509), an I_1 Imidazoline Receptor Selective Ligand with Hypotensive Activity. *J. Med. Chem.* **2001,** *44* (10), 1588–1593.

41. Wildsmith, J. A. W.; Gissen, A. J.; Takman, B.; Covino, B. G. Differential Nerve Blockade: Esters versus Amides and the Influence of PKa. *Br. J. Anaesth.* **1987,** *59* (3), 379–384.

42. Essery, J. M. Preparation and Antibacterial Activity of A-(5-Tetrazolyl)benzylpenicillin. *J. Med. Chem.* **1969,** *12,* 703–705.

43. Glass, R.; Loring, J.; Spencer, J.; Villee, C. The Estrogenic Properties *in Vitro* of Diethylstilbestrol and Substances Related to Estradiol. *Endocrinology* **1961,** *68,* 327–333.

44. a. Lima, L. M.; Barreiro, E. J. Bioisosterism: A Useful Strategy for Molecular Modification and Drug Design. *Curr. Med. Chem.* **2005,** *12,* 23–49.
    b. Sethy, S. P.; Meher, C. P.; Biswal1, S.; Sahoo1, U.; Patro, S. K. The Role of Bioisosterism in Molecular Modification and Drug Design: A Review. *Asian J. Pharm. Sci. Res.* **2013,** *3* (1), 61–87.
    c. Patani, G. A.; LaVoie, E. J. Bioisosterism: A Rational Approach in Drug Design. *Chem. Rev.* **1996,** *96,* 3147–3176.

45. a. Wang, Z.; Fermini, B.; Nattel, S. Evidence for a Novel Delayed Rectifier K^+ Current Similar to Kv1.5 Cloned Channel Currents. *Circ. Res.* **1993,** *73,* 1061–1076.
    b. Fedida, D.; Wible, B.; Wang, Z.; Fermini, B.; Faust, F.; Nattel, S.; Brown, A. M. Identity of a Novel Delayed Rectifier Current from Human Heart with a Cloned Potassium Channel Current. *Circ. Res.* **1993,** *73,* 210–216.
    c. Brendel, J.; Peukert, S. Blockers of the Kv1.5 Channel for the Treatment of Atrial Arrhythmias. *Expert Opin. Ther. Pat.* **2002,** *12* (11), 1589–1598.

46. a. Taglialatela, M.; Castaldo, P.; Pannaccione, A. Human Ether-a-gogo Related Gene (HERG) K Channels as Pharmacological Targets: Present and Future Implications. *Biochem. Pharmacol.* **1998,** *55* (11), 1741–1746.
    b. Vaz, R. J.; Li, Y.; Rampe, D. Human Ether-a-go-go Related Gene (HERG): A Chemist's Perspective. *Prog. Med. Chem.* **2005,** *43,* 1–18.
    c. Kang, J.; Wang, L.; Chen, X. L.; Triggle, D. J.; Rampe, D. Interactions of a Series of Fluoroquinolone Antibacterial Drugs with the Human Cardiac K^+ Channel HERG. *Mol. Pharmacol.* **2001,** *59,* 122–126.

47. Virshup, A. M.; Contreras-García, J.; Wipf, P.; Yang, W.; David, N.; Beratan, D. N. Stochastic Voyages into Uncharted Chemical Space Produce a Representative Library of All Possible Drug-Like Compounds. *J. Am. Chem. Soc.* **2013,** *135,* 7296–7303.

48. a. Reeder, E.; Sternbach, L. H. Process for Preparing 5-phenyl-1,2-dihydro-3H-1,4-benzodiazepnes US3109843, 1963.

    b. Schlager, L. H. Novel 3-hydroxy-1,4-benzodiazepine-2-ones and Process for the Preparation Thereof US4388313, 1983.

    c. Hester, J. B. 6-Phenyl-4H-s-triazolo[4,3-a][1,4]benzodiazepines US3987052, 1976.

    d. Tachikawa, R.; Takagi, H.; Miyadera, T.; Kamioka, T.; Fukunaga, M.; Kawano, Y. *Antidepressant Benzodiazepine, DE 1812252;* 1968.

49. Lipinski, C. A.; Lombardo, F.; Dominy, B. W.; Feeney, P. J. Experimental and Computational Approaches to Estimate Solubility and Permeability in Drug Discovery and Development Settings. *Adv. Drug Deliv. Rev.* **1997,** *23,* 3–25.

50. DiMasi, J. A.; Hansen, R. W.; Grabowski, H. G.; Lasagna, L. Cost of Innovation in the Pharmaceutical Industry. *J. Health Econ.* **1991,** *10,* 107–142.

51. Veber, D. F.; Johnson, S. R.; Cheng, H. Y.; Smith, B. R.; Ward, K. W.; Kopple, K. D. Molecular Properties that Influence the Oral Bioavailability of Drug Candidates. *J. Med. Chem.* **2002,** *45,* 2615–2623.

In vitro ADME and In vivo Pharmacokinetics

In the preceding chapters, a great deal of emphasis has been placed on identifying compounds that potently bind to macromolecular targets in order to induce the desired biological outcome. Structure–activity relationship principles and guidelines have been described from the perspective of potency with the ultimate goal of identifying compounds with optimized interactions with the target. There can be no doubt that biological potency is an important consideration in the process of identifying a new therapeutic entity. At the end of the day, however, there is far more to the drug discovery and development process than simply identifying the most potent compound at a single biological target. As previously discussed, compounds with poor selectivity for their intended targets are likely to have off-target effects that preclude eventual development. Issues of physicochemical properties such as solubility, permeability, and lipophilicity can also have a major impact on the success of a drug discovery program.

In modern drug discovery research, a wide variety of physicochemical properties are examined in the early stages of research programs, but this was not always the case. Prior to 1988, drug discovery and development was heavily focused on identifying active compounds. Issues of such as solubility, stability, pharmacokinetic properties, and toxicity were not examined until a clinical candidate had been identified. By the mid-1980s, however, it had become clear that the clinical failure rate and the associated costs were simply not sustainable. In an effort to understand why clinical candidates were failing to reach the market, Prentis et al. studied the causes of clinical failures in a set of clinical candidates identified by seven companies operating in the United Kingdom between 1964 and 1985.[1] Their study provided the startling revelation that nearly 40% of clinical candidate failures were caused by poor pharmacokinetic properties.

In simple terms, the pharmacokinetic profile of a compound can be defined as what the body does to the compound. Is the compound readily absorbed in the gastrointestinal (GI) tract? Compounds that are not absorbed into the body are unlikely to be drugs. Does the compound distributed broadly throughout the body or does it accumulate in a particular organ or tissue type? If a compound targeting the CNS is unable to pass through the blood–brain barrier, it will never reach its intended target. What is the metabolic fate of the compound and which enzymes are critical to its metabolism? Biologically active compounds that are rapidly metabolized are often removed from the systemic circulation by the liver before they are able to influence pharmacological events. Is the compound rapidly excreted by the kidneys or is it reabsorbed as it passes through the nephrons? Excretion by the kidneys removes the compound from the systemic circulation, preventing it from exerting biological activity. Collectively, these questions address the absorption, distribution, metabolism, and excretion profile of a compound and are often referred to as a compound's *in vivo* ADME properties.

In failing to study the pharmacokinetic profile and *in vivo* ADME properties of candidate compounds, the pharmaceutical industry was missing out on an opportunity to save millions of dollars. Understanding these features of a molecule can provide a great deal of insight as to whether or not a compound will survive the rigors of clinical trials. Since the largest expenditures associated with bringing a new drug to market are associated with clinical development, the ability to predict which compounds are more likely to succeed and which are likely to fail can substantially reduce the risk and cost of development programs.

Fortunately, over the last few decades, the pharmaceutical industry has developed a number of tools and techniques to study *in vivo* ADME and pharmacokinetic properties of compounds in the discovery phase. High throughput robotics, sophisticated detection systems, specialized cell lines, and advanced molecular modeling software have been developed to ascertain *in vitro* ADME properties, which can serve as predictive tools for *in vivo* ADME properties. *In vivo* animal pharmacokinetic studies are also routinely used to predict which compounds are capable of delivering potential drug compounds to their biomolecular target and predict drug performance in human populations. In addition, scientists have developed a greater understanding of the interplay of chemical structure, physicochemical properties, *in vitro* ADME, and *in vivo* pharmacokinetics. Thus, scientists are able to design new molecules with improved biopharmaceutical properties by applying structural modifications in an interactive process in the same manner as that described for optimizing target binding potency. The impact of this is clearly demonstrated by the marked decrease in the contribution of poor PK properties to clinical failures between 1991 and 2000. In 1991, 39% of clinical failures were the results of poor PK properties, but by 2000, this number had fallen to just 8% as the industry increased its emphasis on this important area (Figure 6.1).[2]

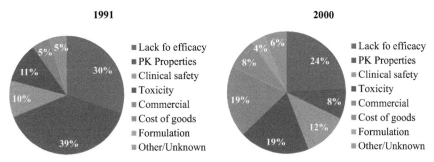

1991
5%
5%
11%
30%
10%
39%
Lack fo efficacy
PK Properties
Clinical safety
Toxicity
Commercial
Cost of goods
Formulation
Other/Unknown
2000
4%
6%
8%
24%
19%
8%
12%
19%
Lack fo efficacy
PK Properties
Clinical safety
Toxicity
Commercial
Cost of goods
Formulation
Other/Unknown

FIGURE 6.1 In 1991, poor pharmacokinetic properties caused an estimated 39% of clinical failures in the pharmaceutical industry. By 2000, an increased focus on designing compounds with improved physicochemical properties resulted in a significant decrease in the contribution of poor pharmacokinetics to drug failure. Only 8% of clinical failures could be attributed to poor pharmacokinetic properties.

In an ideal world, it would be possible to monitor changes in drug concentration as a function of time and dosage in all areas of the body independently. This would provide a real-time understanding of how the drug is distributed across the various tissues and organs as it is absorbed and subsequently removed from the body. In practice, this is neither possible nor practical. There are methods to examine drug levels in a wide array of tissues and organs that are employed in the later stages of a drug discovery and development program, but the expense of running all of them on an early stage compound is simply too high.

Rather than run an extensive array of tissue and organ studies, concentration in the systemic circulation is determined. These measurements are then used to define the pharmacokinetic profile of a compound in terms of its clearance (CL), volume of distribution (V_d), half-life ($t_{1/2}$), and bioavailability (%F). Although a more detailed analysis of these terms is provided later in this chapter, for the purposes of the earlier portions of this chapter, the following definitions will suffice (Figure 6.2). Clearance

1) Clearance (CL): Volume of blood cleared per unit time
2) Volume of Distribution: Compares drug plasma concentration to total drug in the body
3) Half-life ($T_{1/2}$): Time to decrease drug concentration 50%.
4) Bioavailability (%F): % dose reaching systemic circulation

FIGURE 6.2 Key pharmacokinetics terms.

describes how efficiently the body removes a compound from the systemic circulation and is generally defined as a volume of blood cleared per unit of time. Elimination of the compound can occur via excretion of the unchanged drug or metabolic conversion of the drug into a new compound (a metabolite). Volume of distribution is a mathematical term that relates the concentration of the drug in plasma to the total amount of drug in the body as a whole. As will be discussed in the latter half of this chapter, the volume of distribution is not a real volume and can dramatically exceed the actual physical volume of the body. The half-life of a compound is the amount of time required to decrease drug concentration by 50%, and is dependent on both clearance and the volume of distribution. Finally, bioavailability refers to the amount of drug that reaches the systemic circulation as compared to an intravenous injection of the same drug. In general, the oral bioavailability is expressed as a percentage of the corresponding intravenous injection.

It is worth noting at this point that the pharmaceutical industry is heavily focused on the identification of orally active therapeutic agents. While other methods of delivery such as intravenous injection, transdermal patches, and even intranasal delivery systems are available, compounds that cannot be delivered orally are far less likely to be pursued. If the pharmacokinetic parameters of clearance (CL), volume of distribution (V_d), half-life ($t_{1/2}$), and bioavailability (%F) are known, then it is often possible to predict whether or not a particular compound will reach the intended biomolecular target at a concentration high enough to induce a physiological response.

While this information is informative in deciding whether or not to move a single molecule further down the path of drug development, a deeper understanding is required in order to manipulate these parameters across a series of compound. Suppose, for example, a compound with excellent *in vitro* potency and selectivity was identified, but its bioavailability would not support oral dosing. In principle, one could attempt to design new analogs with improved bioavailability, but there are numerous factors that impact bioavailability. An understanding of the processes that drive bioavailability and the ability to design high throughput screens that model these factors would greatly enhance these efforts. The same is true of designing compounds with improved clearance, volume of distribution, and half-life. If these drivers can be identified, then one could correlate changes in chemical structure to changes in the drivers, thereby establishing a structure property relationship data set (SPR data set). This data set could then be used to optimize the desired properties and improve the likelihood that a compound will succeed. This raises the questions, what are the drivers of clearance, volume of distribution, half-life, and bioavailability and how can they be manipulated?

Simply put, the oral pharmacokinetic profile of a given drug is determined by (1) its absorption in the gastrointestinal tract; (2) its distribution through the systemic circulation, extracellular fluids, and tissues; (3) the rate at which it is metabolized; and (4) the rate at which it is excreted by the kidneys. Together, metabolism and excretion eliminate drugs from

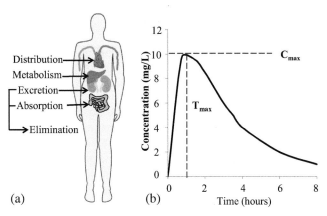

FIGURE 6.3 (a) The oral pharmacokinetic profile of a drug is determined by its absorption, distribution, metabolism, and excretion. (b) Oral delivery of a drug leads to an initial rise in plasma concentration until the maximum concentration (C_{max}) is reached (T_{max}). When the rate of elimination becomes greater than the rate of absorption, plasma concentration decreases. When absorption is complete, the drug is removed through elimination pathways.

the body (Figure 6.3(a)). Oral delivery of a drug leads to an initial rise in plasma concentration as the drug is absorbed. As the drug is distributed through the body, metabolism and excretion eliminate the drug from systemic circulation. In a single dose situation, the maximum drug concentration (C_{max}) is achieved when the rate of absorption is equal to the rate of elimination and is referred to as the T_{max} of the drug. When the rate of elimination is greater than the rate of absorption, plasma concentration begins to decrease. Once absorption is complete, the drug is eventually removed entirely through the elimination pathways (Figure 6.3(b)). Thus, in order to improve the pharmacokinetic profile within a series of compounds through structural changes (the application of medicinal chemistry), it is necessary to understand the fundamental factors that impact absorption, distribution, metabolism, and excretion.

ABSORPTION

As mentioned earlier, the pharmaceutical industry is heavily focused on oral drug delivery, ideally using a once-daily dosing regimen to improve patient compliance. In order for a compound to be orally bioavailable, it must be readily absorbed by the body. The key factors influencing absorption are solubility and permeability (the ability to pass through a biological barrier). Within a given series of compounds, structural modifications designed to optimize these properties can lead to an improvement in absorption, providing improved pharmacokinetic properties.

Solubility

Solid tablets are the preferred method of oral delivery, so the first step in administering an oral therapeutic is dissolution of the compound in the fluids of the gastrointestinal tract. The solubility of a compound, the maximum concentration of compound attainable in a solution, is a critical aspect of oral delivery. Compounds that are not in solution are not readily absorbed by the body. No matter how potent and selective a compound may be, if it cannot be absorbed, it will not be able to reach its intended target and no therapeutic effect will be observed. In other words, compounds with low solubility are likely to have absorption issues and unlikely to become marketed drugs.

In essence, the process of dissolution of any material into a solvent requires that solvent molecules disrupt the interaction between the molecules that comprise the solid. These interactions are then replaced by interactions between the solvent molecules and the molecules of the dissolving compound (Figure 6.4). It stands to reason, therefore, that a compound's

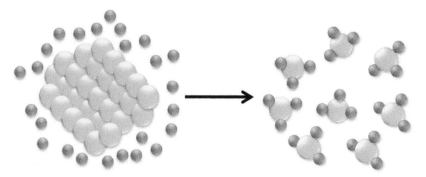

FIGURE 6.4   When a compound (yellow) dissolves in water (blue), the water molecules disrupt the interactions that stabilize the solid form of the material. The solubilized compound forms a favorable interaction with water.

aqueous solubility is a function of its ability to form energetically favorable interaction with water molecules as the solid material breakdown. There are a number of factors that can be manipulated in a given series of compounds to increase aqueous solubility. It has long been known, for example, that as the molecular weight of compounds within a given series increases, the aqueous solubility decreases. This is demonstrated by the change in solubility observed in 2-amino-imidazolidinone Kv1.5 blockers (**1**) and (**2**) (Figure 6.5). Extending the linker chain between the central imidazolidinone and the left-side benzene ring provides a relatively small increase in molecular weight, but that leads to a nearly 5 fold decrease in solubility.

It is also possible to manipulate the aqueous solubility of compounds in a given series by modifying the structures so that they are more polar in nature. This can be accomplished by adding polar groups, increasing the number of hydrogen bond donors and acceptors, or by adding ionizable groups to the compound in question. The addition of even a single hydrogen bond donor or acceptor, for example, can have a significant impact on solubility. Again referring to the series of amino-imidazolidinone Kv1.5 blockers, exchanging a methyl group (**3**) with a methoxy (**4**) more than doubles the observed solubility (Figure 6.5).[3] Increased polarity and the ability

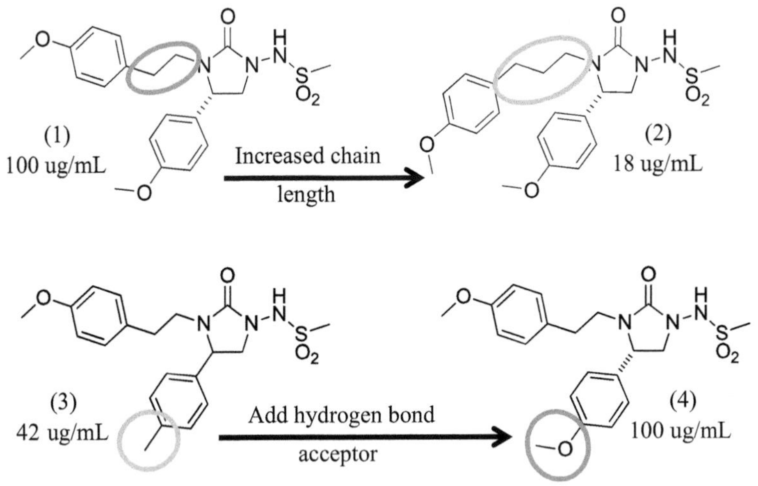

FIGURE 6.5   Increasing the length of the carbon linker by one methylene unit, as seen in (**1**) and (**2**), leads to a 5 fold change in solubility. The addition of a hydrogen bond acceptor (**3** vs **4**) more than double aqueous solubility.

to form additional hydrogen bonds were also factors in improved solubility in the series of benzoimidazol-2-one androgen receptor antagonists (Figure 6.6). Modification of benzoimidazol-2-one (**5**) to include the nitrile as seen in (**6**) increases the overall polarity of the compound and adds an additional hydrogen bond acceptor, which results in a greater than 5 fold increase in solubility. In the same series, modification of (**7**) to incorporate a fluorine atom and an *N*-methyl acetamide (**8**) provides increased polarity and a substituent capable of acting as both a hydrogen bond donor and acceptor. As a result, the aqueous solubility increases by more than 75 fold.[4]

In some cases, dramatic improvements in aqueous solubility can be accomplished with a simple isosteric replacement as seen in the carbon/ nitrogen switch in hexahydropyrazinoquinolines (**9**) and (**10**) (Figure 6.7). This simple change had minimal impact on the desired dopamine D_3 binding (5.1 nM vs 9.7 nM), but provided a 50 fold increase in solubility (from 1 mg/mL to 50 mg/mL, **9** vs **10**). In this case, the pyridine nitrogen installed in (**10**) is both capable of hydrogen bonding interaction and can

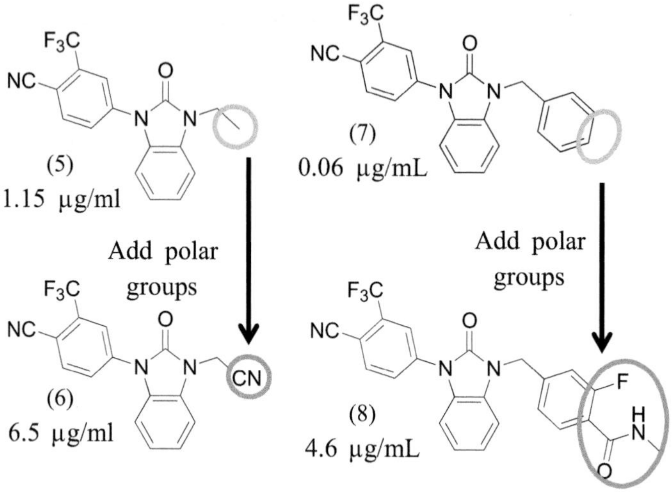

FIGURE 6.6   The increased polarity and additional hydrogen bond acceptor present in (6) as compared to (5) increases solubility by a factor of five. Adding a fluorine atom and an N-methyl acetamide to (7) increases polarity and adds a hydrogen bond donor and a hydrogen bond acceptor. As a result, (8) is over 75 times more soluble.

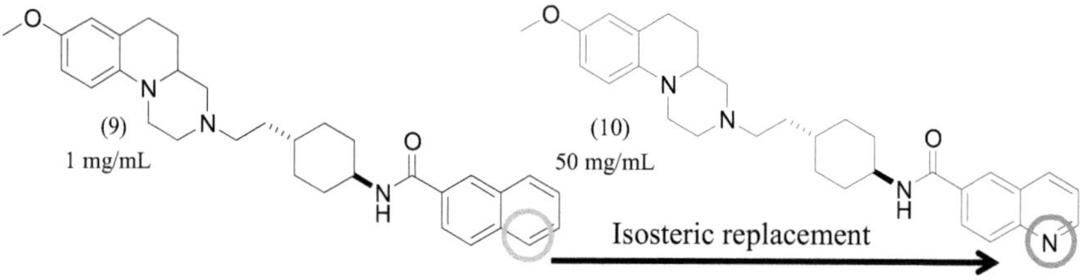

FIGURE 6.7   Isosteric replacement of a carbon (9) with a nitrogen (10) produces a 50 fold increase in aqueous solubility with minimal change to *in vitro* activity at the dopamine D_3 receptor.

be protonated to form the salt, which provides an additional boost in solubility relative to (9).[5]

Modification of a series of compounds to incorporate a carboxylic acid or basic amine is the most effective method of increasing solubility and also the most common. In fact, over 70% of marketed drugs possess a basic amine, and less than 5% are neutral compounds.[6] The presence of an ionizable functional group supports the formation of a charged species in the appropriate pH buffer system, and this can improve solubility by several orders of magnitude. In the development of histone deacylase inhibitors (Figure 6.8), for example, replacing the alcohol substituent in compound (11) with a dimethylamino substituent (12) increased solubility from 5 µg/mL

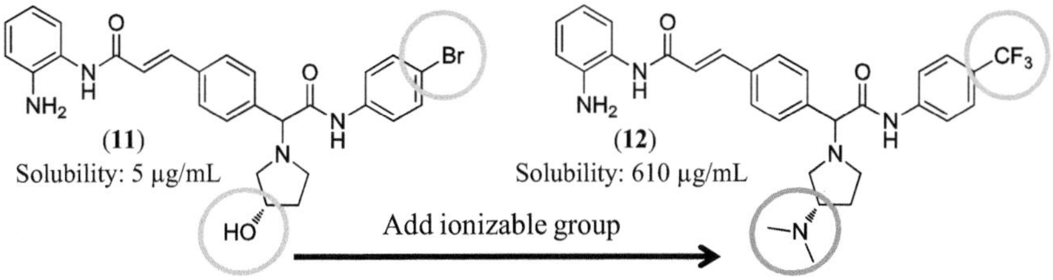

FIGURE 6.8   Although the alcohol functionality of (11) can form hydrogen bonds, replacing it with an *N,N*-dimethylamine substituent (12), an ionizable group, provides a more than 100 fold increase in aqueous solubility.

to 610 μg/mL, a 122 fold increase. Although the alcohol functionality is capable of forming hydrogen bonds, the ionizable character of the dimethylamine dramatically improves aqueous solubility.[7]

A common theme among all of the changes discussed thus far is the overall decrease in lipophilic character of the compounds. In general, in a series of related compounds, aqueous solubility increases as lipophilicity decreases. The lipophilicity of a compound can be assessed by determining its partition ratio in an 1-octanol/water system. Compounds that are more lipophilic will be more heavily concentrated in the 1-octanol layer of the system, while less lipophilic (more polar) compounds will preferentially occupy the water layer. Direct physical measurement of this phenomenon is readily accomplished, however, in most cases the ratio is easily determined with software packages and expressed in a logarithmic scale. It is most often designated as LogP (or cLogP), and refers to the calculated value under neutral conditions (pH = 7).[8a,b] The term LogD is less frequently employed, and refers to the partitioning between 1-octanol and water at pH 7.4, or physiological pH.[9] One could, of course, also view these changes as an increase in hydrophilic character. This feature also plays a role in permeability and will be discussed again in the following pages.

In some cases, it is possible to increase solubility by decreasing the stability of the solid material. Compounds that contain planar regions, for example, will form solid matrices in which the planar regions are effectively stacked on top of each other. This is especially true with aromatic systems in which the stacked π clouds interact in an energetically favorable manner. Adding substituents capable of preventing this stacking or removing the aromatic character of a ring will eliminate these interactions, destabilizing the solid form of the compound, which can lead to improved aqueous solubility. In the identification of hedgehog signaling inhibitors, for example, a nearly 10 fold increase in solubility is observed between the fully aromatic compound (13) and the corresponding cyclohexyl variant (14)

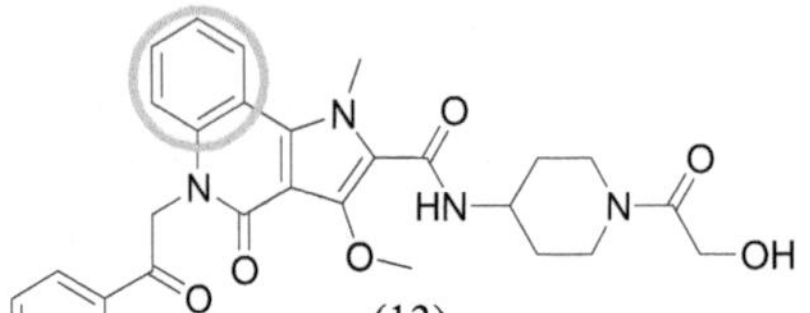

(13)
8.4 ug/mL

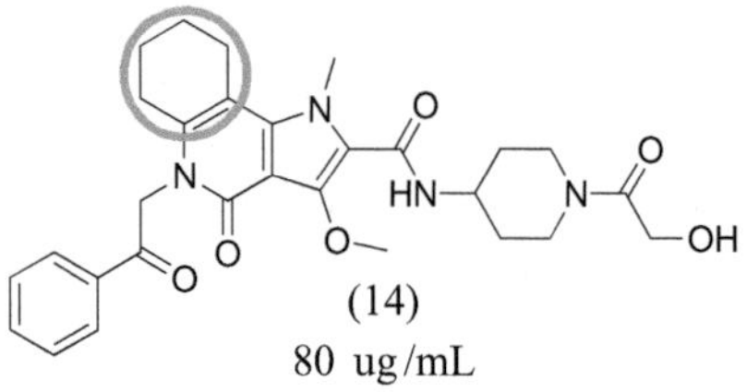

(14)
80 ug/mL

FIGURE 6.9   Eliminating potential π-stacking interactions and decreasing the planarity of (13) by replacing a benzene ring with a cyclohexane ring provides a 10 fold increase in solubility in (14) in a series of hedgehog signaling inhibitors.

(Figure 6.9).[10] Loss of planarity that accompanies this structural alteration decreases the energy of interaction in the solid state, decreasing the energy barrier to solvation by water. Increasing the number of rotatable bonds in a series of compounds can also lead to a similar decrease in the stabilization energy provided by a solid packed form, decreasing the energy required for solvation.

In considering the issue of solubility as it relates to drug discovery, a question often posed by drug discovery scientists is how soluble does a compound need to be in order for it to be considered for advancement in a program? Answering this question is not straight forward, as there are many factors to consider, but an initial assessment may be provided from the molecular weight and potency at the target of interest. Consider, for example, a compound with a molecular weight of 400. If this compound has an IC_{50} of 1.0 μM at its macromolecular target and a plasma concentration of three times the IC_{50} is desired for therapeutic effect, then the minimum aqueous solubility for this compound is 1.2 μg/mL. On the other hand, if the compound has an IC_{50} of 10 nM, then the concentration necessary for the same relative plasma concentration (three times the IC_{50}) is only 0.012 μg/mL, a 100 fold decrease. To a first approximation, as target potency increases, solubility requirements decrease.

Permeability

Once a compound is in solution, it must be able to pass through the cellular barriers that separate the gastrointestinal tract from the systemic circulation. In other words, the compound must be permeable. A compound that is both potent and soluble is unlikely to succeed as an orally delivered therapeutic in the absence of permeability. The compound will not be absorbed, it will not be able to reach its intended target, and no therapeutic effect will be observed. Permeability can also play an important role in determining the utility of a drug once it has entered the systemic circulation, as there are additional cellular barriers that may come into play. If the biomolecular target is inside a specific cell type, a hepatocyte for example, then the compound must be able to move through both the barriers of the gastrointestinal tract and the hepatocyte cell membrane in order to induce a therapeutic response. Differences in permeability between various cell types can also be an issue, as differences in membrane composition, junction tightness (the space between cells), and transporter protein activity can all influence compound permeability. Compounds intended to modulate CNS functions, for example, must traverse the blood–brain barrier, which is far more restrictive in nature. Tight cellular junctions and higher levels of transport protein expression designed to protect the brain from xenobiotics must be overcome in order for CNS targeted compounds to elicit a response.

The observed permeability of any given compound across a biological barrier is the sum of five modes of transport: passive diffusion, active transport, endocytosis, paracellular transport, and efflux (Figure 6.10). In

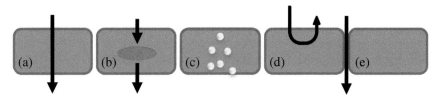

FIGURE 6.10   (a) Passive diffusion; (b) Active transport; (c) Endocytosis; (d) Efflux; (e) Paracellular transport.

considering the transfer of potential therapeutics across the GI tract, the most common mode of transport is passive diffusion. An estimated 95% of all marketed orally available drugs employ this method of transport.[11a,b] Simply put, compounds that are absorbed via passive diffusion move down a concentration gradient from areas of high concentration, such as the GI tract, to areas of low concentration, such as the systemic circulation. Compound polarity can play a major role in passive diffusion, as in order for passive diffusion to occur a compound must exit an aqueous environment, pass through the lipophilic environment of the cell membrane (which is composed of a phospholipid bilayer), and then reenter the aqueous environment on the other side of the biological membrane. It should be no surprise that neutral compounds undergo passive diffusion far more readily than charged species. Absorption in the GI track is further complicated by the requirement that compounds transit through two biological barriers, the apical and basolateral membranes of the GI tract.

Environmental pH also can influence rates of passive diffusion. In acidic regions such as the stomach, basic compounds are largely protonated, so the availability of neutral material capable of passing through a cellular barrier is low. In basic environments, on the other hand, basic compounds are largely neutral and therefore more available for passive diffusion. Compounds that are acidic in nature experience the opposite situations as those described for basic compounds.

A second method available for transporting a compound across a cellular barrier is active transport.[12] The gastrointestinal tract is, of course, designed to absorb nutrients using a variety of active transport system, and drug compounds that can take advantage of one of these systems enter the systemic circulation using these transport systems. Active transport systems in other types of cells can also be exploited for the transport of a compound across a biological membrane. In the majority of cases, a transmembrane protein is involved and energy, typically in the form of ATP, must be expended in order for transport to occur. This method is far less common than passive diffusion.

It is also possible for compounds to move across biological barriers via paracellular transport and endocytosis, but these mechanisms are far less common than passive diffusion. Endocytosis requires compounds to be trapped in a membrane vesicle on the outer surface of the cell, transported through the cell membrane, and then released on the opposite side when the vesicle reopens.[13] Paracellular transport, on the other hand takes advantage of the space between cells in certain types of membranes.[14] Compounds move through the pores between cells of the GI tract, kidney cells, or other "leaky" membranes, thus avoiding the lipophilic environment of the cell membrane. Hydrophilic compounds incapable of diffusing into the lipid layer of a cellular membrane may be transported in this manner, provided they are small enough to fit in the space between the cells (typically 8 Å). Larger compounds (e.g., proteins) are too large to undergo paracellular transport.

Passive diffusion, active transport, endocytosis, and paracellular transport generally act to move a compound from one side of a biological barrier to the other, often moving down a concentration gradient, but not always (active transport). There are, however, systems in place designed to protect the body from xenobiotics. These efflux transporters generally move compounds against a concentration gradient, require the expenditure of energy (e.g., ATP), and their action decreases the ability of a compound to cross a biological barrier. In simple terms, compounds entering the cell membrane are captured by an efflux system protein, and are then ejected from the biological barrier on the same side as they entered. The net effect is an overall decrease in compound permeability through the cell membrane in question.

P-glycoprotein (Pgp) efflux proteins, a member of the ATP-binding cassette family of transporters, are a prototypical example.[15a,b] Unlike many other proteins, Pgps are capable of acting upon a broad range of substrates. They are heavily expressed in tissues such as the brain, liver, kidneys, intestines, and uterus. In principle, their action could decrease passage of compounds from the GI tract into the systemic circulation, but high concentrations generated by oral dosing often far exceed Pgp capacity in the gut. The impact of Pgp and other transport systems is more significant in areas of lower drug concentration, such as the blood–brain barrier. In this scenario, plasma drug concentration is unlikely to exceed Pgp expression and their substrates can be effectively prevented from entering the brain. Programs designed to modulate CNS targets can be significantly impacted by Pgp activity. Other transport systems include the breast cancer resistant protein (BCRP)[16a,b] and the multidrug resistance protein 2 (MDR2).[17] All three of these transporters play a major role in chemotherapy-induced drug resistance in cancer cells.

As mentioned above, the overall permeability of a given compound is the sum of all methods of transmembrane transport, but given that over 95% of drug molecules enter the systemic circulation via passive diffusion, improving this mode of entry is often the focus of efforts to improve overall permeability. Lipophilicity and polarity are key drivers of passive diffusion, but there must be a balance between the two. Compounds that are too polar will be less able to desolvate and enter the highly non-polar region of the phospholipid bilayer that makes up a cell membrane. On the other hand, compounds that are too hydrophobic (non-polar) may permeate the cell membrane, but then fail to exit the membrane as it is the preferred environment for the compound. Different strategies are employed to address problems of relative polarity and hydrophobicity. Highly polar side chains such as carboxylic acids can have a significant, negative impact on permeability. In order for a compound to cross a cellular membrane, it must be neutral, and in many physiologically relevant environments, carboxylic acids are substantially deprotonated. Ciproflaxcin, for example, suffers from low cell permeability as a result of the presence of a carboxylic acid moiety (Figure 6.11). Masking this group as a methyl ester (**16**) provides a 10 fold increase in cell permeability.[18] In a similar manner, GS-4071, an influenza neuraminidase inhibitor (Figure 6.11), is virtually incapable of entering the systemic circulation

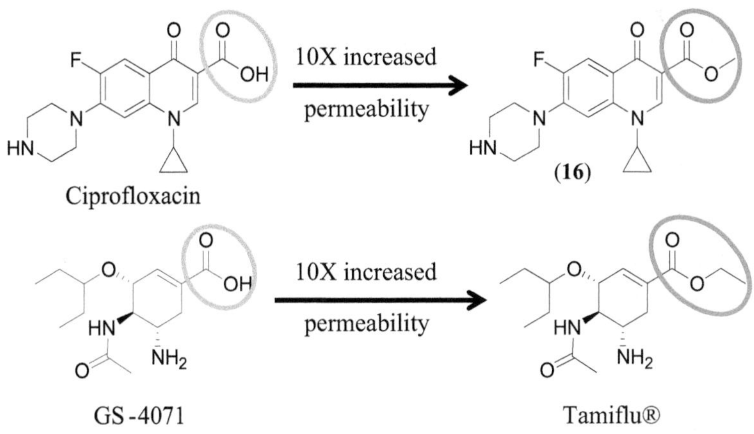

FIGURE 6.11   The cellular permeability of ciprofloxacin and GS-4071 are substantially improved by converting them to the ester prodrugs (**16**) and Tamiflu®.

from the GI tract. Masking the polarity of the carboxylic acid, however, provides a compound capable of entering the systemic circulation, Tamiflu®, which is then converted to the active agent (GS-4071) by enzymes in the plasma. The strategy of forming a prodrug to increase compound permeability is a common theme in drug discovery.[19]

In contrast, altering the basicity of a compound can lead to significant changes in permeability. Once again, the issue of neutrality becomes an important factor in a compound's ability to pass through a biological barrier. Substantially basic compounds may be protonated under physiologically relevant conditions, and the resulting positive charge will limit permeability. In other words, compounds that are overly basic may have low permeability due to the low proportion of free base available to traverse a cell membrane, and adjusting the pKa may increase permeability. In a series of IKKβ inhibitors (Figure 6.12), for example, a 25-fold increase

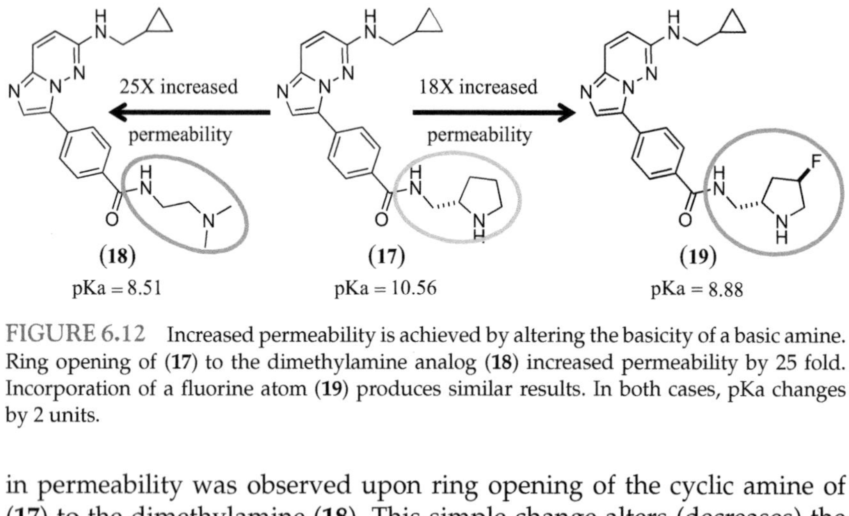

FIGURE 6.12 Increased permeability is achieved by altering the basicity of a basic amine. Ring opening of (17) to the dimethylamine analog (18) increased permeability by 25 fold. Incorporation of a fluorine atom (19) produces similar results. In both cases, pKa changes by 2 units.

in permeability was observed upon ring opening of the cyclic amine of (17) to the dimethylamine (18). This simple change alters (decreases) the pKa of the basic amine by 2 units, substantially altering the amount of free base available for passive diffusion. The more subtle addition of a fluorine atom to the structure (19) produces a similar decrease in pKa and increase in permeability.[20]

As mentioned earlier, lipophilicity, the ability of a compound to enter a non-polar matrix such as a cellular membrane, can also be a key driver of passive diffusion. In theory, one could measure the relative concentration of a compound in membrane fractions exposed to an aqueous solution in order to determine how they would distribute between the two. In practice, however, partitioning between 1-octanol and water is used as a surrogate for the membrane–water interface. The relative concentration can be experimentally determined and the results are reported as a the log of the ratio of the concentration in octanol versus the concentration in water. This value is referred to as LogP, and is readily calculated using modern software packages. The calculated values are designated as cLogP, and both refer to situations in which the aqueous solution is adjusted to pH of 7.[21]

Increasing the permeability of a series of compounds can be accomplished by making structural modification that bring the cLogP closer to that observed for the majority of orally dosed drugs. In general, cLogP between 1 and 3 is considered optimal, although the actual optimal value for any given set of compounds will depend on the specific nature of the series itself. In the case of the renin inhibitors (20) and (21) (Figure 6.13),

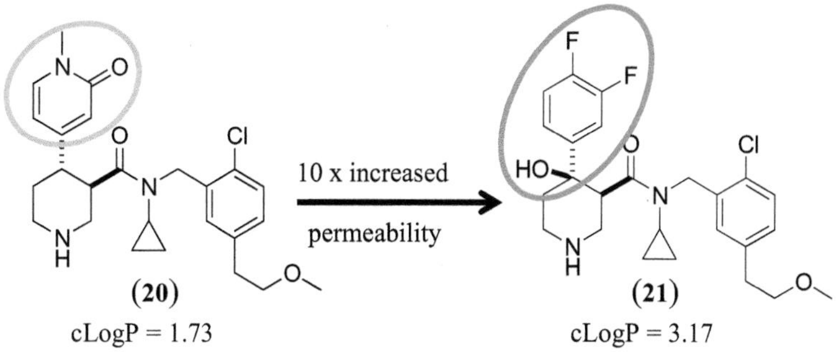

FIGURE 6.13 Decreasing the cLogP of 20 by replacing the pyridone ring with a 3,4-difluorobezene ring and tertiary alcohol (21) produces a 10 fold increase in permeability.

for example, a ten-fold increase in permeability was achieved by increasing compound lipophilicity. Replacing the pyridone moiety of 20 with an alcohol and 3,4-difluorophenyl unit (21) decreased the polarity of the overall compound as indicating by the increase in cLogP from 1.73 to 3.17.[22] This can be contrasted with the development of the series of β-secretase-1 (BACE) inhibitors typified by (22), (23), and (24) (Figure 6.14). In this case, the first compound, (22), is overly lipophilic as indicated by its cLogP of 5.55. Modifying the lipophilic tert-butyl moiety to incorporate a cyano group (23) decreased the lipophilicity of the compound as indicated by the decrease in cLogP to 4.80. This subtle change led to an 8-fold increase in permeability. Decreasing lipophilicity further by replacing the entire tert-butyl group with a cyano group (24) lowers the cLogP even further (cLogP = 3.86) and the permeability doubles as a result.[23]

The relative polarity of a series of compounds can also be expressed as the sum of the surface area of the polar substituents, also referred to as the topological polar surface area (TPSA) of a compound. Adjusting the TPSA of a series of compounds can significantly alter the permeability of a given series of compounds. In the series of hydroxethylamine BACE-1

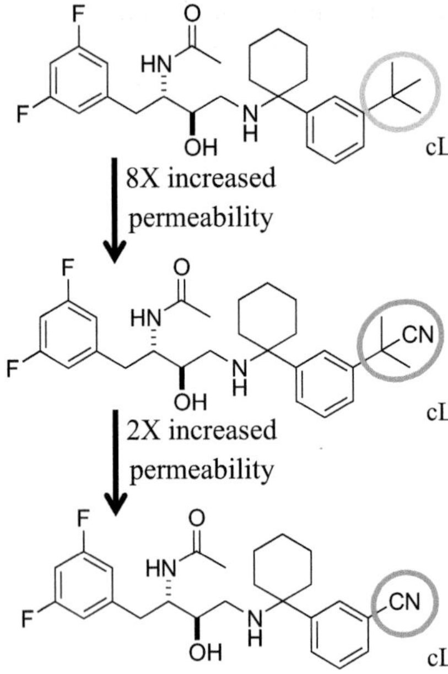

(22)
cLogP = 5.55

(23)
cLogP = 4.80

(24)
cLogP = 3.86

FIGURE 6.14 The permeability of a series of Beta-secretase (BACE) inhibitors is increased by a factor of 16 by decreasing cLogP. Modification of the *t*-butyl group (**22**), first by addition of a cyano (**23**) and then by removal of the flanking methyls (**24**) lowers the cLogP from 5.5 to 3.86.

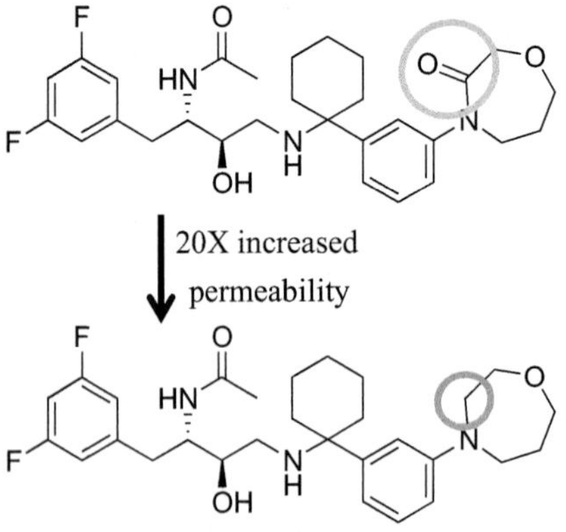

(25)
TPSA = 90.9

(26)
TPSA = 73.8

FIGURE 6.15 Removal of a carbonyl in a series of BACE-1 inhibitors leads to a 20 fold increase in permeability.

inhibitors, for example, the presence (**25**) or absence (**26**) of a carbonyl produces a 20-fold change in permeability (Figure 6.15).[23]

Although modifying polarity, lipophilicity, and TPSA across a series of compounds can lead to significant changes in permeability, there are additional modifications that can be applied to increase permeability.

In many cases, decreasing the number of rotatable bonds and increasing compound rigidity will have a positive impact on permeability. Common methods of accomplishing this include insertion of double bonds, constraining systems into a ring, or adding steric hindrance to slow molecular rotation (Figure 6.16). In each case, the constrained compound

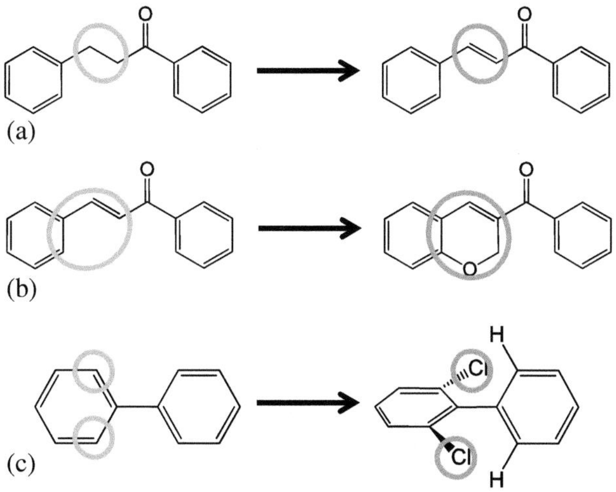

FIGURE 6.16 (a) Incorporation of an alkene lowers the number of rotatable bonds. (b) Formation of a new ring eliminates rotational freedom. (c) Addition of flanking groups restricts rotation.

is subject to less entropy change upon moving from an aqueous environment to the lipid environment of the cell membrane. This lowers the overall energy required for transition between an aqueous environment and the hydrophobic interior of a cell membrane, which increases a compounds overall permeability.

Achieving optimal levels of permeability requires a delicate balance between polarity and lipophilicity. In some cases, structural changes designed to impart greater permeability may have a negative impact on compound solubility. Decreasing compound polarity by removing a carbonyl as seen in Figure 6.15 (decrease in TPSA), for example, has a positive impact on permeability, but the loss of a potential hydrogen bond donor could have a negative impact on solubility. Increasing compound rigidity through the incorporation of points of unsaturation, rings, or slowing molecular rotation (Figure 6.16) is also likely to have a negative impact on aqueous solubility. It is important to realize that in a given series, compounds with the highest solubility may be distinctly different from compounds with highest degree of permeability. Program progression may require a choice between optimizing one property over another, but the choice of which property to optimize must be made carefully, as both properties are important to absorption and oral drug delivery.

FIGURE 6.17 The Biopharmaceutics Classification System (BCS) categorizes compounds based on their solubility and permeability. The likelihood of successful development of an orally delivered drug is best for class 1 compounds and worst for class 4 compounds. Compounds in class 2 and 3 can be used for oral dosing, but may require special formulation or alternative approaches.

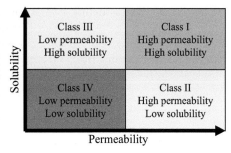

The Biopharmaceutics Classification System (Figure 6.17) was developed to assist scientists in assessing the development risk associated with a compound's absorption as it relates to solubility, permeability, and its likelihood of success as an orally delivered compound. Based on their high aqueous solubility and permeability, BCS class 1 compounds are the most likely to be useful upon oral administration. On the other hand, BCS class 4 compounds represent a significant risk if pursued via oral administration, as their poor aqueous solubility and permeability are likely to significantly limit absorption. Compounds with low aqueous solubility but high permeability (BCS class 2) may require special formulations, while compounds with low permeability but high solubility (BCS class 3) may be suitable candidates for a prodrug approach.[24]

DISTRIBUTION

After a compound enters the systemic circulation, whether through oral absorption, intravenous (IV) injection, intraperitoneal (IP) injection, or any other method, its ability to elicit a biological response will be controlled in part by its ability to reach the intended target. Even the most potent compound will fail to produce the desired outcome if it is unable to reach the macromolecular system it was designed to modulate. Once the compound enters the systemic circulation, it will be distributed throughout the body, but this distribution is often uneven. Some compounds will readily pass into the muscles and organs, while others may be unable to exit the circulatory system. Compounds with the correct physicochemical properties will be able to pass through additional protective barriers around some organs, such as the blood–brain barrier, while others may be excluded and therefore incapable of generating a response in the targeted organ. The presence of off-target side effects is also governed in part by a compounds distribution. If a compound has the potential to cause a side effect, but is unable to reach the offending biological target, then the side effect will not occur. Thus, an overarching question that must be considered as any compound moves into animal models and eventually human studies is where does a compound go when it enters the bloodstream? Stated in another manner, how is the compound distributed?

The distribution of a compound in the body refers to the reversible transfer of a drug between the various tissues, organs, cells, etc. of the body. Compound distribution can be the key to having a successful drug rather than a failed compound as highlighted by nitrofurantoin and desmethylprodine (MPPP). Macrobid® (nitrofurantoin) was originally introduced in 1957 as a treatment for urinary tract infections (Figure 6.18). It

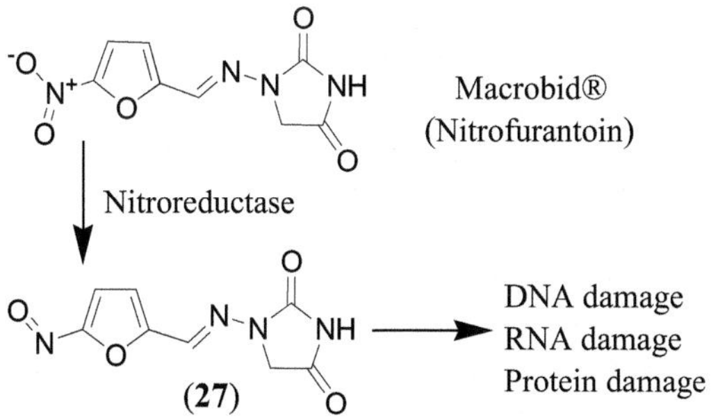

FIGURE 6.18 Macrobid® (nitrofurantoin) is useful for the treatment of urinary tract infections, despite its potential for conversion into (27) by nitroreductase. Its potential for toxicity via DNA damage, RNA damage, and protein damage is mitigated by its rapid excretion into the bladder.

has been highly successful, despite the fact that it contains an aryl nitro group which is usually associated with mutagenicity, teratogenicity, and carcinogenicity.[25a,b] The successful application of this drug is a direct result of its limited distribution in the body. Although it is capable of causing significant DNA, RNA, and protein damage, Nitrofurantoin is rapidly distributed to the bladder and excreted in the urine. This prevents any untoward effect on the body, and places the drug exactly where it needs to be in order to eliminate urinary tract infections.[26]

In a similar manner, but in the opposite sense, the failure of desmethylprodine, also known as MPPP, is in part a result of compound distribution (Figure 6.19). MPPP is an opioid analgesic originally brought to market by Hoffmann-La Roche in the 1940s, but is no longer used in clinical practice. Although MPPP itself is not dangerous, it is metabolized to another compound, 1-methyl-4-phenyl-1,2,3,6-tetrahydropyridine (MPTP), which crosses the blood–brain barrier. Once inside the brain, MPTP is absorbed by glial cells which convert it to 1-methyl-4-phenylpyridinium (MPP$^+$). This charged species is incapable of exiting the brain, is absorbed by the dopaminegic cells of the substantia nigra (the region of the brain that

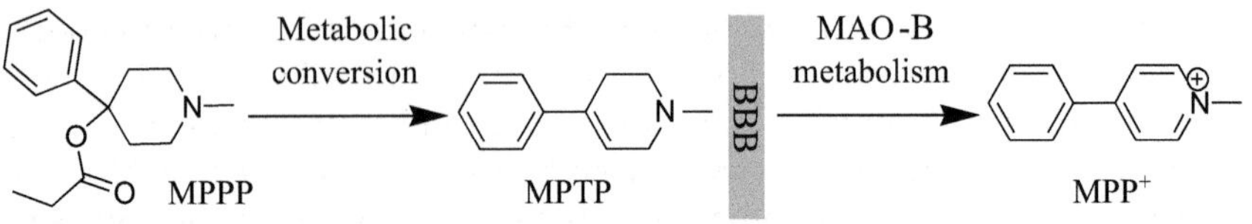

FIGURE 6.19   Desmethylprodine (MPPP) is an opioid analgesic that is no longer marketed due to the formation of the metabolites MPTP and MPP⁺. MPTP enters the brain, where it is converted to MPP⁺ by the enzyme MAO-B. MPP⁺ cannot exit the brain and kill dopamine-producing cells in the brain leading to the rapid development of Parkinson's disease symptoms including paralysis. MPTP, 1-methyl-4-phenyl-1,2,3,6-tetrahydropyridine; MPP⁺, 1-methyl-4-phenylpyridinium; MAO-B, monoamine oxidase B; BBB, blood–brain barrier.

generates dopamine enabling movement), and kills these cells. The death of these cells causes rapid development of Parkinson's disease symptoms and paralysis. In this case, the combination of metabolism and compound distribution leads to terrible outcome that would not occur in the absence of MPPP's ability to distribute into the brain.[27]

As was the case with absorption, issues of permeability can play a major role in determining the extent of distribution of a compound, but there are other factors that can have an impact on this process. Plasma protein binding and transporter proteins also play a significant role in determining the relative concentration of potential drug compounds in various areas of the body. Each of these factors warrants consideration in designing and evaluating potential new therapeutic entities.

Permeability

When a compound is delivered orally, the first biological barrier that it encounters is the lining of the gastrointestinal tract, but this is not the only barrier a compound may be required to cross in order for it to influence a biological system. If the targeted macromolecule is inside a particular cell type such as a cardiac myocyte, an islet cell of the pancreas, or an astrocyte within the brain, then the passage through an additional biological membrane will be required. In order to reach many of these targets, compounds must exit the systemic circulation and then enter the targeted cells. As such, the ability of a given compound to distribute into the proper tissues and cell types will be controlled in part by the compound's permeability through the various biological barriers between bloodstream and the target of interest. A compound that can achieve passage into the systemic circulation via absorption or one that is delivered via a method that bypasses oral absorption (e.g., intravenous injection or similar methods), but cannot distribute to the intended target of interest will have limited utility as a therapeutic agent due to poor target engagement.

In considering a compound's ability to distribute throughout the body and the impact of permeability, it is important to realize that biological barriers are not uniform. The plasma membranes of the cells lining the GI tract are substantially different from those of the pancreas, which are different still from the cell membrane that surrounds the various types of neuronal cells. A compound with a high degree of permeability through the GI tract may still experience distribution issues as a result of poor permeability through cellular barriers encountered between the GI tract and the intended target.

Of course, compounds designed to modulate targets that circulate in the bloodstream do not need to exit the systemic circulation, so permeability beyond the GI tract is not an issue with respect to reaching the biomolecular target. However, a compound's permeability through other tissue types can still play an indirect role. As will be discussed later in this chapter, extraction of compounds by the liver and kidneys can lead to metabolism and excretion. Rates of metabolism by liver enzymes will be impacted by a compound's ability to enter liver cells (liver cell permeability), while excretion rates will be controlled in part by a compound's ability to permeate through the cellular barriers within the kidney. In other words, a compound's liver metabolism and renal excretion will depend in part on its ability to be distributed into these organs.

Distribution of potential drug compounds into the brain is a special case worthy of further elaboration. Compounds that impact the CNS account for a substantial portion of the drug market, and yet they are perhaps the most difficult to identify, as therapeutic action requires that compounds pass through the blood–brain barrier. This protective layer consists of a single monolayer of cells that line the inner surface of the capillaries that run throughout the brain, providing it with nutrients and oxygen, and removing cellular waste products. The endothelial cells of the BBB are packed together tightly (Figure 6.20), preventing

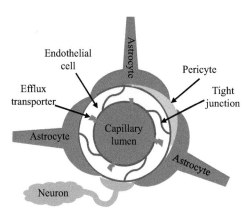

FIGURE 6.20 The blood–brain barrier consists of a monolayer of tightly packed endothelial cells that prevent paracellular transport and endocytosis. Passive diffusion is the main route of entry into the brain and abundant transporter proteins further slow entry of xenobiotics.

paracellular transport, and endocytosis is far more limited than in other cellular barriers. Passive diffusion is the main route of entry for xenobiotics, and the same physical properties that govern passive diffusion in the GI tract also hold sway in the BBB. It is important to note, however, that the composition of the phospholipid bilayer of the endothelial cells of the BBB is significantly different from that of the GI. It is, in fact, far more restrictive in nature, and the presence of high levels of P-glycoprotein transporters on the apical side of the monolayer further complicates entry of compounds into the brain.[28]

Modulating the distribution of a series of compounds by altering cellular permeability may be achieved using the same types of changes capable of influencing absorption in the GI tract. Thus, structural changes designed to alter lipophilicity, polarity, TPSA, rotatable bond count, and compound rigidity can impact compound distribution via changes in cellular permeability. In some cases positive changes designed to increase absorption in the GI tract will also lead to improvements in compound distribution properties, as the changes also improve a compound's ability to reach the intended targets. In other instances, however, structural changes designed to improve absorption though improved permeability may be counterproductive with respect to distribution, as the changes may decrease cellular permeability at the targeted cell, tissue, or organ. It is important that drug discovery scientists be aware of this potential crosscurrent of drug design in order to maximize the benefit to GI tract permeability and permeability into other biologically relevant systems.

Transporters

While passive permeability plays a major role in compound distribution, it is not the only driver of distribution. As discussed earlier, differences in the composition of biological membrane can have a significant impact on passive diffusion, leading to uneven distribution of a compound through the body. In a similar manner, differences in transporter protein expression, both amount and type, will have an impact on a given compound's ability to penetrate a particular part of the body. In general, a specific transporter protein is expressed on only one side of a cell membrane, apical or basolateral, in order to ensure unidirectional movement of the substrate. They are capable of moving compounds against a concentration gradient or transporting compounds that are incapable of passive diffusion across a membrane. These functions require the expenditure of energy, often in the form of ATP/ADP hydrolysis, and transporters can be saturated if substrate concentration exceeds transporter protein capacity. For this reason, transporters have a lesser role in absorption as compared to distribution. Drug concentration in the intestine is high compared to transporter expression leading to saturation of transporters, whereas in other regions of the body (e.g., blood), drug concentrations are generally well below the concentrations required for transporter saturation.[29]

As mentioned earlier, P-glycoproteins (also referred to as multidrug resistant protein 1 (MDR1)) have been extensively studied as a result of their impact on the drug concentrations, especially with respect to the blood–brain barrier, but there are a number of other transporters that have an impact on compound distribution.[30a,b] The breast cancer resistant protein (BCRP) was originally identified as a key mechanism for the development of chemotherapeutically resistant breast tumor cells as the name implies, but it has been subsequently identified in normal hepatocytes, the placenta, and other tissues.[31] Additional examples of transporters include the organic anion transporters (OATs),[32] organic cation transporters (OATs),[33] di/tri peptide transporters,[34a,b] organic anion-transporting polypeptides (OATPs),[35] and monocarboxylic acid transporter.[36]

The impact of a transporter system on a compound's distribution can be either positive or negative depending on the desired outcome. If, for example, one is attempting to develop a compound that is directed towards a target in the brain, high Pgp activity would be a significant barrier to success. On the other hand, if entry into the brain would produce an undesired side effect, then high Pgp activity would be a positive outcome, as distribution into the brain would be limited. In a similar manner, compounds that are substrates for uptake transporters on specific organs can increase compound concentration in these organs, leading to either a positive or negative result, depending on the outcome. If high concentrations created by transporter activity enhance a compound's ability to reach its intended target, then *in vivo* activity may be higher than anticipated by pharmacokinetic studies. On the other hand, if an off-target activity or toxicity is driven by concentration of the compound in a particular organ or tissue type, transporter activity leading to increased concentration in the organ or tissue type can enhance undesired activity or lower the threshold required for toxicity.

It is also worth noting that drugs can be substrates for more than one transporter protein, which further complicates the picture. Expression levels of functional transporter proteins are variable across tissue types, and of course, compound affinities/rates of transport are different for each protein. Crestor® (rosuvastatin), for example, is a substrate for MDR1, multidrug resistance-associated protein 4 (MRP4), and OATP-1B3, as well as several others. Similarly, Gleevec® (imatinib) is a substrate for at least six different transporter proteins, each of which influences its tissue distribution (Figure 6.21).[37] It should be clear from these examples that a complete understanding of the impact of transporter protein on drug distribution requires knowledge of the role of multiple members of this important family of macromolecules. Given the wide variety of transporter proteins, there is no single set of structural changes that can be used as a guide for altering transporter-mediated drug transport. Removing or augmenting

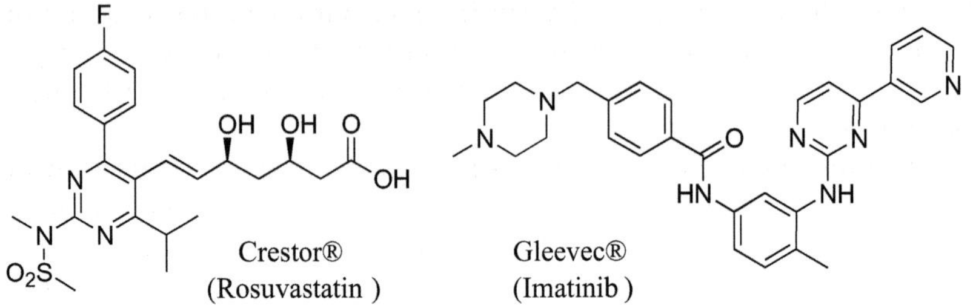

FIGURE 6.21   Crestor® (rosuvastatin) and Gleevec® (imatinib) are blockbuster drugs despite the fact they are transporter protein substrates.

transporter activity is best approached through an SAR-driven paradigm similar to that employed to optimize potency at the intended drug target. As complex as this problem may be, however, significant transporter activity is not necessarily a terminal problem for a drug discovery and development program. Despite their interaction with multiple transporter proteins, both Gleevec® (imatinib) and Crestor® (rosuvastatin) are multibillion dollar products.

Plasma Protein Binding

Once a compound enters the bloodstream, it will interact with the various blood components. Red blood cells, platelets, leukocytes, and various proteins can bind xenobiotics, and these interactions will have an impact on compound distribution and therapeutic action. The most important aspect of these binding interactions are those that occur between compounds and plasma proteins. Drug discovery programs routinely screen compounds to determine the extent of plasma protein binding associated with compounds of interest. While there are a number of plasma proteins, the two main contributors to plasma protein binding are human serum albumin (HSA)[38] and α1-acid glycoprotein (AGP).[39] Both of these proteins are tasked with transporting naturally occurring compounds through the body, so it should come as no surprise that they also undergo binding interactions with xenobiotics. There are, however, significant differences between the two. HSA is significantly more abundant, comprising approximately 60% of the overall plasma protein content, and blood concentrations of HSA are typically on the order of 35–50 mg/mL (500–750 μM). AGP concentration, on the other hand, is typically 0.6–1.2 mg/mL (~15 μM, 1–3% of total plasma protein), but can approach 3 mg/mL in some disease states. The type and number of compounds that can bind to either protein also differs significantly.[40a,b] AGP binds hydrophobic (e.g., steroids) and basic compounds (e.g., amines) primarily via non-specific hydrophobic interactions and contains a single binding site. In contrast, HSA can interact with acid, basic, and neutral compounds in a large number of primary and secondary binding sites. A single molecule of HSA can, for example, bind to at least 10 molecules of imipramine.[41]

The importance of plasma protein binding of drugs is expressed in the "free drug hypothesis"[42] which states that plasma protein bound compounds are incapable of passive diffusion and paracellular transport. Only "free" compounds, those that are not bound to plasma proteins, can cross biological barriers and gain access to tissues and organs outside of the bloodstream (Figure 6.22). In other words, plasma protein binding

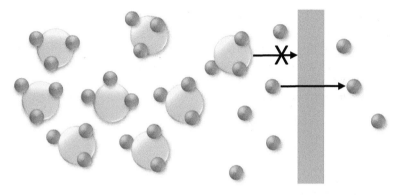

FIGURE 6.22   According to the "free drug hypothesis" compounds (blue) that are bound to plasma protein (yellow) are incapable of crossing a cellular membrane (green).

can have a direct impact on compound distribution. The level of impact will depend on both the extent of binding, often expressed as a percent of compound bound at equilibrium, and the association/disassociation rate. Compounds that are highly protein bound and have a very slow disassociation rate, for example, may be restricted to the bloodstream, as very little of the free drug is available. This can restrict entry of compounds into target tissues, the impact of which can be either positive or negative depending on the perspective. Brain penetration, for example, may be limited, which would be bad for compounds designed to modulate CNS targets, but good for compounds with potential CNS side effects. Liver metabolism and excretion by the kidneys may also be decreased in these situations, leading to improvements in pharmacokinetic properties. On the other hand, compounds that are highly protein bound and have a slow on/off rate are also less available to interact with their intended target, so their therapeutic impact may be less than anticipated. Compounds that possess these properties are "restricted" by plasma protein binding effects.

In contrast, compounds that rapidly disassociate from plasma protein (i.e., have fast kinetics) may not be limited by the impact of plasma protein binding. Even in cases where plasma protein binding is high as a percentage, some level of free drug will always be available to move down a concentration gradient through a plasma membrane. Compounds that have a high plasma protein binding and fast dissociation rates, or have low plasma protein binding (irrespective of the rate of dissociation) are referred to as "permissive" compounds. The importance of dissociation rates as it relates to plasma protein binding becomes evident when considering the currently marketed compounds. A 2002 review of 1500 drugs revealed that over 40% of the compounds are >90% protein bound, and that a high number of anti-inflammatory drugs (26%) displayed >99% protein binding. Clearly, some free drug must be available in order for these compounds to exert a therapeutic effect. In these cases, high on/off rates (dissociation rates) offset the impact of high levels of protein binding.[43] Thus, high plasma protein binding can be tolerated under the correct circumstances (Figure 6.23).

Name	Plasma Protein Binding	Use
Diazepam (Valium®)	99%	Antianxiety agent
Ibuprofen (Advil®)	99%	Antiinflammatory
Lorazepam (Ativan®)	92%	Antianxiety agent
Naproxen (Aleve®)	99%	Antiinflammatory
Amlodipine (Norvasc®)	93%	Antihypertensive
Omeprazole (Prilosec®)	95%	Proton pump inhibitor
Doxycycline (Vibramycin®)	90%	Antibiotic
Efavirenz (Sustiva®)	99%	Antiviral

FIGURE 6.23   Protein binding of known drugs.

ELIMINATION PATHWAYS

The concepts of absorption and distribution describe how a compound enters and moves through the body. At some point, however, compounds that enter the body are removed via one or two basic elimination methods. Molecules can be metabolized to a different compound, effectively removing the original compound from circulation, or they can be excreted unchanged into the urinary tract by the kidney. These pathways are not mutually exclusive, but one pathway may be the dominant mechanism of elimination. The combined action of these mechanisms, however, will ultimately control how long a given compound will reside in the body after a dose is provided. In order to alter the residence time of a compound in the body, it is essential to understand the contribution of these pathways.

Extending the half-life of a compound is often the focus of drug dis-covery programs, but it is important to keep the end goal of a program in mind as a program progresses. While it may be useful to develop a pain medication that is slowly eliminated from the body in order to allow once a day dosing for 24 h relief, the same is not true of a medication designed to cure insomnia. A potential insomnia treatment that last for 24 h could prevent the patient from waking up after the standard 8 h of sleep. While people suffering from insomnia might need to catch up on a significant amount of sleep, it is unlikely that they would want to accomplish this by sleeping for 24 h after taking a sleeping pill. The end goal of a program must be kept in mind when considering the elimination of a compound from systemic circulation.

Metabolism

As soon as a compound enters the body, the various mechanisms in place designed to protect the body from xenobiotics begin to process the foreign material. The biological barriers to the entry into the circulatory system of an orally delivered material, the walls of the intestines and the stomach, are the first line of defense. When a compound has been absorbed in the GI tract, it moves directly into the portal vein, which transports the compound to the liver (Figure 6.24). While metabolic processing of

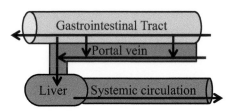

FIGURE 6.24 Upon absorption in the GI tract, compounds enter the portal vein, which delivers them to the liver. The meta-bolic machinery of the liver acts as an addi-tional barrier between xenobiotics and the systemic circulation.

compounds can occur in a variety of different places, the vast majority of drug metabolism occurs in the liver. This organ is specifically designed to chemically modify a wide range of compounds in order to enhance their elimination through the kidney. In order for an orally delivered material to move beyond the liver and exert an influence on bodily function, some fraction of the material must survive the presystemic metabolism of the liver. This is also referred to as "first-pass metabolism" as it is the first time the drug moves through the liver. Compounds that exit the liver enter the systemic circulation and move through the body, but they are continu-ously exposed to the liver with each cycle through the circulatory system, providing repeated opportunities for metabolic processing. Each pass through the liver will decrease the concentration of the compound, and this in turn will have a direct impact on properties such as bioavailability, exposure, and half-life. Higher rates of metabolism lead to decreases in each of these properties, and as such, compound metabolism is often a critical concern in drug discovery programs.

Compound metabolism can be divided into two sets of processes that are referred to as phase 1 and phase 2 metabolism.[44] In general, phase 1 metabolism reactions modify the molecular structure of a compound via processes such as oxidation or dealkylation, often leading to the production of material that possess a "handle" that can be utilized as an attachment point for a polar group that facilitate excretion by the kidneys. Phase 2 metabolism takes advantage of the presence of potential points of attachment, such as hydroxyl groups, by adding polar groups to molecules, such as glucuronic acid or glutathione. These reactions are referred to as conjugation reactions, and although phase 1 metabolism can prepare a compound for conjugation, phase 1 metabolism is not necessarily a prerequisite for phase 2 metabolism. Compounds containing functional groups capable of undergoing conjugation reactions can bypass phase 1 metabolism and act as substrates for the enzymes responsible for phase 2 metabolism.

Chemical modifications that occur in phase 1 metabolism include oxidations, reductions, and dealkylation. While there are a wide range of enzymes that take part in these processes, *monooxygenases* are the most prominent players in these processes. As the name implies, these enzymes generally add an oxygen atom to their substrate (Scheme 6.1). This process

$$R\text{-}H + O_2 + NADPH + H^+ \longrightarrow R\text{-}OH + O_2 + NADP^+ + H_2O$$

SCHEME 6.1   General process of monooxygenase metabolism.

requires the consumption of molecular oxygen and converts NADPH to NADP$^+$ (dihydronicotinamide-adenine dinucleotide phosphate to nicotinamide adenine dinucleotide phosphate).

Monooxygenases are classified based on the cofactors that are required for activity. The cytochrome P450 (CYP450) family of enzymes is the largest family in this class. Functional activity requires the presence of both a heme molecule and an iron atom (Figure 6.25).[45] It is estimated

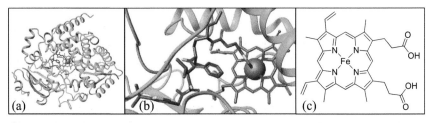

FIGURE 6.25   (a) Crystal structure of CYP3A4 bound to desoxyritonavir analog (RCSB 4K9W). (b) Close up of binding site (RCSB 4K9W). (c) Structure of heme molecule from CYP3A4.

that over 90% of drug metabolism occurs through CYP450-mediated processes, so it should be no surprise that drug discovery programs focus on developing compounds that are resistant to these enzymes. This is not easily accomplished, however, as to date 57 human CYP450 genes have been characterized. The gene products can be further characterized as members of 18 distinct families of CYP450s and 43 subfamilies within these sets.[46] It is also worth noting that these enzymes are capable of functionalizing a wide range of compounds with diverse structures, as the enzyme's pockets are broad and less discriminating than other types of enzymes. This decreases their overall velocity (reaction speed), but allows the CYP450s to broadly protect the body from potentially harmful chemicals. Their variable level of expression and uneven contribution further complicates the picture. In human liver microsomes (Figure 6.26(a)), for example, CYP3A4 is the most abundant

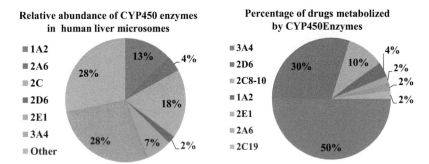

FIGURE 6.26   Relative abundance of CYP450 enzymes in human liver microsomes and their contribution to overall metabolism.

member of the family (28%), followed by the 2C subfamily (18%), and CYP1A2 (13%), while CYP2D6 is expressed as only 2% of the CYP mixture. Abundance, however, does not necessary equate with importance with respect to drug metabolism. Even though CYP2D6 comprises only 2% of the CYPs expressed in the liver, it is responsible for approximately 30% of drug metabolism. In contrast, CYP1A2, which has a significantly higher expression level, is responsible for only 4% of drug metabolism. Given that CYP3A4, 2D6, and the 2C family predominate drug metabolism, drug discovery programs have focused on designing compounds with increasing degrees of metabolic stability in the presence of these important enzymes.

The flavin monooxygenase (FMO)[47] family is another important contributor to the phase 1 metabolism of xenobiotics, but the active site requirements are substantially different from the CYP450s. In these enzymes the heme and iron atom are replaced by a flavin adenine

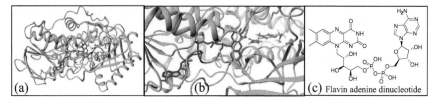

FIGURE 6.27   (a) Human monoamine oxidase B (MAO-B) in complex with rosiglitazone, RCSB file 4A7A. (b) Close up image of binding site of MAO-B in complex with rosiglitazone, RCSB file 4A7A. (c) Structure of flavin adenine dinucleotide.

dinucleotide (Figure 6.27)[48] which serves as a carrier for oxygen that is delivered to FMO substrates. To date, five FMOs, designated FMO1–5 have been identified, and much like the CYP450s, their catalytic sites are designed to accommodate a wide range of structural diversity, which leads to lower substrate conversion rates (decreased enzyme velocity). Additional enzymes that play a role in phase 1 metabolism include aldehyde oxidase (AO),[49] alcohol dehydrogenase (ADH),[50] monoamine oxidases (MAOs),[51] and nitroreductases.[52] Some common phase 1 processes are shown in Figure 6.28. All of these enzyme-mediated reactions can be employed by the body to eliminate potential drug molecules.

FIGURE 6.28   Typical phase 1 metabolic processes.

Hydroxylation of an aliphatic or aromatic carbon is the most commonly observed metabolic product. If further oxidation to an aldehyde, ketone, or carboxylic acid is possible, then additional phase 1 oxidative metabolism may occur. Heteroatoms such as nitrogen and sulfur can also be oxidized to form the corresponding hydroxyl amines, sulfoxides and sulfones. Ether cleavage and deamination reactions are also possible, as enzymatic oxidation of carbon atoms with alpha heteroatom substitution leads to structures that are chemically unstable and decompose to the corresponding by-products.

Initial processing of compounds through phase 1 metabolism generally establishes new nucleophilic sites on a substrate, either via direct functionalization as observed in hydroxylation pathways or via decomposition of unstable compounds such as those generated in oxidative deamination reactions. These new functional groups, most often alcohols and amines, provide an opportunity for further metabolic processing in phase 2 metabolic pathways (conjugation reactions). Of course, if these groups are present in a molecule to start with, then phase 1 metabolism is not necessary for phase 2 conversions to occur. The enzymes that comprise the phase 2 metabolic pathways append groups onto the new sites, generally producing compounds of increased polarity that are more readily excreted into the urine. UDP-glucuronosyl transferases (UGTs),[53] for example, appends glucuronic acid onto compounds with a suitable functional group such as an alcohol or amine (Figure 6.29(a)). This process is often referred to as glucuronidation, while the products are referred to as glucuronides. In a similar fashion, sulfotransferases[54] increase the polarity of phase 2 substrates via sulfation of a suitable functional group. The resulting compounds are substantially more polar and significantly more susceptible to excretion in the urine (Figure 6.29(b)).

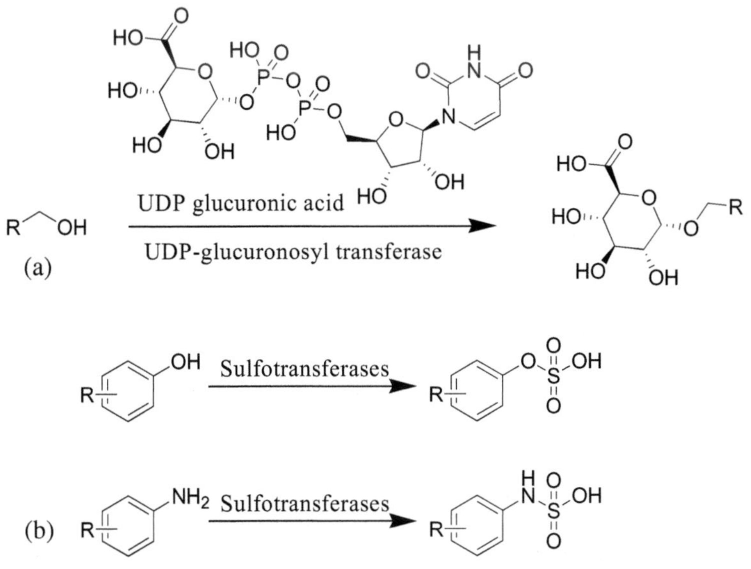

FIGURE 6.29 (a) Glucuronidation of an alcohol. (b) Sulfonylation of an alcohol or amine.

While many phase 2 reactions occur at nucleophilic sites such as, carboxyl (–COOH), hydroxyl (–OH), amino (NH₂), and sulfhydryl (–SH) groups, they can also occur at electrophilic sites such as electron-deficient double bonds and epoxides. Glutathione conjugation, the addition of glutathione, a tripeptide antioxidant designed to protect the body from reactive oxygen species, is also a common phase 2 process. The reaction is mediated by members of the glutathione *S*-transferases (GSTs) family of enzymes, which are capable of appending glutathione on to substrates with suitable electrophilic sites (Figure 6.30).[55]

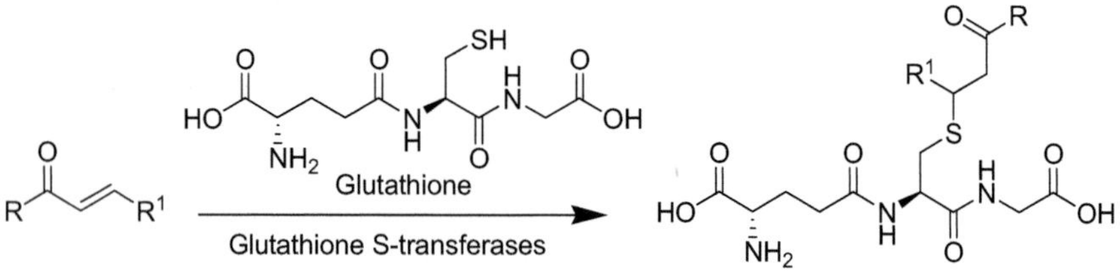

FIGURE 6.30　Addition of glutathione to a Michael acceptor.

Although the preceding discussion on metabolism has been concerned with individual conversions that can occur within the context of compound metabolism, it is important to be aware that individual compounds may be a substrate for multiple metabolic enzymes. A compound that possesses an ether, a benzene ring that can be hydroxylated by an appropriate CYP450 enzyme, and an alcohol that can be glucuronidated will most likely be subject to metabolism at all three sites (Figure 6.31). The relative

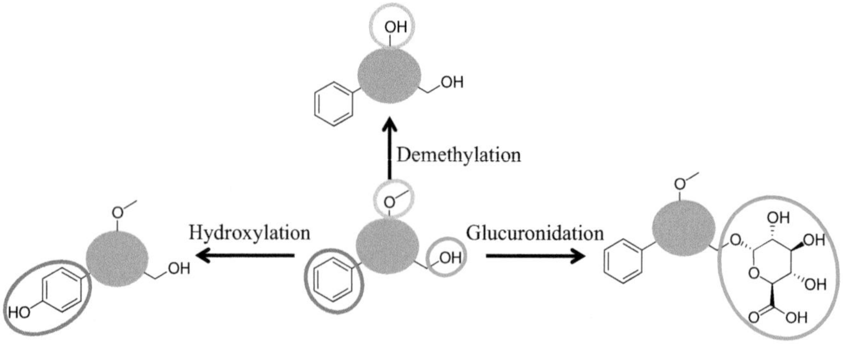

FIGURE 6.31　Compounds can be metabolized at multiple sites in a molecule, and each process occurs at a different rate. If the primary pathway is blocked, a secondary pathway will become the dominant metabolic pathway.

velocity of each metabolic conversion will dictate the ratio of metabolites that are formed as a result of this mixture of activity. Also, if one of these pathways is blocked by structural changes in the molecule or saturated by concentration of the substrate that exceed the metabolic pathway's capacity, then secondary metabolic pathways will become more favorable. This phenomenon is known as metabolic switching.

The metabolism of a given compound can also be impacted by another compound if they are coadministered. If, for example, a drug that is metabolized by a particular CYP450 is administered to a patient and a second drug that inhibits the activity of the same CYP450 is also administered to the same patient, then the metabolism of the first drug will be slowed by the presence of the second drug. In this situation, a drug–drug interaction (DDI) occurs in which the metabolism of the first drug is lower than expected as a result of the presence of the second drug. This change in metabolic profile will lead to changes in the pharmacokinetic profile of the first drug, increasing drug concentration, and possibly increased toxicity. DDIs, especially those that are caused by CYP450 inhibition, are a major concern in the discovery and development of novel therapeutics, as these enzymes play a major role in the metabolism of marketed drugs. The identification of DDIs led to the revocation of marketing approval for previously successful drugs such as Seldane® (terfenadine), which can cause fatal cardiac arrhythmias (torsades de pointes) when coadministered with macrolide antibiotics or ketoconazole (Nizoral®) (Figure 6.32).[56]

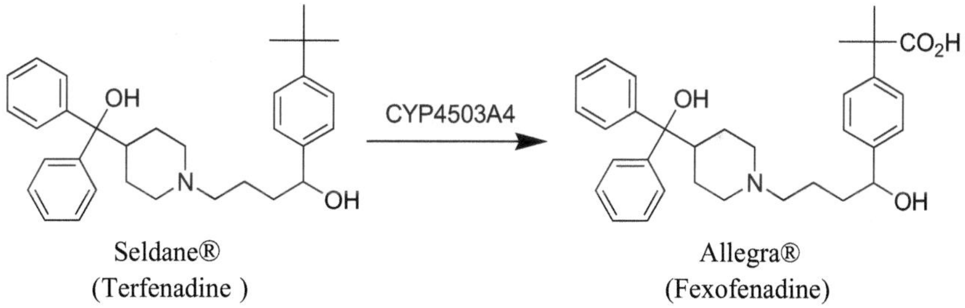

Seldane®
(Terfenadine)

Allegra®
(Fexofenadine)

FIGURE 6.32 Terfenadine is normally metabolized to fexofenadine, but the presence of certain macrolide antibiotics or ketoconazole blocks this process. This increases the systemic concentration of terfenadine to unsafe levels that can cause fatal cardiac arrhythmias.

In the absence of metabolic stability, it is highly unlikely that a compound will proceed beyond initial investigation. Compounds that are highly metabolized are typically removed from the systemic circulation before they are able to exert an impact on a biological system. In many cases, however, compounds with excellent activity at a biological target of interest are rapidly metabolized. In these instances, an understanding of structure–activity properties as they related to enzymatic metabolism can serve as a guide for improving the metabolic profile of candidate compounds in a series of interesting compounds. Designing compounds that are stable to metabolic processes is simply a matter of designing compounds that are no longer substrates for the various metabolizing enzymes. This can be accomplished through the same type of structure–activity analysis designed to optimize biological activity at the target of interest, but with the opposite goal, minimized enzymatic activity. Of course, this has to be accomplished without eliminating the desired biological activity, as a metabolically stable compound with no biological activity is not very useful as a therapeutic agent.

There are essentially three methods of addressing metabolic instability through structural modifications of candidate compounds, remove, replace, and restrict. If a functional group on a compound is a metabolic weak point (site of metabolism), it may be possible to remove the offending functionality in order to increase metabolic stability. Metabolically labile sections of a candidate compound can also be replaced with groups that are less vulnerable to metabolic degradation. Bioisosteric replacement of labile functionality is a common theme in drug discovery. Restricting access to sites of metabolism, either by physically blocking access via steric crowding, conformational restrictions, or changes in the electronic character of an aromatic ring, has all been successfully employed to slow the metabolism of biologically interesting compounds.

Unfortunately, given the wide range of metabolic enzymes and the disparate nature of chemical space and compound classes that have potential utility as drugs, there is no one specific course of action that can be taken to alleviate problems associated with poor metabolic stability while maintaining activity at the intended biological target (not to mention target selectivity, solubility, permeability, and every other property that must be maintained in a successful candidate compound). Each class of compounds must be examined independently with an eye towards balancing the structure–activity relationships that maintain the desired properties, while using structure activity relationships that suppress metabolism to enhance stability. It is, however, worthwhile to consider some examples of how structural changes impact metabolic stability.

Removing functionality that is not necessary for biological activity can be an effective method of improving metabolic stability. In an effort to identify novel heat shock protein 90 (HSP90) inhibitors, Zehnder et al. identified a series of pyrrolodinopyrimidines. Although a phenethyl side chain (Figure 6.33(a)) was tolerated within the context of HSP90 inhibition, its presence was a liability with respect to metabolic stability. Removal of the phenyl ring (Figure 6.33(b)) greatly improved metabolic stability with minimal impact on HSP90 inhibition.[57] In a similar manner, the aromatic character of the pyridine ring of a series of Rho-associated protein kinase-2 (ROCK-2) inhibitors proved to be a metabolic liability that could be removed by simply eliminating the aromatic character of the ring. Replacing the pyridine ring with the corresponding piperidine ring provided a significant boost in metabolic stability

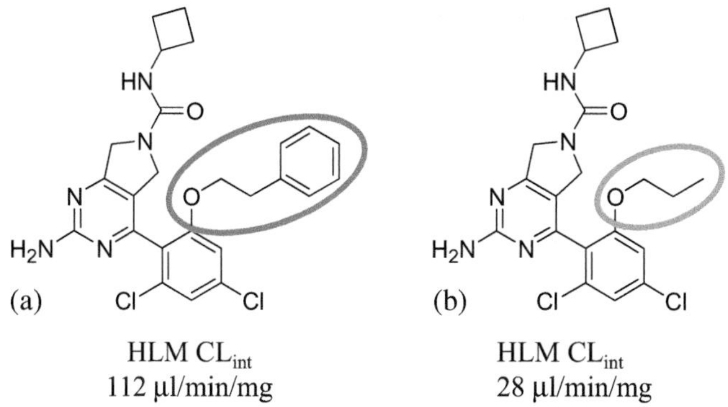

(a) HLM CL$_{int}$
112 µl/min/mg

(b) HLM CL$_{int}$
28 µl/min/mg

FIGURE 6.33 Removal of a benzene ring that is not required for biological activity increases metabolic stability as indicated by the drop in human liver microsome intrinsic clearance (HLM Clint CL$_{int}$) from (a) to (b).

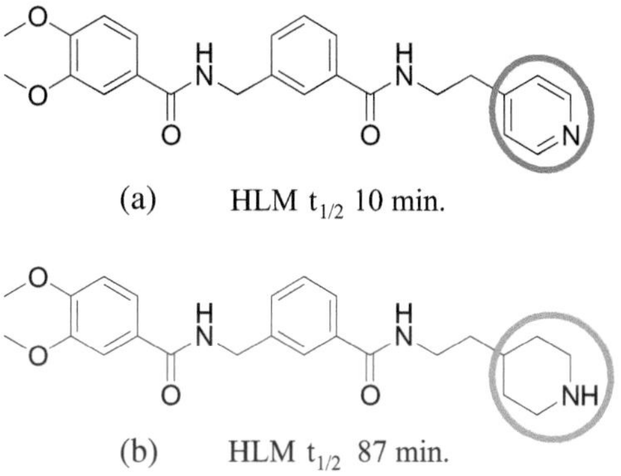

(a) HLM t$_{1/2}$ 10 min.

(b) HLM t$_{1/2}$ 87 min.

FIGURE 6.34 A significant boost in metabolic stability is observed when a pyridine ring (a) is replaced with a piperidine ring (b), as indicated by the increase in human liver microsome half-life (t$_{1/2}$). HLM, human liver microsome.

(Figure 6.34).[58] In both of these cases, removal of a metabolically labile feature of a compound provided a path forward for a drug discovery program.

In some situations, it is not possible to remove metabolically labile groups, as they are required for biological activity. It may, however, be possible to modify a vulnerable site within the labile group so that it is no longer subject to metabolism. This can often be accomplished by replacing hydrogen atoms with fluorine atoms. While hydrogen atoms on an alkyl chain can be metabolically replaced with an alcohol, fluorine atoms cannot. This strategy was successfully employed to increase the metabolic stability of urea transporter B inhibitors. A compound that was substantially metabolized after a 30-min incubation with rat liver microsomes (<5% remaining, Figure 6.35(a)) could be modified to a compound that was minimally metabolized under the same conditions (96% remaining after 30 min) by simply replacing two hydrogen atoms

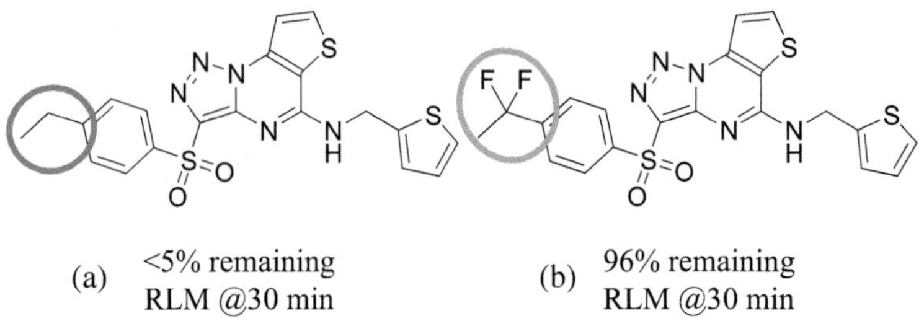

(a) <5% remaining
     RLM @30 min

(b) 96% remaining
     RLM @30 min

FIGURE 6.35   In the absence of fluorine atoms on the ethyl side chain (a) is highly metabolized by rat liver microsomes (RLM). Adding two fluorine atoms to the ethyl side chain (b) greatly enhances metabolic stability.

with two fluorine atoms (Figure 6.35(b)).[59] In a similar manner, replacing two hydrogen atoms in a metabolically labile cyclohexane ring of a chemokine (C-C motif) receptor 1 (CCR1) antagonist with two fluorine atoms led to the identification of a compound with improved metabolic stability with human liver microsomes (Figure 6.36). Further modification of this same set of compounds by replacing the primary amide with a hydroxamic acid led to additional improvements in metabolic stability, suggesting that the amide is also a key player in the metabolism of this compound class (Figure 6.36(c)).[60]

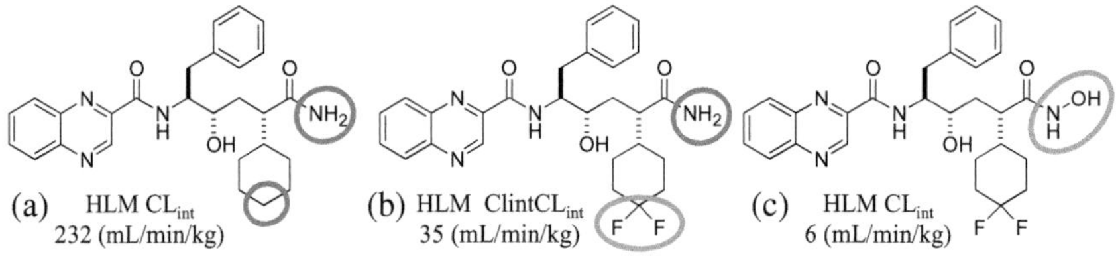

(a) HLM CL$_{int}$
     232 (mL/min/kg)

(b) HLM ClintCL$_{int}$
     35 (mL/min/kg)

(c) HLM CL$_{int}$
     6 (mL/min/kg)

FIGURE 6.36   Human liver microsome (HLM) stability of (a) is low, but incorporation of fluorine atoms in the cyclohexane ring (b) increases metabolic stability. Conversion of the primary amide of (b) into hydroxamic acid (c) further improves microsomal stability in this series of compounds.

Hydroxylation of aromatic ring system is also a common metabolic pathway and is often a problem in the identification of suitable lead compounds. Blocking metabolism by installing aryl substituents in the position that would otherwise be hydroxylated by an enzymatic process has been shown to be an effective means of dealing with issues of metabolic instability. In an effort to identify calcium-sensing receptor antagonists, for example, it was found that replacing an aromatic hydrogen atom with a trifluoromethyl substituent produced a greater than 10 fold increase in stability (Figure 6.37). In this case, the addition of a strongly electron withdrawing group both blocks potential hydroxylation in the 4-position of the benzene ring and significantly decreases the electron

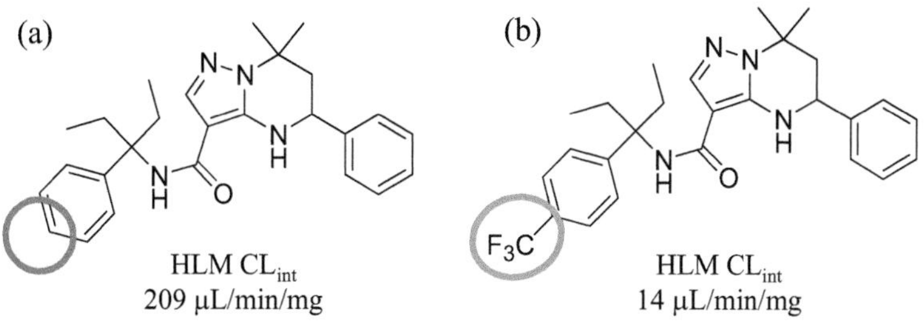

FIGURE 6.37 Adding a CF_3 group to the four position of (a) provides a compound that is significantly more stable to human liver microsomes (b) as indicated by the change in intrinsic clearance. HLM CLint, human liver microsome intrinsic clearance.

density of the aromatic ring as a whole. This change in electronic character further decreases the susceptibility of the benzene ring to metabolic oxidation processes.[61] In a similar manner, the metabolic stability of a series of indole-2-carboxamide antituberculosis agents was substantially improved with the incorporation of a second chlorine atom. Once again, in this case a halogen is used to replace a vulnerable hydrogen atom and decrease the overall electronic character of the benzene ring (Figure 6.38).[62]

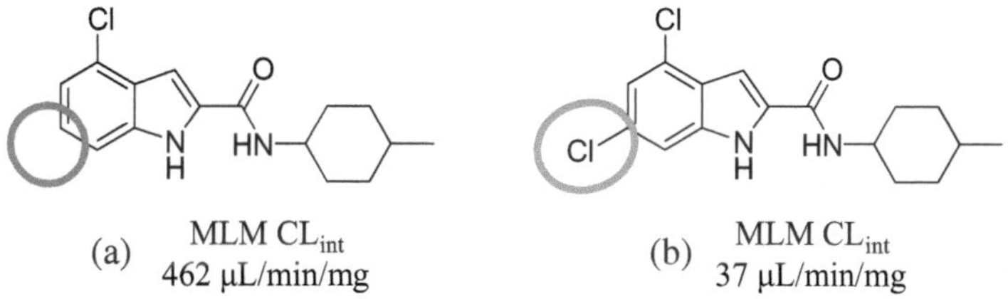

FIGURE 6.38 Metabolism of (a) in mouse liver microsomes (MLM) was suppressed by chlorination of the indole ring (b) as indicated by the change in intrinsic clearance.

In some instances, it is not possible to directly block a metabolically labile site while still preserving the desired biological activity. An alcohol or amine in biologically relevant compound could participate in a key hydrogen-bonding interaction with the target of interest, while also serving as a target for metabolic enzymes. Replacing the offending functionality could be explored, but it may also be possible to restrict metabolic activity at this site by increasing the steric encumbrance around the site. Adding sequential methyl groups to the alcohol side chain of a series of PI3-kinase inhibitors, for example, demonstrated a step-wise increase in metabolic stability as each methyl group was added to the overall framework (Figure 6.39). Each additional methyl group increases the steric hindrance around the alcohol, limiting its ability to act as a

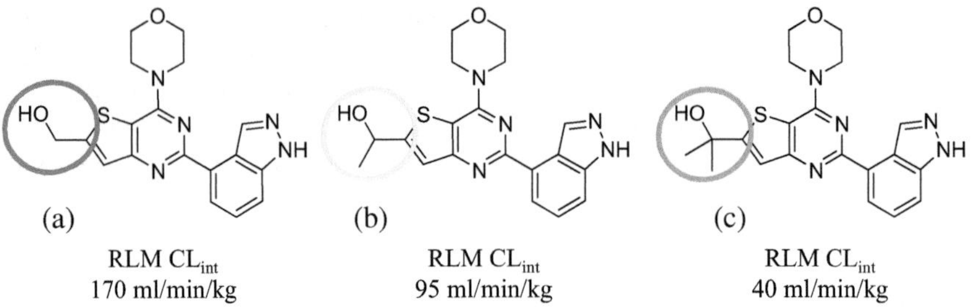

FIGURE 6.39   Adding steric bulk next to the alcohol of (a) decreased metabolism by rat liver microsomes (RLM) as indicated by the change in RLM intrinsic clearance upon incorporation of a first (b) and second (c) methyl group on the alkyl chain.

metabolic substrate, thereby increasing metabolic stability as compared to the unsubstituted analog.[63]

There are, of course, cases in which a metabolically labile site plays a role in the overall structure of a pharmacophore, but does not participate directly in binding with the macromolecular target of interest. In these cases, removing the metabolically labile group would be detrimental to the biological activity of the compound class, as the substituent or functionality plays a role in maintaining the molecule in the proper configuration for biological activity. In these instances, it may be possible to replace the metabolically labile section of the molecule with an alternate group of atoms that can serve the same purpose but that are also more stable to metabolism. Bioisosteric replacements of this type are a common method of suppressing undesired metabolic activity. In the development of CB2 agonists, for example, the metabolically labile piperidine was readily replaced with the corresponding morpholine ring (Figure 6.40). This

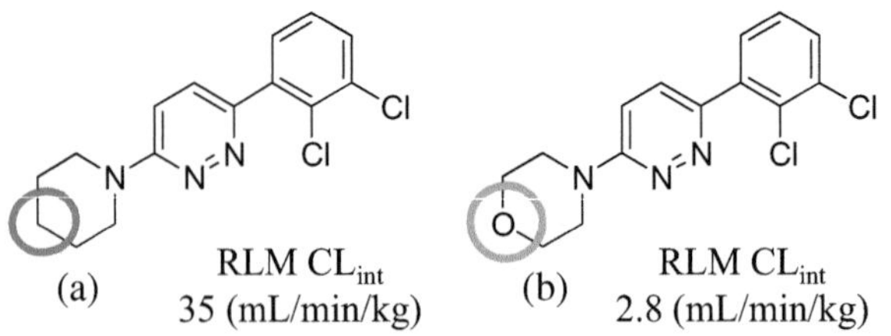

FIGURE 6.40   A carbon/oxygen isosteric exchange to convert a piperidine analog (a) into a morpholine analog (b) increases metabolic stability in rat liver microsomes (RLM).

exchange represents a carbon/oxygen isosteric replacement that produced a greater than 10 fold increase in metabolic stability.[64] Biological activity was not significantly impacted, as crucial binding interaction with the target of interest were not changed as a result of this exchange of atoms.

In some situations, it may be possible to replace large portion of a compound in order to solve metabolic issues. As long as the replacement group can serve as a suitable bioisosteric replacement, metabolic stability can be improved while biological activity is maintained. The metabolic stability of a series of 1,4-diazepames CB2 agonists, for example, was significantly enhanced when a thiazole ring was replaced with a similarly substituted isoxazole ring (Figure 6.41). Bioisosteric replacement is a common tactic employed in an effort to improve the metabolic profile of potential lead compounds.[65]

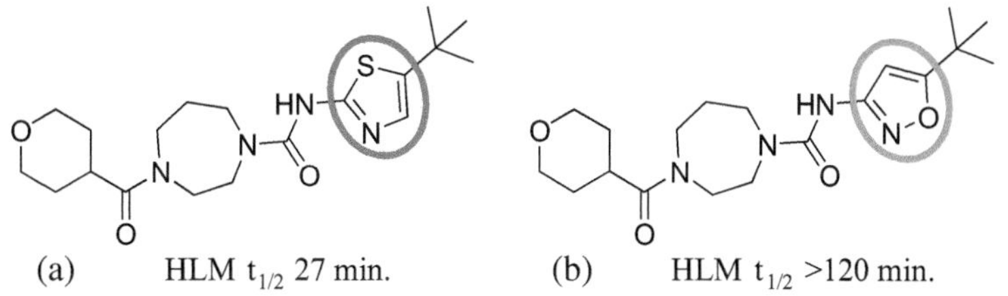

(a)   HLM $t_{1/2}$ 27 min.   (b)   HLM $t_{1/2}$ >120 min.

FIGURE 6.41   Human liver microsomal stability in a series of 1,4-diazepames is increased when a thiazole ring (a) is replaced with an isoxazole ring (b) as indicated by the increase in half-life ($t_{1/2}$). HLM, human liver microsome.

There are many additional methods that can be employed to improve the metabolic stability of candidate compounds. Cyclization to form a ring system, changing the size of a ring, inversion of a chiral center, or altering lipophilicity may have a positive impact on the metabolic profile within a given series of compounds. Metabolism itself is an enzymatic process, and as such, all of the methods one might employ to modulate active site binding through structure activity relationships are available for the purposes of modulating metabolism. There are many reviews on the topic that those with a deeper interest in the area are encouraged to read.[66a,b,c,d]

Excretion

Although metabolic processes can be an effective mechanism for the elimination of xenobiotics from the systemic circulation, they are not the only method available for clearing materials from the body. The rate of elimination of compounds from the body is also impacted by their rate of excretion. Metabolic by-products are also cleared from the body by excretion, which serves to prevent the build up of undesired metabolites. In general, compounds, both unchanged drugs and their metabolites, can be excreted into urine, sweat, bile, breast milk, or even into expired air as it leaves the lungs (e.g., anesthetic gases). The two most important methods are excretion into the urine by the kidneys and deposition into bile fluids by the liver.

In order to understand the mechanism of renal excretion, also referred to as renal elimination, it is important to have a basic understanding of the functional units of the kidney that carries out this process. The working unit of the kidney that functions to regulate the concentration of soluble material

FIGURE 6.42   The nephron is the working unit of the kidney. Its interlocking system of blood vessels and renal tubes regulates body water level and the concentration of soluble material.

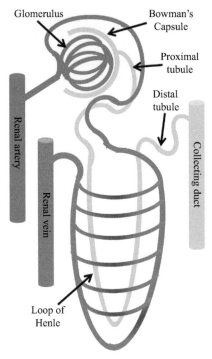

in the blood and water levels in the body is the nephron (Figure 6.42). In simple terms, the nephron can be described as an interlocking system of blood vessel and the renal tubules. The proximal portion of the nephron contains an intricate network of blood vessels called the glomerulus, which is surrounded by Bowman's capsule, a cup-like sack that is the beginning of the tubular portion of the nephron. The tubules of the nephron extend off of Bowman's capsule to form the proximal tubule, which feeds into the loop of Henle, the distal tubule, and eventually the cortical collecting duct that carries fluid out of the nephron and towards the bladder. Each of these sections of the nephron is paralleled by blood vessels.

As blood flows through the nephrons of the kidneys, approximately 10% of the blood flow (120–130 mL/min, the glomerular filtration rate) is diverted into the glomerulus. The membranes of the glomerulus are porous enough to allow diffusion of water and dissolved material to filter into the Bowman's capsule that surrounds the glomerulus. Plasma water and compounds with a molecular weight below 60,000 AMU are effectively filtered out of the blood in this process. Proteins and protein-bound drugs are not filtered by this process, so compounds with higher protein binding will be less susceptible to this process. The filtered fluid in Bowman's capsule collects and moves through the proximal tubule, which runs in parallel to the blood vessels that exit the glomerulus and rejoin the blood vessels that extend through the rest of the nephron. Further compound elimination can occur in the proximal tubule via active tubular secretion. Transport proteins capable of actively transporting various drugs, primarily weakly acidic and weakly basic compounds, out of the blood and into the urine can further increase the renal rate of elimination of compounds.

If these were the only processes that occurred within the nephron, everything that exited the blood via glomerular filtration or active transport would exit the body through the urine. While this would be acceptable for undesirable material or metabolic waste products, the same could not be said of essential materials such as glucose. Fortunately, many compounds, both good and bad, are reabsorbed into the blood in the loop of Henle and the distal tubule. Neutral compounds can be reabsorbed via passive diffusion, so lipid soluble materials reenter the systemic circulation. Compounds that are highly charged in the urine, on the other hand, are not capable of passive diffusion, but can be reabsorbed via active transport. Some vitamins, electrolytes, and amino acids are actively reabsorbed in this region of the nephron, and compounds capable of accessing the available transport proteins can be reabsorbed through the same pathways. Tubular reabsorption can also be affected by urine flow rate, as changes in this rate will alter the tubular residence time of potentially reabsorbed compounds. In summary, renal elimination is a function of glomerular filtration, passive diffusion (both into and out of the renal tubules), active secretion, and active reabsorption.[67]

Compounds can also be removed from the body by secretion into bile fluids that are produced by the liver, stored in the gall bladder, and eventually released into the intestinal tract. This material is composed of a mixture of water, various electrolytes, cholesterol, phospholipids, and bile acids (Figure 6.43). An average adult produces between 400 and 800 mL

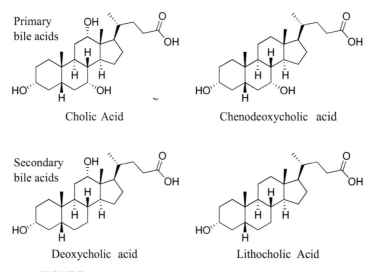

FIGURE 6.43   Examples of primary and secondary bile acids.

of bile each day. This material has two major functions. First, bile acids contained within this fluid are critically important for the digestion and absorption of lipids, lipid-soluble material in the intestinal tract. Its second major function is to facilitate transport of waste materials, such as bilirubin (a degradation product of hemoglobin), into the intestinal tract so that they may be eliminated in fecal material.

As mentioned earlier, compounds that are absorbed into the systemic circulation are brought to the liver immediately after absorption. The metabolic pathways available in the liver may convert compounds to metabolites as they pass through the liver, they may exit into the systemic circulation unchanged, or they may be secreted into the bile fluids as either the unchanged compound or a metabolite thereof. Any material that is deposited into the bile fluids will eventually reenter the intestinal tract when it is emptied from its temporary storage location in the gall bladder. If nothing further occurred, then any unchanged drug or metabolites in the bile would exit the body in feces. It is possible, however, for both unchanged compounds and their metabolites to be reabsorbed by the intestines, providing both with the opportunity to reenter the systemic circulation. The reentry of compounds into the systemic circulation in this manner is referred to as enterohepatic recirculation (Figure 6.44). Material that is capable of using this mechanism

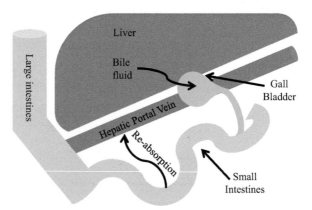

FIGURE 6.44 Compounds and their metabolites can be excreted into the gall bladder along with bile fluid produced by the liver. The gall bladder releases bile fluids into the small intestines, providing an opportunity for reabsorption of compounds in the bile fluid. This process is referred to as enterohepatic recirculation.

of absorption can cycle through this pathway many times, prolonging the exposure to of the body to the material. In addition, compounds that undergo phase 2 metabolism to form drug conjugates such as glucuronides, can be deconjugated by intestinal enzymes and bacteria that reside in the gut, providing additional opportunity for the compounds to be reabsorbed into the body.[68]

IN VITRO ADME SCREENING METHODS

Ideally, it would be possible to examine the *in vivo* ADME characteristics of large numbers of compounds in a given program in order to select the compounds with best overall ADME properties. In practice, however, this is simply not possible as *in vivo* experiments are both expensive and time-consuming. Fortunately, a number of *in vitro* assays that are useful predictors of *in vivo* ADME properties have been developed. Insight into compound absorption, for example, can be obtained by determining a compound's solubility in aqueous media. High-throughput solubility assays can be performed using commercially available 96-well plate platforms. Measuring the solubility of a compound in a range of biologically relevant buffer system will provide insight into how soluble a compound will be in the various section of the gastrointestinal tract.

Absorption is also driven by compound permeability, and *in vitro* assays have been developed to model this as well. Permeability across a biological barrier can be assessed in an *in vitro* setting in a parallel artificial membrane permeability assay, also known as PAMPA (Figure 6.45).[69] In

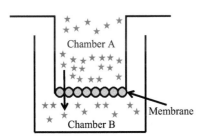

FIGURE 6.45 In parallel artificial membrane permeability assay, an artificial membrane (orange) separates two compartments. The test compound is placed in one compartment and after an incubation period, the concentration in each compartment is measured to assess membrane permeability.

this experiment, two compartments are separated by an artificial membrane and the test compound is placed in one side of the two compartments. After an incubation period, the amount of compound in each of the two compartments is measured. The change in compound concentration in each chamber provides direct insight into its ability to passively diffuse across a biological barrier.

Transporter activity, which as discussed earlier can impact absorption, distribution, and excretion, can be assessed in a similar manner

FIGURE 6.46 Measuring transporter protein activity can be accomplished by measuring the relative rate of migration of compounds across a monolayer of cells (Caco-2 or MDCK). Transporter activity is indicated if there is a difference between "A to B" and "B to A" transfer of compound.

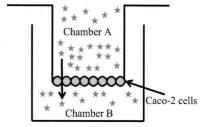

A→B = B→A Passive transport only

A→B > B→A Membrane Transport Activity

A→B < B→A Membrane Transport Activity

(Figure 6.46). In this case, the artificial membrane is replaced by a monolayer of cells that are known to possess transporters. The most commonly used cells for this purpose are CACO-2 cells, a cell line developed from human epithelial colorectal adenocarcinoma cells,[70] and the Madin–Darby canine kidney epithelial cell line, also referred to as MDCK cells.[71] Both of these cell lines contain membrane transport proteins and can be employed to assess a compound's susceptibility to active transport. These assays use the same two chamber arrangement designed to measure passive permeability, but in this case, two experiments are required. The migration of a compound moving across a monolayer of cells is measured from both sides of the monolayer in two independent experiments. If the rate of migration across the monolayer is the same irrespective of the starting point of the test compound, then compound permeability is not significantly impacted by transporter activity. If, however, the rates of migration are different, then active transport is occurring, and the difference in migration rates provides insight into the impact of membrane transporter activity on compound permeability.

In vivo metabolism can also be predicted with *in vitro* models.[72] Liver microsomes, vesicles formed from pieces of the endoplasmic reticulum (ER) of liver cells via differential centrifugation, from a variety of species are commercially available and contain the majority of enzymes responsible for xenobiotic metabolism. Incubation of a known concentration of a test compound with liver microsomes can be used as an initial assessment of a compound's susceptibility to phase 1 metabolism. Compound concentration measurements over a defined time course provides a liver microsomal half-life. In addition, metabolic by-products can be identified from the incubation, and this information can be used to design compounds that are less likely to be metabolized. Phase 2 metabolism can be predicted in similar assays using additional cellular components generated during the differential centrifugation or fully intact hepatocytes which contain all of the metabolic machinery of the liver. Metabolic half-lives determined using these methods have been demonstrated to be useful predictors of *in vivo* metabolic half-lives.

Additional *in vitro* methods designed to predict *in vivo* ADME properties such as plasma stability,[73] plasma protein binding,[74] CYP450 inhibition,[75] and blood–brain barrier permeability[76] have also been developed. All of these assays were developed with the ultimate goal of improving the odds of selecting successful compounds for *in vivo* studies. In theory, this would decrease the time and costs required to identify compounds suitable for *in vivo* study, but at the end of the day, each of these *in vitro* systems is only a model of the *in vivo* situation. Eventually, compounds must be tested in an *in vivo* setting in order for a program to move forward. It is at this point that drug discovery scientists must confront to the complexities created by the overlapping aspects of ADME in an *in vivo* setting in order to determine the *in vivo* pharmacokinetic properties of a test compound.

IN VIVO PHARMACOKINETICS

As stated earlier, the simplest definition of the term pharmacokinetics is that it is a measure of what the body does to a compound. If one were to view the body as a set of complex, overlapping, and integrated systems, similar to a factory, determining the PK properties of a single compound would be an extremely complex problem. The PK properties of a single compound are determined by its physical, chemical, and biochemical properties, and when the body is viewed as a "factory," the number of experiments required to derive a *complete* understanding of a compound's PK properties rapidly becomes overwhelming. Clearance of a given compound, for example, is determined by liver metabolism, excretion into the urine, removal of compound into bile fluids, and any other process that removes the compound from the systemic circulation. Understanding the distribution across the individual organs and tissue types in the body independently would also be quite a task. A series of experiments monitoring changes in compound concentration over time for each tissue and organ would quickly become very time-consuming and expensive.

Fortunately, a "factory" view of the body is not necessary in order to gain a significant amount of insight into how a compound behaves in the body. Instead, one could view the body as a single compartment that a compound enters via a defined route of administration (orally, intravenously, etc.) and exits via the various processes described earlier in this chapter (metabolism and excretion). The body is essentially a fluid-filled bucket with all tissues, organs, and systems treated as equal in the sense that when a compound enters the body, it is rapidly distributed throughout the system. A given compound is also viewed as "exiting" the bucket through a single "drain" that represents the sum total of all methods of

FIGURE 6.47 Determining a compound's pharmacokinetic properties can be substantially simplified if the body is viewed as a single compartment such as a fluid-filled bucket. Administration of a compound leads to rapid equilibrium distribution, and loss of the compound is treated collectively, much like a single drain at the bottom of the bucket. Plasma concentration measurements are used to monitor compound concentration.

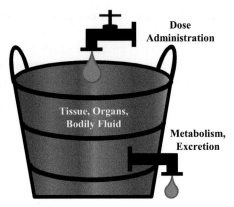

elimination (Figure 6.47). This scenario greatly simplifies PK property determination, as the concentration of a drug in the plasma can be used as a tool for determining the PK properties of a compound.

There are, of course, more complex scenarios in which the body cannot be viewed as a single compartment. When a compound is unequally distributed across two different tissue types, for example, then two "compartments" are required (Figure 6.48). A common example of this is the

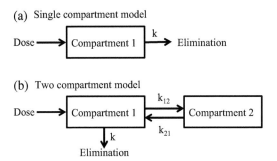

FIGURE 6.48   (a) The single compartment model of compound pharmacokinetics has a single rate constant, k, which represents the rate of elimination of the compound. (b) The two compartment model requires three rate constants, one for the elimination of compounds (k), one for movement of compound from compartment 1 to compartment 2 (k_{12}), and one for compound movement from compartment 2 to compartment 1 (k_{21}). The model assumes that equilibrium between the compartments is achieved rapidly.

unequal distribution of compound into the brain as compared to the rest of the body. In this example, the brain is treated as a "compartment" that is distinct, but reversibly linked to a separate compartment that consists of the rest of the body. A compound can move freely between the two compartments across a concentration gradient, but the kinetic properties of each "compartment" are different from each other. Multicompartment models of PK properties make the assumption that groups of tissues that behave the same with respect to a compound's PK properties and kinetics are indistinguishable from each other, irrespective of the tissue type. Clearly, as the number of "compartments" increases, the mathematics required to understand the PK of individual compounds becomes increasing complex, so for the purposes of the remainder of this chapter, emphasis will be placed on the single compartment model of PK. Keeping in mind the single compartment model, the concepts of volume of distribution, clearance, half-life, and bioavailability can be further elaborated upon.

Volume of Distribution

In practice, the PK properties of a given compound are most often determined by using a single dose *in vivo* animal model. If this dose is applied as an IV injection, various barriers to absorption into the systemic circulation such as absorption in the gastrointestinal tract and first-pass metabolism are no longer an issue. A compound injected in this manner is rapidly distributed into the bloodstream as a consequence of the rapid circulation of the blood, allowing the compound to move throughout the body quickly. Determining the plasma concentration at T_0, which can extrapolated from multiple plasma concentration measurements after the drug is injected, allows one to calculate the volume of distribution (V_d) of a compound in the body. This mathematically derived volume is not a real volume, but rather an assessment of the compound's ability to distribute through the body.

Suppose, for example, 10 mg of a compound were injected into a person weighing 70 kg, and a measurement of the plasma concentration indicated that the compound was present in the plasma at 0.25 mg/L. In order for the compounds to be present at this concentration, then the theoretical "compartment" would have a volume of 40 L, and the compound would have a V_d of 0.57 L/kg (Figure 49(a)). If the same experiment were performed with

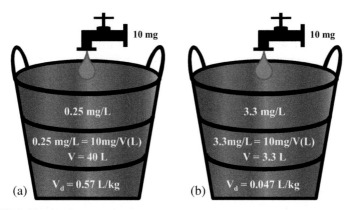

FIGURE 6.49 (a) If 10 mg of compound produces a plasma concentration of 0.25 mg/L, then the volume of distribution (Vd) is 0.57 L/kg. (b) If the same compound produces a plasma concentration of 3.3 mg/L, then the volume of distribution (Vd) is 0.047 L/kg.

a different compound and the plasma concentration was determined to be 3.3 mg/L, then the same "compartment" would have a volume of only 3.3 L, and the second compound would have a V_d of 0.047 L/kg (Figure 49(b)). In this instance, the "compartment" is a human with a fixed fluid volume that consists of approximately 40–46 L of total body water and 8 L of lipophilic material. If the "compartment" does not change in size, why does the volume of distribution change and what can be inferred from this change?

The difference between the two volumes of distribution provides insight into the ability of each compound to move beyond the systemic circulation and distribute into the various tissue and organs of the body. While the human body may contain a total of 48–54 L of fluid, total plasma volume is approximately 3 L.[77] In order for the second compound described about to have a V_d of 0.047 L/kg, it must be largely confined to the plasma. This is typical of compounds that are highly hydrophilic or those that are highly and tightly bound to plasma proteins. Their distribution is limited, indicating that they are less capable of penetrating the tissues and organs of the body. On the other hand, the first compound's V_d of 0.57 L/kg is consistent with wider distribution across the body, as a substantial amount of the drug must exit the systemic circulation in order for the plasma concentration to reach the observed levels. This is typical of compounds with moderate plasma protein binding and moderate lipophilicity.

There are cases in which the volume of distribution can be very high. Tamoxifen, for example, has a V_d of 50–60 L/kg,[78] which would equate to a "compartment" volume of 3500–4200 L. Clearly, this is well beyond the total fluid volume of a 70-kg human. Viewed in another manner, a 10-mg dose of tamoxifen delivered by IV would produce a plasma concentration of 0.0028–0.0023 mg/L. Since plasma volume is only 3 L, only 8.4–6.9 µg of tamoxifen resides in the plasma. The vast majority of the 10-mg dose is distributed into the tissues and organs of the body. This is typical of compounds that are highly lipophilic, as they preferentially bind to the lipophilic components of the body (lipids, proteins, cellular membranes, etc.) and have very low concentration in the plasma. Their distribution is very broad.

Knowledge of a compound's V_d can be a useful tool in determining its potential therapeutic utility. If, for example, a compound is being designed to modulate the activity of a macromolecule that is present in the plasma, then it would be useful to have a compound with a very low V_d, which is indicative of the compound being restricted to the systemic circulation. A compound with a high V_d would be less effective, as it would distribute into the rest of the body, decreasing its concentration in the plasma, the location of the target of interest. On the other hand, a compound whose intended target is outside of the systemic circulation needs to be distributed outside of the bloodstream, which would be indicated by a higher V_d. Selecting a compound with the proper V_d can be the difference between efficacy and failure. It is also worth noting that a compound's V_d is tied to its permeability, lipophilicity, and plasma protein binding. Structural alteration designed to alter these properties, such as those described in this and earlier chapters will have an impact on the V_d of a series of compounds, allowing one to predict changes in V_d across a series of related compounds.

Clearance

Volume of distribution is one of the two independent PK parameters that determine the *in vivo* half-life of a compound. The other independent parameter is clearance (CL), which describes the rate at which a compound is removed from the systemic circulation irrespective of the method. Systemic clearance (CL_s) is the sum total of all methods of elimination of a compound (Eqn (6.1)), and is usually described with units of mL/min/kg. The largest contributors to CL_s are excretion into the urine by the kidneys (renal clearance designated as CL_r) and metabolic loss of the compound via liver metabolism (hepatic clearance designated as CL_h). Other methods of elimination (CL_o) also contribute to CL_s, such as loss of the compound into sweat and saliva, but their contributions to the process is very small when compared to CL_r and CL_h.

$$CL_s = CL_r + CL_h + CL_o \qquad (6.1)$$

$$CL = Q * E_r \qquad (6.2)$$

The upper limit of clearance by any organ or tissue is defined by the blood flow into the organ (Q) and the extraction rate (E_r) of a compound by that particular organ (Eqn (6.2)). Organ blood flow rates have been determined for a wide range of species, and the extraction ratio is a constant for a given compound. The extraction ratio indicates the portion of a compound that is removed from a particular organ each time blood passes through the organ in question. If a compound is rapidly and efficiently removed from the blood as it passes through an organ, then the extraction value will approach its maximum value of 1 (100% of compound is removed). On the other hand, if an organ has very little impact on the blood concentration of a compound as it passes through the organ, then its extraction ratio will approach its minimum value of 0. As an example, consider a compound passing through the liver of a rat. Hepatic blood flow in the average rat is 55 mL/min/kg.[79] If 45% of the compound is removed with each passage of blood through the liver, its extraction ratio is 0.45 and its hepatic clearance, CL_h is 25 mL/min/kg.

Since hepatic clearance and renal clearance are the major factors in determining systemic clearance, it stands to reason that in a series of related compounds, structural changes that impact the ability of the kidney and liver to process the compounds can have a significant impact on the CL_s. The vast majority of the liver's contribution to CL_s is driven by compound metabolism, so changes within a compound class that change the rate of metabolic loss of a parent compound will alter its CL_h, thereby altering CL_s. As a compound's ability to resist liver metabolism increases, its CL_h will decrease, leading to an overall lowering of CL_s. Several strategies designed to increase a compound's metabolic stability were described earlier in this chapter.

In a similar fashion, structural changes within a series of compounds that prevent excretion by the nephrons of the kidney will decrease CL_r, thereby lowering CL_s. Structural changes designed to increase the kidney's ability to reabsorb a compound after glomerular filtration will decrease CL_r, as would changes that suppress active tubular secretion of compounds into the proximal tubule of the nephron by membrane transport proteins.

Half-life

The *in vivo* half-life ($t_{1/2}$) of a compound is a major determinant of its success or failure as it moves down the path toward clinical study and eventual commercialization. Simply stated, the *in vivo* half-life of a compound is the time required for removal of 50% of the compound from the body. Dosing regimens are directly impacted by $t_{1/2}$, as compounds that have a short $t_{1/2}$ are quickly removed from the body and require more frequent dosing. If the *in vivo* half-life is too low, it may not be possible for the compound to remain in the systemic circulation long enough for the compound to have an impact on a biological system. If, for example, a compound has a $t_{1/2}$ of 15 min, then 1 h after dosing, over 90% of the compound would be cleared from the body, and in 90 min less than 2% would remain in the systemic circulation. On the other hand, compounds with high *in vivo* half-lives will require less frequent dosing in order to maintain systemic availability. A compound whose $t_{1/2}$ is 200 h, such as the antimalarial drug chloroquine,[80] would require over 33 days in order to achieve 94% elimination from the systemic circulation. In general, most drug discovery and development programs focus on designing compounds capable of once or twice daily dosing, but as discussed earlier, the end goal of the program should be kept in mind in determining the required half-life. The requirements for a sleeping pill will be different from those of a pain medication or cancer treatment.

As previously mentioned, the *in vivo* half-life is dependent on its clearance and volume of distribution. Understanding the relationship of these concepts can provide a great deal of insight into how to improve the characteristic of candidate compounds. In most cases, the *in vivo* half-life of a given compound is assessed by measuring the plasma concentration of the compound after a single IV dose. Multiple concentration measurements are taken over a set of time period, usually 12–24 h. A plot of the log of the compound concentration over time can be used to determine the elimination

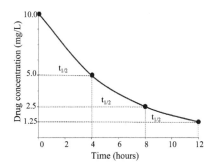 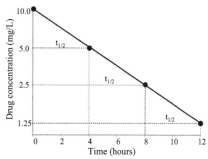

FIGURE 6.50 Plasma concentration of a compound administered by intravenous injection plotted as a function of time. The elimination rate constant (k) of a compound is determined using a semilog plot (right).

rate constant (k) for the compound (Figure 6.50). The *in vivo* half-life of the compound is then calculated using Eqn (6.3). An alternate mathematical definition of the *in vivo* half-life shown in Eqn (6.4) describes the relationship between *in vivo* half-life, clearance, and volume of distribution.

$$t_{1/2} = 0.693/k \tag{6.3}$$

$$t_{1/2} = 0.693 * (V_d/Cl) \tag{6.4}$$

As indicated in Eqn (6.4), the *in vivo* half-life is directly related to both clearance and V_d. *In vivo* half-life is indirectly proportional to clearance, but directly proportional to V_d. In other words, as clearance increases, $t_{1/2}$ falls and as V_d increases, $t_{1/2}$ increases. Since clearance is driven by the rate of elimination (metabolism, excretion, etc.), it is easy to understand why *in vivo* half-life decreases as clearance increases. Higher clearance values indicate that a compound is more rapidly removed from the body by any available mechanism, which logically shortens the time required to remove 50% of the compound from the systemic circulation. This phenomenon is readily apparent in the observed half-lives of the drugs ethosuximide[81] and flucytosine[82] (Figure 6.51). The volume of distribution of both compounds is 49 L, but ethosuximide's $t_{1/2}$ is 48 h, while flucytosine's $t_{1/2}$ is only 4.2 h. This nearly 10 fold difference is a reflection of the nearly 10 fold difference in clearance (0.7 L/h vs 8.0 L/h). Since clearance is a representation of the sum total of all mechanisms available for removal of the compound from the body, it stands to reason that efforts to slow down these mechanisms would lead to decreased clearance and therefore a longer $t_{1/2}$. In other words, improving the metabolic stability of a compound or decreasing renal excretion can have a direct impact on $t_{1/2}$ by decreasing clearance.

The impact of volume of distribution on $t_{1/2}$ is not as readily apparent. As discussed earlier, volume of distribution is an imaginary volume used to describe how widely a compound is distributed through the body. Compounds that are readily distributed beyond the systemic circulation will have a large volume of distribution, while compounds that are more restricted to the blood will have a lower volume of distribution. The impact of volume of distribution on $t_{1/2}$ becomes readily apparent in a comparison of flucytosine and digoxin[83] (Figure 6.51), as there is a 10 fold difference in $t_{1/2}$, even though

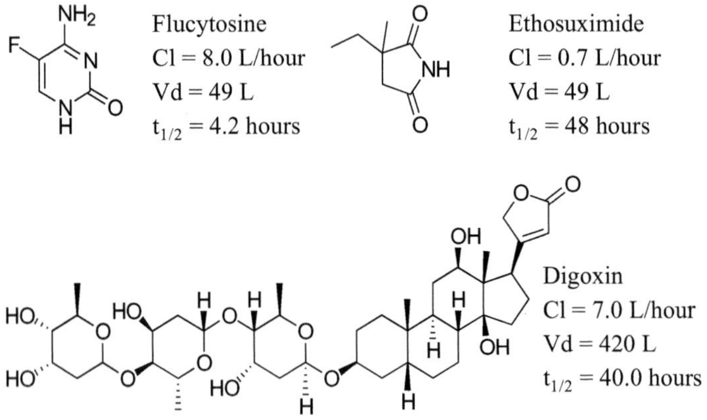

Flucytosine
Cl = 8.0 L/hour
Vd = 49 L
$t_{1/2}$ = 4.2 hours

Ethosuximide
Cl = 0.7 L/hour
Vd = 49 L
$t_{1/2}$ = 48 hours

Digoxin
Cl = 7.0 L/hour
Vd = 420 L
$t_{1/2}$ = 40.0 hours

FIGURE 6.51   The structure and pharmacokinetic properties of flucytosine, ethosuximide, and digoxin. Differences in half-life are a result of differences in clearance and volume of distribution.

drug clearance is nearly identical. In this case, the difference in half-life is mirrored by a 10 fold difference in volume of distribution, and digoxin, which has a significantly larger volume of distribution, has a longer half-life. This phenomenon can be explained by considering the relative concentration of a compound in the blood versus tissues. Compounds that are highly distributed into the tissues and organs (those that have a higher volume of distribution) are less concentrated in the blood. This means that less compound is available for processing by the liver and kidney with each cycling of the blood supply through these organs. Thus, in a series of related compounds, changes in physical properties that alter volume of distribution, such as lipophilicity and transporter protein activity, will influence half-life. Changes that increase the volume of distribution will increase half-life, while changes that decrease the volume of distribution will decrease half-life.

Bioavailability

The majority of drug discovery and development programs are directed towards the identification and commercialization of orally active drugs. Although it is clearly possible to deliver medication using other delivery mechanisms (e.g., intravenous injection, intraperitoneal injection, etc.), the ability to hand a patient a pill greatly simplifies the situation. In order to be able to achieve this, it is necessary to understand how much of the orally delivered dose actually reaches the systemic circulation. The term bioavailability (designated as F or %F) describes the fraction of an orally delivered dose that reaches the systemic circulation as compared to an equivalent dose delivered intravenously. In principle, an intravenously delivered dose bypasses all of the barriers encountered by an orally delivered compound and is therefore the maximum possible dose achievable upon oral delivery.

The overall oral bioavailability of a compound is the summation of the impact of the biological barriers between the GI tract and the systemic circulation. Consider, for example, oral delivery of 100 mg of a compound. The first barrier encountered upon oral delivery is the lining of the gastrointestinal tract. If only 70 mg of the GI dose is capable of crossing this barrier, then the epithelial lining of the tract limits the compound to 70% ($F_g = 0.7$). As discussed earlier, once a compound passes through the wall of the GI tract, it enters the portal vein that leads directly to the liver. If further processing by the liver removes 35 mg of the 70 mg entering the liver, then only 50% of the compound that enters the liver will reach the systemic circulation, so liver contribution to the overall bioavailability of the compound is 0.5 ($F_h = 0.5$). Of the initial dose, only 35 mg actually reached the systemic circulation. Therefore, the compound's bioavailability is 35%, which is the product of F_g and F_h (Figure 6.52).

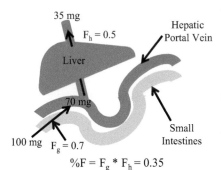

35 mg
$F_h = 0.5$
Hepatic
Portal Vein
Liver
70 mg
Small
Intestines
100 mg $F_g = 0.7$
$\%F = F_g * F_h = 0.35$

FIGURE 6.52 The oral bioavailability of a compound is determined by the fraction of compound that is absorbed by the gastrointestinal (GI) tract (F_g) and the fraction of compound that escapes liver metabolism (F_h). If the compound were also subject to metabolism by enzyme in either the systemic circulation or within the GI tract, the fraction of compound that escaped these processes would also need to be considered.

Since the bioavailability of a compound is partly driven by absorption in the GI tract, it stands to reason that within a given series, changes that improve the ability of a compound to move across the walls of the GI tract (the compound's absorption) will improve the overall bioavailability of a compound. Improving a compound's solubility through the methods described earlier in this chapter, for example, can have a positive impact on bioavailability, as more of the delivered dose is in solution and available for absorption. Similarly, structural changes in a compound series that improve permeability will promote absorption across the GI wall, and can improve oral bioavailability. Thus, modifying features such as a compound's polarity, lipophilicity, TPSA, and even the number of rotatable bonds can influence oral bioavailability.

In a similar manner, structural changes in a series of compounds that decrease the liver's ability to extract compounds from the systemic circulation can lead to improvements in oral bioavailability. Improving the metabolic stability in a class of compounds through structural modifications that prevent liver metabolism, for example, can have a positive impact on oral bioavailability. There are many examples in the literature describing significant improvements in oral bioavailability that have been driven by the suppression of liver metabolism.

From a practical standpoint, bioavailability is most often measured by comparing the concentration of compound available over time in two groups of animals, one dosed via IV injection, the other dosed orally. Plotting the changes in plasma concentration over time for both delivery methods (Figure 6.53) provides the total exposure of the animal to the drug

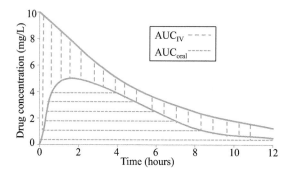

FIGURE 6.53   Total plasma exposure after a single oral dose of a compound (AUC$_{oral}$, blue area) can be compared to the total plasma exposure after a single intravenous dose (AUC$_{IV}$, green area) in order to determine a compound's bioavailability. AUC, area under the concentration curve.

as the area under the concentration curve (AUC). The IV dose is defined as the 100% mark, and the bioavailability is the fraction of the oral dose that reaches the systemic circulation relative to the IV dose. If the oral dose and IV dose are different, then this difference is incorporated into the calculation with a multiplying factor that represents the relative increase in AUC that would occur if the IV and oral doses matched. In other words, if the IV dose is 5 mg/kg and the oral dose is 15 mg/kg, then the IV AUC should be multiplied by a factor of three in order to assess the oral bioavailability of the 15 mg/kg oral dose.

Bioavailability data can be used to predict the total exposure of animal to a compound as the dose is increased if there is a linear relationship between the IV and oral doses. When a linear relationship exists, then doubling the oral dose will lead to double the exposure (AUC).

Dose	$AUC_{0\text{-inf}}$ (ng*hr/mL)
5 mg/kg IV	746
15 mg/kg PO	1989
50 mg/kg PO	33562

(a) $\%F\ 15\text{mg/kg PO} = \dfrac{15\ \text{mg/kg PO}}{3 \times (1\text{mg/kg IV})} = \dfrac{1989\ (\text{ng*hr/mL})}{2238\ (\text{ng*hr/mL})} = 82\%$

(b) $\%F\ 50\text{mg/kg PO} = \dfrac{50\ \text{mg/kg PO}}{10 \times (1\text{mg/kg IV})} = \dfrac{33562\ (\text{ng*hr/mL})}{7460\ (\text{ng*hr/mL})} = ?\ 479\%\ ?$

FIGURE 6.54 (a) The bioavailability of a compound dosed orally at 15 mg/kg is determined using the AUC values provided in the table. (b) The apparent bioavailability of a compound is determined based on the AUC values provided in the table, but the greater than dose proportional increase in AUC indicates that the compound has non-linear pharmacokinetic properties at 50 mg/kg. This suggests saturation of an elimination pathway. AUC, area under the concentration curve.

In other words, if the 15 mg/kg-dose in Figure 6.54 where to be doubled to 30 mg/kg, the expected AUC would be 3978 ng*h/mL based on the 82% bioavailability determined at 15 mg/kg. When the relationship is not linear, however, the comparison between an IV dose and a higher oral dose may lead to an apparent %F that exceeds 100%. If, for example, a compound dosed at 5-mg/kg IV and 50-mg/kg oral produces AUCs of 746 ng*h/mL and 33,562 ng*h/mL, then the apparent bioavailability of the 50-mg/kg oral dose is 479% (Figure 6.54(b)). The predicted exposure based on a linear relationship would be 7460 ng*h/mL. The difference between the expected and observed results suggests saturation of an elimination pathway.

In the absence of saturation events, compound exposure will rise proportionally with the dose. Doubling the dose will double the AUC. When an elimination pathway, such as a metabolic enzyme or a renal elimination method, is saturated, increases in compound dose will produce larger than expected increases in exposure. This situation is referred to as non-linear pharmacokinetics and can represent a significant problem in determining the proper dose in a given situation. Saturation of the elimination pathways will happen for most compounds, if the dose is raised to a high enough level. Ideally, the dose required for efficacy will be substantially removed from the concentration at which non-linear PK becomes an issue.

SPECIES SELECTION

In considering *in vivo* pharmacokinetics, it is important to consider the overall goals of the program, especially with regard to efficacy studies. While it may be useful to know and understand the pharmacokinetics of a given compound in rats, this information is not very helpful if the *in vivo* efficacy models for the project are run in dogs. There are often substantial differences between the pharmacokinetic parameters of a given compound in different species, making it challenging to judge dosing requirements for one species using the pharmacokinetic profile in a second species. In most cases, *in vivo* pharmacokinetic studies will be performed in multiple species in order to establish dosing levels for efficacy and safety studies of potential lead compounds. One rodent and one non-rodent species, usually a dog, but in some cases a primate, are generally performed for compounds that are being considered for clinical evaluation, as safety studies require assessments in one rodent and one non-rodent species. Predicting optimal dosing and expected compound exposure is an important aspect of this process and cannot be done in the absence of proper pharmacokinetic studies.

QUESTIONS

1. What are three methods employed to increase aqueous solubility?
2. Of the two compounds shown below, which one is likely be the more water soluble and why?

3. How does the lipophilicity of a compound effect its solubility in water?
4. Provide a possible explanation for the ten-fold difference in solubility between the two compounds shown below.

5. What are the five major methods by which compounds are moved across a cellular barrier?
6. How do *p*-glycoproteins (Pgps) impact the permeability of compounds?
7. How do changes in the cLogP and TPSA of a compound impact its permeability?
8. What is the "free drug hypothesis"?
9. What general processes occur during phase 1 and phase 2 metabolism?

10. What are the basic strategies employed to increase the metabolic stability of a candidate compound?
11. Of the two compounds shown below, (1) is substantially more stable than (2). Provide a plausible explanation.

(1) <5% remaining
     RLM @30 min

(2) 96% remaining
     RLM @30 min

12. Explain the process of enterohepatic recirculation.
13. If a compound has a very high volume of distribution, what can be inferred about its concentration in the systemic circulation?
14. What are the two independent parameters that determine the *in vivo* half-life of a compound?
15. What two factors determine the clearance of a compound by an organ?
16. If two candidate compounds have the same clearance, but one has a five-fold greater *in vivo* half-life than the other, what does this indicate about their relative volumes of distribution?
17. What does an oral bioavailability of greater than 100% suggest about a compound?

References

1. Prentis, R. A.; Lis, Y.; Walker, S. R. Pharmaceutical Innovation by the Seven UK-owned Pharmaceutical Companies (1964–1985). *Br. J. Clin. Pharmacol.* **1988,** *25,* 387–396.
2. Meanwell, N. A. Improving Drug Candidates by Design: A Focus on Physicochemical Properties as a Means of Improving Compound Disposition and Safety. *Chem. Res. Toxicol.* **2011,** *24,* 1420–1456.
3. Blass, B. E.; Fensome, A.; Trybulski, E.; Magolda, R.; Gardell, S.; Liu, K.; Samuel, M.; Feingold, I.; Huselton, C.; Jackson, C.; Djandjighian, L.; Ho, D.; Hennan, J.; Janusz, J. Selective Kv1.5 Blockers: Development of KVI-020/WYE-160020 as a Potential Treatment for Atrial Arrhythmia. *J. Med. Chem.* **2009,** *52* (21), 6531–6534.
4. Guo, C.; Pairish, M.; Linton, A.; Kephart, S.; Ornelas, M.; Nagata, A.; Burke, B.; Dong, L.; Engebretsen, J.; Fanjul, A. N. Design of Oxobenzimidazoles and Oxindoles as Novel Androgen Receptor Antagonists. *Bioorg. Med. Chem. Lett.* **2012,** *22* (7), 2572–2578.
5. Chen, J.; Ding, K.; Levant, B.; Wang, S. Design of Novel Hexahydropyrazinoquinolines as Potent and Selective Dopamine D3 Receptor Ligands with Improved Solubility. *Bioorg. Med. Chem. Lett.* **2006,** *16,* 443–446.
6. Manallack, D. T. The PKa Distribution of Drugs: Application to Drug Discovery. *Perspect. Med. Chem.* **2007,** *1,* 25–38.

7. Wong, J. C.; Tang, G.; Wu, X.; Liang, C.; Zhang, Z.; Guo, L.; Peng, Z.; Zhang, W.; Lin, X.; Wang, Z.; Mei, J.; Chen, J.; Pan, S.; Zhang, N.; Liu, Y.; Zhou, M.; Feng, L.; Zhao, W.; Li, S.; Zhang, C.; Zhang, M.; Rong, Y.; Jin, T. G.; Zhang, X.; Ren, S.; Ji, Y.; Zhao, R.; She, J.; Ren, Y.; Xu, C.; Chen, D.; Cai, J.; Shan, S.; Pan, D.; Ning, Z.; Lu, X.; Chen, T.; He, Y.; Chen, L. Pharmacokinetic Optimization of Class-selective Histone Deacetylase Inhibitors and Identification of Associated Candidate Predictive Biomarkers of Hepatocellular Carcinoma Tumor Response. *J. Med. Chem.* **2012**, *55* (20), 8903–8925.

8. a. Moriguchi, I.; Hirono, S.; Liu, Q.; Nakagome, I.; Matsushita, Y. Simple Method of Calculating Octanol/Water Partition Coefficient. *Chem. Pharm. Bull.* **1992**, *40* (1), 127–130.

   b. Ghose, A. K.; Viswanadhan, V. N.; Wendoloski, J. J. Prediction of Hydrophobic (Lipophilic) Properties of Small Organic Molecules Using Fragmental Methods: An Analysis of AlogP and ClogP Methods. *J. Phys. Chem. A* **1998**, *102* (21), 3762–3772.

9. Scherrer, R. A.; Howard, S. M. Use of Distribution Coefficients in Quantitative Structure–Activity Relationships. *J. Med. Chem.* **1977**, *20* (1), 53–58.

10. Ohashi, T.; Oguro, Y.; Tanaka, T.; Shiokawa, Z.; Tanaka, Y.; Shibata, S.; Sato, Y.; Yamakawa, H.; Hattori, H.; Yamamoto, Y.; Kondo, S.; Miyamoto, M.; Nishihara, M.; Ishimura, Y.; Tojo, H.; Baba, A.; Sasaki, S. Discovery of the Investigational Drug TAK-441, a Pyrrolo[3,2-c]pyridine Derivative, as a Highly Potent and Orally Active Hedgehog Signaling Inhibitor: Modification of the Core Skeleton for Improved Solubility. *Bioorg. Med. Chem.* **2012**, *20* (18), 5507–5517.

11. a. Mandagere, A. K.; Thompson, T. N.; Hwang, K. K. Graphical Model for Estimating Oral Bioavailability of Drugs in Humans and Other Species from Their Caco-2 Permeability and *In vitro* Liver Enzyme Metabolic Stability Rates. *J. Med. Chem.* **2002**, *45*, 304–311.

    b. Kerns, E. D.; Di, L. *Drug like Properties: Concepts Structure, Design, and Methods. From ADME to Toxicity;* Elsevier Inc.: Burlington, MA, 2008.

12. Hediger, M. A.; Romero, M. F.; Peng, J. B.; Rolfs, A.; Takanaga, H.; Bruford, E. A. The ABCs of Solute Carriers: Physiological, Pathological and Therapeutic Implications of Human Membrane Transport Proteins. *Pflügers Arch.* **2004**, *447* (5), 465–468.

13. Marsh, M.; McMahon, H. T. The Structural Era of Endocytosis. *Science* **1999**, *285* (5425), 215–220.

14. Van Itallie, C. M.; Anderson, J. M. Claudins and Epithelial Paracellular Transport. *Annu. Rev. Physiol.* **2006**, *68*, 403–429.

15. a. Holland, I. B.; Blight, M. A. ABC-ATPases, Adaptable Energy Generators Fuelling Transmembrane Movement of a Variety of Molecules in Organisms from Bacteria to Humans. *J. Mol. Biol.* **1999**, *293* (2), 381–399.

    b. Al-Shawi, M. K.; Omote, H. The Remarkable Transport Mechanism of P-glycoprotein; a Multidrug Transporter. *J. Bioenerg. Biomembr.* **2005**, *37* (6), 489–496.

16. a. Doyle, L. A.; Yang, W.; Abruzzo, L. V.; Krogmann, T.; Gao, Y.; Rishi, A. K.; Ross, D. D. A Multidrug Resistance Transporter from Human MCF-7 Breast cancer Cells. *Proc. Natl. Acad. Sci. U.S.A.* **1998**, *95* (26), 15665–15670.

    b. Hazai, E.; Bikadi, Z. Homology Modeling of Breast cancer Resistance Protein (ABCG2). *J. Struct. Biol.* **2008**, *162* (1), 63–74.

17. Borst, P.; Evers, R.; Kool, M.; Wijnholds, J. The Multidrug Resistance Protein Family. *Biochim. Biophys. Acta – Biomembr.* **1999**, *1461* (2), 347–357.

18. Tehler, U.; Fagerberg, J. H.; Svensson, R.; Larhed, M.; Artursson, P.; Bergström, C. A. Optimizing Solubility and Permeability of a Biopharmaceutics, Classification System (BCS) Class 4 Antibiotic Drug Using Lipophilic Fragments Disturbing the Crystal Lattice. *J. Med. Chem.* **2013**, *56*, 2690–2694.

19. Li, W.; Escarpe, P. A.; Eisenberg, E. J.; Cundy, K. C.; Sweet, C.; Jakeman, K. J.; Merson, J.; Lew, W.; Williams, M.; Zhang, L.; Kim, C. U.; Bischofberger, N.; Chen, M. S.; Mendel, D. B. Identification of GS 4104 as an Orally Bioavailable Prodrug of the Influenza Virus Neuraminidase Inhibitor GS 4071. *Antimicrob. Agents Chemother.* **1998**, *42* (3), 647–653.

20. Shimizu, H.; Yasumatsu, I.; Hamada, T.; Yoneda, Y.; Yamasaki, T.; Tanaka, S.; Toki, T.; Yokoyama, M.; Morishita, K.; Iimura, S. Discovery of Imidazo[1,2-b]pyridazines as IKKβ Inhibitors. Part 2: Improvement of Potency *In vitro* and *In vivo*. *Bioorg. Med. Chem. Lett.* **2011,** *21,* 904–908.

21. Hansch, C.; Leo, A.; Hoekman, D. *Exploring QSAR. Fundamentals and Applications in Chemistry and Biology, Vol. 1. Hydrophobic, Electronic, and Steric Constants, Vol. 2;* New York Oxford University Press: New York, 1995.

22. Lévesque, J. F.; Bleasby, K.; Chefson, A.; Chen, A.; Dubé, D.; Ducharme, Y.; Fournier, P. A.; Gagné, S.; Gallant, M.; Grimm, E.; Hafey, M.; Han, Y.; Houle, R.; Lacombe, P.; Laliberté, S.; MacDonald, D.; Mackay, B.; Papp, R.; Tschirret-Guth, R. Impact of Passive Permeability and Gut Efflux Transport on the Oral Bioavailability of Novel Series of Piperidine-based Renin Inhibitors in Rodents. *Bioorg. Med. Chem. Lett.* **2011,** *21,* 5547–5551.

23. Truong, A. P.; Probst, G. D.; Aquino, J.; Fang, L.; Brogley, L.; Sealy, J. M.; Homa, R. K.; Tucker, J. A.; John, V.; Tung, J. S.; Pleiss, M. A.; Konradi, A. W.; Sham, H. L.; Dappen, M. S.; Tóth, G.; Yao, N.; Brecht, E.; Pan, H.; Artis, D. R.; Ruslim, L.; Bova, M. P.; Sinha, S.; Yednock, T. A.; Zmolek, W.; Quinn, K. P.; Sauer, J. M. Improving the Permeability of the Hydroxyethylamine BACE-1 Inhibitors: Structure–Activity Relationship of P20 Substituents. *Bioorg. Med. Chem. Lett.* **2010,** *20,* 4789–4794.

24. Amidon, G. L.; Lennernäs, H.; Shah, V. P.; Crison, J. R. A Theoretical Basis for a Biopharmaceutic Drug Classification: the Correlation of *In vitro* Drug Product Dissolution and *In vivo* Bioavailability. *Pharm. Res.* **1995,** *12* (3), 413–420.

25. a. Neumann, H. G. Monocyclic Aromatic Amino and Nitro Compounds: Toxicity, Genotoxicity and Carcinogenicity, Classification in a Carcinogen Category. *MAK Collect. Occup. Health Saf.* **2005,** *21,* 3–45.

    b. Letelier, M. E.; Izquierdo, P.; Godoy, L.; Lepe, A. M.; Faúndez, M. Liver Microsomal Biotransformation of Nitro-aryl Drugs: Mechanism for Potential Oxidative Stress Induction. *J. Appl. Toxicol.* **2004,** *24,* 519–525.

26. Cunha, B. A. Nitrofurantoin – Current Concepts. *Urology* **1988,** *32* (1), 67–71.

27. Langston, W. J.; Palfreman, J. *The Case of the Frozen Addicts: How the Solution of an Extraordinary Medical Mystery Spawned a Revolution in the Understanding and Treatment of Parkinson's Disease;* Pantheon: New York, 1996.

28. Ballabh, P.; Braun, A.; Nedergaard, M. The Blood–Brain Barrier: An Overview: Structure, Regulation, and Clinical Implications. *Neurobiol. Dis.* **2004,** *16* (1), 1–13.

29. Krajcsi, P. Drug-transporter Interaction Testing in Drug Discovery and Development. *World J. Pharmacol.* **2013,** *2* (1), 35–46.

30. a. Schinkel, A. H. P-Glycoprotein, a Gatekeeper in the Blood–Brain Barrier. *Adv. Drug Deliv. Rev.* **1999,** *199* (36), 179–194.

    b. Hennessy, M.; Spiers, J. P. A Primer on the Mechanics of P-glycoprotein the Multidrug Transporter. *Pharmacol. Res.* **2007,** *55,* 1–15.

31. Maliepaard, M.; Scheffer, G. L.; Faneyte, I. F.; van Gastelen, M. A.; Pijnenborg, A. C. L.; Schinkel, A. H.; van de Vijver, M. J.; Scheper, R. J.; Schellens, J. H. M. Subcellular Localization and Distribution of the Breast Cancer Resistance Protein Transporter in Normal Human Tissues. *Cancer Res.* **2001,** *61,* 3458–3464.

32. Sekine, T.; Cha, S. H.; Endou, H. The Multispecific Organic Anion Transporter (OAT) Family. *Pflügers Archiv – Eur. J. Physiol.* **2000,** *440* (3), 337–350.

33. Ciarimboli, G. Organic Cation Transporters. *Xenobiotica* **2008,** *38* (7–8), 936–971.

34. a. Vig, B. S.; Stouch, T. R.; Timoszyk, J. K.; Quan, Y.; Wall, D. A.; Smith, R. L.; Faria, T. N. Human PEPT1 Pharmacophore Distinguishes between Dipeptide Transport and Binding. *J. Med. Chem.* **2006,** *49,* 3636–3644.

    b. Zhang, E. Y.; Emerick, R. M.; Pak, Y. A.; Wrighton, S. A.; Hillgren, K. M. Comparison of Human and Monkey Peptide Transporters: PEPT1 and PEPT2. *Mol. Pharm.* **2004,** *1* (3), 201–210.

35. Hagenbuch, B.; Meier, P. J.; Organic anion transporting polypeptides of the OATP/ SLC21 family: phylogenetic classification as OATP/ SLCO superfamily, new nomenclature and molecular/functional properties. *Pflugers Archv European Journal of Physiology*, **2004**, *447* (5), 653–665.

36. Sai, Y.; Tsuji, A. Transporter-Mediated Drug Delivery: Recent Progress and Experimental Approaches. *Drug Discovery Today* **2004**, *9*, 712–720.

37. Kell, D. B.; Dobson, P. D.; Bilsland, E.; Oliver, S. G. The Promiscuous Binding of Pharmaceutical Drugs and Their Transporter-Mediated Uptake into Cells: What We (Need to) Know and How We Can Do So. *Drug Discovery Today* **2013**, *18* (5/6), 218–239.

38. Ascenzi, P.; Fasano, M. Allostery in a Monomeric Protein: The Case of Human Serum Albumin. *Biophys. Chem.* **2010**, *148*, 16–22.

39. Kremer, J. M. H.; Wilting, J.; Janssen, L. H. M. Drug Binding to Human Alpha-1-acid Glycoprotein in Health and Disease. *Pharmacolo. Rev.* **1988**, *40* (1), 1–47.

40. a. Talbert, A. M.; Tranter, G. E.; Holmes, E.; Francis, P. L. Determination of Drug-plasma Protein Binding Kinetics and Equilibria by Chromatographic Profiling: Exemplification of the Method Using L-tryptophan and Albumin. *Anal. Chem.* **2002**, *74*, 446–452.

   b. Colombo, S.; Buclin, T.; Décosterd, L. A.; Telenti, A.; Furrer, H.; Lee, B. L.; Biollaz, J.; Eap, C. B. Orosomucoid (A1-acid Glycoprotein) Plasma Concentration and Genetic Variants: Effects on Human Immunodeficiency Virus Protease Inhibitor Clearance and Cellular Accumulation. *Clin. Pharmacol. Ther.* **2006**, *80* (4), 307–318.

41. Yoo, M. J.; Smith, Q. R.; Hage, D. S. Studies of Imipramine Binding to Human Serum Albumin by High-performance Affinity Chromatography. *J. Chromatogr. B* **2009**, *877*, 1149–1154.

42. Smith, D. A.; Di, L.; Kerns, E. H. The Effect of Plasma Protein Binding on *in Vivo* Efficacy: Misconceptions in Drug Discovery. *Nat. Rev.: Drug Discov.* **2010**, *9*, 929–939.

43. Kratochwil, N. A.; Huber, W.; Müller, F.; Kansy, M.; Gerber, P. R. Predicting Plasma Protein Binding of Drugs: a New Approach. *Biochem. Pharmacol.* **2002**, *64* (9), 1355–1374.

44. Lu, C., Li, A. P., Eds. *Enzyme Inhibition in Drug Discovery and Development: The Good and the Bad;* John Wiley & Sons, Inc.: New York, 2010. (Chapter 5, Bohnert, T.; Gan, L. S. The Role of Drug Metabolism in Drug Discovery; pp. 91–176.

45. Sevrioukova, I. F.; Poulos, T. L. Dissecting Cytochrome P450 3A4-ligand Interactions Using Ritonavir Analogues. *Biochemistry* **2013**, *52*, 4474–4481.

46. Lewis, D. F. V. 57 Varieties: The Human Cytochromes P450. *Pharmacogenomics* **2004**, *5* (3), 305–318.

47. Cashman, J. R.; Zhang, J. Human Flavin-containing Monooxygenases. *Annu. Rev. Pharmacol. Toxicol.* **2006**, *46*, 65–100.

48. Binda, C.; Aldeco, M.; Geldenhuys, W. J.; Tortorici, M.; Mattevi, A.; Edmondson, D. E. Molecular Insights into Human Monoamine Oxidase B Inhibition by the Glitazone Anti-Diabetes Drugs. *ACS Med. Chem. Lett.* **2012**, *3*, 39–42.

49. Gordon, A. H.; Green, D. E.; Subrahmanyan, V. Liver Aldehyde Oxidase. *Biochem. J.* **1940**, *34* (5), 764–774.

50. Theorell, H.; McKee, J. S. Mechanism of Action of Liver Alcohol Dehydrogenase. *Nature* **1961**, *192* (4797), 47–50.

51. Edmondson, D. E.; Mattevi, A.; Binda, C.; Li, M.; Hubálek, F. Structure and Mechanism of Monoamine Oxidase. *Curr. Med. Chem.* **2004**, *11* (15), 1983–1993.

52. Green, M. N.; Josimovich, J. B.; Tsou, K. C.; Seligman, A. M. Nitroreductase Activity of Animal Tissues and of Normal and Neoplastic Human Tissues. *Cancer* **1956**, *9* (1), 176–182.

53. King, C.; Rios, G.; Green, M.; Tephly, T. UDP-glucuronosyltransferases. *Curr. Drug Metab.* **2000**, *1* (2), 143–161.

54. Negishi, M.; Pedersen, L. G.; Petrotchenko, E.; Shevtsov, S.; Gorokhov, A.; Kakuta, Y.; Pedersen, L. C. Structure and Function of Sulfotransferases. *Arch. Biochem. Biophys.* **2001**, *390* (2), 149–157.

55. Hayes, J. D.; Flanagan, J. U.; Jowsey, I. R. Glutathione Transferases. *Annu. Rev. Pharmacol. Toxicol.* **2005**, *45*, 51–88.

56. Thompson, D.; Oster, G. Use of Terfenadine and Contraindicated Drugs. *J. Am. Med. Assoc.* **1996**, *275* (17), 1339–1341.

57. Zehnder, L.; Bennett, M.; Meng, J.; Huang, B.; Ninkovic, S.; Wang, F.; Braganza, J.; Tatlock, J.; Jewell, T.; Zhou, J. Z.; Burke, B.; Wang, J.; Maegley, K.; Mehta, P. P.; Yin, M. J.; Gajiwala, K. S.; Hickey, M. J.; Yamazaki, S.; Smith, E.; Kang, P.; Sistla, A.; Dovalsantos, E.; Gehring, M. R.; Kania, R.; Wythes, M.; Kung, P. P. Optimization of Potent, Selective, and Orally Bioavailable Pyrrolodinopyrimidine-containing Inhibitors of Heat Shock Protein 90. Identification of Development Candidate 2-amino-4-{4-chloro-2-[2-(4-fluoro-1H-pyrazol-1-yl)ethoxy]-6-methylphenyl}-n-(2,2-difluoropropyl)-5,7-dihydro-6H-pyrrolo[3,4-d]pyrimidine-6-carboxamide. *J. Med. Chem.* **2011**, *54*, 3368–3385.

58. Morwick, T.; Büttner, F. H.; Cywin, C. L.; Dahmann, G.; Hickey, E.; Jakes, S.; Kaplita, P.; Kashem, M. A.; Kerr, S.; Kugler, S.; Mao, W.; Marshall, D.; Paw, Z.; Shih, C.-K.; Wu, F.; Young, E. Hit to Lead Account of the Discovery of Bisbenzamide and Related Ureidobenzamide Inhibitors of Rho Kinase. *J. Med. Chem.* **2010**, *53*, 759–777.

59. Anderson, M. O.; Zhang, J.; Liu, Y.; Yao, C.; Phuan, P. W.; Verkman, A. S. Nanomolar Potency and Metabolically Stable Inhibitors of Kidney, Urea Transporter UT-B. *J. Med. Chem.* **2012**, *55*, 5942–5950.

60. Brown, M. F.; Avery, M.; Brissette, W. H.; Chang, J. H.; Colizza, K.; Conklyn, M.; DiRico, A. P.; Gladue, R. P.; Kath, J. C.; Krueger, S. S.; Lira, P. D.; Lillie, B. M.; Lundquist, G. D.; Mairs, E. N.; McElroy, E. B.; McGlynn, M. A.; Paradis, T. J.; Poss, C. S.; Rossulek, M. I.; Shepard, R. M.; Sims, J.; Strelevitz, T. J.; Truesdell, S.; Tylaska, L. A.; Yoon, K.; Zheng, D. Novel CCR1 Antagonists with Improved Metabolic Stability. *Bioorg. Med. Chem. Lett.* **2004**, *14*, 2175–2179.

61. Yoshida, M.; Mori, A.; Kotani, E.; Oka, M.; Makino, H.; Fujita, H.; Ban, J.; Ikeda, Y.; Kawamoto, T.; Goto, M.; Kimura, H.; Baba, A.; Yasuma, T. Discovery of Novel and Potent Orally Active Calcium-sensing Receptor Antagonists that Stimulate Pulse like Parathyroid Hormone Secretion: Synthesis and Structure-activity Relationships of Tetrahydropyrazolo-pyrimidine Derivatives. *J. Med. Chem.* **2011**, *54*, 1430–1440.

62. Kondreddi, R. R.; Jiricek, J.; Rao, S. P. S.; Lakshminarayana, S. B.; Camacho, L. R.; Rao, R.; Herve, M.; Bifani, P.; Ma, N. L.; Kuhen, K.; Goh, A.; Chatterjee, A. K.; Dick, T.; Diagana, T. T.; Manjunatha, U. H.; Smith, P. W. Design, Synthesis, and Biological Evaluation of Indole-2-carboxamides: A Promising Class of Antituberculosis Agents. *J. Med. Chem.* **2013**, *56*, 8849–8859.

63. Sutherlin, D. P.; Sampath, D.; Berry, M.; Castanedo, G.; Chang, Z.; Chuckowree, I.; Dotson, J.; Folkes, A.; Friedman, L.; Goldsmith, R.; Heffron, T.; Lee, L.; Lesnick, J.; Lewis, C.; Mathieu, S.; Nonomiya, J.; Olivero, A.; Pang, J.; Prior, W. W.; Salphati, L.; Sideris, S.; Tian, Q.; Tsui, V.; Wan, N. C.; Wang, S.; Wiesmann, C.; Wong, S.; Zhu, B. Y. Discovery of (Thienopyrimidin-2-yl)aminopyrimidines as Potent, Selective, and Orally Available Pan-pi3-kinase and Dual Pan-pi3-kinase/mtor Inhibitors for the Treatment of Cancer. *J. Med. Chem.* **2010**, *53*, 1086–1097.

64. Gleave, R. J.; Beswick, P. J.; Brown, A. J.; Giblin, G. M. P.; Goldsmith, P.; Haslam, C. P.; Mitchell, W. L.; Nicholson, N. H.; Page, L. W.; Patel, S.; Roomans, S.; Slingsby, B. P.; Swarbrick, M. E. Synthesis and Evaluation of 3-amino-6-aryl-pyridazines as Selective CB2 Agonists for the Treatment of Inflammatory Pain. *Bioorg. Med. Chem. Lett.* **2010**, *20*, 465–468.

65. Riether, D.; Wu, L.; Cirillo, P. F.; Berry, A.; Walker, E. R.; Ermann, M.; Noya-Marino, B.; Jenkins, J. E.; Albaugh, D.; Albrecht, C.; Fisher, M.; Gemkow, M. J.; Grbic, H.; Löbbe, S.; Möller, C.; O'Shea, K.; Sauer, A.; Shih, D.-T.; Thomson, D. S. 1,4-Diazepane Compounds as Potent and Selective CB2 Agonists: Optimization of Metabolic Stability. *Bioorg. Med. Chem. Lett.* **2011**, *21*, 2011–2016.

66. a. St Jean, D. J.; Fotsch, C. Mitigating Heterocycle Metabolism in Drug Discovery. *J. Med. Chem.* **2012,** *55,* 6002–6020.

    b. Thompson, T. N. Optimization of Metabolic Stability as a Goal of Modern Drug Design. *Med. Res. Rev.* **2001,** *21,* 412–449.

    c. Kerns, E. H.; Di, L. *Drug-like Properties: Concepts, Structure, Design and Methods;* Academic Press: San Diego, CA, 2008.

    d. Smith, D. A. Discovery and ADMET: Where Are We Now? *Curr. Top. Med. Chem.* **2011,** *11,* 467–481.

67. Dipiro, J. P., Ed. *Concepts in Clinical Pharmacokinetics,* 5th ed.; American Society of Health System Pharmacists: Bethesda, Maryland, 2010.

68. Roberts, M. S.; Magnusson, B. M.; Burczynski, F. J.; Weiss, M. Enterohepatic Circulation: Physiological, Pharmacokinetic and Clinical Implications. *Clin. Pharmacokinet.* **2002,** *41* (10), 751–790.

69. Kansy, M.; Senner, F.; Gubernator, K. Physicochemical High Throughput Screening: Parallel Artificial Membrane Permeability Assay in the Description of Passive Absorption Processes. *J. Med. Chem.* **1998,** *41* (7), 1007–1010.

70. Hidalgo, I. J.; Raub, T. J.; Borchardt, R. T. Characterization of the Human Colon Carcinoma Cell Line (Caco-2) as a Model System for Intestinal Epithelial Permeability. *Gastroenterology* **1989,** *96* (3), 736–749.

71. Irvine, J. D.; Takahashi, L.; Lockhart, K.; Cheong, J.; Tolan, J. W.; Selick, H. E.; Grove, J. R. MDCK (Madin-Darby Canine Kidney) Cells: A Tool for Membrane Permeability Screening. *J. Pharm. Sci.* **1999,** *88* (1), 28–33.

72. Iwatsubo, T.; Hirota, N.; Ooie, T.; Suzuki, H.; Shimada, N.; Chiba, K.; Ishizaki, T.; Green, C. E.; Tyson, C. A.; Sugiyama, Y. Prediction of *In vivo* Drug Metabolism in the Human Liver from *In vitro* Metabolism Data. *Pharmacol. Ther.* **1997,** *73* (2), 147–171.

73. Di, L.; Kerns, E. H.; Hong, Y.; Chen, H. Development and Application of High Throughput Plasma Stability Assay for Drug Discovery. *Int. J. Pharm.* **2005,** *297* (1–2), 110–119.

74. Yasgar, A.; Furdas, S. D.; Maloney, D. J.; Jadhav, A.; Jung, M.; Simeonov, A. High-throughput 1,536-Well Fluorescence Polarization Assays for α1-Acid Glycoprotein and Human Serum Albumin Binding. *PLoS One* **2012,** *7* (9), e45594.

75. Lin, T.; Pan, K.; Mordenti, J.; Pan, L. *In vitro* Assessment of Cytochrome P450 Inhibition: Strategies for Increasing LC/MS-based Assay Throughput Using a One-point IC_{50} Method and Multiplexing High-performance Liquid Chromatography. *J. Pharm. Sci.* **2007,** *96* (9), 2485–2493.

76. Di, L.; Kerns, E. H.; Fan, K.; McConnell, O. J.; Carter, G. T. High Throughput Artificial Membrane Permeability Assay for Blood–Brain Barrier. *Eur. J. Med. Chem.* **2003,** *38* (3), 223–232.

77. Rhoades, R.; Bell, D. R. *Medical Physiology: Principles of Clinical Medicine,* 4th ed.; Lippincott Williams & Wilkins: Baltimore, Maryland, 2013.

78. Lien, E. A.; Solheim, E.; Ueland, P. M. Distribution of Tamoxifen and its Metabolites in Rat and Human Tissues during Steady-State Treatment. *Cancer Res.* **1991,** *51,* 4837–4844.

79. Davies, B.; Morris, T. Physiological Parameters in Laboratory Animals and Humans. *Pharm. Res.* **1993,** *10,* 1093–1095.

80. Moore, B. R.; Page-Sharp, M.; Stoney, J. R.; Ilett, K. F.; Jago, J. D.; Batty, K. T. Pharmacokinetics, Pharmacodynamics, and Allometric Scaling of Chloroquine in a Murine Malaria Model. *Antimicrob. Agents Chemother.* **2011,** *55* (8), 3899–3907.

81. Livingston, S.; Pauli, L.; Najmabadi, A. Ethosuximide in the Treatment of Epilepsy. Preliminary Report. *J. Am. Med. Assoc.* **1962,** *180,* 822–825.

82. Vermes, A.; Guchelaar, H. J.; Dankert, J. Flucytosine: a Review of its Pharmacology, Clinical Indications, Pharmacokinetics, Toxicity and Drug Interactions. *J. Antimicrob. Chemother.* **2000,** *46* (2), 171–179.

83. Hauptman, P. J.; Kelly, R. A. Digitalis. *Circulation* **1999,** *99,* 1265–1270.

7

Animal Models of Disease States

In the early stages of a drug discovery and development program, there is heavy emphasis on the application of *in vitro* screening technology. These efforts are focused on identifying compounds with the desired biological and physicochemical properties. Throughout this process, it is assumed that *in vitro* assays will be predictive of activity in an *in vivo* setting. In fact, the pharmaceutical industry and academic community have spent a significant amount of time and money (billions of dollars and decades of research time) focused on improving the utility and predictive power of *in vitro* screening methods. As discussed in the previous chapters, it is now possible to assess the ability of compounds to impact a wide range of macromolecules. Physicochemical properties of compounds, such as their propensity to undergo metabolism, inhibit the metabolism of other compounds, or move across biological barrier can be predicted using *in vitro* models. Given the wide range of biological and physical properties that can be measured in an *in vitro* setting, why is it necessary to study potential drug compounds in animal models?

A simplistic answer to this question is that the FDA and other regulatory agencies require both proof of efficacy and safety in animal models before they will allow a compound to be studied in humans. Of course, from scientific standpoint, this answer is not a satisfactory explanation. A more satisfying answer comes from considering the limitation of *in vitro* assay systems with respect to whole animal models. While *in vitro* assays have proven to be essential tools in the hunt for new therapeutic agents, they cannot be used as a substitute for *in vivo* studies. At the end of the day, an *in vitro* assay, irrespective of whether it is at the molecular, cellular, or tissue level, is only an incomplete model of a far more complicated system. In whole animals, the net biological impact of a compound is the sum total of the compound's effect on all of the macromolecules, tissues, and organs that it comes into contact with. This includes a compound's ability to modulate the activity of potential drug targets, as well as the impact the body has on the compound itself (e.g., protein binding, metabolism, etc.). An intact animal, whether it is a mouse, dog, monkey, or human, is a complex web of overlapping and interacting biological systems. Currently, there are no *in vitro* assays or combination of *in vitro* assays capable of recapitulating the complexities of a whole animal. This makes it critical to investigate the impact of potential new therapeutic agents in animal models that are predictive of the human condition or disease of interest.

The first step in the process of transitioning from *in vitro* assays to animal models is determining a compound's pharmacokinetic profile. As discussed in Chapter 6, the pharmacokinetic profile of a compound provides a wealth of information on what happens to a compound when it enters the body. These studies, however, are not designed to determine if a compound will provide the desired biological impact. Assessing the biological impact of a compound in an intact animal requires more complex models that are designed to determine the pharmacodynamics (PD) of a compound. In simple terms, pharmacodynamics is the study of what a compound does to the body. Pharmacodynamic investigations are typically designed to determine the capacity of a compound to produce a desired biological endpoint, referred to as its efficacy. In addition, pharmacodynamic studies determine the amount of compound required to produce an effect of a given intensity, referred to as the compound's potency. Of course, efficacy and potency are not finite values like temperature or length, but rather they are specific to the biological endpoint under examination.

Determining the dose of a compound required to achieve a particular effect in an *in vivo* assay is an important aspect of pharmacodynamic studies. In practice, pharmacokinetic data are used to set compound doses that ensure the plasma concentration of the compound of interest is above the IC_{50} (or EC_{50}) of the primary therapeutic target for an extended period of time. Consider, for example, a compound designed to modulate the 5-HT_7 serotonin receptor with an EC_{50} of 100nM. A dosing regimen that fails to provide a minimum plasma concentration of 100nM is unlikely to allow the compound to reach its intended target at a high enough concentration to elicit a biological response. Even if a dose produces plasma concentration in excess of 100nM, in the absence of a second dose, elimination pathways will remove the compound from the systemic circulation. In most cases, as the plasma concentration drops, the biological response will fade, tracking with the half-life of the compound. Irrespective of the nature of the *in vivo* model, it is critical to design *in vivo* studies so that data are collected while the compound is still "on board." This is, of course, only possible if the pharmacokinetic profile of a compound is available for the same species that will be used in the *in vivo* efficacy model. Although mice and rats have some similarities, the PK profile of a given compound can be very different in the two species. The rat PK profile of a compound is of limited value in determining the proper dosing strategy for a mouse efficacy study and vice versa.

The main goal of an *in vivo* efficacy model is, of course, to determine whether or not a compound is capable of producing a desired physiological response, to compare the impact of a range of doses of the same compound, and to compare different compounds to each other. The nature of the data produced in any given series of experiments will depend on the study design, but the reporting method is generally the same. Efficacy in an *in vivo* model is typically conveyed as a combination of the dose provided and the percent effect relative to the maximal effect that can be achieved with the compound under consideration. The most commonly reported value for an *in vivo* study is its ED_{50}, the dose at which 50% of the maximum effect is observed. These values are useful for rank-ordering compounds in a particular *in vivo* model.

In general, the first *in vivo* efficacy model that a compound encounters is most likely geared towards the primary endpoint of the program. A compound developed to lower cholesterol will almost certainly be assessed in an animal model designed to assess cholesterol levels. It is well known, however, that a given compound can generate multiple biological responses. It is often necessary to monitor more than one physiological response, and the ED_{50}'s for each physiological response may be different. These determinations are often used to assess the propensity of a compound to produce undesired effects as well as the desired pharmacological endpoint. While the desired activity of a compound is likely due to the successful targeting of one or more biological targets, there are many other targets available in the body that can produce a response. Pharmacological studies designed to determine the dose dependency of an undesired biological response are also very useful. Knowledge of the dose dependency of both the desired and undesired biological responses allows the determination of therapeutic window in which the positive effect is elicited while the negative effect is minimized. The ratio of these two doses is referred to as the therapeutic index. In general, although there may be a number of possible negative effects associated with increasing doses of a compound, the therapeutic index of a compound is generally established with the dose of the first negative consequence that occurs. Thus, if a compound lowers cholesterol effectively at 1 mg/kg, but causes heart rate to drop at 10 mg/kg, then the therapeutic index is 10. If the same compound causes hair loss at 100 mg/kg, then the therapeutic index for this effect is 100, but as stated above, the lowest therapeutic index is generally considered the most important in a drug discovery setting.

SOURCES OF ANIMAL MODELS

While animal models are as diverse as the diseases and conditions they are designed to mimic, the methods of developing new animal models is fairly limited. Animal models have been developed based on spontaneously occurring variations within an animal population, selective breeding designed to develop a population that expresses (or do not express) a trait, and with the advent of biotechnology, via the insertion or deletion of genes. *In vivo* models have also been developed using drug effects, behavioral training and responses, and surgical/physical alterations to an intact animal.

The earliest animals available were the ones that occurred spontaneously or that were generated via selective breeding experiments. In these situations, random mutations provided animals that possessed attributes that made them suitable models for human diseases or conditions. Selective breeding experiments could then be used to increase the population of animals possessing the desired traits, and entire breeding colonies could be established to generate a steady supply of the animal in question. The leptin-deficient (Lep$^{Ob/Ob}$) mouse, for example, was identified in the Jackson Laboratory in 1950.[1] Although the importance, and in fact the existence of leptin was not known at the time, it was clear that the animal in question ate voraciously, experienced significant weight gain, and eventually developed conditions strikingly similar to morbid obesity and type 2 diabetes in humans.[2] Additional examples of animal models developed from naturally occurring mutations and/or selective breeding experiments include the nude mouse,[3] the spontaneously hypertensive rat,[4] and the non-obese diabetic mouse.[5] These and other important animal models were developed in the absence of modern biotechnological methods and highlight the important role of natural mutations in animal models.

Of course, it is not always practical or possible to wait for nature to provide a genetic framework (animal) suitable for *in vivo* modeling of potential therapeutics. Fortunately, modern science has provided an alternative to natural mutations and selective breeding. As discussed in Chapter 2, the biotechnology revolution of the late twentieth century provided scientists with the ability to manipulate the genetic material of living cells. The breakthrough science of this era has allowed scientists to create "designer animals." Insertion of non-native DNA or the suppression of normal gene function (knockout) via transgenic technology has been used to develop numerous animal models that would have been unavailable otherwise. The study of amyotrophic lateral sclerosis (ALS), also known as Lou Gehrig's disease, for example, has been significantly enhanced by the development of the SOD1-G93A transgenic mouse model. The incorporation and overexpression of the human gene for superoxide dismutase 1 containing a mutation at the 93rd codon (a glycine/alanine switch) in mice produces animals that display neurodegeneration and symptomology consistent with ALS. In the absence of this model, it would be difficult to study this disease.[6] Many other genetically modified animal models are commercially available. The ability to generate new animal models via transgenic technology is primarily limited by whether or not the modification to the animal's genome is an inherently lethal change.

Although genetically produced animal models, natural or artificially manipulated, are very useful, it is not always necessary, or even possible, to use an animal model that is "hard wired" to produce a specific disease state or condition. In many cases, a model of a disease state or condition can be induced temporarily. Drug-induced animal models, for example, utilize pharmacological intervention in order to establish the models. In general, an otherwise healthy animal is treated with a compound known to induce a disease condition or symptoms that closely mimic the disease. The MPTP model of Parkinson's disease, for example, creates a Parkinsonian type condition in primates[7] and mice.[8] Administration of 1-methyl-4-phenyl-1,2,3,6-tetrahydropyridine (MPTP) leads to rapid destruction of dopamine-synthesizing neurons in the substantia nigra region of the brain. This triggers the rapid development of Parkinson's disease symptoms in a manner consistent with the human condition. Interestingly, administration of MPTP to rats has no effect, which highlights the importance of species selection in animal models.

Animal models can also be produced by physical means (e.g., mechanical, surgical, etc.). Ischemic events can be surgically created by limiting or blocking blood flow in order to study the impact of test compounds on stroke survival[9] or cardiac reperfusion after a heart attack.[10] *In vivo* models designed to study the ability of a compound to suppress pain sensation can also be generated using physical means. The ability of an animal to tolerate contact with a heated surface, such as a hotplate underneath a paw, could be used to identify compounds that suppress pain.[11]

It is also possible to use an animal's environment to create a situation suitable for assessing the pharmacological impact of test compounds. This is especially useful in the CNS arena, as changes in behavioral response to environmental condition in the presence or absence of a test compound may be the only meaningful data available. Consider, for example, animal models designed to identify compounds that may be useful for the treatment of depression. Although it is not possible to determine if a mouse is experiencing depression, it is possible to study potential antidepressants in mice by observing their responses to various situations. In the Porsolt forced swimming test, for example, a mouse is placed in a glass cylinder containing water too deep to stand in and with walls too high to allow escape. Eventually, mice in this situation will stop attempting to reach the top of the cylinder to escape and simply float on the surface. Mice that are treated with antidepressants spend a significantly greater time attempting to find an escape than untreated mice.[12] This model has been effectively employed to identify clinically useful antidepressants, even though there is no clear relationship between a mouse's inability to escape a situation and human depression.

VALIDITY OF ANIMAL MODELS

Although it is possible to argue the merits and drawbacks of each method of model development, from the standpoint of drug discovery, the source of a model is not nearly as important how well the model recapitulates the human condition. An animal model that does not correlate with a human disease or condition is of little value for the development of novel therapies. In general, animal models can be divided into three categories based on how well they reproduce the human disease or condition, homologous, isomorphic, and predictive.[13] Homologous animal models are the most desirable, as they have the same causes, symptoms, and treatment options available for humans. They are the rarest type, as they are difficult to achieve, but unlike other models they have construct validity. Typical examples include the non-obese diabetic mouse model of type 1 diabetes[5] and numerous bacterial infection models.[14] The majority of animal models are not homologous models.

Isomorphic animal models, on the other hand, are far more common. This type of animal model has the same symptomology as the human conditions and treatment options are generally the same. However, the root cause of the disease or condition in the animal model is not the same as that observed in humans. Consider, for example, animal models of stroke. The ischemic damage caused by a stroke can be experimentally created in an animal model by interfering with brain blood flow.[9] The impact of this interference will certainly be similar to an occlusive event caused by a blood clot in a blood vessel of the human brain, and treatment options will be similar, but the root cause of the stroke is not the same. Similarly, the degenerative damage associated with osteoarthritis can be mimicked by injecting sodium iodoacetate into the joints of an animal. Osteoarthritis begins to develop in 2–4 months, and this model can be used to study the antidegenerative properties of test compounds.[15] This is clearly not how this condition develops in humans, but the symptoms and disease progression make it a suitable model of the human condition. Isomorphic models are considered to have "face validity." In other words, the models have the same overall appearance of the disease state that they are modeling, but are not actually the same.

When homologous and isomorphic models are not available, predictive animal models are employed instead. This class of animal models is most often used when the disease or condition is poorly understood or simply does not occur in animals. In some cases, the animal model itself may have little or no obvious resemblance to the human condition, but there are facets of the model that allow researchers to use it as a predictive tool. The impact of potential therapeutic intervention in the model can be correlated to how humans will respond to the same compound. The model displays the treatment characteristics of the disease or condition and is said to have predictive validity. Consider, for example, schizophrenia. Determining whether or not an animal is suffering from schizophrenia is not generally possible due to the significant communication barrier. In addition, the mechanisms through which this condition develops are poorly understood at the best. Despite these gaps in scientific knowledge, there are several *in vivo* models available for the study of schizophrenia, most of which depend upon the administration of drugs that exacerbate or induce symptoms in humans.[16]

SPECIES SELECTION

In considering which species to use for a particular *in vivo* study, it is important to be aware of the limitations of each possible choice. *In vivo* efficacy experiments are key decision points in the life of a program, as they are often viewed as proof of concept experiments. Failure of an *in vivo* experiment can lead to termination of a program, especially if the therapeutic target is novel (e.g., no marketed drugs using the proposed mechanism of action). It is, therefore, absolutely critical that the species employed gives the best correlation possible with the human condition or disease.

Non-human primates are, of course, the closest animal to humans overall, but they are rarely used in animal trials. There are very few non-human primates available for study, and non-human primates are both difficult and expensive to maintain. Their large size also directly impacts compound supply issues, as potential therapeutics are most often dosed on a milligram per kilogram basis. Larger animals require larger amounts of compounds, further driving up the expense of non-human primate studies. Additional ethical considerations also come into play. Non-human primates are typically used only when no other option is available.

Smaller species such as rats, mice, and dogs are far more common choices for *in vivo* experiments. Each species has its pluses and minuses, but the most important aspect in species selection is ensuring that the selected animal model correlates with the human condition. A rat model of atrial arrhythmia, for example, may be easier to work with than a more sophisticated model that employs beagles. If, however, the rat model does not correlate with the human condition, the data provided in running the rat model will have very little values in a drug discovery and development program. Although the beagle model may be more difficult and time-consuming, it is the appropriate choice as it will provide insight into the utility of a potential therapeutic agent in humans, whereas the rat model will not. It is critical to keep the end goal in mind when selecting the species for an *in vivo* experiment. It is also worth noting that PK studies (described in Chapter 6) should be performed in the same species as planned for the pharmacodynamics experiments.

NUMBER OF ANIMALS

Once an animal model has been selected, the number of animals necessary to demonstrate efficacy must be considered. In theory, the animal model could provide a determination using a single animal. The compound either produces the desired result or it fails to do so. Unfortunately, animal models are often not capable of producing yes and no answers. Just like any other measurement, the "signal" produced in an animal model is subject to errors. The errors can be the result of differences within the animal population selected (e.g., genetic difference that occur within a population of otherwise homogenous animals such as those described above), inaccuracies inherent to the means of measuring the "signal," intensity of the "signal" relative to the untreated population (a control is always necessary) as well as several other factors. In order for the results of an animal study to be considered meaningful, the "signal" generated in the study must be statistically significant. In other words, the "signal" strength must exceed the margin of error inherent to the model itself. The number of animals required to accomplish this will depend on the strength of the expected "signal" relative to the expected margin of error. If the expected "signal" is strong and the expected margin of error is low, then fewer animals will be required. In contrast, if the expected "signal" is weak and the margin for error is large, then more animals will be required in order for the average "signal" to be outside of the margin of error for the model. Statistical models that are beyond the scope of this text are routinely employed in order to determine the number of animals required for a study to have a reasonable chance of demonstrating a statistically significant result.

EXEMPLARY ANIMAL MODELS BY DISEASE CATEGORY

There are literally thousands of animal models available for the study of diseases, conditions, and novel therapies. Although it is certainly useful for drug discovery scientists to have an in-depth understanding of the animal models associated with their ongoing programs, it is simply not possible to enumerate them all in this text, or any other single text. The ability to generate animal models is primarily limited by the creativity of drug discovery scientists. Given their inherent creativity, a large number of animal models are available. It is instructive to review a limited number of examples of animal models from a range of disease states so that one can develop an appreciation for the effort required to generate the result of an animal study. Those interested in an in-depth description of animal models in a particular disease area are encouraged to consult the scientific literature for additional information.

Animal Models in Neuroscience

The Forced Swimming Test: A Model of Depression[12,17]

Depression is a clinically important indication, but it is difficult to model in animals. It is described in humans as a pathological complex of somatic, neuroendocrine, and of course, psychological symptoms. Replicating these symptoms in animal model, rodents or otherwise, is not possible, and as such, drug discovery programs aimed toward the development of novel antidepressant depend on specific measurable behaviors that are predictive of antidepressant activity in humans. The forced swimming test, originally described by Porsolt et al. has been effectively employed to identify novel antidepressants by monitoring the behavior of mice placed in an inescapable cylinder of water. When initially placed in the cylinder, a mouse will swim and search for a means of escape. Eventually, the mouse will stop attempting to find a way out of the cylinder, adopt a characteristic floating position, and move only enough to keep its head above water. In the original paper, Porsolt et al. were able to correlate an increase in activity in mice treated with known antidepressants as compared to untreated mice, establishing a correlation between mouse behavior and human depression (Figure 7.1).

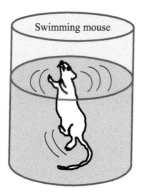

FIGURE 7.1   In the Porsolt forced swimming test, compounds with antidepressant properties will extend the length of time a mouse will continue to swim attempting to find an exit to the chamber. This experimental model has been successfully employed to identify novel antidepressants.

The Elevated Plus Maze: A Measure of Anxiety[18]

Similar to depression, anxiety is a complex disorder in humans that is not easily recapitulated in animal models. Several models have been developed to assess the antianxiolytic activity of novel compounds. The elevated plus maze is a very simple model that has been validated with known antianxiolytics, such as benzodiazepines, and can provide insight into the efficacy of a test compound. In this model, a mouse or rat is placed at the center of an elevated plus-shaped platform. One set of oppositely facing arms of the platform is enclosed and dark, while the other two are open to the air. The rodent's propensity to spend time in the two different types of arms of the platform can be assessed using a video camera and motion tracking software. In the absence of an effective compound, rodents will preferentially spend time in the closed spaces in order to avoid the anxiety of height and open spaces. Compounds capable of suppressing anxiety will increase the amount of time a rodent will spend in the open arms of the platform. This information can be used to assess the anxiety-like behavior of the rodent in the presence or absence of test compounds (Figure 7.2).

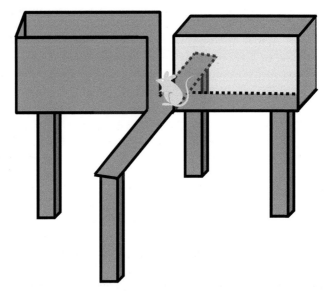

FIGURE 7.2   The elevated plus maze tracks the amount of time a mouse spends in enclosed areas versus open sections of the platform using video cameras and motion sensors. The time difference has been correlated to antianxiolytic efficacy.

The Novel Object Recognition Test: A Model of Memory and Cognition[19]

Another important aspect of central nervous system drug discovery is the identification of compounds capable of restoring or enhancing cognition and memory functions. As the population ages, memory impairment, whether as the result of diseases such as Alzheimer's disease or as a natural consequence of aging, is a serious health issue and has been the focus of many drug discovery programs. The novel object recognition model takes advantage of the natural curiosity of rodents. When placed in the presence of a familiar object and a new object, rodents will spend significantly more time exploring the novel object. This sense of familiarity with the familiar object reflects the animal's ability to remember and recognize objects, providing an opportunity to quantify cognition and memory changes (Figure 7.3).

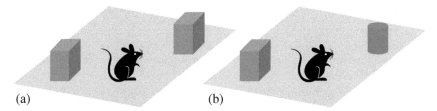

(a)   (b)

FIGURE 7.3   The novel object recognition model takes advantage of the natural curiosity of mice and can be used to determine the impact of candidate compounds on memory and learning. (a) Mice are initially familiarized with an environment and two objects. (b) After the acclimation period, one of the known objects is replaced with a new object. Candidate compounds capable of impacting memory formation (positively or negatively) will influence the amount of time a mouse spends examining the novel object relative to the familiar object.

In practice, a rodent is first familiarized with an arena containing two identical objects set at an equal distance from each other. After the rodent has acclimated to the presence of the two known objects, typically 24 h, one of the two objects is replaced by a new object. The time spent with each of the two objects can be quantified to generate a discrimination index that indicates the rodent's preference for the new object. If a rodent's ability to form memories is impaired, it will be less likely to remember the familiar object, leading to a decreased tendency to explore the novel object. Compounds that are capable of improving cognition and memory can be identified using this model in combination with rodents that would normally suffer from cognitive and memory impairment. Animals treated with successful test compounds would show an increased tendency to explore the new object as compared to the untreated control, which would be expressed as a higher discrimination index. Alternatively, this model could also be used to identify compounds that impair cognition and memory function. Compounds that suppress memory and cognition would prevent the rodent from forming a memory of the first object during the acclimation period. When the new object is added, the rodent will show less preference for the new object, leading to a decreased discrimination index.

Contextual Fear Conditioning Model: A Model of Contextual Learning[20]

Memory function and the impact of novel compounds can also be assessed by taking advantage of Pavlovian responses in rodents.

The contextual fear conditioning model is designed to train rodents to expect a negative sensation, such as a mild electric shock, when a specific cue occurs such as the ringing of a bell. Once the rodents have been trained to expect the negative sensation when the cue occurs, they will freeze when the cue occurs in anticipation of the negative sensation. In practice, this can be accomplished by placing a rodent in a cage in which the floor has been attached to a low power electric shock generator. The rodent is trained to associate a specific sound with a low electric shock in the first day, and then 24h later, the same procedure is repeated, but the shock is not delivered when the sound occurs. In the absence of memory and learning impairment, the rodent will remember the sound and freeze immediately, as they are expecting a shock. Memory and learning impairment is indicated by a decreased tendency to freeze upon cue, as the rodents are less likely to remember the cue from the previous day's training sessions. Novel compounds that have the potential to enhance or restore memory can be assessed for efficacy in this model by using rodents that experience cognitive decline, either as the result of a disease (e.g., Alzheimer's disease models rodents) or ischemic events (e.g., stroke). Alternatively, this model can be used to assess a compound's ability to cause memory and learning deficits. Compounds that impair memory and learning will also decrease a rodent's freezing response upon cue (Figure 7.4).

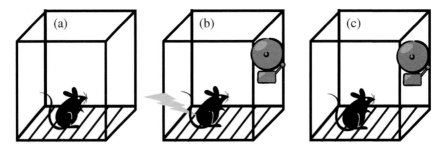

FIGURE 7.4   The contextual fear model can be used to assess the impact of candidate compounds on memory and learning. (a) A mouse is acclimated to a cage equipped with a low-power shock generator attached to the floor. (b) The mouse is then trained to associate an electric shock with the ringing of a bell. (c) Once trained, mice will freeze upon hearing the bell in anticipation of an electric shock. Candidate compounds can be assessed for their ability to impact learning and memory using this model.

The Morris Water Maze: A Model of Spatial Learning and Memory[21]

Although both of the previously described animal models of memory and learning are useful, neither are well suited to probing spatial learning and memory. These aspects of brain function are more appropriately assessed in the Morris water maze. In this model, a rodent's ability to remember the spatial relationships of object within a previously encountered environment is measured. The model itself consists of a pool of water in which the animal must swim. A single platform is placed in the pool below the water surface, but high enough that the rodent can stand on the platform rather than swim in the water. Rodents are initially trained in the pool with clear water so that they can see the platform below the surface of the water. Once the rodents are trained, which generally requires several days, the pool water is made opaque with a coloring agent. When the rodents are placed in the pool, the time required to find the hidden platform is measured, providing an assessment of their ability to form spatial memories regarding the location of the platform. Alternatively, the platform can be removed from the opaque pool, and the amount of time the rodent spends in the quadrant that previously contained the platform can be measured. Rodents suffering from cognitive decline, such as transgenic mice designed to model Alzheimer's disease, will require longer periods of time to find the hidden platform. If the platform has been removed, impaired rodent will spend less time in the appropriate quadrant of the pool as compared to normal rodents. The ability of a test compound to prevent cognitive decline, restore, or enhance memory can be tested in this model. Compounds that either protect or restore cognitive function would be expected to decrease the time required for a rodent to locate the hidden platform or increase the time spent in the proper quadrant of the pool if the platform has been removed as compared to control animals (Figure 7.5).

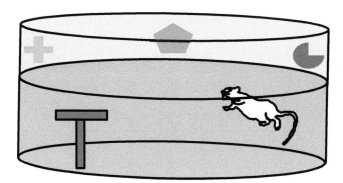

FIGURE 7.5 The Morris water maze provides an assessment of a candidate compound's impact on learning and memory. Rodents are trained to locate a submerged platform in a clear pool. Their ability to remember the location of the raised platform is then challenged by replacing the clear water with an opaque solution that hides the submerged platform. Candidate compounds that impact learning and memory will affect the rodent's ability to locate the submerged platform when the liquid is opaque.

Animal Models of Neurodegeneration

Although memory loss and cognitive decline are important aspects of human health, there are other diseases of the central nervous system that do not impact effect memory and learning. Neurodegenerative diseases such as amyotrophic lateral sclerosis (ALS), Parkinson's disease, and Huntington's disease are linked to death or dysfunction of nerve cells associated with movement. Models that recapitulate the symptoms of movement-associated neurodegenerative diseases have been developed using both pharmacological intervention and transgenic science. It is worth noting, however, that these models are generally isomorphic in nature, as the cause of the human conditions are unknown. The loss of motor control associated with Parkinson's disease, for example, is the result of the death of dopamine-generating cells in the substantia nigra region of the brain, but the cause of the cell death remains a mystery. These models have face validity, making them useful for exploring pathophysiology and possible treatments for the associated conditions.

The SOD1-G93A Mouse of Amyotrophic Lateral Sclerosis[6,22]

Amyotrophic lateral sclerosis is a debilitating neurodegenerative disease characterized by the death of both upper and lower motor neurons in the motor cortex of the brain, the brain stem, and the spinal cord. The progressive denervation of muscle tissue causes muscular weakness, paralysis, and atrophy that become more extensive through the course of the disease. Although the exact cause of ALS is unknown, the SOD1-G93A mouse model suffers from the same progressive motor neuron death. These transgenic mice overexpress mutated human copper–zinc superoxide dismutase (SOD1) in which the glycine in the 93rd position is replaced with an alanine residue. The pathology created in this transgenic mouse model correlates well with the human condition, as symptoms such as paralysis begin to occur at approximately 90 days old, and in the absence of therapeutic intervention, the mice die in approximately 135 days.

Novel compounds that are capable of slowing or stopping the progression of this neurodegenerative disease can be studied in this animal model by observing physical capabilities and changes in compound-treated animals as compared to untreated control animals. Grip strength, body weight, righting reflex (the ability of an animal to right itself after being placed on its back), and rotarod performance (Figure 7.6) can all be observed as the treated and untreated animals age. Compounds capable of slowing motor neuron degeneration will demonstrate improved responses in these measures for treated animals. Overall mortality will also improve with compounds that are efficacious.

An important aspect of this model is the length of time and resources required to complete an experiment. Unlike the previously described

FIGURE 7.6   The rotarod performance test employs a rotating horizontal cylinder suspended above the cage floor. Under normal conditions, rodents placed on the rotating rod will attempt to stay on the rod for as long as possible. Neurodegenerative diseases and candidate compounds that impact balance, coordination, motor skills, and wakefulness (sedatives) will have a negative impact on the length of time a rodent is able to stay on the rod. Candidate compounds that delay the progression of neurodegenerative diseases can be studied in this model. In addition, this model has been used to identify compounds that cause sedation.

animal models of memory and cognition which can be completed in a relatively short period of time (5–7 days at most), SOD1-G93A mouse studies are long and labor intensive. Consider an experiment with a single daily dose of a novel compound and a control group of untreated mice. If there are 10 mice per group, then 20 mice will have to be cared for, treated with the test compound, and assessed for at least 19 weeks (the point at which all untreated mice would be dead) in order to demonstrate improvements in overall mortality. Ideally, the treated group would show significant improvement, possibly increasing the length of the trial even further. This highlights the importance of the non-mortality driven measurements (grip strength and body weight) that can provide an early indication as to the success or failure of the test compound. In general, it is a good idea to have early secondary indication in long-term models, as these may provide guidance as to whether or not the study should be terminated early due to lack of efficacy. Terminating a study early for this reason is not a desirable outcome, but it can preserve resources for future experiments.

The MPTP Model of Parkinson's Disease[23]

In some cases, it is possible to mimic human disease by using pharmacological tools (e.g., chemicals) to create a pathology in animals that is consistent with the human condition. Parkinson's disease-type symptoms and pathology, for example, can be induced in monkeys

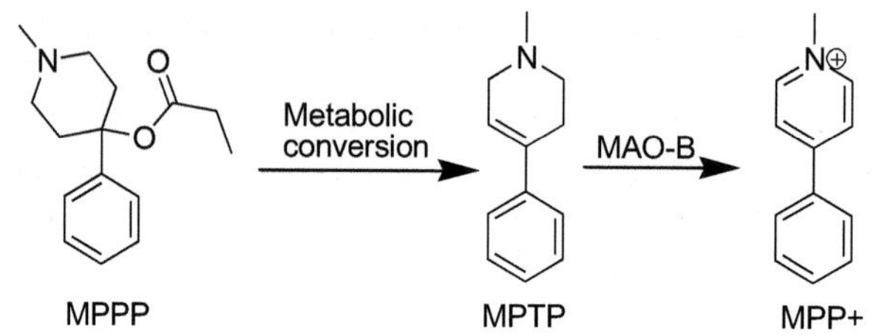

MPPP　　　　　　　　　　MPTP　　　　　MPP+

FIGURE 7.7　Desmethylprodine (1-methyl-4-phenyl-4-propionoxypiperidine (MPPP)), an opioid analgesic that was identified by scientists at Hoffmann-La Roche in the 1940s, is metabolically converted to 1-methyl-4-phenyl-1,2,3,6-tetrahydropyridine (MPTP). Metabolic processing of MPTP by monoamine oxidase-B (MAO-B) in the brain produces the cationic species 1-methyl-4-phenylpyridinium (MPP+), which enters the dopaminergic cells found in the substantia nigra region of the brain and kills them. Dopaminergic cells are the only cells in the brain that produce dopamine, a chemical required for movement. As these cells die, Parkinson's disease symptoms begin to emerge and become more pronounced as the dopaminergic cell population decreases.

and mice with MPTP (1-methyl-4-phenyl-1,2,3,6-tetrahydropyridine) (Figure 7.7). Although the exact cause of Parkinson's disease remains a mystery, it is clear that this chronic and progressive movement disorder results from the loss of dopamine producing cells in the substantia nigra region of the brain. The decline of these cells leads to the onset of uncontrollable tremors, rigidity of limbs, bradykinesia (slowness of movement), impaired balance, and diminished physical coordination. In the late 1980s, it was accidently discovered that MPTP causes rapid onset of Parkinson's disease symptoms, including death of dopamine-producing cells in humans (the details of this discovery will be discussed in Chapter 13), and as mentioned above, administering MPTP to primates[7] and mice[8] has the same effect. The dopamine-producing cells in the brains of both primates and mice die as a result of exposure to MPTP, creating a situation nearly identical to Parkinson's disease. Onset of symptoms occurs within 6–9 days of MPTP injection and in the absence of therapeutic intervention, symptoms persist just as they do in the human disease. The MPTP model of Parkinson's disease is often considered to have construct validity. The main criticism of this model has been the absence of Lewy bodies, an abnormal aggregation of proteins inside the neurons, which are a hallmark of Parkinson's disease.

The use of MPTP to induce a Parkinson's disease condition in an animal model provides an opportunity to study the associate neurodegeneration in a relatively short time frame and from two different perspectives. Compounds with the potential to prevent MPTP-induced neurodegeneration (neuroprotectants) can be assessed by dosing the animals with test compounds prior to administration of MPTP. Compounds that have efficacy as neuroprotective agents within the context of Parkinson's disease will prevent the development of symptoms associated with MPTP exposure. Alternatively, compounds capable of preserving dopamine function can be identified by dosing the animals with test compounds after MPTP administration, but prior to symptom onset. Compounds capable of preserving dopamine function will demonstrate a decreased level of parkinsonian-type symptoms.

Animal Models of Cardiovascular Disease

Despite decades of research, cardiovascular disease remains the number one cause of death. Cardiovascular disease as a whole contributes to a larger number of deaths than that of HIV and cancer combined. There are, of course, many aspects and contributing factors to the development of cardiovascular disease as well as many different subtypes of cardiovascular disease. Multiple animal models have been developed in order to study, understand, and develop therapies for the treatment of the different facets of cardiovascular disease. The following exemplary models provide a glimpse of the complex and often time-consuming animal models employed in the search for novel treatments.

Models of Hypertension

Of the various risk factors associated with myocardial infarction, heart failure, and strokes, the most important is hypertension, a chronic condition in which blood pressure is elevated (greater than 140/90 mmHg). It is estimated that in 2008, complications associated with hypertension accounted for approximately 9.4 million deaths, and that over 1 billion people were living with this condition, diagnosed or otherwise.[24] Given the importance of this condition, it should come as no surprise that a substantial amount of effort has been devoted to the identification of novel antihypertensive agents. Animal models of hypertension have been developed using surgical methods, pharmacological induction, and genetic manipulation. In the surgical model, hypertension can be established by constriction of the renal artery (the 2K1C model, Figure 7.8). This leads to a chronic increase in blood pressure that plateaus in 2–3 weeks. Hypertension in mice, rats, rabbits, dogs, pigs, and non-human primates have been studied using surgical intervention of this type.[25]

Pharmacologically established animal models of hypertension, on the other hand, are often produced using chronic administration of mineralocorticoids,

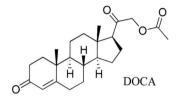

DOCA

FIGURE 7.9 Prolonged exposure to mineralocorticoids such as deoxycorticosterone acetate (DOCA) produces hypertension in several useful animal models.

FIGURE 7.8 In the 2K1C model of hypertension, surgical constriction of the renal artery leads to chronically increased blood pressure (hypertension) that plateaus 2–3 weeks after surgery.

particularly deoxycorticosterone acetate (DOCA, Figure 7.9). Prolonged exposure to mineralocorticoids (2–4 weeks) causes hypertension in rats, dogs, and pigs that are characterized by increased cardiac output and volume expansion. In some cases, animal model performance can be improved by feeding them a high-salt diet. Similar results can be obtained in mice if the mineralocorticoids are replaced with glucocorticoids.[25] Simple dietary intervention can also be used to induce hypertension under the correct circumstances. The Dahl salt-sensitive rat, for example, will develop hypertension in as little as 3 weeks if fed a high-salt diet.[26]

Genetic manipulation, either via selective breeding or transgenic science, has also provided a number of animal models of hypertension. The most recognized of these is the spontaneously hypertensive rat (SHR). This model of hypertension was developed by selective breeding of Wistar rats that naturally demonstrated high blood pressure. Hypertension begins to develop 5–6 weeks of age, and cardiovascular disease becomes apparent at 40–50 weeks. The SHR has been extensively used in the study of hypertension.[4,27]

Irrespective of how an animal model of hypertension is generated, a method of reliably measuring blood pressure that will not in itself cause blood pressure to spike (e.g., fear or anxiety related increases) must be established. In the absence of such a method, a compound's ability to lower blood pressure might be masked by an increase in blood pressure related to the measurement itself. Training of animals to tolerate the application of a blood pressure cuff, which often requires restraining the animal, requires an initial training period of 1–2 weeks. In rats and mice, a tail blood pressure cuff is employed, while larger animals can be assessed on limbs. Once the animals are trained, blood pressure readings can be obtained in the presence and absence of a test compound to assess efficacy.[28]

As an alternative, blood pressure monitoring equipment capable of transmitting data over radio frequencies can be surgically implanted into various animal species. In this case, training the animals to tolerate the use of a blood pressure cuff is unnecessary, but significant healing time (2–4 weeks) is required in order for the animals to recover from the surgical implantation. After the animals have healed and become acclimated to their surroundings, the radio transmitting equipment allows 24h monitoring of blood pressure in the animals. This telemetry-driven animal system has been successfully employed in a wide range of animal species.[29]

Once the animals are ready (e.g., trained, healed, acclimated, etc.), dosing with test compound can be initiated. The typical efficacy study for a potential antihypertensive agent will require extended dosing periods (1–2 weeks) during which changes in blood pressure are monitored. Other aspects of cardiovascular performance such as heart rate, contractility, and ejection fraction are often monitored to ensure that the remaining aspects or cardiovascular performance are not negatively impacted. On the whole, testing a potential antihypertensive agent in a suitable animal model can require several weeks, possibly months of effort by the appropriately trained staff, making it very important that only molecules that are properly vetted reach this stage.

Models of Hyperlipidemia and High Cholesterol

It is well established that increased levels of cholesterol (Figure 7.10(a)) and low-density lipoprotein (LDL, Figure 7.10(c)) are associated with significantly increased risk of atherosclerosis and related cardiovascular

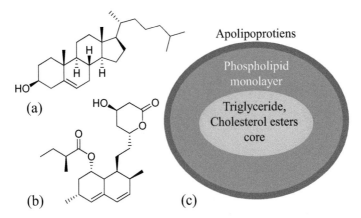

FIGURE 7.10   (a) Cholesterol (b) Mevacor® (lovastatin), an HMGCoA reductase inhibitor. (c) Low-density lipoprotein (LDL) particles are composed of a triglyceride and cholesterol ester core surrounded by a phospholipid monolayer. Apolipoproteins are often embedded in the surface of the LDL particle.

diseases.[30] As a result, a number of animal models have been designed to study the impact of potential therapeutic agents on their plasma concentration. New Zealand White rabbits, for example, can each serve as test animals for potential cholesterol-lowering agents, provided they are fed the appropriate diet. Hypercholesterolemia can be induced in these animals by maintaining them on an appropriate diet for 5–6 weeks (normal cholesterol levels of approximately 70 mg/dL are elevated to approximately 310 mg/ dL). Once hypercholesterolemia is established, the animals can be treated with novel compounds designed to lower circulating cholesterol levels for an extended period of time (1–2 weeks). An analysis of blood samples to determine systemic concentration of the cholesterol and LDL during and at the end of the dosing period can be used to identify compounds that have a positive impact (e.g., lowering the cholesterol level closer to the normal levels). Statins such as Mevacor® (lovastatin, Figure 7.10(b)), inhibit HMGCoA reductase, the rate-limiting enzymes in the biosynthesis of cholesterol. They were identified using this type of animal model.[31]

Although the diet-induced model of hypercholesterolemia has been an effective tool in the development of cholesterol-lowering agents, it is not a perfect model. The ability to induce high levels of cholesterol and LDL does not necessarily lead to the formation of atherosclerosis in a manner similar to that observed in human. Atherosclerotic plaques in the rabbit model, for example, are significantly different from their human counterparts. The difference in plaque structure is not relevant in the identification of compounds capable of lowering cholesterol levels. It is, however, a limiting factor for the identification of compounds capable of altering the progression of atherosclerotic plaques. The rabbit model is not well suited to this task.

Fortunately, the apolipoprotein E-deficient (apoE$^{-/-}$) mouse model does not have this limitation. This transgenic animal has been genetically altered so that it does not express apolipoprotein E (Figure 7.11), which

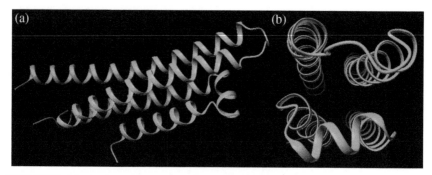

FIGURE 7.11   Structure of apolipoprotien E4 (a) viewed across the four α-helices and (b) viewed down the length of the four α-helices (RCSB 1b68).

is a constituent of all lipoproteins except low-density lipoproteins. As a result, apoE$^{-/-}$ mice exhibit high levels of cholesterol (>500 mg/dL) on a normal diet, and cholesterol levels increase by a factor of four when they are fed a Western diet (high fat). Importantly, these mice also begin to develop fatty streaks in blood vessels as early as 10 weeks of age. The atherosclerotic lesions progress in a manner consistent with the development of atherosclerotic plaques in humans. The impact of potential therapeutic entities on the progression of atherosclerotic plaques can be assessed in the apoE$^{-/-}$ mouse model by direct inspection of the plaque structure, rather than depending on circulating cholesterol levels as a surrogate maker for disease progression.[32]

Models of Atrial Fibrillation

Although not generally life threatening, atrial fibrillation (AF) patients are at a substantially higher risk of heart failure, ischemia, morbidity, and mortality. The condition is characterized by an asynchronous contraction

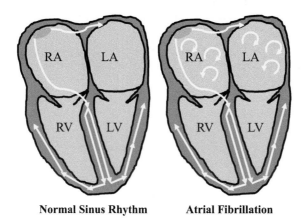

Normal Sinus Rhythm   **Atrial Fibrillation**

FIGURE 7.12   In normal sinus rhythm, an electrical impulse (yellow arrows) (left) is initiated in the sinoatrial (SA) node (orange). It moves across interatrial tracts that spread the signal across the left and right atria. The electrical impulse then passes through the atrioventricular (AV) node and into the ventricles, where it spreads across interventricular tracts. In atrial fibrillation (right), the interatrial tracts of the atrial chambers have become distorted, creating opportunities for aberrant electrical impulses (short circuits) that disrupt the normal sinus rhythm. The disruption presents clinically as an irregular and often rapid heartbeat in the atrial chambers of the heart. RA, right auricle; RV, right ventricle; LA, left auricle; LV, left ventricle.

of hundreds of regions of the atrial chamber of the heart (Figure 7.12), which decreases atrial blood transport and increases the rate of ventricular response. Globally, over 33 million people suffer from this condition, and the development of novel antiarrhythmic agents remains a challenge.[33]

Progression of AF is associated with physical and electrical remodeling of the heart, and currently available animal models of AF are dominated by large animals such as dogs, goats, sheep, and pigs. In fact, historically, arrhythmias were studied in larger mammals, as it was believed that arrhythmia did not occur in mice due to their lack of critical cardiac mass[34] (Janse and Rosen, 2006). Additionally, the hearts of larger animals (e.g., non-human primates, dogs, etc.) are more similar to human hearts than those of mice and rats. In 1999, however, the theory of critical cardiac mass was disproven,[35] and since that time, a number of transgenic mouse models of AF have been reported in the literature. These models have been useful in studying the underlying mechanism of AF, but their utility in the identification of compounds capable of treating AF has as yet to be validated.[36]

The majority of AF drug discovery research employs large animal models such as the sterile pericarditis dog model (Figure 7.13). This dog model

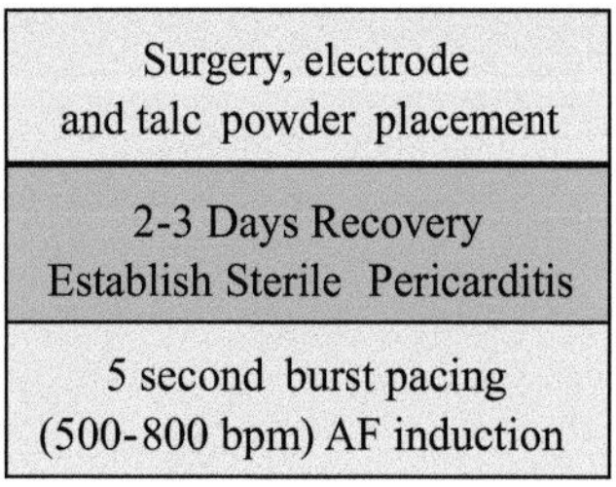

FIGURE 7.13 In the sterile pericarditis dog model, the presence of talc powder in the pericardium causes inflammation. Atrial fibrillation (AF) can be induced with a burst pacing protocol 2–3 days after surgery.

Surgery, electrode and talc powder placement
2-3 Days Recovery Establish Sterile Pericarditis
5 second burst pacing (500-800 bpm) AF induction

of AF exhibits a high incidence of sustained AF with strong similarity to clinically observed AF after an open-heart surgery. In this model, surgical ablation of the pacemaker tissue in the heart, implantation of a pacemaker, and insertion of talc powder into the pericardium is followed by a 2- to 3-day recovery period. The presence of talc causes pericarditis, an inflammation of the pericardium, and this, in turn, renders the dog susceptible to induction of AF via burst pacing (5–10s of 500–800 beats per minutes via implanted pacemaker). Severity of AF can be measured in terms of the average duration of AF induced via burst pacing and the overall AF inducibility (the number of AF episodes lasting longer than 60s after a set number of attempts). In the presence of compounds with antiarrhythmic properties, the average duration of AF and the AF inducibility will decrease when compared with vehicle treatment.

An alternative model that is particularly useful for evaluating drugs that may prevent electrical remodeling can be created by subjecting an animal, usually a dog, to extended periods of high heart rates (Figure 7.14). In this pacing model, surgical implantation of monitoring electrodes and a pacemaker is followed by a 2-week recovery period. Once the animal is

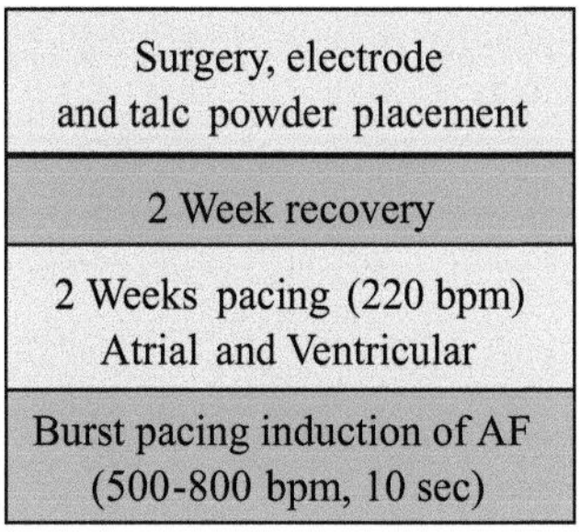

Surgery, electrode and talc powder placement
2 Week recovery
2 Weeks pacing (220 bpm) Atrial and Ventricular
Burst pacing induction of AF (500-800 bpm, 10 sec)

FIGURE 7.14 The pacing model of atrial fibrillation (AF) is similar to the sterile pericarditis model, but an additional 2 weeks of rapid pacing is added to the protocol. This induces extensive physical and electrical remodeling of the heart that promotes atrial fibrillation.

fully recovered from surgery, the pacemaker is adjusted to approximately 220 beats per minute (tachycardia), and this heart rate is sustained for 2–3 weeks. Tachycardia-induced physical and electrical remodeling of the chambers of the heart occurs during this time. This includes alteration of ionic currents and gene expression of ion channels that propagate electrical signals in the heart which promotes the occurrence of AF. At the end of this 2- to 3-week remodeling period, severity of AF can be assessed in a similar manner as that described for the sterile pericarditis model of AF. Once again, compounds with antiarrhythmic properties will demonstrate a decrease in the average duration of AF and the AF inducibility as compared to vehicle treatment.

Models of Heart Failure

An estimated 23 million patients worldwide[37] suffer from heart failure (HF), also referred to as congestive heart failure (CHF), a condition that is the final stage of many diseases of the cardiovascular system. In simple terms, cardiac output (volume of blood pumped by the heart in 1 min) is insufficient to meet the needs of the body. Conditions that decrease myocardial efficiency, such as ischemic events, can lead to the development of HF. In fact, ischemic heart disease is thought to be the most important risk factor for HF,[38] but other conditions such as hypertension and amyloidosis (deposition of protein in the heart muscle) can also contribute to the development of this condition. Over time, decreased cardiovascular efficiency leads to detrimental changes in the heart itself as it attempts to compensate for the increased workload. The heart undergoes hypertrophy (increase in size) as the muscles of the heart increase in size in an effort to improve contractility and increase cardiac output. At the same time, heart rate increases in an attempt to compensate for decreased efficiency, blood vessels narrow to maintain blood pressure, and the production of important effectors of the cardiovascular system such as renin, vasopressin, angiotensin, and aldosterone are altered as part of the body's compensatory mechanism

(a)

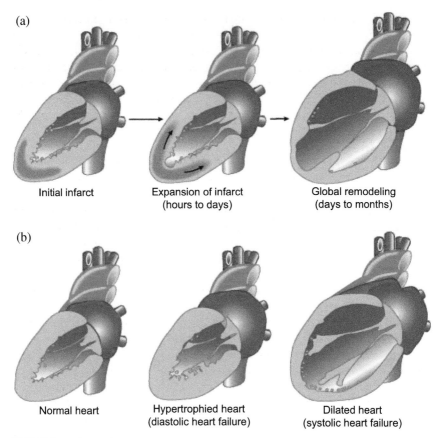

Initial infarct   Expansion of infarct   Global remodeling
                    (hours to days)   (days to months)

(b)

Normal heart   Hypertrophied heart   Dilated heart
                 (diastolic heart failure)   (systolic heart failure)

FIGURE 7.15   Patients suffering from heart failure have enlarged hearts that pump blood with significantly less efficiency than required. In some cases, (a) left ventricle remodeling occurs in response to an ischemic event. The heart expands and its walls become thinner leading to global systolic dysfunction. (b) Reduction in left ventricle compliance can also cause heart failure. Hypertrophy of the heart is followed by expansion of the heart, thinning of the walls of the heart, and a significant decrease in cardiac efficiency. *Source: Reprinted from Ginsberg, F.; Parrillo, J. E. Critical Care Medicine: Principles of Diagnosis and Management in the Adult. Severe Heart Failure, Chapter 30, pages 559–587, copyright 2008, with permission from Elsevier.*

(Figure 7.15). Patients can be asymptomatic at the onset of HF, but eventually develop exercise intolerance that becomes progressively worse until even normal activity (e.g., walking, getting out of bed) becomes a significant challenge. In advanced cases, hospital-supported care is required.

Although the pathology of HF is complex, both surgical and genetic animal models have been developed. As previously discussed, ischemic events promote the development of HF. Myocardial infarction can also be surgically induced by ligation of the coronary artery in a variety of species (rats, dogs, pigs, sheep). As with humans, the damage created by the ischemic event will heal, but the animal's heart will remodel itself in order to compensate for the loss of efficiency. In other words, the animal will develop HF type conditions that can be used to assess the efficacy of test compounds.[39] In practice, surgical induction of an ischemic event (and implantation of cardiovascular monitoring equipment) is followed by a recovery period (3–7 days). Dosing with test compounds is then initiated and cardiovascular function is monitored over a period of weeks to months in order to determine if the test compound is preventing progression into HF. Factors such as mean aortic pressure, contractility, and survival can be used as determinants of efficacy of test compounds relative to untreated animals.[40]

As an alternative, it is also possible to induce HF conditions using a rapid pacing protocol similar to that described for the creation of a model suitable for the study of atrial fibrillation. In the tachycardia-induced HF model, the application of a rapid pacing protocol is used to induce dilated cardiomyopathy (DCM) and left ventricle heart failure (LV HF). These conditions are characterized by LV dilation, an increase in the LV chamber radius to wall thickness ratio, and a resulting increase in LV wall stress. The physical remodeling of the heart is consistent with that observed in human HF. In addition, changes in the neurohormonal system that regulates cardiovascular function (e.g., the renin-angiotensin-aldosterone system) occur in this model in a time-dependent manner consistent with clinical observation. To date, this model has been successfully developed in dogs, pigs, and sheep for the study of HF progression and potential disease modifying therapies.[41]

In practice, the model is generated in a similar manner to the tachycardia-induced AF model. Surgical implantation of a pacemaker and electrical monitoring equipment is followed by a recovery period (2–3 weeks). Rapid pacing (approximately 220 beats per minute) is applied over an additional 2–3 weeks, and this leads to DCM and the associated LV HF described above. Unlike the tachycardia-induced AF model, however, burst pacing is not required for productive use of this model. Animal can be dosed with test compounds during the pacing period in order to assess their ability to slow remodeling of the heart. Efficacy can be determined by monitoring for changes in cardiovascular functions (e.g., contractility, LV ejection fraction, mean arterial pressure, LV systolic pressure, LV diastolic pressure, etc.) over time in the presence of test compounds. In addition, changes in concentration of circulating neurohormones (e.g., norepinephrine, atrial natriuretic peptide, aldosterone, etc.) and renin activity can be assessed via blood sampling. Finally, at the end of the study, myocytes can be isolated from test animals in order to determine the potential protective effect of test compounds. Once isolated, the myocytes can be studied structurally and biochemically to assess their overall health and contractility measurements provide additional insight regarding progression toward HF. Compounds capable of treating HF (suppressing disease progression) will show improvements in some, if not all of these areas.[42]

Non-surgical rodent models of HF are also available. The Dahl salt-sensitive rat, for example, develops heart failure when placed on a high-salt diet, but HF induction requires an extended time period. Initial hypertrophy of the left ventricle develops after 4–6 weeks, while significant HF-associated remodeling is observable after 15–20 weeks. Similarly, a modified version of the spontaneously hypertensive that is prone to heart failure develops left ventricular hypertrophy that progresses into a useful model of HF by the age of 12 months. The gradual development of symptoms and onset of remodeling is more closely related to the progression of cardiovascular disease into HF than the surgical models, but the significant costs associated with maintaining animal colonies for extended periods of time (6–12 months) is a serious drawback.[43]

ANIMAL MODELS OF INFECTIOUS DISEASE

There is no question that infectious disease is a significant health challenge. Despite the development of a wide range of anti-infective agents, preventing, controlling, and treating infectious disease remains an important issue. The appearance of resistant strains of various infectious organisms (e.g., methicillin-resistant *Staphylococcus aureus* (MRSA)) has highlighted the importance of continuing research efforts in this area. The concept of curing a systemic infection with drug therapy was first introduced in 1935 when Domagk demonstrated that sulphonamidocrysoidine (Prontosil) could be used to successfully treat mice suffering from a pneumococcal infection.[44] Since this discovery, hundreds of animal models of infection have been described in the literature. This should come as no surprise, given the wide range of infectious agents that exist in nature. An examination of a few exemplary animal models of infectious disease is useful, but a comprehensive review is well beyond the scope of this text. Those interested in a thorough review of this area are encouraged to consult specialized texts such as The Handbook of Animal Models of Infection.[45]

Murine Thigh Infection Model

The murine thigh infection model (Figure 7.16) is one of many animal models of infectious diseases that have been employed to identify therapeutically useful compounds. In this model, mice are first treated with cyclophosphamide over a 4-day period to render them neutropenic (<100 neutrophils/mm^3) to decrease their ability to mount an immune response. Once neutropenia is established, mice are injected in the thigh

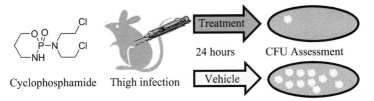

FIGURE 7.16 In the murine thigh infection model, mice are pretreated with cyclophosphamide to render them neutropenic. Introduction of a bacterial infection in the thigh is followed by a candidate compound or vehicle. After 24 h, the mice are sacrificed, and the number of colony-forming unit in treated versus untreated mice can be used to assess antibacterial efficacy of candidate compounds. CFU, colony-forming unit.

with a bacterial suspension (typically in a 10:1 dilution of the growth media) in order to establish infection. After a defined time interval, often 2–4 h, treatment with potential antibacterial agents is initiated. Both single- and multiple-dose protocols have been described in the literature. In both cases, upon completion of the dosing protocol and experimental time frame, the mice are sacrificed, and the number of colony-forming unit (CFUs) in the thigh is assessed. Efficacious compounds will decrease the number of CFUs in the thigh of treated mice relative to the untreated, control mice. Although this model is not a model of "natural" infections (pyomyositis is a very rare condition), this model is simple, versatile, and inexpensive. In addition, infection can be established with a high degree of accuracy, decreasing the number of animals required for each study, lowering the overall cost of this model.[46]

Murine Model of Systemic Infection

The murine model of systemic infection is another important infectious disease model that can be used to identify potential antibacterial agents. There are numerous variations of this procedure, but as with the thigh infection model, neutropenic mice are a typical starting point for this model. Once neutropenia is established, a lethal systemic infection is generated by intraperitoneal injection of 0.5 mL of an inoculum of an infectious bacteria (e.g., *Staphylococcus aureus*, *Listeria monocytogenes*) containing 10^7–10^8 CFU/mL. A single dose of a candidate compound is administered 1 h after inoculation and the mice are then monitored for 48 h. In the absence of effective treatment, the lethal dose of bacteria will kill the mice. Test compounds with the desired antibacterial activity will "rescue" the mice. Repeating the study with multiple doses across different groups of mice can be used to determine the ED_{50} (the dose required to rescue 50% of the mice) of candidate compounds. This

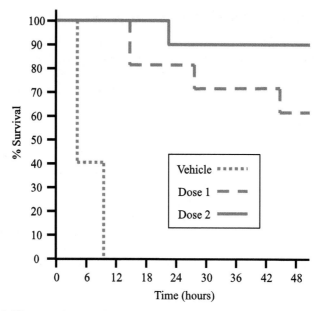

FIGURE 7.17   A Kaplan–Meier survival curves can be used to determine the efficacy of various doses of a candidate compounds in an infection model. In this graph, infected mice treated with the vehicle die faster than mice treated with a candidate compound at two different doses.

information can be used to compare the efficacy of multiple compounds or dose levels[47] and is generally displayed as Kaplan–Meier survival curve (Figure 7.17).

The Mouse Model of Influenza Virus Infection

Animal models of viral infections are also available to support the development of novel antiviral agents. The influenza virus, for example, can be studied using a variety of laboratory mice, as most of them are susceptible to infection with the influenza virus. The mice are typically infected with a strain of the influenza virus intranasally. The infection is allowed to become established over a brief time period (e.g., 4h), and then the mice are treated with candidate compounds on a set schedule. Compound efficacy can be determined by sacrificing the mice at fixed time points after dosing and determining the viral load and cytokine activity in the lung tissue of the animals. Candidate compounds with the desired antiviral activity will demonstrate decreased viral load and cytokine activity as compared to control mice. Alternatively, survival of the mice postinfection can be monitored and compared to survival of untreated animals as a measure of compound efficacy.[48]

Limitations of Animal Models of Infection

While the utility of the animal models described above, as well as hundreds of other, has been made clear by the identification of numerous therapeutic agents, there are some important limitations that should be kept in mind. First, many infectious diseases are host limited. Species-specific interactions between microbial ligands and host receptors are a significant part of the disease process that is often not replicated in an animal model of human infectious disease. Transgenic science is capable of "humanizing" animals to improve their correlation with the human condition, but they are still only an approximation of a human host. In a similar manner, there may be differences between the clinical and laboratory strains of an infectious agent. Although laboratory strains generally originate from primary clinical isolates, careful growth, enrichment, and selection are required to produce biological samples suitable for laboratory experiments. This process may evoke changes in gene expression and causes the infectious agent to adapt to growth in an artificial media. Laboratory monocultures can be significantly different from the clinical isolates that grow in a complex and dynamic environment (the human body).

Finally, while transmission pathways can play an important role in disease progression, very few animal models take this into account. Unique pathogenic phenotypes may result from the residence of an infectious agent in a particular microenvironment. Naturally occurring infectious diseases begin with an initial introduction of an infectious organism (via ingestion, inhalation, etc.), continue with a period of replication establishing the infection in the host, and then an exit from the host for transmission to the next host. In contrast, most animal models of infectious disease introduce an infectious agent at a high dose with a needle and syringe. Clearly, this method of introduction is very different from the natural life cycle of infectious organisms and could have a significant impact on the validity of the data acquired in an animal study.[49]

ANIMAL MODELS OF ONCOLOGY

According to the International Agency for Research on Cancer (IARC), there were 12.7 million new cancer cases and 7.6 million cancer-related deaths in 2008 worldwide. These figures are expected to grow to 21.4 million new cases and 13.2 million cancer-related deaths by 2030 as the population grows and ages.[50] The treatment and prevention of cancer has been and will likely continue to be an important component of the pharmaceutical industry's pipeline of products and research efforts. Much like infectious diseases, however, cancer is not a single disease, but rather a multitude of related diseases that share some common feature. As a result, there are a large number of animal models of cancer (the National Cancer Institute's cancer model database contains over 6000 models as of 2013[51]). They have some overlapping characteristics, but they are not interchangeable. Selecting the proper animal model is critical to the success of an oncology program.

Mouse Xenograft Tumor Model

One of the most widely used types of animal models of cancer is the mouse xenograft tumor model (Figure 7.18). This model takes advantage of the inability of specific types of mice, notably the athymic nude mice or severely compromised immunodeficient (SCID) mice, to mount an immune

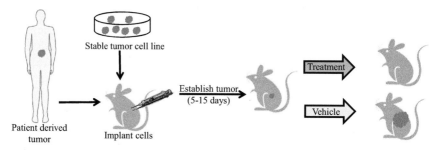

FIGURE 7.18   Mouse xenograft models can be used to determine the antitumor efficacy of candidate compounds. A tumor is established in an immunocompromised mouse using either patient-derived tumor cells or a stable tumor cell line. Once the tumor is established, treatment with a candidate compound or vehicle is initiated. Changes in the tumor size are monitored to determine the efficacy of candidate compounds.

response to foreign (non-native) cells. Human tumor cells derived from stable cell lines, such as those available from the National Cancer Institute's NCI-60 panel,[52] are often used as they are readily available and standardized for laboratory examination. A potential drawback to the use of stable cell lines, however, is that the selection process required to establish a propagating cell line may select for clones that are not necessarily representative of the original, clinically relevant tumors from which they were derived. Also, stable cell lines are typically grown using conditions that are vastly different from those that tumor cells would experience in a patient. Despite these issues, xenograft models using standardized cell lines has been a workhorse of cancer drug discovery programs for several decades.

As an alternative, patient-derived tumor cells can be used in order to establish a xenograft mouse model. In this approach, a small portion of a patient's tumor is removed from the patient and inserted into the immunocompromised mouse. Since the tumors are taken directly from patients, the issues of long term propagation and growth conditions are eliminated from the equation, providing a model that is closer to the clinical situation. The adoption of this model, however, has been slowed by the significant expense and technical difficulties associated with using patient-derived material.[53]

Irrespective of the source of the tumor cells, in practice, tumor cells (or a tumor section if the material is patient-derived) are inserted under the skin of a mouse. The tumor cells (or tumor fragment if patient-derived) are allowed to establish themselves over a short period of time (typically 5–15 days), and then treatment with a test compound is initiated. Changes in tumor volume over time can be used to determine whether or not compounds display any degree of efficacy as antitumor agents. In the absence of an effective treatment, tumors will grow without restriction, while compounds capable of halting or slowing disease progression will slow or stop tumor growth. It is worth noting that these experiments are time consuming and costly, as differences in tumor size and growth rates are typically measured over weeks or months (up to 4 months, sometimes longer), not hours or days.[54]

Mouse Allograft Tumor Model

One of the drawbacks of the mouse xenograft model is the use of cross-species transplantation (human tumor cells to a mouse host). The experiment requires the use of immunocompromised mice in order to ensure that the human tumor cells will not be rejected by an immune response mounted by the host animal. Of course, this is not consistent with the natural progression of cancer in humans. The allograft mouse model provides an opportunity to study potential antitumor agents in the presence of a normal immune system. In this model, immunocompetent mice (those with an intact immune system) are subjected to experimental conditions similar to those employed in the xenograft model, but the human tumor cells are replaced with mouse tumor cell lines (an intraspecies transplantation rather than an interspecies transplantation). Candidate compounds can be introduced in the same manner as described for the mouse xenograft model, and their efficacy can also be determined in the same manner (changes in tumor size over time). The presence of a fully functioning immune system makes the allograft model more similar to the human condition than the xenograft model.

Of course, the use of mouse cancer cells in and of itself is a significant issue. Candidate compounds that show positive efficacy in a mouse allograft model have only demonstrated that they are capable of treating cancer in mice, not humans. Subtle differences between the mouse and human variant of a compound's intended macromolecular target may allow it to perform the desired function in mice, but not in humans. As a result of this issue, allograft models are less utilized than the xenograft models in cancer drug discovery programs.

Genetically Engineered Mouse Models of Cancer

Although xenograft and allograft animal models are staples of cancer drug discovery and have provided a wealth of information on cancer progression, they are far from perfect models. Both of these models require the introduction of tumorigenic material into an animal from cultured cells. This has provided scientist with a means of improving their understanding of tumor growth and the various factors that effect tumor growth, but they have limitations. The microenvironment of naturally occurring tumors is very different from the artificial conditions required to grow tumor cells for implantation in the xenograft and allograft models. Factors that would influence natural tumor cell growth and tumor formation are simply not present in these models. As a result, these models cannot be relied upon to predict the role of a tumor's microenvironment on cancer progression.

The genetically engineered mouse (GEM) model of cancer is an alternative model that has gained in popularity since its introduction. In GEM models, genes that are suspected of participating in the transformation of normal cell into malignant cells and tumors are targeted for mutation, overexpression, or deletion. The resulting mice can then be studied to determine the impact of the genetic alteration on their propensity to develop tumors over time (Figure 7.19). In addition, it is possible to study the various stages

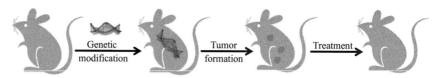

FIGURE 7.19   Genetically engineered mice are created using transgenic techniques. In one method, genes that promote spontaneous tumor formation are inserted into otherwise normal mice. Candidate compounds with *in vivo* antitumor properties will cause the tumors to shrink and/or decreased in number.

of tumor progression and determine the impact of therapeutic agents at each of these stages. Also, unlike xenograft models, GEM models are developed in mice with a fully intact immune system (the mice are immunocompetent). As a result, it has been postulated that the tumor microenvironment of GEM models more closely mimics natural cancer progression.

The ability to replicate specific genetic abnormalities known to produce tumors in humans also provides an opportunity to study specific tumor development pathways with the appropriately designed GEM model.

GEM models of cancer are, of course, not without their disadvantages. As with most genetically engineered animal model, developing a GEM model is a costly and time-consuming process and can require several years to validate. There are commercially available GEM models that can be used if they are appropriate for the cancer to be studied. It is also important to keep in mind that GEM model tumors are not necessarily comparable to actual human tumors. Unlike GEM tumors which are typically the result of limited alteration of the mouse genome, human tumors are heterogeneous in nature. Multiple mutations may be present in a clinical setting, and GEM models do not adequately recapitulate this aspect of disease progression. Finally, and perhaps the most importantly, tumors that develop in GEM models are mouse tumors, not human tumors. As a result, efficacy in a GEM model is not necessarily predictive of what will happen in a clinical setting.[55]

Selecting an appropriate animal model is a key aspect of a drug discovery program. Choosing the wrong animal model can lead to termination of viable programs based on erroneous data, or worse the progression of a compound into clinical trials based on results that are not truly correlated to the human condition. In many cases, discovery programs will employ multiple animal models in order to develop a more comprehensive understanding of potential clinical candidates. Additional animal models may be utilized to assess potential safety and toxicity risks. The animal models discussed in this chapter are just a small fraction of those available for the identification of novel therapeutics. Transgenic and knockout animal models have greatly enhanced the development of novel animal models, and it is likely that the number of viable animal models will continue to increase over time.

QUESTIONS

1. What is the definition of the term therapeutic index?
2. What is a homologous animal model?
3. What is an isomorphic animal model?
4. What is a predictive animal model?
5. Why is it necessary to use more than a single animal in an *in vivo* experiment?
6. A mouse treated with a candidate compound spends more time in the open arms of an elevated plus maze as compared to an untreated mouse. What does this suggest about the candidate compound?

7. In the novel object recognition model, what does it mean if a mouse treated with a candidate compound spends less time with a new object than untreated animals?
8. Describe the rotarod assay.
9. What disease can be modeled in animals by chronic administration of deoxycorticosterone acetate (DOCA)?
10. What are the limitations of diet-induced models of hypercholesterolemia?
11. What is a Kaplan–Meier survival curve?
12. What are some of the limitations of animal models of infectious disease?
13. What is the difference between a xenograft tumor model and an allograft tumor model?

References

1. Ingalls, A. M.; Dickie, M. M.; Snell, G. D. Obese, A New Mutation in the House Mouse. *J. Hered.* **1950**, *41* (12), 317–318.
2. Friedman, J. M.; Leibel, R. L.; Siegel, D. S.; Walsh, J.; Bahary, N. Molecular Mapping of the Mouse Ob Mutation. *Genomics* **1991**, *11* (4), 1054–1062.
3. a. Giovanella, B. C.; Fogh, J. The Nude Mouse in Cancer Research. *Adv. Cancer Res.* **1985**, *44*, 70–120.
   b. Flanagan, S. P. 'Nude', a New Hairless Gene with Pleiotropic Effects in the Mouse. *Genet. Res.* **1966**, *8*, 295–309.
4. Okamoto, K.; Aoki, K. Development of a Strain of Spontaneously Hypertensive Rat. *Jpn. Circ. J.* **1963**, *27*, 282–293.
5. Kachapati, K.; Adams, D.; Bednar, K.; Ridgway, W. M. The Non-obese Diabetic (NOD) Mouse as a Model of Human Type 1 Diabetes. *Methods Mol. Biol.* **2012**, *933*, 3–16.
6. Hegedus, J.; Putman, C. T.; Tyreman, N.; Gordon, T. Preferential Motor Unit Loss in the SOD1-G93A Transgenic Mouse Model of Amyotrophic Lateral Sclerosis. *J. Physiol.* **2008**, *586* (14), 3337–3351.
7. Wichmann, T.; DeLong, M. R. Pathophysiology of Parkinson's Disease: the MPTP Primate Model of the Human Disorder. *Ann. N. Y. Acad. Sci.* **2003**, *991*, 199–213.
8. Lewis, V. J.; Serge Przedborski, S. Protocol for the MPTP Mouse Model of Parkinson's Disease. *Nat. Protoc.* **2007**, *2*, 141–151.
9. Casals, J. B.; Pieri, N. C. G.; Feitosa, M. L. T.; Ercolin, A. C. M.; Roballo, K. C. S.; Barreto, R. S. N.; Bressan, F. F.; Martins, D. S.; Miglino, M. A.; Ambrósio, C. E. The Use of Animal Models for Stroke Research: A Review. *Comp. Med.* **2011**, *61* (4), 305–313.
10. Michael, L. H.; Entman, M. L.; Hartley, C. J.; Youker, K. A.; Zhu, J.; Hall, S. R.; Hawkins, H. K.; Berens, K.; Ballantyne, C. M. Myocardial Ischemia and Reperfusion: a Murine Model. *Am. J. Physiol.* **1995**, *269* (6), H2147–H2154.
11. Le Bars, D.; Gozariu, M.; Cadden, S. W. Animal Models of Nociception. *Pharmacol. Rev.* **2001**, *53* (4), 597–652.
12. Petit-Demouliere, B.; Chenu, F.; Bourin, M. Forced Swimming Test in Mice: a Review of Antidepressant Activity. *Psychopharmacology* **2005**, *177* (3), 245–255.
13. Conn, P. M., Ed. *Animal Models for the Study of Human Disease*; Academic Press: Waltham, MA, USA, 2013; p 354.
14. Zak, O., Sande, M. A., Eds. *Handbook of Animal Models of Infection: Experimental Models in Antimicrobial Chemotherapy*; Academic Press: San Diego, CA, US, 1999.

15. Kalbhen, D. A. Chemical Model of Osteoarthritis–a Pharmacological Evaluation. *J. Rheumatol.* **1987,** *14,* 130–131.

16. Marcotte, E. R.; Pearson, D. M.; Srivastava, L. K. Animal Models of Schizophrenia: a Critical Review. *J. Psychiatry Neurosci.* **2001,** *26* (5), 395–410.

17. Porsolt, R. D.; Anton, G.; Blavet, N.; Jalfre, M. Behavioural Despair in Rats: a New Model Sensitive to Antidepressant Treatments. *Eur. J. Pharmacol.* **1978,** *47* (4), 379–391.

18. Walf, A. A.; Frye, C. A. The Use of the Elevated Plus Maze as an Assay of Anxiety-related Behavior in Rodents. *Nat. Protoc.* **2007,** *2,* 322–328.

19. Antunes, M.; Biala, G. The Novel Object Recognition Memory: Neurobiology, Test Procedure, and its Modifications. *Cognit. Process.* **2012,** *13* (2), 93–110.

20. Wehner, J. M.; Radcliffe, R. A. Cued and Contextual Fear Conditioning in Mice. *Curr. Protoc. Neurosci.* **2004.** 8.5C.1–8.5C.14.

21. D'Hooge, R.; De Deyn, P. P. Applications of the Morris Water Maze in the Study of Learning and Memory. *Brain Res. Rev.* **2001,** *36* (1), 60–90.

22. Gurney, M. E.; Pu, H.; Chiu, A. Y.; Dal Canto, M. C.; Polchow, C. Y.; Alexander, D. D.; Caliendo, J.; Hentati, A.; Kwon, Y. W.; Deng, H. X.; Chen, W.; Zhai, P.; Sufit, R. L.; Siddique, T. Motor Neuron Degeneration in Mice that Express A Human Cu,Zn Superoxide Dismutase Mutation. *Science* **1994,** *264* (5166), 1772–1775.

23. Porras, G.; Li, Q.; Bezard, E. Modeling Parkinson's Disease in Primates: The MPTP Model. *Cold Spring Harb. Perspect. Med.* **2012,** *2* (3), 1–10.

24. *A Global Brief on Hypertension: Silent Killer, Global Public Health Crisis" World Health Day 2013;* World Health Organization: Geneva, Switzerland, 2013.

25. Lerman, L. O.; Chade, R. A.; Sica, V.; Napoli, C. Animal Models of Hypertension: An Overview. *J. Lab. Clin. Med.* **2005,** *146* (3), 160–173.

26. Li, J.; Wang, D. H. Role of TRPV1 in Renal Hemodynamics and Function in Dahl Saltsensitive Hypertensive. *Exp. Physiol.* **2008,** *93* (8), 945–953.

27. Conrad, C. H.; Brooks, W. W.; Hayes, J. A.; Sen, S.; Robinson, K. G.; Bing, O. H. Myocardial Fibrosis and Stiffness with Hypertrophy at Heart Failure in the Spontaneously Hypertensive Rat. *Circulation* **1995,** *91* (1), 161–170.

28. a. Kubota, Y.; Umegaki, K.; Kagota, S.; Tanaka, N.; Nakamura, K.; Kunitomo, M.; Shinozuka, K. Evaluation of Blood Pressure Measured by Tail-cuff Methods (Without Heating) in Spontaneously Hypertensive Rats. *Biol. Pharm. Bull.* **2006,** *29* (8), 1756–1758.

    b. Nariai, T.; Fujita, K.; Mori, M.; Katayama, S.; Hori, S.; Matsui, K. SM-368229, a Novel Promising Mineralocorticoid Receptor Antagonist, Shows Antihypertensive Efficacy with Minimal Effect on Serum Potassium Level in Rats. *J. Cardiovasc. Pharmacol.* **2012,** *59* (5), 458–464.

29. a. Braga, V. A.; Burmeister, M. A. Applications of Telemetry in Small Laboratory Animals for Studying Cardiovascular Diseases Chapter 9. In *Modern Telemetry;* Krejcar, O., Ed.; Intech: Rijeka, Croatia, 2011.

    b. Wood, J. M.; Maibaum, J.; Rahuel, J.; Markus, G.; Grutter, M. G.; Cohen, N. C.; Rasetti, V.; Heinrich Ruger, H.; Goschke, R.; Stutz, S.; Fuhrer, W.; Schilling, W.; Rigollier, P.; Yamaguchi, Y.; Cumin, F.; Baum, H. P.; Schnell, C. R.; Herold, P.; Mah, R.; Jensen, C.; O"Brien, E.; Stanton, A.; Bedigianj, M. P. Structure-based Design of Aliskiren, a Novel Orally Effective Renin Inhibitor. *Biochem. Biophys. Res. Commun.* **2003,** *308,* 698–705.

30. Carmena, R.; Duriez, P.; Fruchart, J. C. Atherogenic Lipoprotein Particles in Atherosclerosis. *Circulation* **2004,** *109,* III-2–III-7.

31. a. Krause, B. R.; Newton, R. S. Lipid-lowering Activity of Atorvastatin and Lovastatin in Rodent Species: Triglyceride-Lowering in Rats Correlates with Efficacy in LDL Animal Models. *Atherosclerosis* **1995,** *117,* 237–244.

    b. Kroon, P. A.; Hand, K. M.; Huff, J. W.; Alberts, A. W. The Effects of Mevinolin on Serum Cholesterol Levels of Rabbits with Endogenous Hypercholesterolemia. *Atherosclerosis* **1982,** *44,* 41–48.

32. Meir, K. S.; Leitersdorf, E. Atherosclerosis in the Apolipoprotein E–deficient Mouse: A Decade of Progress. *Arterioscler., Thromb., Vasc. Biol.* **2004**, *24*, 1006–1014.

33. Chugh, S. S.; Havmoeller, R.; Narayanan, K.; Singh, D.; Rienstra, M.; Benjamin, E. J.; Gillum, R. F.; Kim, Y. H.; McAnulty, J. H., Jr; Zheng, Z. J.; Forouzanfar, M. H.; Naghavi, M.; Mensah, G. A.; Ezzati, M.; Murray, C. J. L. Worldwide Epidemiology of Atrial Fibrillation: A Global Burden of Disease 2010 Study. *Circulation* **2014**, *129*, 837–847.

34. Janse, M. J.; Rosen, M. R. History of Arrhythmias. *Handb. Exp. Pharmacol.* **2006**, *171*, 1–39.

35. Vaidya, D.; Morley, G. E.; Samie, F. H.; Jalife, J. Reentry and Fibrillation in the Mouse Heart: A Challenge to the Critical Mass Hypothesis. *Circulation Res.* **1999**, *85*, 174–181.

36. Riley, G.; Syeda, F.; Kirchhof, P.; Fabritz, L. An Introduction to Murine Models of Atrial Fibrillation. *Front. Physiol.* **2012**, *3* (296), 1.

37. McMurray, J. J.; Petrie, M. C.; Murdoch, D. R.; Davie, A. P. Clinical Epidemiology of Heart Failure: Public and Private Health Burden. *Eur. Heart J.* **1998**, *19*, P9–P16.

38. Loehr, L. R.; Rosamond, W. D.; Chang, P. P.; Folsom, A. R.; Chambless, L. E. Heart Failure Incidence and Survival (From the Atherosclerosis Risk in Communities Study). *Am. J. Cardiol.* **2008**, *101*, 1016–1022.

39. Dixon, J. A.; Spinale, F. G. Large Animal Models of Heart Failure: A Critical Link in the Translation of Basic Science to Clinical Practice. *Circ.: Heart Failure* **2009**, *2*, 262–271.

40. a. Yarbrough, W. M.; Mukherjee, R.; Escobar, G. P.; Mingoia, J. T.; Sample, J. A.; Hendrick, J. W.; Dowdy, K. B.; McLean, J. E.; Lowry, A. S.; O'Neill, T. P.; Spinale, F. G. Selective Targeting and Timing of Matrix Metalloproteinase Inhibition in Post–Myocardial Infarction Remodeling. *Circulation* **2003**, *108*, 1753–1759.

    b. Sakai, S.; Miyauchi, T.; Kobayashi, M.; Yamaguchi, I.; Goto, K.; Sugishita, Y. Inhibition of Myocardial Endothelin Pathway Improves Long-Term Survival in Heart Failure. *Nature* **1996**, *384*, 353–355.

41. a. Dixon, J. A.; Spinale, F. G. Large Animal Models of Heart Failure: A Critical Link in the Translation of Basic Science to Clinical Practice. *Circ.: Heart Failure* **2009**, *2*, 262–271.

    b. Moea, G. W.; Armstrong, P. Pacing-induced Heart Failure: a Model to Study the Mechanism of Disease Progression and Novel Therapy in Heart Failure. *Cardiovasc. Res.* **1999**, *42*, 591–599.

42. Spinale, F. G.; Holzgrefe, H. H.; Mukherjee, R.; Hird, R. B.; Walker, J. D.; Arnim-Barker, A.; Powell, J. R.; Koster, W. H. Angiotensin-converting Enzyme Inhibition and the Progression of Congestive Cardiomyopathy: Effects on Left Ventricular and Myocyte Structure and Function. *Circulation* **1995**, *92*, 562–578.

43. Patten, R. D.; Hall-Porter, M. R. Small Animal Models of Heart Failure: Development of Novel Therapies, Past and Present. *Circ.: Heart Failure* **2009**, *2*, 138–144.

44. Domagk, G. Ein Beitrag zur Chemotherapie der bakteriellen Infektionen. *Dtsch. Med. Wochenschr.* **1935**, *61*, 250–253.

45. Sande, M. A., Zak, O., Eds. *Handbook of Animal Models of Infection;* , 1998. London, UK.

46. a. Vogelman, B.; Gudmundsson, S.; Leggett, J.; Turnidge, J.; Ebert, S.; Craig, W. A. Correlation of Antimicrobial Pharmacokinetic Parameters with Therapeutic Efficacy in an Animal Model. *J. Infect. Dis.* **1988**, *158* (4), 831–847.

    b. Gudmundsson, S.; Erlendsdottir, S. Murine Thigh Infection Model. In *From Handbook of Animal Models of Infection;* Sande, M. A., Zak, O., Eds.; 1998; pp 137–144. London, UK.

47. a. Xin, Q.; Fan, H.; Guo, B.; He, H.; Gao, S.; Wang, H.; Huang, Y.; Yang, Y. Design, Synthesis, and Structure–Activity Relationship Studies of Highly Potent Novel Benzoxazinyl-oxazolidinone Antibacterial Agents. *J. Med. Chem.* **2011**, *54*, 7493–7502.

    b. Jang, W. S.; Lee, S. C.; Lee, Y. S.; Shin, Y. P.; Shin, K. H.; Sung, B. H.; Kim, B. S.; Lee, S. H.; Lee, I. H. Antimicrobial Effect of Halocidin-derived Peptide in a Mouse Model of Listeria Infection. *Antimicrob. Agents Chemother.* **2007**, *51* (11), 4148–4156.

48. a. Sidwell, R. W. The Mouse Model of Influenza Virus Infection. In *From Handbook of Animal Models of Infection;* Sande, M. A., Zak, O., Eds.; 1998; pp 981–987. London, UK.

   b. Ohgitani, E.; Kita, M.; Mazda, O.; Imanishi, J. Combined Administration of Oselta-
       mivir and Hochu-ekki-to (TJ-41) Dramatically Decreases the Viral Load in Lungs of
       Senescence-accelerated Mice during Influenza Virus Infection. *Arch. Virol.* **2014,** *159,*
       267–275.
49. Wiles, S.; Hanage, W. P.; Frankel, G.; Robertson, B. Modelling Infectious Disease – Time
    to Think outside the Box? *Nat. Rev. Microbiol.* **2006,** *4,* 307–312.
50. Ferlay, J.; Shin, H. R.; Bray, F.; Forman, D.; Mathers, C. D.; Parkin, D. *GLOBOCAN 2008,
    Cancer Incidence and Mortality Worldwide: IARC CancerBase No.10;* International Agency
    for Research on Cancer: Lyon, France, 2010.
51. https://cancermodels.nci.nih.gov/camod/.
52. http://dtp.nci.nih.gov/branches/btb/ivclsp.html.
53. Morton, C. L.; Houghton, P. J. Establishment of Human Tumor Xenografts in Immunode-
    ficient Mice. *Nat. Protoc.* **2007,** *2* (2), 247–250.
54. a. Gangjee, A.; Zhao, Y.; Raghavan, S.; Rohena, C. C.; Mooberry, S. L.; Hamel, E.
       Structure Activity Relationship and *in Vitro* and *in Vivo* Evaluation of the Potent
       Cytotoxic   Anti-microtubule   Agent   N(4-methoxyphenyl)n,2,6-trimethyl-6,7-
       dihydro-5Hcyclopenta[d]pyrimidin-4-aminium Chloride and its Analogues as
       Antitumor Agents. *J. Med. Chem.* **2013,** *56,* 6829–6844.
    b. Hennessy, E. J.; Adam, A.; Aquila, B. M.; Castriotta, L. M.; Cook, D.; Hattersley, M.;
       Hird, A. ,W.; Huntington, C.; Kamhi, V. M.; Laing, N. M.; Li, D.; MacIntyre, T.; Omer,
       C. A.; Oza, V.; Patterson, T.; Repik, G.; Rooney, M. T.; Saeh, J. C.; Sha, L.; Vasbinder, M.
       M.; Wang, H.; Whitston, D. Discovery of a Novel Class of Dimeric Smac Mimetics as
       Potent IAP Antagonists Resulting in a Clinical Candidate for the Treatment of Cancer
       (AZD5582). *J. Med. Chem.* **2013,** *56,* 9897–9919.
55. Richmond, A.; Su, Y. Mouse Xenograft Models Vs GEM Models for Human cancer Thera-
    peutics. *Dis. Models & Mech.* **2008,** *1,* 78–82.

Safety and Toxicology

The preceding chapters of this text have been primarily focused on the identification of candidate compounds that are capable of eliciting a desired biological effect for a sustained period of time in order to treat a condition or disease. A compound capable of achieving these end points, however, is not necessarily a good drug. In order for a candidate compound to matriculate into a marketed drug, it must produce a benefit that substantially outweighs the risk of not only leaving the condition or disease untreated, but also any risk associated with the compound itself. In other words, safety and toxicity associated with candidate compounds must be minimized in order to limit risks to patients. Thus, safety pharmacology, the study of a candidate compound's impact on normal physiological function, is an important aspect of drug discovery and development.

The negative impact of a candidate compound is caused by its interaction with biomolecules (e.g., enzymes, GPCR, ion channels, etc.) in the same manner as the desired effects. This means that there will be a measurable relationship between the dose of compound provided and the intensity of the negative response. This dose response relationship can be compared to the dose response relationship for the desired biological effect in order to identify treatment levels capable of producing the desired effect without eliciting a negative response. The distance between these two doses is referred to as a safety window, and the ratio of the two doses is the therapeutic index (Figure 8.1). For example, a compound capable of treating a bacterial infection with a plasma concentration of 1.0 nM that also elicits kidney damage at 250 nM has a therapeutic index of 250 with respect to kidney damage.

Another important measure of safety and toxicity is the highest dose or exposure at which no toxicity is observed, the NOEL (no observed effect level) of a compound. It is determined in animal studies that employ an ascending series of doses and subsequent analysis to determine the point at which any toxic event occurs, no matter how small it may be. Hair loss is considered equal to ischemic events with regard to determining an NOEL dose. Of course, in reality there is a significant difference between these

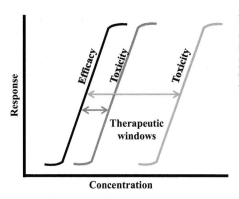

FIGURE 8.1 The therapeutic window defined by the red toxic event is substantially narrower than the therapeutic window defined by the green toxic event. It is easier to manage patient risk with a wider therapeutic window.

two types of toxicity, as one is far more manageable than the other, which leads to the concept of the no observed adverse effect level (NOAEL) of a drug. This is the maximum dose or exposure level of a compound that can be employed without creating unmanageable toxicity. This is also referred to as the maximum tolerable dose (MTD).[1]

When considering safety and toxicity, a question that inevitably arises is how wide does the therapeutic window need to be in order for a compound to be considered safe enough to proceed to market? In other words, how safe is safe enough? Ideally, candidate compounds would be designed to eliminate all possible negative effects. Drug advertisements are, however, filled with provisos regarding potential side effects, toxicities, and warnings regarding the possibility of negative consequences of taking a given drug. Clearly, marketed drugs carry a risk, even though they have been deemed suitable for clinical use. The safety requirements for a new drug are largely dependent upon the nature of the disease or condition that they are designed to treat. Compounds designed to treat disease and conditions that would be lethal if left untreated often have lower safety requirements, especially if there is no effective treatment available. Amyotrophic lateral sclerosis (ALS), for example, has an 80% 5 year mortality rate and very limited treatment options. Rilutek® (Riluzole, Figure 8.2) is the only available treatment for ALS, but it increases life expectancy by only 2–3 months after 15 months of treatment.[2] If a curative agent was identified, but it increased the risk of myocardial infarction by a factor of 4, this might be considered tolerable, as the alternative is almost certain death within 5 years. On the other hand the same fourfold increase in risk would be intolerable in a novel pain medication, as there are many pain medications available. While pain is undesirable, it does not generally qualify as a lethal condition. The painkiller Vioxx® (Rofecoxib, Figure 8.2) was removed from the market after it was discovered that it increased the risk of heart attacks by a factor of 4.[3]

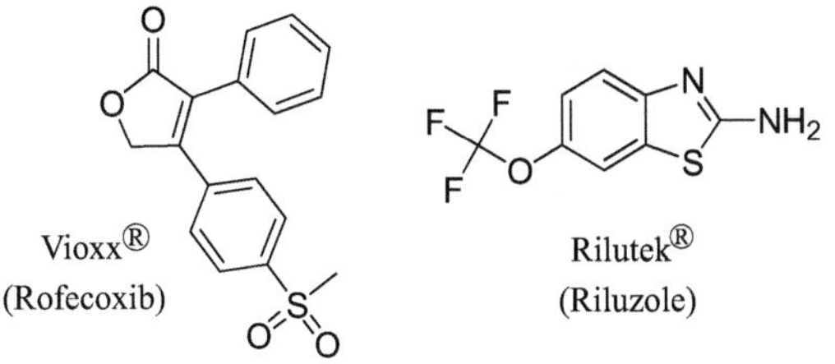

FIGURE 8.2 Vioxx® (Rofecoxib, left) was removed from market due to increased risk of cardiac ischemia. Rilutek® (Riluzole, right) is the only available treatment for ALS.

SOURCES OF TOXICITY

Although it is not always possible to determine the underlying biochemical mechanism of toxicity associated with a candidate compound, if this information is available, it can be very helpful in determining how to eliminate safety issues. In some cases, toxicity or safety issues are associated with the therapeutic target itself. This is often referred to as target-based or mechanism-based toxicity and can be described as unintended effects associated with modulation of the activity of the macromolecular target itself.[4] Inhibition of matrix metalloproteinases (MMPs), for example, had been viewed as a viable method of slowing the progression of various forms of arthritis.[5] Unfortunately, clinical study of various MMP inhibitors demonstrated that inhibition of this class of enzymes causes a musculoskeletal syndrome in which matrix protein builds up and severely restricted patient movement.[6] Compounds and programs that are definitively associated with mechanism based toxicity are difficult to manage, as increasing target activity is likely to increase toxicity. Discovery and development programs are often terminated when mechanism-based toxicity or safety issues are identified.

Safety and toxicity issues can also be caused by interference with the normal biological activity of biomolecules other than the therapeutic target of interest. In other words, a lack of target selectivity can give rise to undesirable side effects which are classified as "off-target effects." Unlike mechanism-based issues, off-target effects are not tied to desired mechanism of action of the candidate compound. Instead, they can be traced back to a lack of target selectivity.[4] As discussed in earlier chapters, in an ideal world, candidate compounds would interact with only the target or targets necessary to exert the desired biological influence and nothing else. Unfortunately, this is often not the case. The vast majority of candidate compounds are capable of interacting with a wide range of biomolecules with varying degrees of potency. Fortunately, it is possible to develop structure–activity relationships for these interactions in the same manner as achieved for the target of interest. This, in turn, provides an opportunity for scientists to design compounds that are less prone to off-target effects.

Large pharmaceutical companies typically have the ability to run a number of off-target selectivity assays. Typically, a set of "nearest neighbor" targets are selected to serve as a preliminary assessment of risk. If, for example, a program were targeting the 5-HT_{1a} serotonin receptor, a selectivity panel designed to assess compounds for off-target side effects would most likely include the remaining serotonin receptors (5-HT_2 through 5-HT_7), as they have a high degree of homology with the target of interest. This type of assessment often occurs early in the *in vitro* screening process and can be used to prioritize candidate compounds. Ultimately, however, a candidate compound must be assessed across a wide range of targets in order to assess its propensity for side effects, safety, and toxicity issues. Even the largest of pharmaceutical companies do not have the capacity to establish *in vitro* screens for the multitude of possible targets that could produce untoward effects. There are, however, a number of organizations that specialize in this particular aspect of the drug discovery and development process. These institutions (e.g., EMD Millipore,[7] Perkin Elmer,[8] Cerep,[9] National Institute of Mental Health's Psychoactive Drug Screening Program,[10] etc.) maintain a diverse set of *in vitro* assays covering important enzymes, ion channels, GPCRs, and other biomolecules that can provide additional insight into potential side effects.

In some cases, safety and toxicology problems associated with a candidate compound are not directly associated with the compound itself. When compounds enter the body, they are subject to the metabolic processes designed to remove them from the circulation. Modification of the candidate compound by metabolic enzyme will produce new compounds that have properties different from the original. If these new compounds are, in turn, capable of binding to biomolecules associated with negative effects, then safety and toxicity issue may arise. Consider, for example, a candidate compound that has negligible binding to the hERG channel, but is metabolized into a compound with strong hERG binding (Figure 8.3). Even though

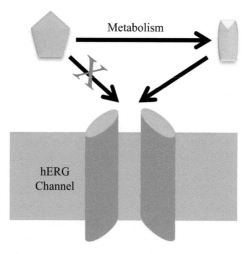

FIGURE 8.3   The drug (green) does not interact with the hERG channel. It is converted to a metabolite (yellow) that blocks the hERG channel.

the parent compound is unable to induce hERG-related cardiac toxicity, its metabolic by-product creates a very serious safety concern that would likely halt progression of the compound in question. In a similar manner, metabolism of a candidate compound can also produce reactive metabolites. Nucleophilic or electrophilic centers created by the action of metabolic processes normally designed to protect the body creates compounds capable of forming covalent bonds with endogenous biomolecules. The resulting adducts will no longer function properly. Normal function can only be restored by the generation of replacement protein/enzyme and removal of the damaged material. In the meantime, loss of function of the damage macromolecule can lead to significant issues. Consider, for example, the conversion of a candidate compound into a molecule capable of covalently binding to cytochrome P450 3A4 (CYP3A4), a key metabolic enzyme of the liver. Inactivation of this enzyme will impair the metabolic processing of *all* compounds that are eliminated through the CYP3A4 pathway, potentially producing toxic concentrations of compounds that would otherwise have been removed by this important metabolic enzyme. The antidepressant Dutonin® (Nefazodone), for example, is converted to a quinoneiminium ion by the liver metabolism enzyme, which reacts with CYP3A4 and deactivates it (Figure 8.4).[11]

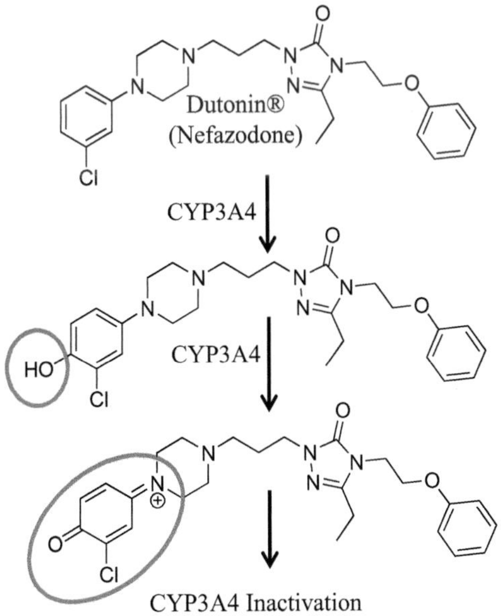

FIGURE 8.4 The antidepressant Dutonin® (Nefazodone) is hydroxylated by CYP3A4 (middle structure, red circle). This compound is subsequently oxidized to the quinoneiminium ion (bottom compound, red circle), which reacts with CYP3A4, deactivating the enzyme. Deactivation of CYP3A4 by this process will impede the metabolism of all compounds that require CYP3A4 for metabolic processing.

There are some safeguards in the body capable of capturing reactive metabolites before they cause damage, such as glutathione, a scavenger of electrophiles, but these systems are far from perfect and can be overwhelmed by candidate compounds, especially when cells are in a state of oxidative stress.

Alternatively, a compound or metabolite may be capable of forming an adduct with a biomolecule that will elicit an immune response. Compounds of this type are referred to as haptens. In the absence of a carrier protein, a hapten will not evoke an immune response, but once it is linked a suitable protein, the resulting adduct is no longer recognized by the body. Antibodies for the hapten–protein adduct are generated, and an immune response is launched targeting the "foreign" compound. The clinical manifestation of this type of response can range from a simple skin rash to anaphylaxis and death. Allergic reactions to β-lactam are a well-known example of the interaction of a hapten with normal proteins leading to an adverse reaction (Figure 8.5).[12]

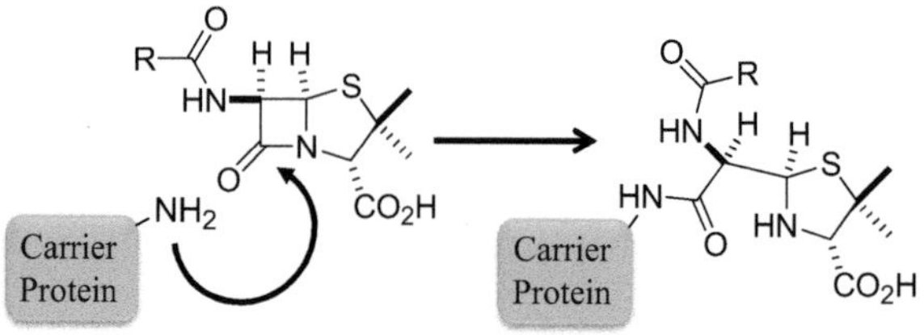

FIGURE 8.5   The reaction of β-lactam containing compounds with a protein is a well-known example of how a small molecule can act as a hapten and elicit an immune response. Nucleophilic functional groups on a protein (e.g., amines) can react with a β-lactam. The resulting hapten–protein adduct is not be recognized as "self" leading to an immune response.

Clearly, understanding the metabolism of a candidate compound can play a key role in minimizing the risk of safety and toxicity issues. Metabolic process, however, often produces multiple metabolites, and determining which compound is responsible for observed negative effects is often problematic. Preparation and testing of the proposed metabolite can provide insight into its potential role in safety and toxicity, but this can be a time and resource intensive process. Fortunately, a great deal of information is available regarding the metabolism of a wide range of functional groups and substructures that may initiate toxicity through the formation of reactive intermediates. Aryl nitro groups, for example, are well known for their propensity to undergo metabolic activation. Tasmar® (Tolcapone), an inhibitor of catechol-O-methyl transferase (COMT) useful for the Parkinson's disease, for example, is converted to a quinone imine that reacts with available nucleophiles (Figure 8.6(a)).[13] As a result, this

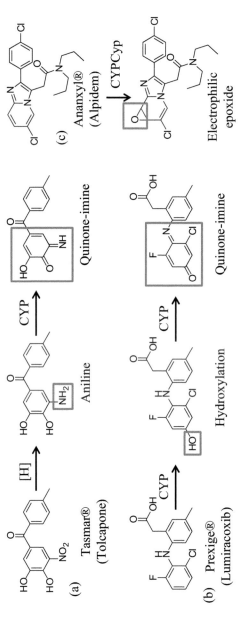

FIGURE 8.6 (a) Tasmar® (Tolcapone) is metabolically processed first to the amine and then to the reactive quinone imine. (b) Metabolic processing of Prexige® (Lumiracoxib) produces a hydroxylated product that is then converted to a reactive quinone imine. (c) Ananxyl® (Alpidem) is metabolically converted to a reactive epoxide.

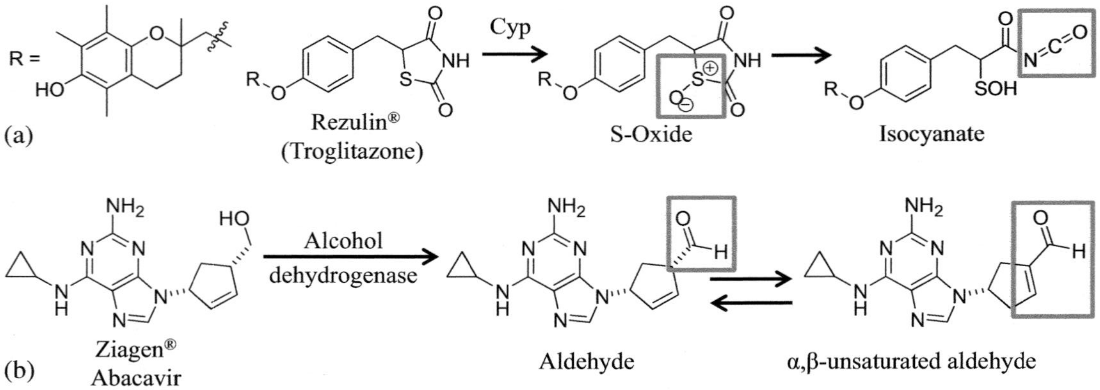

FIGURE 8.7 (a) Oxidation of the sulfur atom of Rezulin® (troglitazone) leads to an unraveling of the thiazolidine-2,4-dione and formation of the corresponding reactive isocyanate. (b) Oxidation of the primary alcohol of Ziagen® (Abacavir) produces the corresponding aldehyde. The aldehyde can react with available nucleophiles or isomerize to the α,β-unsaturated aldehyde, which can also reacted with available nucleophiles.

compound is associated with liver toxicity. The cyclooxygenase-2 (COX-2) inhibitor Prexige® (Lumiracoxib) is also metabolically activated by conversion into a quinone imine, but its metabolic pathway is different. In this case, hydroxylation of an aromatic ring precedes metabolic activation. The resulting reactive metabolite will react with available nucleophiles, which can lead to toxicity (Figure 8.6(b)).[14] Direct oxidation of an aromatic ring can also lead to the generation of an epoxide, as seen in the CYP-mediated metabolism of Ananxyl® (Alpidem), an antianxiety agent that was withdrawn from the market based on reports of severe liver damage in patients (Figure 8.6(c)).[15]

The formation of reactive metabolites is not limited to the activation of aromatic systems. Oxidation of heteroatoms is another common metabolic process. In the case of Rezulin® (troglitazone), an antidiabetic drug withdrawn from the market in 2000, oxidation of a sulfur atom leads to liver toxicity. In this case, CYP oxidation of the sulfur atom of the thiazolidine-2,4-dione ring leads to ring opening and the formation of an isocyanate that readily reacts with available nucleophiles (Figure 8.7(a)).[16] It is also important to keep in mind that CYP enzymes are not the only pathway toward metabolic activation. There are many other metabolic processes that can produce potentially toxic materials from otherwise unreactive compounds. Alcohol dehydrogenase, for example, oxidizes the primary alcohol of Ziagen® (Abacavir), a reverse transcriptase inhibitor useful for the treatment of HIV infection. In this instance, the resulting aldehyde is in equilibrium with the corresponding α,β-unsaturated aldehyde, both of which can react with various biomolecules to produce toxic effects (Figure 8.7(b)).[17]

In an effort to avoid potential reactive metabolites, many drug discovery and development programs will attempt to design analogs and compounds that are devoid of functionality known to produce reactive metabolites. Candidate compounds that contain suspicious functionality are not always problematic, but given the perceived risk and wide range of compounds that can be made, they are often given a lower priority. There are a number of reviews on the metabolic activation of functional groups and structural classes that provide excellent guidance on points to consider in compound design and priority relative of the risk of reactive metabolite formation.[18]

ACUTE VERSUS CHRONIC TOXICITY

Given the wide range of biochemical reaction that must occur in order to sustain life, it should come as no surprise that there are many different types of toxicity. The end results can be as mild as a skin rash or as severe as death, depending on the importance and influence of the molecular mechanism that is being impacted in an undesirable manner. In general, compound toxicity can be divided into categories based on the number of doses required to elicit toxic effects. Compounds that produce toxicity after a single dose are referred to as possessing acute toxicity, while compounds whose toxicity is associated with long-term dosing are said to have chronic toxicity. Acutely toxic compounds are very rarely employed as drugs, although there are exceptions. Arsenic trioxide, for example, is marketed as a second line treatment for certain types of leukemia under the trade name Trisenox®,[19] despite its well-known acute toxicity (Arsenic trioxide is routinely employed as a rat poison.). Modern compound screening methods, both *in vitro* and *in vivo*, are employed to identify and eliminate from consideration compounds that are acutely toxic.

Chronically toxic compounds, on the other hand, require prolonged or repeated exposure in order to elicit untoward effects. Whether or not chronic toxicity can be tolerated in a candidate compound depends on the nature of toxicity, the severity of the condition being treated, and the duration of exposure required to elicit toxicity. Consider, for example, the musculoskeletal syndrome associated with MMP inhibitors described above.[6] A compound capable of inducing reversible musculoskeletal syndrome over the course of several months of exposure is unlikely to be a viable treatment for chronic conditions such as osteoarthritis, neuropathic pain, or high blood pressure. On the other hand, use of the same compound as treatment for a bacterial infection might be acceptable. In this case, patient exposure is limited to a short time frame, typically 5–14 days, which is not long enough to elicit musculoskeletal syndrome. Chronic toxicity associated with increased risk of conditions such as cancer, cardiovascular disease, ischemic disease, and irreversible nerve damage can significantly limited the likelihood of the successful commercialization of candidate compounds.

CYTOTOXICITY

Compounds that exhibit cytotoxicity are capable of killing otherwise healthy cells. With the exception of programs aimed at the identification of anti-cancer agents, cytotoxicity is not generally tolerated in candidate compounds. Fortunately, *in vitro* screening methods have been designed to identify cytotoxic compound early in the drug discovery process. The MTT human hepatotoxicity assay is one commonly used method. In this assay, changes in cell viability are monitored by measuring the rate of conversion of the yellow dye 3-[4,5-dimethylthiazol-2-yl]-2,5-diphenyltetrazolium bromide (MTT) into formazan, a

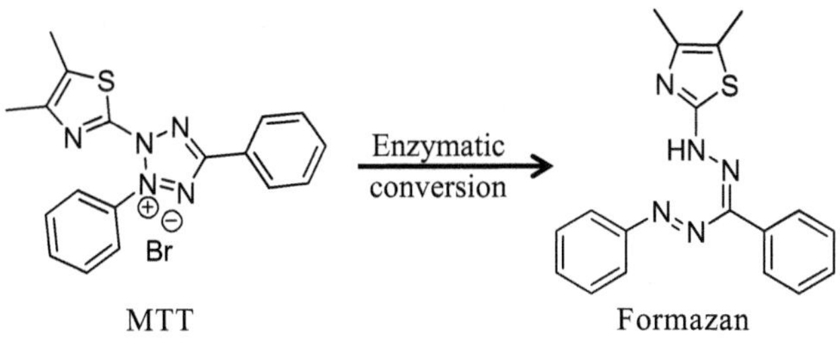

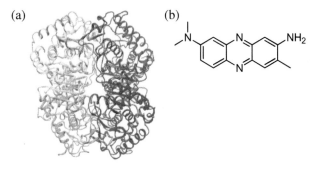

FIGURE 8.8 The MTT human hepatotoxicity assay monitors the rate of conversion of MTT to formazan in a cell culture. Compounds that are cytotoxic decrease the rate of production of formazan from MTT.

purple compound (Figure 8.8). This reaction occurs in the mitochondria of cells at a known rate, and concentrations of both MTT and formazan can be measured spectroscopically. Cytotoxic compound will decrease the rate of conversion, making it possible to screen for this potential safety issue. In addition, since this assay uses hepatocytes (e.g., liver cells), compounds that are converted to cytotoxic metabolites will also be identified in this assay.[20]

The lactate dehydrogenase (LDH) assay[21] and the neutral red assay[22] are also popular methods of identifying compounds that represent a cytotoxicity risk. Both of these assays can be run with hepatocytes, preserving the ability to identify compounds that produce cytotoxic metabolites in the liver. In the LDH assay, the amount of LDH (Figure 8.9(a)) released by dead cells after an incubation period with candidate compounds is used as an indicator of cytotoxicity. Quantitation of the production of formazan by LDH can be accomplished using commercially available kits (Roche, catalog number 11644793001). Since LDH is not released by living cells, increased concentration of LDH in the cell growth media is an indication of cytotoxicity. In a similar manner, healthy hepatocytes absorb the dye neutral red (Figure 8.9(b)) and sequester it in lysosomes.

FIGURE 8.9 (a) Lactate dehydrogenase (LDH) (RCSB PDB file 1T2F). (b) Neutral red.

Measuring the uptake of neutral red by a cell culture after incubation with a candidate compound will provide insight into the candidate compound's impact on cell viability. Compounds that are cytotoxic will decrease the absorption of neutral red as compared to cells grown in the absence of the compounds.

Although each of these assays can be used to identify cytotoxic compounds at a very early stage of a program, they do not provide information on the biochemical causes of the cytotoxicity. In order to determine the root cause, additional studies would be required. The value of gaining an understanding of the mechanistic underpinnings of cytotoxicity of a particular candidate compound should be carefully assessed.

CARCINOGENICITY, GENOTOXICITY, AND MUTAGENICITY

The interrelated concepts of carcinogenicity,[23] genotoxicity,[24] and mutagenicity[25] are exceptionally important in the identification of novel therapeutics. In general, carcinogenic compounds cause cancer through a variety of different mechanisms such as alteration in cellular metabolism or DNA damage that cause uncontrolled proliferation of malignant cells. Compounds that are genotoxic damage the genetic information within a cell. Changes to DNA created by genotoxic compounds can be in the form of single strand DNA breaks, double stranded DNA breaks, or mutation of the DNA. In some cases, genotoxicity leads to apopotosis (programmed cell death), but it can also lead to the formation of malignant cells (cancer). Mutagenicity is a subset of genotoxicity in which the DNA is mutated. In this case, damage caused by a genotoxin is improperly repaired, permanently altering the DNA. Compounds with these properties represent a potential risk to patient health. While there are some marketed drugs that possess these attributes, modern drug discovery programs actively eliminate suspect candidates through a variety of *in vitro* screening methods.

The Ames assay is one of the most widely employed methods of identifying potential carcinogenic compounds (Figure 8.10). Originally described by Bruce Ames in the early 1970s,[26] this test is specifically designed to identify compounds that have mutagenic properties. Although not all mutagens are carcinogenic, a positive signal in an Ames assay is viewed as an indication of high risk. Compounds that are "Ames positive" are rarely progressed further. In practice, the assay monitors the growth of a specially designed bacterial strain, usually *Salmonella typhimurium*, which cannot grow in the absence of histidine. Mutations in the genes that control the histidine synthesis prevent the bacteria from producing histidine on their own, so they will only grow when histidine is added to the growth media. The bacteria are grown in the presence of a test compound and a limited supply of histidinde. Once the histidine supply

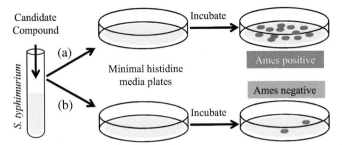

FIGURE 8.10 In the Ames assay, cells genetically engineered to be incapable of producing histidine (such as *Salmonella typhimurium*) are cultured in the presence of candidate compounds and a limited supply of histidine. Once the histidine is depleted, only cells with mutations that restore histidine synthesis pathways will survive. (a) Candidate compounds that promote mutations (mutagenic compounds) will lead to an increase in cell survival in the absence of a histidine-rich media. This is referred to as an Ames positive result. (b) Compounds that do not increase cell survival in this assay are referred to as being Ames negative.

is exhausted, only bacteria that have mutated to reactivate the histidine synthesis pathway will survive. If the test compound is mutagenic, it will produce mutations that restore the histidine synthesis pathway, allowing more bacteria to survive after the histidine supply is exhausted (as compared to the same bacteria grown in the absence of the test compound). This can be quantified by counting the number of bacterial colonies present after a set time interval. The assay can also be run in the presence of rat liver extract (specifically the S9 fraction) in order to determine if a metabolite of the test compound has mutagenic properties. Like all assays, the Ames assay is not perfect, as both false positives and false negatives have been reported. However, the speed at which this assay can be run and the overall cost/time saving as compared to *in vivo* screening in rats or mice are substantial advantages. Also, regulatory bodies have provided specific guidance regarding acquiring Ames data on candidate compounds.

Another important method used to identify potentially risky compounds is the Micronucleus Assay (Figure 8.11).[27] Unlike the Ames assay, this

FIGURE 8.11 (a) Cytochalasin B. (b) In the Micronucleus Assay, Chinese hamster ovarian cells (CHO cells) are grown in the presence of a candidate compound for a set period of time and then exposed to cytochalasin B. This blocks cytokinesis, leading to a buildup of binucleated cells. The presence of micronuclei indicates that a candidate compound can cause chromosomal damage.

8. SAFETY AND TOXICOLOGY

method uses normal cells (e.g., Chinese hamster ovary cells (CHO cells)) and is designed to identify compounds that damage chromosomes or the cell division system. When these systems are damaged, membrane bound DNA fragments, referred to as micronuclei, form during the cell division process due to improper migration of chromosomes during mitosis. Morphologically, micronuclei are identical to the main nuclei of a cell, but they are substantially smaller, making them readily identifiable when present. In practice, cells are grown in the presence of candidate compounds for a brief period of time to provide an opportunity for chromosomal damage. Upon completion of this time period, the cytokinesis-blocking agent cytochalasin B (Figure 8.11(a)) is added, causing the accumulation of cells that have completed one nuclear division to become binucleated cells. These cells are then examined and scored for the presence of micronuclei. A "positive" result is indicated when a compound induces a dose dependent increase in the formation of micronuclei in this assay, suggesting that the compound has genotoxic properties.

Examining cells for structural changes in the chromosomes using the Chromosomal Aberration Assay (Figure 8.12)[28] has also proven to be a

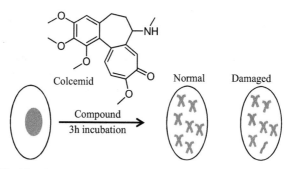

FIGURE 8.12 The chromosomal aberration assay is used to identify compounds that damage genetic material. Incubation of a cell culture in the presence of a candidate compound for a short period of time is followed by the introduction of colcemid. This compound stops the cell cycle at the metaphase, a point at which the condensed and highly coiled chromosomes are aligned in the middle of the cell and visible with a microscope. Cell fixation and microscopic examination provides an opportunity to assess DNA damage that may have been caused by the candidate compound.

useful tool in identifying potentially genotoxic compounds. DNA damage associated with genotoxic compounds can lead to breaks in the chromosomal material (e.g., single and double strand breaks). These breaks can be repaired correctly, rejoined together incorrectly, or not rejoined at all. The first scenario repairs the cell to its normal state, but the second and third scenarios create changes in the overall structure and appearance of chromosomal material that can be observed under a microscope when cells are in the metaphase portion of the cell cycle. The observation of gross changes in chromosomal structure serves as an indicator that significant DNA damage/mutation has occurred and deleterious consequences are likely (e.g., cancer, genetic diseases). In practice, a cell culture is incubated for 3h with a candidate compound and then a compound capable of arresting the cell cycle at metaphase (e.g., colcemid) is added. At this stage of mitosis, the condensed and highly coiled chromosomes are aligned in the middle of the cell. The cells are then fixed and microscopically examined to assess for chromosomal aberrations. Compounds that form metabolites capable of causing chromosomal aberrations can also be identified by performing the experiments in the presence of rat liver extract (specifically the S9 fraction).

As discussed above, some genotoxic compounds damage DNA by causing strand breaks. Compounds that cause this kind of DNA damage can be identified using a single-cell gel electrophoresis (SCGE) method referred to as the Comet assay (Figure 8.13).[29] Originally developed by

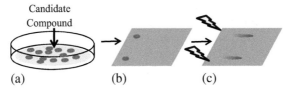

FIGURE 8.13 The Comet assay can detect compounds that cause DNA strand breaks. (a) Cells are incubated with a candidate compound for a defined time period. (b) A single cell is embedded in a gel electrophoresis matrix (agarose matrix) lysed, and then (c) an electric field is applied across the gel. If the candidate compound causes DNA strand breaks, staining of the gel upon completion of the experiment will produce a comet-shaped image as a result of differential rates of migration of the full DNA as compared to the DNA fragments produced by strand breaks.

Peter Cook,[30] this assay takes advantage of differences in migration rates of DNA strands in an electrophoresis gel. DNA fragments created by strand breaks and relaxed chromatin migrate through electrophoresis gel at a faster rate as compared to unchanged DNA. Upon staining and visualization, compounds that are capable of inducing DNA strand breaks will produce an image that resembles the tailing of a comet, which is the rational for the name of the assay. Typically, a cell culture is treated with a test compound for a sustained time period, and then single cells are embedded in an agarose matrix on a microscope slide. Lysis of the cells under mildly basic conditions releases the DNA and this is followed by the application of an electric field across the gel. This causes the released DNA to migrate through the gel at a rate based on their size. The presence of a comet shape upon visualization indicates a positive result in the Comet assay and suggests that the compound in question has genotoxic properties.

DRUG–DRUG INTERACTIONS

In considering the potential safety and toxicity issues that may be associated with a candidate compound, it is important to look beyond the compound itself. As discussed earlier, the metabolic processes designed to clear xenobiotics from the body can produce compounds (e.g., metabolites) that possess undesirable characteristics. It is also possible that the presence of one compound can cause a second compound to produce side effects that would not occur if it were used alone. In this case, co-administration of the two compounds causes a change in the pharmacokinetic profile of one of the two compounds. These changes could present themselves as higher systemic exposure of the parent compound (e.g., decreased metabolism of the parent compound) or increased production of metabolites (e.g., increased metabolism of the parent compound). In both of these scenarios, a "drug–drug interaction" (DDI) creates an opportunity for side effects to appear.

Consider, for example, a compound that is converted to two metabolites. One metabolic pathway converts 99.99% of the compound to a benign metabolite, while the second metabolic pathway converts the remaining 0.01% to a compound with toxic properties. If a second compound capable of blocking the major metabolic pathway is co-administered, then metabolism of the first compound will be forced into the minor pathway, leading to higher concentration of the toxic metabolite (Figure 8.14). Alternatively,

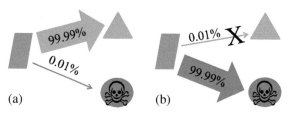

FIGURE 8.14   (a) A compound with the potential to metabolize into a dangerous compound (red) may be safe to use clinically if an alternative, dominant pathway converts it to a safe metabolite (green). (b) Blockade of the metabolic pathway leading to a safe metabolite by another compound, however, can lead to unexpected toxicity as a result of an increase in production of the toxic metabolite (red).

consider a compound that is rapidly metabolized by one metabolic pathway and more slowly by a second pathway. Co-administration of a compound that blocks the fast pathway will lead to a buildup in concentration of the first compound potentially to concentration beyond the established safety window.

Predicting the propensity of a compound to block metabolic pathways can provide substantial guidance as to the likelihood of drug–drug interaction becoming an issue. Compounds that inhibit CYP3A4, a major metabolic enzyme of the liver, for example, are likely to undergo a drug–drug interaction with compounds that are substantially metabolized by this enzyme. Compounds can be screened for their ability to block metabolic enzymes using *in vitro* assays designed to detect changes in the rate of metabolism of a known substrate. Changes in the rate of conversion of 7-benzyloxy-4-trifluoromethylcoumarin to 7-hydroxy-4-trifluoromethylcoumarin by CYP3A4 in the presence of a varying concentrations of a candidate compound, for example, provides an IC_{50} for the candidate compound's interaction with CYP3A4 (Figure 8.15).[31] In theory, it would be possible to generate IC_{50}s for a candidate compound's ability to block all of the metabolic CYP enzymes (Table 8.1), and this information could be used to predict which compounds

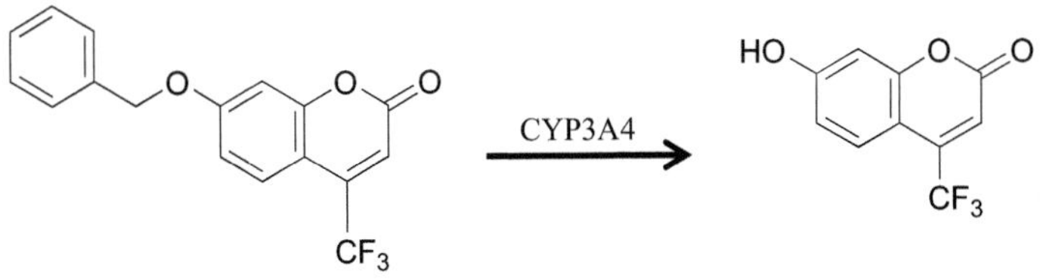

FIGURE 8.15 Monitoring the rate of conversion of 7-benzyloxy-4-trifluoromethylcoumarin into 7-hydroxy-4-trifluoromethyl coumarin by CYP3A4 is an effective means of identifying compounds that block CYP3A4 metabolism and represent a risk of drug–drug interactions.

TABLE 8.1 CYP Enzymes, Drug Substrates, and Metabolites that can be Used to Detect Possible Drug–Drug Interactions

CYP isozyme	Drug substrate	Drug metabolite
3A4	Midazolam	1'-Hydroxymidazolam
2D6	Bufuralol	1'-Hydroxybufuralol
2C9	Diclofenac	4'-Hydroxydiclofenac
1A2	Ethoxyresorufin	Resorfin
2C19	S-Mephenytoin	4'-Hydroxymephenytoin
2A6	Coumarin	7'-Hydroxycoumarin
2C8	Paclitaxel	6α-Hydroxypaclitaxel

would be at risk for drug–drug interactions. In practice, however, it is not cost-effective to screen compounds against all known metabolic enzymes. Instead, typical drug discovery programs often screen compounds of interest against the major CYP enzymes CYP3A4, CYP2D6, and CYP2C9 in order to obtain a preliminary sense of associated risks. As a candidate compound advances toward human trials, its ability to inhibit additional metabolic enzymes (e.g., CYP1A2, monoamine oxidase, (MAO) etc.) may be assessed in order to further determine DDI risks.

Another pathway that can lead to drug–drug interaction is changes in expression of metabolic enzymes. There are many reported cases of compounds that induce increases in expression of CYP enzymes, an effect referred to as CYP induction. This can have a significant impact on the metabolic fate of candidate compounds. Increased CYP expression will lead to increased metabolism of susceptible compounds, decreasing their potential for efficacy, while at the same time increasing the rate of formation of metabolites. If any of the metabolites are associated with safety or toxicity issues, the increased concentration caused by CYP induction may lead to negative effects that would not have occurred in absence of CYP induction (e.g., drug–drug interaction). Identifying CYP inducers early in a drug discovery program can be accomplished using *in vitro* techniques similar to those used to identify compounds that inhibit CYP enzymes. Hepatocytes are exposed to candidate compounds for a sustained period to provide an opportunity for CYP induction and then assessed for their ability to metabolize substrates known to be specific to a single CYP isozyme (e.g., CYP3A4, CYP2D6, CYP2C9, etc.). If the rate of metabolism is increased relative to control hepatocytes (those not exposed to the candidate compound), then CYP induction is indicated. It is also possible quantify the amount of enzyme or mRNA present using biochemical means.[32]

CARDIOVASCULAR SAFETY AND TOXICOLOGY STUDIES

The manifestation of cardiovascular side effects, even in a very small portion of the patient population, can lead to significant marketing restrictions or even removal from the market. The painkiller Vioxx®, for example, had 2003 sales of $2.5 billion, but was removed from the market when it was determined that it increased the rate of cardiovascular events from 0.78 per 100 patients to 1.50 per 100 patients.[33] In order to minimize the risk to both the patient and sponsor company, the cardiovascular safety profile of candidate compounds is thoroughly examined prior to the initiation of clinical trials. It is, of course, cost prohibitive to assess large numbers of compounds in full animal models of cardiovascular safety. Fortunately, there are a number of assays available that can be used to identify and eliminate compounds with the potential to negatively impact cardiovascular function.

Assays designed to detect blockade of the hERG channel (also known as $K_v11.1$) are generally the first step in determining whether or not a candidate compound is at risk for cardiovascular side effects. This channel is an important component of the electrical system that allows the heart to beat in the rhythmic fashion required to move blood through the body. Specifically, the hERG channel conducts potassium ions out of cardiac myocytes (the "rapid" delayed rectifier current (I_{Kr})) during the repolarization phase of the cardiac action potential (Figure 8.16(a)), allowing the cells to regain their ability to

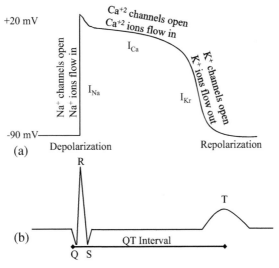

FIGURE 8.16 (a) At the cellular level, an electrical signal causes the sodium channels of a cardiac myocyte to open. Sodium atoms flow into the cell, causing a rapid depolarization. Maximum depolarization causes calcium channels to open, and the flow of calcium atoms into the cell begins the repolarization process. Potassium channels open in response and complete the repolarization process, bringing the myocyte back to its resting state. (b) An electrocardiogram (ECG) provides a view of the electrical activity on the surface of the heart, and correlates with myocyte action potentials. The QT interval represents the amount of time between depolarization and repolarization of the ventricles and is critical indicator of cardiovascular activity and health.

contract when the next electrical signal that arrives. Briefly, a cardiac action potential is initiated by an electrical signal that triggers the opening of sodium (Na$^+$) channels. The resulting rapid influx of Na$^+$ leads to depolarization of the myocyte, changing the electrical gradient across the cell membrane from the resting potential of −90 mV to about +20 mV. The depolarized state is maintained for a short period of time by the opening of calcium (Ca^{+2}) channels, but eventually repolarization back to −90 mV is accomplished by the flow of potassium (K$^+$) out of the myocyte through a number of potassium channels. The hERG channel is the most important part of the repolarization event.[34]

At the organ level, individual myocyte action potentials contribute to the electrical activity of the heart as a whole. An electrocardiogram (ECG) provides a view of the electrical activity on the surface of the heart, which correlates with myocyte action potentials. In an ECG, the distance between the Q wave and the end of T wave, referred to as the QT interval, represents the amount of time between depolarization and repolarization of the ventricles. This time interval can be mapped to the myocyte action potential (Figure 8.16(b)).[35] It has been established that increases in the QT

FIGURE 8.17 Compounds that block the hERG channel increase the QT interval of the heart and raise the risk of ventricular arrhythmia, Torsades de pointes, and sudden cardiac death. (a) Cellular level electronic signaling. (b) Electrocardiogram view of electrical signaling. The black line represents normal electrical activity, while the red line represents electrical activity in the presence of a hERG channel blocker.

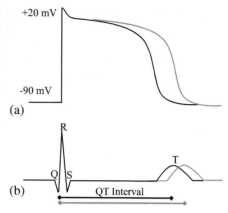

interval (Figure 8.17) are linked to increased risk for ventricular arrhythmia, Torsades de pointes (ventricular tachycardia), and sudden cardiac death. These dangerous events have, in turn, been linked blockade of the hERG channel. Decreased hERG activity as a result of channel blockade increases the duration of the action potential, lengthens the time required for repolarization, and increases the likelihood of dangerous side effects.[36]

Given the significant risk associated with hERG channel, it is clear that monitoring candidate compounds for activity at this "anti-target" is very important. At the *in vitro* level, it is possible to assess compounds for hERG activity in competitive binding assays (Figure 8.18). In these assays, a potent, radiolabeled hERG channel binder such as

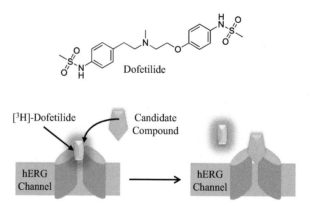

FIGURE 8.18 Displacement of [³H]-dofetilide from it binding site on the hERG channel by candidate compounds can be used to assess the level of hERG risk associated with candidate compounds. This method will only identify compounds that bind to the dofetilide binding site, but there are other binding sites on the hERG channel. Additional assessments using other methods are often performed as a candidate compound advances in the drug discovery process.

[^{3}H]-dofetilide can be applied to cells expressing the hERG channel or a membrane preparation containing the hERG channel. Candidate compounds are then examined for their ability to displace the radiolabel from the channel using the appropriate detection methods (e.g., scintillation counters).[37] While this method is relatively low cost and available in high-throughput mode, it is has an important draw back. The assay will only identify compounds that target the dofetilide binding site. Since there are multiple compound binding sites on the hERG channel, a negative result (no binding) in this assay is not necessarily an indication that the compound in question is clear of hERG issues. Also, this assay is not capable of providing information on the electrophysiological impact of a compound on targeted cells.

As an alternative, hERG activity can be assessed using a rubidium (Rb^{+}) efflux assay (Figure 8.19). Rubidium can move through the hERG channel,

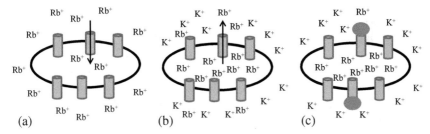

FIGURE 8.19 Rubidium (Rb^{+}) ions are capable of moving through potassium channels such as the hERG channel (blue) as a result of the similar size and charge. The rubidium (Rb^{+}) efflux assay takes advantage of this phenomenon. (a) Cells expressing the hERG channel are loaded with Rb^{+} by placing them in a media rich in the ion. (b) Replacing the Rb^{+} with K^{+} will cause the cells to release Rb^{+}, which can be measured using atomic absorption spectroscopy or scintillation counting (^{86}Rb^{+}). (c) Compounds that block the hERG channel (red) will slow the exit of Rb^{+}, indicating a potential hERG liability.

as it is has the same charge and size as K^{+}. Since it is not normally present in cells or media, background Rb^{+} concentration is negligible. When cells expressing the hERG channel, such as CHO cells that have been transfected with the hERG gene, are exposed to Rb^{+} containing media, they will absorb the Rb^{+} until equilibrium has been achieved between the inside and outside of the cells. Replacing the media with Rb^{+}-free media containing high concentrations of the K^{+} will cause the hERG channels to open, allowing the Rb^{+} to escape the cells. Measuring the concentration of Rb^{+} that enters the media over a fixed period of time provides insight into hERG channel activity. Compounds can be screened for hERG channel activity by including the test compound in the K^{+}-containing media. Compounds that block hERG channel activity will prevent Rb^{+} from leaving the cells, leading to decreased concentration in the media. Rb^{+} concentration inside the cells can also be assessed at the end of the assay by lysing the cells. In either case, Rb^{+}

concentration can be measured using either scintillation counting methods to detect $^{86}Rb^+$ or atomic absorption spectroscopy methods.[38]

The Rb^+ efflux assay and hERG binding assays are extensively used in the pharmaceutical industry, but ultimately they are both indirect methods of determining ion channel activity. As discussed in Chapter 4, electrophysiological patch clamp is the "gold standard" in determining the impact of candidate compounds on an ion channel. Experiments of this type can be conducted using an appropriately engineered cell line (hERG-expressing CHO or HEK293 cells), but this method is generally only used on advanced candidates. Electrophysiological patch clamp studies remain time and labor intensive processes that are not amenable to high throughput methods, and this significantly limits the number of compounds that can be screened for hERG activity using this method.

While hERG channel activity is an important aspect of cardiovascular safety, it is by no means the only issue that needs to be considered in determining cardiac safety risks associated with a candidate compound. Changes in factors such as blood pressure, heart rate, contractility, and ejection fraction must also be considered as alteration in any of these aspects of cardiovascular function can also lead to serious consequences. The impact of candidate compounds could be studied in a variety of animal models capable of providing direct insight into cardiovascular function, but it is not practical to study large numbers of compounds in this manner. The time and cost associated with these studies is simply too high to allow broad screening of candidate compounds in an *in vivo* setting. Fortunately, it is possible to identify compounds that may have an impact on cardiovascular function using *in vitro* screening systems. The cardiovascular system is exceedingly complex and there are a wide variety of enzymes, ion channels, GPCR, transporters, and other biomolecules that modulate its activity. *In vitro* screening methods designed to identify compounds that interact with these targets have been developed to identify useful drugs for the treatment of cardiovascular diseases such as hypertension and congestive heart failure. These same assays can be used as counterscreens to identify compounds that could induce cardiovascular side effects. SAR patterns can be determined in order to minimize the interaction of candidate compounds with biomolecules that control cardiovascular function. If a program is targeting a biomolecule that is closely related to a second biomolecule that is involved in cardiac function, then an assay focused on activity of the second biomolecule is often part of the screening paradigm used to identify compounds suitable for advancement. Consider, for instance, a program whose goal is the identification of 5-HT_7 antagonists for the treatment of irritable bowel disease (IBD).[39] The 5-HT_{2b} receptor shares a high degree of homology with 5-HT_7, and agonist activity at 5-HT_{2b} is associated with cardiovascular side effects.[40] As a result, the risk of designing a compound with activity at both of these receptors is relatively high. In order to mitigate this risk, compounds identified as active at the 5-HT_7 receptor would likely be screened in a secondary assay to assess their capacity for modulating 5-HT_{2b} activity. Compounds that represent a risk for cardiovascular safety via 5-HT_{2b} activity can be eliminated early in the process.

Of course, the concept of screening candidate compound for activity at "nearest neighbor" targets is insufficient on its own. Screening the hypothetical 5-HT_7 modulators for 5-HT_{2b} activity will provide no insight on a candidate compound's capacity for interacting with the myriad of other potential cardiovascular targets that exist (Table 8.2). As discussed earlier,

TABLE 8.2   Exemplary Cardiovascular Drug Targets

Aldosterone receptor	Mineralocorticoid receptor
Alpha adrenergic receptor	Neutral endopeptidase
Angiotensin-converting enzyme	Nicotinic acid receptor
Beta adrenergic receptor	Prostaglandin E2 receptor
Endothelin A receptor	Vasopressin V1a receptor
L-type calcium channel	T-type calcium channel

even the largest of pharmaceutical companies do not maintain the enormous arrays of *in vitro* screens necessary to monitor for all possible untoward effects. Although the number of targets relevant to cardiovascular disease is only a subset of the total, it would still be a very large undertaking to have all of the necessary assays available in a single pharmaceutical company (The hERG assay is a notable exception. Given its importance, many companies choose to run this assay internally). Advanced candidate compounds are often screened using *in vitro* panels of cardiovascular targets at contract research organizations such as EMD Millipore,[7] Perkin Elmer,[8] and Cerep.[9] *In vitro* cardiovascular safety panels are often run on advanced candidate compounds as part of a wider *in vitro* safety assessment discussed earlier.

Failure to identify activity at in a series of *in vitro* screens designed to identify potential cardiovascular risk is good start, but it is not sufficient to fully derisk a candidate compound. *In vitro* screening, by definition, is not the same as screening a compound in an animal model. Data from *in vitro* screening assays are not a suitable substitute for testing a compound in a living system. In some organizations, the first step toward *in vivo* assessment of cardiovascular risk is the Langendorff preparation, an isolated heart perfusion model originally reported in 1898.[41] Essentially, this model consists of a heart that has been isolated from a terminally anaesthetized animal and attached to a perfusion system capable of delivering a constant flow of fluid. The perfusion fluid is typically maintained at 37 °C, contains a physiological salt mixture that mirrors plasma (pH 7.4), and is gassed with a 95%/5% O_2/CO_2 mixture (Figure 8.20). These conditions allow the heart to

FIGURE 8.20 Langendorff preparations provide an opportunity to study cardiovascular function in the presence of a candidate compound. A perfusion fluid containing a candidate compound is pumped through an isolated heart while monitoring for changes in cardiovascular function.

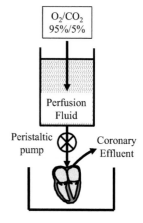

remain viable for several hours, providing an opportunity to study cardiovascular function in the presence of a test compound. Test compounds are simply added to the perfusion fluid at the desired concentration and the isolated heart is monitored for changes in function (contractility, heart rate, etc.). Since the heart has been isolated from all other organs and bodily system, the results of a Langendorrf model are independent of any noncardiac system. Neurohormonal modulation of the cardiovascular system in response to outside influences, for example, is not part of the equation in the Langendorrf system. Any effects observed are directly related to the heart itself. This model has been successfully used with hearts from a number of different species including mice, rats, rabbits, dogs, pigs, and primates.[42]

As useful as the Langendorff model may be, it is not without its disadvantages. First and foremost, no matter how well the model is established, it must still be considered a "dying" preparation. Cardiac function will naturally decay over time, and the heart will become less and less viable as time moves on (5–10% degradation of function per hour). It is, therefore, essential that Langendorff studies be accomplished quickly. In addition the advantage of working with an isolated heart is also a disadvantage, as it decreases the clinical relevancy of the model. Compensatory mechanisms from neurohormonal systems are absent, and compounds that might exert their effect through neurohormonal modulation may not elicit a change in cardiac function in this model. Although the Langendorff model can provide a wide range of information, it is not a substitute for *in vivo* study of candidate compounds.

An *in vivo* study designed to assess the level of risk associated with a particular compound under various dosing regimens typically employs telemetry equipped animals. Surgical implantation of telemetry equipment capable of detecting changes in parameters such as heart rate, blood pressure, contractile force, and ejection fraction is followed by a recovery period (2–4 weeks) to allow healing. Once the animals have healed, they can be treated with various doses of the compound in question to determine whether or not the candidate compound has any impact of cardiovascular function. Animals can be subjected to a single dose of a candidate compound to in an effort to identify acute cardiovascular safety issue. Alternatively, animals can be dosed with multiple doses of the same compound over an extended period of time in order to assess the risk associated with chronic exposure. In either event, the implanted telemetry equipment will provide significant insight into the cardiovascular risks associated with the candidate compound under review.[43] Cardiovascular studies of this type can take weeks, or even months to complete, and throughout the study the test subjects (the lab animals) must be properly housed, fed, and treated humanely. Given the length of time required, the significant cost, and the difficulties associated with properly caring for study animals, it should come as no surprise that *in vivo* cardiovascular safety studies are reserved for only the most promising candidate compounds.

CENTRAL NERVOUS SYSTEM SAFETY AND TOXICOLOGY STUDIES

There are many drugs on the market specifically designed to modulated central nervous systems functions, such as antidepressant, antipsychotic agents, and antiseizure medication. The lives of millions of patients have been improved as a result of advances in the treatment of CNS diseases. It is important to understand, however, that modulation of CNS functions can produce a range of negative effects that may preclude advancement of candidate compounds. Off-target CNS-mediated effects can include hallucinations, insomnia, sedation, seizures, increases sensitivity to pain, impaired memory function, depression, and even an increased risk of suicide. Popular drugs such as Prozac® (fluoxetine, antidepressant, Figure 8.21(a)),[44] Effexor® (venlafaxine, antidepressant, Figure 8.21(b)),[45] Chantix® (Varenicline, smoking cessation, Figure 8.21(c)),[46] and Accutane® (Isotretinoin, acne treatment, Figure 8.21(d))[47] have been linked to an increased risk of suicidal thoughts and tendencies. While the level of risk associated with each of these drugs was not high enough to warrant withdrawal from the market, these examples highlight the importance of monitoring for CNS-mediated side effects.

CNS-mediated side effects can also complicate the drug discovery process. Consider, for example, a program focused on the development of

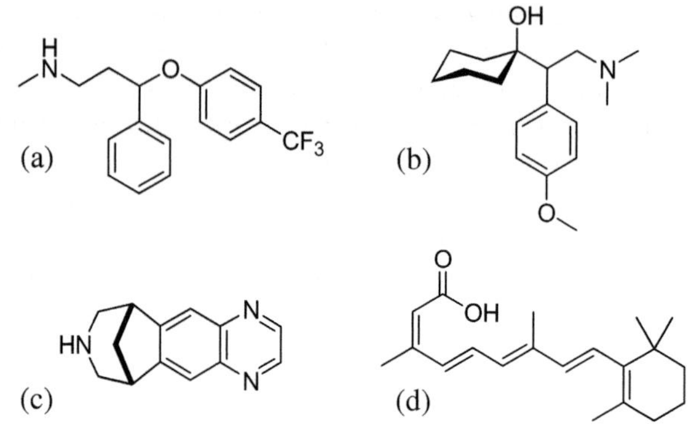

FIGURE 8.21 (a) Prozac® (fluoxetine), (b) Effexor® (venlafaxine), (c) Chantix® (Vareni-cline), (d) Accutane® (Isotretinoin).

novel pain relievers. An *in vivo* model designed to measure an animal's response to pain would be significantly impacted if a candidate compound is capable of acting as a sedative. Decreased response to pain stimulus might be driven by sedation, rather than an increased threshold to the pain. Understanding the impact of candidate compounds on the CNS can be critical to the success or failure of drug discovery and development programs.

As was the case for cardiovascular safety and toxicology studies, the number of possible targets potentially associated with negative CNS effects is quite extensive. Unless there is significant level of homology between the therapeutic target of interest and a potential CNS "antitargets," most companies do not have the resources to maintain *in vitro* screening systems for the wide range biomolecules that can impact CNS function (Exemplary CNS targets are listed in Table 8.3). Typically, advanced compounds are screened

TABLE 8.3   Exemplary Central Nervous System Targets

Adenosine receptor A2a	Nicotinic acetylcholine receptor
α-adrenergic receptor	NMDA receptor
β-2 adrenergic receptor	Norepinephrine transporter
Cannabinoid receptor 1	Opioid receptor
Dopamine receptor	Serotonin receptor
GABA receptor	Serotonin transporter
Histamine H1 receptor	Substance-P receptor
Monoamine oxidase A	Voltage-gated sodium channel
Muscarinic acetylcholine receptor	Voltage-gated potassium channel

in CNS panels at contract research organizations that specialize in *in vitro* screening services such as EMD Millipore,[7] Perkin Elmer,[8] and Cerep.[9] The National Institute of Mental Health also provides *in vitro* screening services specifically focused on CNS targets through the Psychoactive Drug Screening Program at the University of North Carolina at Chapel Hill.[48]

At the *in vivo* stage, candidate compounds can be assessed for their impact on central nervous system using several of the animal models described in Chapter 7. The novel object recognition model,[49] the contextual fear conditioning model,[50] and the Morris water maze[51] are all capable of identifying compounds that have a deleterious effect on memory formation. Separately, compounds capable of acting as a sedative can be identified with the rotarod test.[52] Animals that are under the influence of a sedative are more likely to fall off of the rotating wheel than those that are not suffering from the lethargy produced by a sedative. These and other *in vivo* CNS safety and toxicology studies are, of course, significantly more resource intensive than their *in vitro* counterparts.

IMMUNE SYSTEM MEDIATED SAFETY ISSUES

Safety and toxicity issues can also occur as a result of a candidate compound's interaction with the immune system or macromolecules that regulate the immune system. In some cases, an immune response is initiated as a result of the formation of covalent bond between a candidate compound and an otherwise normal biomolecule. Although neither the candidate compound, which is referred to in this context as a hapten (a small molecule that can elicit an immune response, but only when bound to a carrier protein), nor the carrier macromolecule are noticed by the immune system, the combination of the two creates a hapten-carrier adduct that is recognized by the immune system as foreign material. In response, the immune system generates antibodies and launches an allergic response. The severity of the response will increase with each exposure to the candidate compound and can be as severe as anaphylactic shock.[53] Penicillin allergy[12a,54] is a well-known example (Figure 8.22),

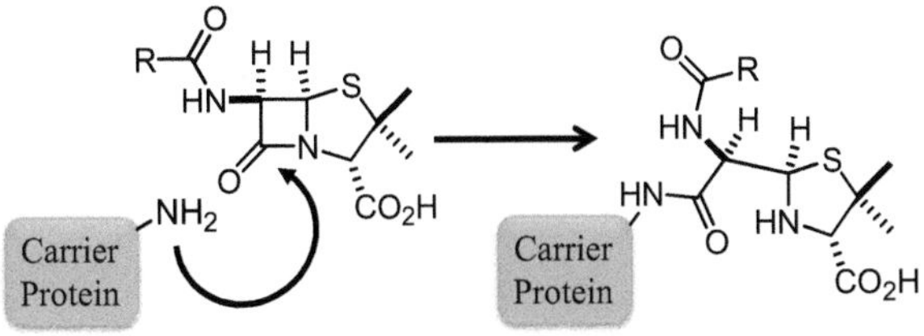

FIGURE 8.22 The reaction of β-lactam containing compounds with a protein is a well-known example of how a small molecule can act as a hapten and elicit an immune response. Nucleophilic functional groups on a protein (e.g., amines) can react with a β-lactam. The resulting hapten–protein adduct may not be recognized as "self" leading to an immune response.

but any compound capable of forming a covalent bond with a normal protein has the potential to initiate a hapten-carrier adduct immune response. This includes not just candidate compounds, but also their metabolites. Reactive metabolites formed as a result of the normal processing of a xenobiotic could produce compounds capable of acting as a hapten. Understanding the metabolic fate of candidate compounds can provide insight into how to prevent the formation of haptens. The simplest way to avoid hapten-carrier adduct mediate issues, of course, is to eliminate potential reactive sites in candidate compounds and their metabolites.

Immunosuppression is another serious safety issue that should be considered as drug discovery and development program progress. The immune system has a number of different components, and molecules that interfere with the normal function of any one of these components can suppress the immune system, decreasing the body's ability to defend itself from foreign entities (e.g., viruses, bacteria, etc.). Within the context of organ transplant patients, immunosuppression is required to prevent rejection of the transplanted organ, but in most other disease states, immunosuppression is a serious side effect. It is, of course, not feasible to assess all program compounds using an appropriate *in vivo* model, as the cost and time required are simply too high. There are, however, *in vitro* assays available for a wide range of target that are known to play a role in the immune response, and candidate compounds can be screened in these assays in an effort to gauge the likelihood of immunosuppression. If the therapeutic target of interest is closely related to a biomolecule that is linked to immunosuppression, then an assay to identify off-target effects in the closely related biomolecule will likely be an early part of the screening paradigm designed to derisk candidate compounds. If not, then candidate compounds of particular interest can be assessed in *in vitro* assays provided by companies such as EMD Millipore,[7] Perkin Elmer,[8] and Cerep[9] to screen for potential issues.

Autoimmune diseases such as rheumatoid arthritis, psoriasis, inflammatory bowel disease, and multiple sclerosis represent an interesting challenge for safety issues. In general, autoimmune diseases arise from an abnormal immune response in which the body targets normal tissues and substances as if they were foreign material.[55] The resulting tissue damage can lead to symptoms such as skin rashes (psoriasis), joint pain (rheumatoid arthritis), or even movement disorders (multiple sclerosis). It has been demonstrated that immunosuppression therapy through a number of different targets can mitigate the symptoms associated with these conditions, but the inherent safety risk of immunosuppression must be carefully weighed against the benefit derived from treatment. The janus kinase (JAK) inhibitor Xeljanz®

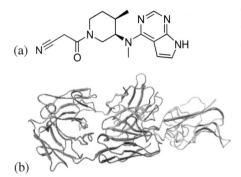

FIGURE 8.23 (a) Xeljanz® (tofacitinib), (b) Humira® (adalimumab) RCSB file 3WD5.

(tofacitinib, Figure 8.23(a)), for example, is marketed for the treatment of rheumatoid arthritis, and is known to modulate both the innate and adaptive immune response.[56] In a similar manner, Humira® (adalimumab, Figure 8.23(b)), a human monoclonal antibody that targets tumor necrosis factor-alpha (TNFα), is used clinically for the treatment of rheumatoid arthritis. Humira® (adalimumab) binds to TNFα, preventing it from binding with TNF receptors, dampening associated inflammatory response and mitigating symptoms.[57] However, TNF system is part of the larger immune system, so there is an inherent risk of immune-related safety issues such as increased risk of infection, reemergence of latent infections (e.g., tuberculosis), decreased immune response, and in rare cases increased cancer risk.[58] Both of these cases could be described as a mechanism-based safety issue, as the drug's mode of action is also the means through which safety issues, in this case decreased immune response, can arise.

TERATOGENICITY

Prior to the early 1960s, determining the impact of drugs and candidate compounds on a developing fetus was rarely considered. As discussed in Chapter 2, the prevailing theory at the time was that the placenta protected a developing fetus from drugs and toxins ingested by the mother. The thalidomide disaster,[59] however, made it clear that fetal development could be significantly altered by xenobiotics. The link that was established between thalidomide and severe birth defects was undeniable and led to a significant expansion of efforts to identify teratogenic compounds. If there was any doubt as to the importance of understanding the potential teratogenicity of candidate compounds, these doubts were eliminated by the identification of fetal alcohol syndrome in 1973.[60]

The term teratogen refers to a material that is capable of blocking the normal growth and development of a fetus. As with other types of pharmacological effects, the impact of exposure to a teratogen is a dose-dependent phenomenon. The severity of deviant development will increase as the exposure increases. Teratogen exposure can lead to malformations, growth retardation, functional deficits (mental or physical), and even death of the embryo/fetus. In addition, the impact of a teratogen is dependent on the developmental stage of an embryo/fetus. The staged nature of development from embryo to fetus to child dictates that different developmental processes will be ongoing in different stages of the process. The first trimester is generally considered to be the highest risk time frame, as differentiation of the major organs occurs within the first 2–12 weeks of gestation.[61] Fetal vulnerability decreases as the fetus develops, but the thalidomide story makes it clear that teratogens can have severe impacts at later stages of development.

The complexity of fetal development has made it challenging to create model systems capable of identifying compounds that have teratogenic potential. There are, however, some assays that can be employed to mitigate the risk. The embryonic stem cell test, for example, monitors the conversion of mouse embryonic stem cells in cardiomyocytes. Conversion of the stem cells into cardiac myocytes will lead to the production of α-actinin and sarcomeric myosin heavy chain. Potentially teratogenic compounds will prevent differentiation of the stem cells, and the impact of test compounds can be measured by assessing the amount of α-actinin and sarcomeric myosin heavy chain at the end of an experiment. Compounds that block differentiation (possible teratogens) will exhibit decreased levels of both of these proteins. Alternatively, the number of contracting myocytes can be assessed using microscopy techniques to determine the impact of the test compounds.[62]

The mammalian micromass (MM) test is an alternative procedure that is also capable of identifying potential teratogens. It is based on the observation that undifferentiated mesenchyme cell from chick embryos differentiate into chondrocytes when grown in a small volume at a high density. Since its initial disclosure, the procedure has been extended to mouse and rat embryonic tissues, and to embryonic CNS cell, which form aggregates that differentiate into neurons under similar conditions. The mammalian micromass test has been adapted to a 96-well plate format and cell behavior typical of embryogenesis such as cell adhesion, movement, communication, division, and differentiation can be observed in the presence or absence of candidate compounds. Potential teratogens will demonstrate decreases in the aforementioned embryogenic features, providing an opportunity to exclude at risk compounds from further progression.[63]

The zebrafish assay is a relatively recent addition to the battery of test capable of identifying potential teratogens. The development of zebrafish from fertilization to maturity is well characterized, and occurs over a relatively short time frame. The time between formation of a zygote and hatching is approximately 48–72 h. In addition, the egg remains transparent through much of the process (beyond pharyngulation), which provides an unobstructed view of morphological changes. Changes in the development process that are induced by the presence of candidate compounds can be monitored to gain insight into their teratogenic potential. The sensitivity of fish to chemical exposure during embryo development is well documented. Further, zebrafish embryos that are malformed, missing organs, or have dysfunctional organs generally survive beyond the time at which normal organ function would occur under ideal conditions. Upon hatching, evaluations of teratogenicity can continue by monitoring the resulting larvae for skeletal deformities, the ability to swim properly, and body position. Abnormalities in any of these features are considered an indication of teratogenicity.[64]

IN VIVO TOXICITY AND SAFETY STUDIES

As a drug discovery program moves toward the identification of a single compound suitable for clinical evaluation, the emphasis on safety and toxicity intensifies. Ultimately, candidate compounds must be evaluate in animal studies in order to gauge the potential risk to patients in clinical trials. The primary goals of preclinical *in vivo* toxicity and safety studies are to predict hazards to patients, define the phase I dosing regimens, and identify toxicity and safety markers to monitor in clinical trials. Organ-specific impact is often determined, toxic metabolites may be identified, and drug responses that cannot be assessed in humans can be studied during the course of preclinical evaluation.[65]

In many organizations, initial *in vivo* animal safety studies are performed under non-Good Laboratory Practices (GLP) conditions in order to decrease costs and increase throughput. Compounds destined for clinical study, however, must be assessed under GLP conditions prior to first in human (FIH) phase I clinical trials as part of an overall preclinical safety assessment that is included in an Investigational New Drug (IND) application. Safety data from two species, one rodent and one non-rodent (typically dog, but non-human primates can be used if necessary), are required in order to meet IND application requirements.

In practice, *in vivo* safety assessments are designed to monitor a variety of aspects of animal health such as physical appearance, body weight, food consumption, eye function, electrocardiography, blood chemistry, urine, and organ weight. Medical examinations and test designed to evaluate system and organ function (e.g., central nervous system, cardiovascular (including radiotelemetry), respiratory, gastro-intestinal systems, and kidney) are also performed. Microscopic histological examination of a wide range of tissue samples from dosed animals is another important aspect of these studies, as they may identify safety issues before overt symptoms appear. Study guidelines for many of these procedures are defined by the International Conference on Harmonization of Technical Requirements for Registration of Pharmaceuticals for Human Use.[66]

The first round of *in vivo* safety experiments are often performed as a series of single dose assessments (single ascending dose study: SAD study) to determine the maximum tolerated acute dose (the acute MTD), and to define both the NOEL and NOAEL doses in an acute setting. If these experiments demonstrate an acceptable therapeutic index, they are followed chronic dosing studies designed to identify the MTD, NOEL, NOAEL, and therapeutic index for extended administration of the candidate compound. Chronic dosing studies are typically 2–14 weeks in length. Both of these studies are paired with pharmacokinetic studies so that toxicity and safety issues can be correlated to exposure levels of the candidate compound. These toxicokinetic studies are critical to determining the therapeutic index of a compound as a function of systemic concentration. If the therapeutic index is too low in either the acute or chronic dosing studies or if the toxicity identified is too severe, the progression of a candidate toward clinical study will be terminated.

If a candidate compound survives preclinical safety evaluation, an IND is approved, and clinical trial are initiated, additional animal safety studies are often performed in order to gain a greater understanding of the level and types of risk associated with a clinical candidate. Additional chronic dosing studies lasting between 3 and 12 months are especially important for candidate compounds designed to treat chronic conditions such as hypertension or arthritis. Additional long-term animal studies that may be considered to identify potential issues as a candidate compound progresses toward market approval include reproductive health studies (to predict changes in fertility, mating habits, estrous cycle, and sperm formation), embryonic and development studies (monitors survival, fetal development, offspring growth and health), immunotoxicity studies,[67] and oncology studies. The results from each of these studies can be used to predict problems in the human population, and enable more informed decisions in clinical trial design and execution.

QUESTIONS

1. Define the terms NOEL and NOAEL.
2. What is the maximum tolerable dose (MTD) of a compound?
3. What is a mechanism-based toxicity?
4. Why is a compound's activity at the hERG channel an important aspect of its safety profile? How can changes in metabolism impact the importance of hERG channel activity?
5. What is a hapten? How does it evoke a biochemical response?
6. What is the difference between acutely toxic compounds and chronically toxic compounds?
7. What is the purpose of the MTT human hepatotoxicity assay?

8. What is the purpose of the AMES assay?
9. What is the purpose of the micronucleus assay?
10. What does the comet assay detect?
11. How do drug–drug interactions occur?
12. What are the limitations of the [^{3}H]-dofetilide hERG assay?
13. What are some of the major cardiovascular parameters that are routinely examined in safety assessments of lead compounds?
14. What animal model of CNS function can also be used to identify compounds with sedation as a risk?
15. If a candidate compound is described as being teratogenic, what inherent safety risks are associated with it?

References

1. a. Duffus, J. H.; Nordberg, M.; Templeton, D. M. Glossary of Terms Used in Toxicology, 2nd Edition (IUPAC Recommendations 2007). *Pure Appl. Chem.* **2007,** *79* (7), 1153–1344.
   b. U.S. Department of Health and Human Services, Environmental Health and Toxicology Specialized Information Services IUPAC Glossary of Terms Used in Toxicology, http://sis.nlm.nih.gov/enviro/iupacglossary/frontmatter.html.
2. a. ALS Association "Facts You Should Know" Web page, http://www.alsa.org/about-als/facts-you-should-know.html.
   b. Bensimon, G.; Lacomblez, L.; Meininger, V. A Controlled Trial of Riluzole in Amyotrophic Lateral Sclerosis. ALS/Riluzole Study Group. *N. Engl. J. Med.* **1994,** *330,* 585–591.
3. Karha, J.; Topol, E. J. The Sad Story of Vioxx, and What We Should Learn from It. *Cleve. Clin. J. Med.* **2004,** *71* (12), 934–939.
4. Rudmann, D. G. On-Target and Off-Target-Based Toxicologic Effects. *Toxicol. Pathol.* **2013,** *41* (2), 310–314.
5. Close, D. R. Matrix Metalloproteinase Inhibitors in Rheumatic Diseases. *Ann. Rheum. Dis.* **2001,** *60,* iii62–iii67.
6. Fingleton, B. MMPs as Therapeutic Targets—Still a Viable Option? *Semin. Cell Dev. Biol.* **2008,** *19* (1), 61–68.
7. https://www.millipore.com/life_sciences/flx4/drug-discovery-solutions-research-capability.
8. http://www.perkinelmer.com/Catalog/Category/ID/Profile%20Programs.
9. http://www.cerep.fr/cerep/users/index.asp.
10. http://pdsp.med.unc.edu/indexR.html.
11. Kalgutkar, A. S.; Vaz, A. D. N.; Lame, M. E.; Henne, K. R.; Soglia, J.; Zhao, S. X.; Abramov, Y. A.; Lombardo, F.; Collin, C.; Hendsch, Z. S.; et al. Bioactivation of the Nontricyclic Antidepressant Nefazodone to a Reactive Quinone-Imine Species in Human Liver Microsomes and Recombinant Cytochrome P450 3A4. *Drug Metab. Dispos.* **2005,** *33* (2), 243–253.
12. a. Perez-Inestrosa, E.; Suau, R.; Montanez, M. I.; Rodriguez, R.; Mayorga, C.; Torres, M. J.; Blanca, M. Cephalosporin Chemical Reactivity and Its Immunological Implications. *Curr. Opin. Allergy Clin. Immunol.* **2005,** *5,* 323–330.
   b. Elisabetta Padovan, E.; Baue, T.; Tongio, M. M.; Kalbache, H.; Weltzien, H. U. Penicilloyl Peptides Are Recognized as T Cell Antigenic Determinants in Penicillin Allergy. *Eur. J. Immunol.* **1997,** *27,* 1303–1307.

13. Jorga, K.; Fotteler, B.; Heizmann, P.; Gasser, R. Metabolism and Excretion of Tolcapone, a Novel Inhibitor of Catechol-*O*-methyltransferase. *Br. J. Clin. Pharmacol.* **1999**, *48*, 513–520.

14. a. Kang, P.; Dalvie, D.; Smith, E.; Renner, M. Bioactivation of Lumiracoxib by Peroxidases and Human Liver Microsomes: Identification of Multiple Quinone Imine Intermediates and GSH Adducts. *Chem. Res. Toxicol.* **2009**, *22*, 106–117.

    b. Li, Y.; Slatter, G.; Zhang, Z.; Li, Y.; Doss, G. A.; Braun, M. P.; Stearns, R. A.; Dean, D. C.; Baillie, T. A.; Tang, W. *In Vitro* Metabolic Activation of Lumiracoxib in Rat and Human Liver Preparations. *Drug Metab. Dispos.* **2008**, *36*, 469–473.

15. a. Durand, A.; Thenot, J. P.; Bianchetti, G.; Morselli, P. L. Comparative Pharmacokinetic Profile of Two Imidazopyridine Drugs: Zolpidem and Alpidem. *Drug Metab. Rev.* **1992**, *24*, 239–266.

    b. Usui, T.; Mise, M.; Hashizume, T.; Yabuki, M.; Komuro, S. Evaluation of the Potential for Drug-Induced Liver Injury Based on *In Vitro* Covalent Binding to Human Liver Proteins. *Drug Metab. Dispos.* **2009**, *37*, 2383–2392.

16. Kassahun, K.; Pearson, P. G.; Tang, W.; McIntosh, I.; Leung, K.; Elmore, C.; Dean, D.; Wang, R.; Doss, G.; Baillie, T. A. Studies on the Metabolism of Troglitazone to Reactive Intermediates *In Vitro* and *In Vivo*. Evidence for Novel Biotransformation Pathways Involving Quinone Methide Formation and Thiazolidinedione Ring Scission. *Chem. Res. Toxicol.* **2001**, *14*, 62–70.

17. Walsh, J. S.; Reese, M. J.; Thurmond, L. M. The Metabolic Activation of Abacavir by Human Liver Cytosol and Expressed Human Alcohol Dehydrogenase Isozymes. *Chem. Biol. Interact.* **2002**, *142*, 135–154.

18. a. Stepan, A. F.; Walker, D. P.; Bauman, J.; Price, D. A.; Baillie, T. A.; Kalgutkar, A. S.; Aleo, M. D. Structural Alert/Reactive Metabolite Concept as Applied in Medicinal Chemistry to Mitigate the Risk of Idiosyncratic Drug Toxicity: A Perspective Based on the Critical Examination of Trends in the Top 200 Drugs Marketed in the United States. *Chem. Res. Toxicol.* **2011**, *24*, 1345–1410.

    b. Park, B. K.; Laverty, H.; Srivastava, A.; Antoine, D. J.; Naisbitt, D.; Williams, D. P. Drug Bioactivation and Protein Adduct Formation in the Pathogenesis of Drug-Induced Toxicity. *Chem. Biol. Interact.* **2011**, *192*, 30–36.

    c. Kalgutkar, A. S.; Gardner, I.; Obach, R. S.; Shaffer, C. L.; Callegari, E.; Henne, K. R.; Mutlib, A. E.; Dalvie, D. K.; Lee, J. S.; Nakai, Y.; et al. A Comprehensive Listing of Bioactivation Pathways of Organic Functional Groups. *Curr. Drug Metab.* **2005**, *6*, 161–225.

    d. Kalgutkar, A. S.; Didiuk, M. T. Structural Alerts, Reactive Metabolites, and Protein Covalent Binding: How Reliable Are These Attributes as Predictors of Drug Toxicity? *Chem. Biodivers.* **2009**, *6* (11), 2115–2137.

19. Soignet, S. L.; Frankel, S. R.; Douer, D.; Tallman, M. S.; Kantarjian, H.; Calleja, E.; Stone, R. M.; Kalaycio, M.; Scheinberg, D. A.; et al. United States Multicenter Study of Arsenic Trioxide in Relapsed Acute Promyelocytic Leukemia. *J. Clin. Oncol.* **2001**, *19* (18), 3852–3860.

20. a. Mosmann, T. Rapid Colorimetric Assay for Cellular Growth and Survival: Application to Proliferation and Cytotoxicity Assays. *J. Immunol. Methods* **1983**, *65* (1–2), 55–63.

    b. Berridge, M. V.; Herst, P. M.; Tan, A. S. Tetrazolium Dyes as Tools in Cell Biology: New Insights into Their Cellular Reduction. *Biotechnol. Annu. Rev.* **2005**, *11*, 127–152.

21. Decker, T.; Lohmann-Matthes, M. L. A Quick and Simple Method for the Quantitation of Lactate Dehydrogenase Release in Measurements of Cellular Cytotoxicity and Tumor Necrosis Factor (TNF) Activity. *J. Immunol. Methods* **1998**, *115* (1), 61–69.

22. Repetto, G.; del Peso, A.; Zurita, J. L. Neutral Red Uptake Assay for the Estimation of Cell Viability/Cytotoxicity. *Nat. Protoc.* **2008**, *3* (7), 1125–1131.

23. Dorland. Dorland's Medical Dictionary: Dorland's Illustrated Medical Dictionary (32nd ed.). Elsevier Health Sciences; 2011.

24. Nagarathna, P. K. M.; Wesley, M. J.; Reddy, P. S.; Reena, K. Review on Genotoxicity, its Molecular Mechanisms and Prevention, International journal of pharmaceutical sciences review and research. **2013**, *22* (1), 236–243.

25. Benigni, R.; Bossa, C. Mechanisms of Chemical Carcinogenicity and Mutagenicity: A Review with Implications for Predictive Toxicology. *Chem. Rev.* **2011**, *111* (4), 2507–2536.

26. Ames, B. N.; Durston, W. E.; Yamasaki, E.; Lee, F. D. Carcinogens Are Mutagens: A Simple Test System Combining Liver Homogenates for Activation and Bacteria for Detection. *Proc. Natl. Acad. Sci. U.S.A.* **1973**, *70* (8), 2281–2285.

27. a. Fenech, M. The Cytokinesis-Block Micronucleus Technique and Its Application to Genotoxicity Studies in Human Populations. *Environ. Health Perspect. Suppl.* **1993**, *101* (S3), 101–107.

 b. Doherty, A. T. The *In Vitro* Micronucleus Assay. *Methods Mol. Biol.* **2012**, *817*, 121–141.

28. Galloway, M. A. Cytotoxicity and Chromosome Aberrations *In Vitro*: Experience in Industry and the Case for an Upper Limit on Toxicity in the Aberration Assay. *Environ. Mol. Mutagen.* **2000**, *35*, 191–201.

29. Collins, A. R. The Comet Assay for DNA Damage and Repair: Principles, Applications, and Limitations. *Mol. Biotechnol.* **2004**, *26* (3), 249–261.

30. Cook, P. R.; Brazell, I. A.; Jost, E. Characterization of Nuclear Structures Containing Superhelical DNA. *J. Cell Sci.* **1976**, *22*, 303–324.

31. a. Walsky, R. L.; Obach, R. S. Validated Assays for Human Cytochrome P450 Activities. *Drug Metab. Dispos.* **2004**, *32*, 647–660.

 b. Bjornsson, T. D.; Callaghan, J. T.; Einolf, H. J.; Fischer, V.; Gan, L.; Grimm, S.; Kao, J.; King, S. P.; Miwa, G.; Ni, L.; et al. The Conduct of *In Vitro* and *In Vivo* Drug–Drug Interaction Studies: A Pharmaceutical and Research Manufacturers of America (PhRMA) Perspective. *Drug Metab. Dispos.* **2003**, *31*, 815–832.

 c. Obach, R. S.; Walsky, R. L.; Venkatakrishnan, K.; Houston, J. B.; Tremaine, L. M. *In Vitro* Cytochrome P450 Inhibition Data and the Prediction of Drug–Drug Interactions: Qualitative Relationships, Quantitative Predictions, and the Rank-Order Approach. *Clin. Pharmacol. Ther.* **2005**, *78* (6), 582.

32. a. Li, A. P. Primary Hepatocyte Cultures as an *In Vitro* Experimental Model for the Evaluation of Pharmacokinetic Drug–Drug Interactions. *Adv. Pharmacol.* **1997**, *43*, 103–130.

 b. Moore, J. T.; Kliewer, S. A. Use of the Nuclear Receptor PXR to Predict Drug Interactions. *Toxicology* **2000**, *153*, 1–10.

33. Bresalier, R.; Sandler, R.; Quan, H.; Bolognese, J.; Oxenius, B.; Horgan, K.; Lines, C.; Riddell, R.; Morton, D.; Lanas, A.; et al. Cardiovascular Events Associated with Rofecoxib in a Colorectal Adenoma Chemoprevention Trial. *N. Engl. J. Med.* **2005**, *352* (11), 1092–1102.

34. Grant, A. O. Cardiac Ion Channels. *Circ. Arrhythm. Electrophysiol.* **2009**, *2*, 185–194.

35. a. Bazett, H. C. An Analysis of the Time-Relations of Electrocardiograms. *Heart* **1920**, *7*, 353–370.

 b. Sagie, A.; Larson, M. G.; Goldberg, R. J.; Bengston, J. R.; Levy, D. An Improved Method for Adjusting the QT Interval for Heart Rate (The Framingham Heart Study). *Am. J. Cardiol.* **1992**, *70* (7), 797–801.

36. a. Sanguinetti, M. C.; Jiang, C.; Curran, M. E.; Keating, M. T. A Mechanistic Link between an Inherited and an Acquired Cardiac Arrhythmia: HERG Encodes the I_{Kr} Potassium Channel. *Cell* **1995**, *81* (2), 299–307.

 b. Sanguinetti, M. C.; Tristani-Firouzi, M. hERG Potassium Channels and Cardiac Arrhythmia. *Nature* **2006**, *440* (7083), 463–469.

37. Finlayson, K.; Turnbull, L.; January, C. T.; Sharkey, J.; Kelly, J. S. [^{3}H]Dofetilide Binding to HERG Transfected Membranes: A Potential High Throughput Preclinical Screen. *Eur. J. Pharmacol.* **2001**, *430*, 147–148.

38. Chaudhary, K. W; O'Neal, J. M.; Mo, Z. L.; Fermini, B.; Gallavan, R. H.; Bahinski, A. Evaluation of the Rubidium Efflux Assay for Preclinical Identification of HERG Blockade. *Assay Drug Dev. Technol.* **2006,** *4* (1), 73–82.

39. Kim, J. J.; Bridle, B. W.; Ghia, J. E.; Wang, H.; Syed, S. N.; Manocha, M. M.; Rengasamy, P.; Shajib, M. S.; Wan, Y.; Hedlund, P. B.; et al. Targeted Inhibition of Serotonin Type 7 (5-HT$_7$) Receptor Function Modulates Immune Responses and Reduces the Severity of Intestinal Inflammation. *J. Immunol.* **2013,** *190,* 4795–4804.

40. Rothman, R. B.; Baumann, M. H.; Savage, J. E.; Rauser, L.; McBride, A.; Hufeisen, S. J.; Roth, B. L. Evidence for Possible Involvement of 5-HT$_{2B}$ Receptors in the Cardiac Valvulopathy Associated with Fenfluramine and Other Serotonergic Medications. *Circulation* **2000,** *102,* 2836–2841.

41. Langendorff, O. Untersuchungen Am Überlebenden Säugetierherzen. *Pflügers Arch.* **1898,** *61,* 291–332.

42. Bell, R. M.; Mocanu, M. M.; Yellon, D. M. Retrograde Heart Perfusion: The Langendorff Technique of Isolated Heart Perfusion. *J. Mol. Cell. Cardiol.* **2011,** *50,* 940–950.

43. Guth, B. D. Preclinical Cardiovascular Risk Assessment in Modern Drug Development. *Toxicol. Sci.* **2007,** *97* (1), 4–20.

44. Beasley, C. M., Jr.; Dornseif, B. E.; Bosomworth, J. C.; Sayler, M. E.; Rampey, A. H., Jr.; Heiligenstein, J. H.; Thompson, V. L.; Murphy, D. J.; Masica, D. N. Fluoxetine and Suicide: A Meta-Analysis of Controlled Trials of Treatment for Depression. *Br. Med. J.* **1991,** *303* (6804), 685–692.

45. Emslie, G. J.; Findling, R. L.; Yeung, P. P.; Kunz, N. R.; Li, Y. Venlafaxine ER for the Treatment of Pediatric Subjects with Depression: Results of Two Placebo-Controlled Trials. *J. Am. Acad. Child Adolesc. Psychiatry* **2007,** *46* (4), 479–488.

46. Serena Tonstad, S.; Davies, S.; Flammer, M.; Russ, C.; Hughes, J. Psychiatric Adverse Events in Randomized, Double-Blind, Placebo-Controlled Clinical Trials of Varenicline. *Drug Saf.* **2010,** *33* (4), 289–301.

47. Kontaxakis, V. P.; Skourides, D.; Ferentinos, P.; Havaki-Kontaxaki, B. J.; Papadimitriou, G. N. Isotretinoin and Psychopathology: A Review. *Ann. Gen. Psychiatry* **2009,** *8* (2), 1–8.

48. NIH Contract # HHSN-271-2008-025C (NIHM PDSP), http://pdsp.med.unc.edu/index R.html.

49. Antunes, M.; Biala, G. The Novel Object Recognition Memory: Neurobiology, Test Procedure, and Its Modifications. *Cognit. Process.* **2012,** *13* (2), 93–110.

50. Wehner, J. M.; Radcliffe, R. A. Cued and Contextual Fear Conditioning in Mice. *Curr. Protoc. Neurosci.* **2004.** 8.5C.1–8.5C.14.

51. D'Hooge, R.; De Deyn, P. P. Applications of the Morris Water Maze in the Study of Learning and Memory. *Brain Res. Rev.* **2001,** *36* (1), 60–90.

52. Bogo, V.; Hill, T. A.; Young, R. W. Comparison of Accelerod and Rotarod Sensitivity in Detecting Ethanol- and Acrylamide-Induced Performance Decrement in Rats: Review of Experimental Considerations of Rotating Rod Systems. *Neurotoxicology* **1981,** *2* (4), 765–787.

53. Lemus, R.; Karol, M. H. Conjugation of Haptens. *Methods Mol. Med.* **2008,** *138,* 167–182.

54. Weltzien, H. U.; Padovan, E. Molecular Features of Penicillin Allergy. *J. Invest. Dermatol.* **1998,** *110* (3), 203–206.

55. Rose, N. R.; Bona, Constantin Defining Criteria for Autoimmune Diseases (Witebsky's Postulates Revisited). *Immunol. Today* **1993,** *14* (9), 426–430.

56. Ghoreschi, K.; Jesson, M. I.; Li, X.; Lee, J. L.; Ghosh, S.; Alsup, J. W.; Warner, J. D.; Tanaka, M.; Steward-Tharp, S. M.; Gadina, M.; et al. Modulation of Innate and Adaptive Immune Responses by Tofacitinib (CP-690,550). *J. Immunol.* **2011,** *186* (7), 4234–4243.

57. a. Kempeni, J. Preliminary Results of Early Clinical Trials with the Fully Human Anti-TNFα Monoclonal Antibody D2E7. *Ann. Rheum. Dis.* **January 1999,** *58* (S1), I70–I72.
    b. Scheinfeld, N. Adalimumab (HUMIRA): A Review. *J. Drugs Dermatol.* **2003,** *2* (4), 375–377.

58. a. Burmester, G. R.; Matucci-Cerinic, M.; Mariette, X.; Navarro-Blasco, F.; Kary, S.; Unnebrink, K.; Kupper, H. Safety and Effectiveness of Adalimumab in Patients with Rheumatoid Arthritis over 5 Years of Therapy in a Phase 3b and Subsequent Postmarketing Observational Study. *Arthritis Res. Ther.* **2014,** *16* (R24), 1–11.

    b. Bender, N. K.; Heilig, C. E.; Dröll, B.; Wohlgemuth, J.; Armbruster, F. P.; Heilig, B. Immunogenicity, Efficacy and Adverse Events of Adalimumab in RA Patients. *Rheumatol. Int.* **2007,** *27* (3), 269–274.

59. Kim, J. H.; Scialli, A. R. Thalidomide: The Tragedy of Birth Defects and the Effective Treatment of Disease. *Toxicol. Sci.* **2011,** *122* (1), 1–6.

60. Jones, K. L.; Smith, D. W.; Ulleland, C. N.; Streissguth, A. P. Pattern of Malformation in Offspring of Chronic Alcoholic Mothers. *Lancet* **1973,** *301* (7815), 1267–1271.

61. *Mosby's Medical Dictionary on My Desk.*

62. Seiler, A.; Visan, A.; Buesen, R.; Genschow, E.; Spielmann, H. Improvement of an *In Vitro* Stem Cell Assay for Developmental Toxicity: The Use of Molecular Endpoints in the Embryonic Stem Cell Test. *Reprod. Toxicol.* **2004,** *18,* 231–240.

63. Flint, O. P. *In Vitro* Tests for Teratogens: Desirable Endpoints, Test Batteries and Current Status of the Micromass Teratogen Test. *Reprod. Toxicol.* **1993,** *7* (S1), 103–111.

64. a. Selderslaghsa, I. W. T.; Van Rompaya, A. R.; De Coenb, W.; Witters, H. E. Development of a Screening Assay to Identify Teratogenic and Embryotoxic Chemicals Using the Zebrafish Embryo. *Reprod. Toxicol.* **2009,** *28,* 308–320.

    b. Teixidó, E.; Piqué, E.; Gómez-Catalán, J.; Llobet, J. M. Assessment of Developmental Delay in the Zebrafish Embryo Teratogenicity Assay. *Toxicol. In Vitro* **2013,** *27,* 469–478.

65. Jones, T.W. Pre-clinical Safety Assessment: It's No Longer Just a Development Activity; Drug Discovery Technology (R) and Development World Conference, Boston, MA, August 8–10, 2006.

66. http://www.ich.org/.

67. Dean, J. H.; Cornacoff, J. B.; Haley, P. J.; Hincks, J. R. The Integration of Immunotoxicology in Drug Discovery and Development: Investigative and *In Vitro* Possibilities. *Toxicol. In Vitro* **1994,** *8,* 939–944.

Basics of Clinical Trials

The Food, Drug, and Cosmetic Act of 1938[1] requires an FDA safety review of all new drugs before they can be approved for commercial use in the United States, and the Kefauver–Harris Amendment of 1962[2] added proof of efficacy requirement components to the regulatory review process. While these laws do not directly impact other jurisdictions, similar laws are in place, making proof of safety and efficacy a global requirement. The concept of studying the impact of potentially therapeutic material on people is, however, significantly older than the laws that govern the modern marketing approval system. As discussed in Chapter 2, attempts to identify useful medicines can be traced to very early points in human history. Documentation of early attempts can be found in texts such as the Mesopotamians "Treatise of Medical Diagnosis and Prognoses" (1700 BC),[3] the Ebers Papyrus of ancient Egypt (1550 BC),[4] and Pen-tsao Kang-mu, the compendium written by Li Shih-chen in 1857 that documents traditional Chinese medicine used for nearly 2000 years.[5] Many important medicines were developed prior to the advent of the modern clinical trials system, but definitive proof of efficacy and safety was virtually non-existent. In the absence of clinical trials, medicines and medical procedures were employed based on circumstantial evidence or, in many cases, no evidence at all. Bleeding, for example, was viewed as a viable treatment for yellow fever[6] and pneumonia.[7] Prior to 1747, the medical community had not recognized the importance of standardization, objectivity, control subjects, or any of the other features that are required to develop truly useful clinical data.

The earliest documented clinical trial is attributed to Dr. James Lind, a physician in the Scottish navy, who was attempting to identify a means of treating or preventing scurvy (Figure 9.1). This condition was common in sailors, especially those who had been at sea for an extended period of time. In 1747, Dr. Lind selected 12 sailors with similar cases of scurvy during an ocean voyage for an experiment. The men were fed the same basic diet and housed in the same area of the ship, as Dr. Lind recognized the importance of standardizing as many aspects of the experiment as possible. They were

Basic Principles of Drug Discovery and Development
http://dx.doi.org/10.1016/B978-0-12-411508-8.00009-8

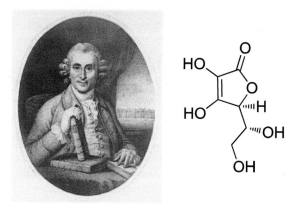

FIGURE 9.1   Dr. James Lind (left) conducted the earliest documented clinical trial in an effort to identify a method of treating and preventing scurvy. He determined that consumption of oranges and lemons prevented this condition. It was later determined that scurvy is caused by a vitamin C (right) deficiency. *Source: Reprinted from Magiorkinis, E.; Beloukas, A.; Diamantis, A. Scurvy: Past, present and future. Eur. J. Intern. Med., 22 (2), 147-152, copyright 2011 with permission from Elsevier.*

divided into six groups and each group was given one of six potential treatments for scurvy, (1) cider, (2) diluted sulfuric acid, (3) vinegar, (4) seawater, (5) a mixture of several foods including nutmeg and garlic, (6) and oranges and lemons. After 1 week, the 2 sailors consuming oranges and lemons showed little sign of scurvy, while the remaining 10 sailor's conditions remained unchanged. Although he did not know that scurvy was caused by a vitamin C deficiency, Dr. Lind had successfully identified a method of treating and preventing scurvy.[8]

Dr. Lind's successful but limited study of scurvy did not, however, lead to a broader understanding of the importance of clinical trials. The foundation of modern clinical trials was not put into place until nearly 90 years later when Dr. Pierre Charles Alexandre Louis defined the "numerical method" of assessing therapeutic interventions. Dr. Louis recognized the importance of utilizing patient populations and the average effect of a potential therapy across the group, rather than using individual patients. He argued that difference between patients in the groups would "average out" provided that the patient population was large enough. In addition, he stressed the importance of understanding the natural progression of untreated subjects (controls), precisely defining the disease or condition prior to initiation of treatment, careful and objective observation of patient outcomes, and an awareness of any deviations from the treatment regimen under examination. Dr. Louis successfully applied these methods to disprove the utility of bleeding as a treatment for pneumonia (Figure 9.2). Although his approach initially met with strong resistance from his contemporaries, the improvements in patient outcome could not be ignored, and Dr. Louis' methods eventually became the foundation of the modern clinical trial system.[9]

FIGURE 9.2   Dr. Pierre Charles Alexandre Louis was the first to recognize the importance of analyzing clinical data from groups of patient. He disproved the utility of bloodletting as a treatment for pneumonia. *Source: acquired by public domain at: http://en.wikipedia.org/wiki/ File:BloodlettingPhoto.jpg, http://en.wikipedia.org/wiki/File:Pierre-Charles_Alexandre_Louis.jpg*

The concept of randomization of patient populations and blinded groups are notably absent from Dr. Louis' work. Patient randomization was first introduced by Greenwood and Yule in 1915 as part of their investigations of potential treatments for cholera and typhoid.[10] A little more than a decade later, Ferguson and his colleagues described what is believed to be the first clinical investigation using a blinded group. In their study of vaccines for the common cold, patients did not know if they were receiving a vaccine or saline, but the researchers were aware of the treatment course for each patient, making this a single-blinded study.[11] The first clinical trial that incorporated the majority of features of modern clinical trials, including properly randomized control groups and fully blinded data analysis, was performed in 1948 in an effort to determine the utility of streptomycin for the treatment of *mycobacterium tuberculosis* infection (Figure 9.3).[12]

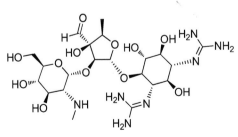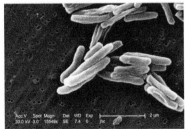

FIGURE 9.3   The first clinical trial that incorporated properly randomized control groups and fully blind data analysis was performed in 1948 in order to determine the utility of Streptomycin (left) as a treatment for *mycobacterium tuberculosis* (right) infection. *Source: CDC public image library, http://phil.cdc.gov/phil/details.asp?pid=8438*

Within the context of the modern pharmaceutical industry, a clinical trial can be defined as a biomedical or behavioral experiment conducted on population of humans that is designed to answer key questions about a potential therapeutic agent. (Medical device and medical procedure can also be the subject of clinical study, but for the remainder of this text, only drug therapies will be considered.) The studies are designed to generate safety and efficacy data that can be presented to the appropriate regulatory bodies in order to gain market approval. In general terms, clinical trials are divided into 4 phases (Phase I to Phase IV), each with its own purpose. Phase I studies define safety margins and the pharmacokinetic profile of a candidate compound. Phase II studies are pilot studies designed to provide an initial assessment of efficacy and safety in the target patient population. Phase III studies are broad efficacy and safety studies designed to determine the risk to benefit ratio of a candidate compound in the target population. Finally, phase IV studies, also referred to as post-approval studies, monitor a drug's efficacy and safety after it has gained marketing approval. Each of these phases and the associated trial designs will be examined in greater detail later in this chapter.[13] It is important to understand, however, that there are some important hurdles that must be crossed in between the identification of a potential clinical candidate and the initiation of clinical trials.

BEFORE THE CLINIC

The identification of a candidate compound that has the desired efficacy in an appropriate *in vivo* animal model and a sufficient safety window in two animal species (rodent and non-rodent) represents the final step in the majority of drug discovery programs. Moving a candidate compound into clinical study, however, requires a significant amount of additional work. There are a number of issues that must be addressed before human studies will be permitted by the various regulatory bodies. Compound manufacture (drug supply), dosing method, and formulation must be determined before an Investigational New Drug Application (IND) would be approved for execution. There are, of course, numerous other areas that must be explored, both scientific and economic, but for the purpose of exemplifying the difficulties associated with moving from discovery to clinical development, we will consider these three key issues.

DRUG SUPPLY

Production methods capable of generating pharmaceutical grade active pharmaceutical ingredient (API, the candidate compound) under good manufacturing practice (GMP) conditions must be established in order to initiate a clinical program. In an ideal world, this would simply be a matter of increasing the scale of the synthetic methods employed in the original preparation of the candidate compound. In practice, however, this is almost always impossible, as there are many procedures and methods that are impractical to replicate on an industrial scale. Consider, for example, a synthetic sequence that requires a temperature of −78 °C. In a laboratory focused on preparing small amounts of material (>20 g), one can simply cool a reaction vessel with a dry ice/acetone bath. Cooling an industrial scale reactor designed to make 200 kg of the same material (a 10,000-fold increase in scale, Figure 9.4) would be challenging and expensive. Purification of

FIGURE 9.4   The center red circle represents 20 g. The black circle represents a 100-fold increase in size, while the green circle represents a 10,000-fold increase in size.

intermediates and the final product can also be an issue. Chromatography methods suitable for laboratory scale synthesis, for example, are generally not feasible for industrial scale production and must be replaced with recrystallizations, precipitations, or distillations as appropriate.

The concept of polymorphism can also create unexpected issues in moving from laboratory scale to GMP production. Polymorphism refers to the ability of a solid material, such as a candidate compound, to exist in more than one crystalline form.[14] If multiple crystal forms are possible, they may have different physical properties, which may have an impact on pharmacokinetic properties. If, for example, one crystal form of a compound (a polymorph) dissolves more slowly than a second crystal form (a different polymorph) of the same compound, the bioavailability of the compound may be significantly lower if the second crystal form is employed. Norvir® (Ritonavir, Figure 9.5), an HIV protease inhibitor, for example, was

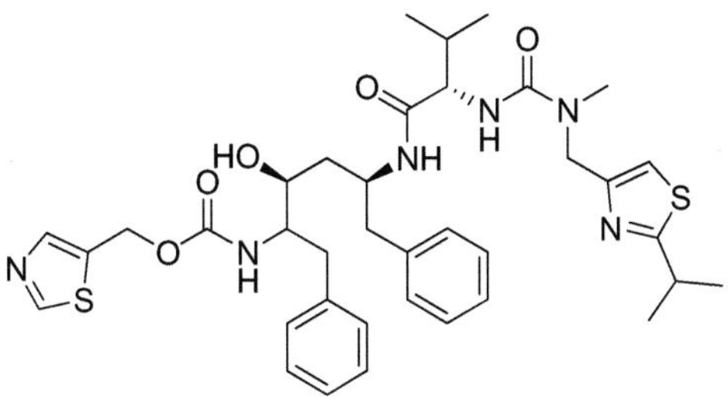

FIGURE 9.5   Norvir® (Ritonavir).

originally approved for the treatment of HIV infection in 1996. By 1998, however, it became clear that manufacturing processes were producing a previously unknown polymorph that did not meet the solubility specification approved for clinical application. The decreased solubility led to a substantial decrease in oral bioavailability and eventually forced the manufacturer (Abbott Labs, now Abbvie) to temporarily withdraw the Norvir® (Ritonavir) from the market while a new formulation was developed and tested.[15] As this example clearly demonstrates, it is important to ensure that manufacturing processes lead to not only the desired compound, but also that they reliably produce the same crystal form every time and that the crystal form is stable under the proposed storage conditions.

Environmental and waste disposal issues also become a major concern. Laboratory-scale reaction can employ solvents that have known safety and waste disposal issues such as methylene chloride (carcinogen, highly volatile), ethyl ether (highly volatile, highly flammable, peroxide former), and benzene (carcinogen, toxic, highly volatile, highly flammable). Industrial-scale procedures designed to prepare material for human consumption (i.e., medication), however, are often limited to more benign solvents (e.g., ethanol, 1,4-dioxane, toluene, ethyl acetate). The sheer volume of waste produced is also significantly higher and must be considered as a process is moved to an industrial scale. A 10,000-fold increase in production scale would lead to a 10,000-fold increase in waste generation if no changes were made on moving from the lab scale to manufacturing. If a lab-scale procedure requires 5 L of solvent to produce 20 g of material, a 10,000-fold increase to produce 200 kg would require 50,000 L of solvent in the absence of procedural changes. All of this solvent would be considered hazardous waste and would require proper disposal. Modifications to minimize solvent utilization can significantly reduce the environmental impact. These factors will also reduce production costs, which will, in turn, lower the cost of the product once it reaches the market. Minimizing solvent use is environmentally important and can help contain the cost of new medicines as they approach the market.

There are additional factors beyond scalability of the chemistry that must also be considered. Changes in synthetic pathways often result in the formation of different impurities or side products with unknown properties, even if the change is as simple as producing a new salt form of the parent compound. Consider, for example, a new synthetic method for the production of a candidate compound that also generates a small but quantifiable amount of an Ames positive impurity. The presence of this new impurity would represent a major issue for the new route. Identification, characterization, and monitoring methods to detect impurities in the final products must be established in order to ensure that the clinical material is truly suitable for human study.

DELIVERY METHODS

As a candidate compound progresses from the discovery stage to the development stage and clinical study, an appropriate method of delivery must be identified. The most common method of dosing is oral delivery (PO). Nearly 70% of marketed medications are delivered to patients in this manner[16] as it is the most convenient, economic, and safe method of administration. This non-invasive method, however, requires a high degree of patient compliance, as efficacy is dependent upon the patient following the proper dosing strategy. In most cases, an oral medication is given without the direct supervision of medical personnel (unless the patient is in a hospital or other medical facility), and there is no guarantee that the patient will take the medicine as directed. The emergence of bacteria that are resistant to modern antibiotics (e.g., Methicillin-resistant *Staphylococcus aureus* (MRSA)[17]) is at least in part due to non-compliance with drug regimens. Patients that feel better part way through the dosing regimen may stop taking the antibiotic, providing an opportunity for bacterial resistance to develop (Interestingly, Alexander Fleming, warned of this possibility in his lecture upon receiving the Nobel prize in 1945[18]). In order to simplify dosing regimens and maximize the likelihood of patient compliance, the majority of pharmaceutical companies target once daily dosing protocols.

It is, of course, possible to move forward with a drug that must be delivered more than once a day, but increasing the number of times a pill must be ingested through the course of a 24 h period generally decrease the patient compliance. Twice daily pills are taken every 12 h, but a pill that is administered four times a day must be taken every 6 h in order to obtain an even drug exposure across a 24 h period. If the first pill is taken at 9:00 AM, then the fourth pill must be taken at 3:00 AM to maintain the schedule (Figure 9.6). Very few people want to get up at 3:00 AM to take a pill.

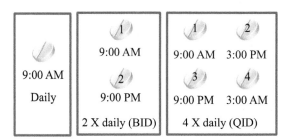

FIGURE 9.6   Dosing regimen scheduled if spaced evenly across a 24-h period.

There are many instances, however, in which oral delivery is simply not an option, as there are physicochemical issues that preclude this route of administration. If the medical need is high enough or the benefit conferred is significant enough that the targeted patient population will accept an alternate form of delivery, there are a number of additional options that could be considered. Intravenous (IV), intraperitoneal (IP), subcutaneous (SC), intramuscular injections (IM), transdermal, and intranasal delivery[19] are some of the options available for candidate compounds that cannot be delivered orally. Insulin, for example, is an absolute necessity for insulin-dependent diabetics, but peptides like insulin are not typically orally bio-available due to their poor absorption. Similarly, patients suffering from migraine headaches were very willing to accept injections of Imitrex® (Sumatriptan, Figure 9.7) when it was originally introduced in 1991 as the

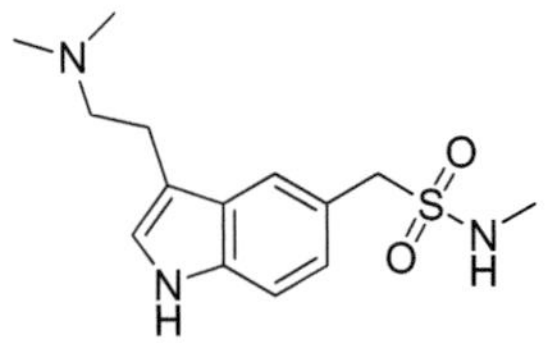

FIGURE 9.7   Imitrex® (Sumatriptan).

first effective treatment for migraine headaches.[20] In this case, the pain was less tolerable than the needle. The benefits of Reclast® (zoledronic acid)[21] also allowed it to reach the market in a non-oral formulation. In this case, the drug is given to patients once a year as a 5 mg intravenous infusion for the treatment and prevention of osteoporosis. There are several competing oral medications, such as Fosamax® (Alendronic acid),[22] Actonel® (Risedronic acid),[23] and Boniva® (Ibandronic acid),[24] but the clinical utility of Reclast® is significant enough to warrant an IV formulation (Figure 9.8).

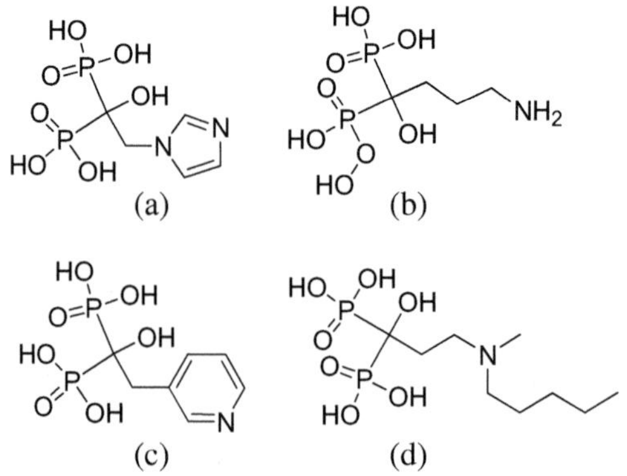

FIGURE 9.8   (a) Reclast® (zoledronic acid) (b) Fosaamax® (Alendronic acid) (c) Actonel® (Risedronic acid) (d) Boniva® (Ibandronic acid).

While there are a many examples of the successful commercialization of drugs that are delivered using non-oral routes of administration, the choice to move down this path must be carefully considered. An IV-deliv-ered drug may reach the market, but if a competitor develops an orally delivered medication that serves the same purpose, the oral drug will most likely supplant the IV drug for most purposes. This could make recouping the costs associated with the discovery and development of the IV drug very difficult. There is also the possibility that the patients will not be willing to accept the new method of delivery. Inhalable insulin sold under the name Exubera®, for example, was developed in an attempt to replace insulin injections and brought to market by Pfizer and Nektar Therapeutics in 2006. Appropriate clinical studies demonstrated that it was safe and effective, but the patients did not migrate to the new protocol, making the drug far less profitable than expected. The drug was pulled from the market by Pfizer in mid-2007, which cost the company an estimated $2.8 billion.[25] Choosing the wrong delivery method can be a costly mistake.

FORMULATION

The ability to effectively deliver a therapeutically useful compound to a patient is at least in part driven by the ability to formulate the compound in a useful delivery vehicle. In theory, a candidate compound with suitable PK properties could be provided to patients as a dry powder in a bottle and they could simply ingest the material according to a physician's or pharmacist's instructions. In practice, this is neither practical nor desirable. Consider, for example, the highly potent antianxiety medication, Ativan® (Lorazepam, Figure 9.9).[26] This well-known benzodiazepine is a white powder with approved dosing levels of 0.5, 1.0, and 2.0 mg. In practice, it

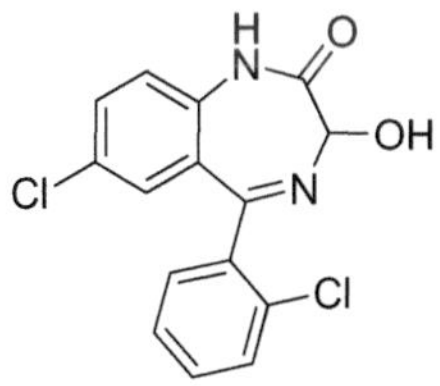

FIGURE 9.9   Ativan® (Lorazepam).

would be difficult to ensure that a patient received the correct dose if he/she were simply given a bottle of dry powder. In addition, a tablet or pill that contains only 0.5–2.0 mg of material would be very small and difficult to produce.

Of course, tablets and pills typically contain a number of additional components beyond the API. In some cases, the added materials, also referred to as excipients (Table 9.1),[27] are simply fillers that enables the production of a pill of suitable size that can be easily manufactured. A reasonably sized pill can improve patient compliance by making it more convenient for the patient to use the medication. In other cases, excipients

TABLE 9.1   Representative Excipients Listed by Class

Excipient Class	Example	Excipient Class	Example
Fillers	Carbohydrates (e.g., glucose, lactose)	Dry binders	Cellulose
	Calcium phosphate		Methyl cellulose
	Calcium carbonate		Polyvinyl pyrrolidone
	Cellulose		Polyethylene glycol
	Starch	Glidants	Silica
	Cellulose		Magnesium stearate
	Polyvinyl pyrrolidone		Talc
Disintegrants	Sodium starch glycolate	Lubricants	Magnesium stearate
	Sodium carboxymethyl cellulose		Stearic acid
	Gelatin		Polyethylene glycol
	Polyvinyl pyrrolidone		Sodium lauryl sulfate
	Cellullose derivatives		Paraffin
	Polyethylene glycol		Talc
Solution binders	Sucrose	Antiadherents	Starch
	Starch		Cellulose

serve an expressed purpose beyond simply adding bulk. Glidants, lubricants, antiadherents, and binding agents may be required to facilitate the manufacturing process. Flavors may be required in order to mask a bitter taste that would reduce patient compliance (especially in pediatric formulations). Colorants and antioxidants may be necessary, if the API is sensitive to air oxidation or exposure to light.

Excipients can also have an impact on compound pharmacokinetics. They can, for example, alter dissolution rates, reduce potential toxicity, change the elimination half-life ($t_{1/2}$), or alter time at which a candidate compound reaches its maximum concentration in the systemic circulation. If, for example, a potential therapeutic agent is readily soluble, but a slow and sustained delivery is required, the candidate compound could be embedded in polymer that dissolves over a longer period of time. As the polymer dissolves or disintegrates, API is released and dissolved.

Special coatings can also be applied to allow a candidate compound to survive an unfavorable environment such as the highly acidic environment of the stomach. Enteric-coated tablets,[28] for example, are passed through the stomach unchanged, as the coating is unaffected by the acidic environment. On the other hand, the enteric coating dissolves in the basic environment of the intestines. The drug is released and then free to pass into the systemic circulation.

Excipients can also have a positive impact on the permeability of a compound. As discussed in previous chapters, passive diffusion plays a major role in permeability. There are a number of excipients referred to as bioenhancers that can be incorporated into a solid dosage form for the purposes of increasing compound permeability. Typical examples include Tween®-80 (Polysorbate-80), sodium lauryl sulfate (sodium dodecyl sulfate), Labrafil® (PEGylated oils), and Labrasol® (PEGylated caprylic/capric glycerides). In some cases, these agents form mixed micelles with membrane lipids or increase the fluidity of the cell membrane, making compound permeation easier and faster. In other cases, excipients interact with lipids at the membrane surface increasing the hydrophobicity on the membrane surface, which allows a compound to enter the membrane more easily.[29]

The choice of excipients is only one aspect of formulation that must be considered prior to beginning a clinical program. Particle size of the API can have a significant impact on whether or not a drug can be effectively delivered, especially when the rate of dissolution is a limiting factor. As the particle size of a therapeutic agent decreases, the overall surface area available for dissolution increases. Decreasing the particle size of a drug product from 10 μm to 200 nM, for example, increase the overall surface area by a factor of 50.[30] The significant increase in surface area can increase the rate of dissolution, which can, in turn, increase the bioavailability and systemic exposure of a candidate compound. The bioavailability of the synthetic steroid Danocrine® (Danazol, Figure 9.10), for example, jumps from 5.1% in a

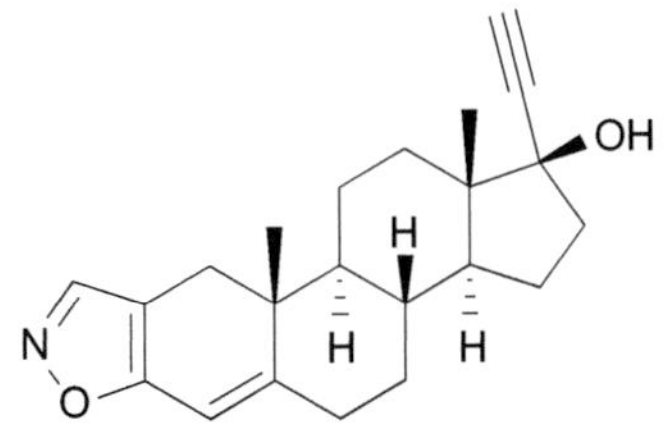

FIGURE 9.10   Danocrine® (Danazol).

conventional formulation to 82.3% when nanoparticles are employed in an oral formulation (average particle size of 169 nM). The C_{max} also increases, jumping from 0.2 μg/mL to 3.01 μg/mL.[31] A number of different milling processes have been developed to facilitate micro- and nanoparticle production (e.g., ball mill, fluid energy mill, cutter mill, and hammer mill).[32] It is important to keep in mind, however, that particle size reduction does not affect the absolute solubility of a given compound. Solubility is a physicochemical property of a compound that cannot be changed by altering the formulation employed for dosing.

Additional formulation options that can be explored in an attempt to design delivery methods capable to producing the desired clinical outcome include specialty tablets and capsules. Multilayer tablets that produce an initial burst of the candidate compound followed by a slow, sustained release can be created by embedding the API into two different matrices with different dissolution rates (one fast, one slow, Figure 9.11).[33] Osmotic pumps

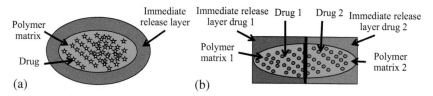

FIGURE 9.11 (a) An immediate release outer layer provides an immediate burst of drug substance. Once the outer layer is gone, the core of the tablet is exposed. The polymer matrix containing the drug is dissolved slowly, releasing the remainder of the drug over an extended period of time. (b) Two different medications coat the outside of the tablet, both of which are released to provide an immediate burst of drug substance. Exposure of the inner core leads to slow release of the two medications from their respective polymer matrices. The rate of release of each drug will be determined by the nature of each polymer matrix.

systems that encapsulate the API in a semipermeable membrane can also be an effective tool in delivering a drug. In a simple system, water enters the capsule through a semipermeable membrane as a result in the difference in osmotic pressure between the inside and the outside of the pill. The water dissolves the API, which then moves out of the capsule through microdrilled orifices in the walls of the pill (Figure 9.12(a)).

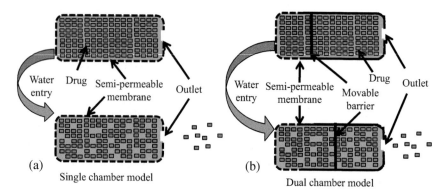

FIGURE 9.12 (a) In the single chamber osmotic pump system, differences in osmotic pressure lead to release of the drug substance (red) through microdrilled holes. (b) In the two chambered system, drug release (red) is promoted by movement of barrier between the two chambers of the pill. Osmotic pressure differences causes water to enter the "push chamber," which expands, forcing drug substance out of the drug chamber.

In a two-chambered osmotic pump system, the drug substance chamber is surrounded by an impermeable membrane with microdrilled orifices. The second chamber, also called the "push chamber" is encapsulated by a semipermeable membrane, and the two chambers are separated by a moveable wall. Water enters the "push chamber" as a result of the difference in osmotic pressure, causing the material in this chamber to expand. This action pushes the moveable wall, which forces drug substance out of the capsule through the microdrilled orifices in the walls of the drug chamber.[34]

There are many other aspects to consider in determining the proper formulation for a clinical candidate that must be finalized before initiating clinical trials. Issues such as hardness, thickness, and friability (the ability of a solid to breakdown into smaller pieces) are unique to oral formulations, whereas issues such as dose uniformity, impurity profiles, and manufacturing methods cut across all types of formulations. Irrespective of the final dosing method and form, decisions on the final drug formulation must be made prior to entering clinical trials.

INVESTIGATIONAL NEW DRUG APPLICATION

Once all of the scientific, manufacturing, and technical hurdles have been overcome, an organization must gain approval before they will be authorized to move forward with human studies. Government agencies such as The Food and Drug Administration (US), the European Medicines Agency (European Union), the Pharmaceuticals and Medical Agency (Japan), and similar agencies across the globe must receive and approve a clinical plan before any potential therapy can be tested in human populations. An Investigational New Drug Application (IND), also referred to as a Clinical Trial Authorization (CTA) or Clinical Trial Notification (CTN), is filed in order to solicit the approval of these agency. The document contains information describing three broad areas of research, (1) Animal pharmacology, safety, and toxicology study data, (2) Chemistry, Manufacturing, and Controls (CMC), and (3) Clinical trial protocols including investigator information.[35]

The animal pharmacology, safety, and toxicology section of an IND describes pharmacological studies that provide proof of efficacy in an accepted animal model of the relevant disease state. This section also includes preclinical safety study data that the regulators will use to determine whether or not the potential new therapeutic agent is safe enough to initiate clinical trials. *In vivo* safety studies from two animal species, one rodent and one non-rodent, run under GLP (Good Laboratory Practices) conditions must also be included. This section also typically includes information derived from many of the safety studies described in Chapter 8 such as Ames testing, micronucleus assay data, chromosomal aberration assessments, and cardiovascular safety assessments. In addition, any data from previous use of the candidate compound in humans is described in this section. Human data may be available from studies performed in other jurisdictions or from previous attempts to gain marketing approval in different disease state.

The second section provides detailed information on Chemistry, Manufacturing, and Controls (CMC). A detailed description of the manufacturing methods including synthetic methods for producing the API, as well as information on all of the excipients and how the clinical formulation will be prepared must be included. It must also include information on impurities that are produced as part of the process, methods for assessing purity of each batch that is prepared, and evidence that equivalent drug batches can be produced in a reliable and consistent manner. Of course, in order to have this information, manufacturing capacity must already be in place and sufficiently tested to demonstrate its capacity to produce the drug product.

The third and final section covers the clinical protocols and the individuals involved. Clinical studies must be described in sufficient detail such that the regulatory agencies can assess whether or not the program will expose the subject to unnecessary risks. In addition, information on the study centers (most clinical trials employ multiple study centers) and the qualifications of the people running the programs and overseeing administration of the clinical candidate (most often physicians) is provided as evidence that the facilities and personnel are sufficient to ensure the safety of the study participants. Adherence to regulatory guidelines, methods of obtaining informed consent of all study participants, and the establishment of an Institutional Review Board (IRB) are also described in this section of the document. An IRB, also referred to as an independent ethics committee or ethical review board, is a committee that is tasked with ensuring that clinical trials are run in an appropriate manner. They are formally charged with the tasks of monitoring, reviewing, and approving all aspects of the clinical programs. This includes any changes to protocols that may be required as a result of new data that emerges through the course of the trials. Risk to benefit analysis may also be performed by the IRB as the clinical trials move forward in order to determine whether or not it is safe to continue the trials.

It should come as no surprise that INDs are exceptionally large documents that require a significant amount of effort to prepare. They are generally written by a large team of sufficiently qualified individuals. Typically, the investigators will meet with representative of the appropriate regulatory body (e.g., FDA, EMA, etc.) in order to ensure that the clinical program design will be sufficient. At the end of the day, all clinical programs must be approved by the regulatory agencies before they can be initiated. It is in the best interest of the trial sponsors (usually, although not always a pharmaceutical company) to ensure they are crafting a document that will be approved. Once the IND is approved, human trial can be initiated.

PHASE I CLINICAL TRIALS

The primary goal of phase I trials, also referred to as First-in-Human (FIH) trials, is to determine whether or not the candidate compound is safe for clinical study. While there may be ample evidence from animal models suggesting that a candidate compound is safe for clinical use, there are significant interspecies differences that cannot be accounted for in animal safety studies. Great care must be taken to ensure proper monitoring the subject of a phase I clinical trial. When the phase I trial is complete, the maximum tolerated dose (MTD) will have been identified, along with dose limiting toxicities (DLT) that are associated with the potential therapeutic agent. The human pharmacokinetic profile will also be determined, and this information will be used to select the doses used in phase II trials. It may also be possible to gain some insight into the pharmacodynamics of the candidate compound, and possibly an early indication of efficacy, depending on the trial design.

In most cases, phase I clinical trial participants are healthy volunteers, and a typical phase I trial will require between 20 and 100 individuals. The patients can be divided into multiple cohorts (dosing groups) and, although it is not required, some studies employ a placebo control group. Since phase I is often the first use of the candidate compound in human subjects, dose selection for earliest portions of phase I trials must be based on the animal safety data developed in the preclinical phase of development. Allometric scaling,[36] a mathematical calculation used to estimate interspecies dosing changes based on known PK parameters in a single species, based on the NOAEL (no observed adverse effect level) determined in animal studies is used as a guide in determining initial dosing levels. The actual starting dose is often lower than that determined by allometric scaling in order to ensure the safety of the volunteers.

A typical phase I clinical trial begins with a single ascending dose (SAD) study (Figure 9.13) in order to determine the MTD and identify any DLTs that may be associated with the candidate compound. In one simple scenario, often referred to as a "3 + 3" study,[37] patients are divided into groups of three and given a single dose of the drug. If no adverse reactions are observed, the dose is double. This doubling of the dose is continued until adverse events occurs, which defines the MTD for a single dose

FIGURE 9.13 A typical phase I trial exposes a small group of patients to increasing amounts of clinical candidate while monitoring for adverse events. The trial is ended when the MTD is determined. The data from phase I trials are used to set dose levels for subsequent trials. DLT, dose limiting toxicities; MTD, maximum tolerated dose.

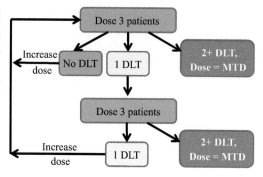

exposure. It may also be necessary to perform a separate study in which the subjects receive multiple doses that increase in each stage of the study. This is often referred to as multiple ascending dose (MAD) study and can provide a significant amount of information on the pharmacokinetics of a candidate compound. In this scenario, patient groups receive multiple doses of the candidate compound at fixed time intervals. The dose is escalated with each successive round of exposures until a predetermined maximum (often set by the MTD of the SAD studies) is reached.

In practice, all of the patients must be carefully monitored through a variety of means. Full physical exams and monitoring of cardiovascular function (e.g., BP, heart rate) and other vital signs are conducted through the course of the study. In addition, multiple types of biological specimens (e.g., blood, urine) are collected. The biological samples are used to determine the *in vivo* ADME properties of the candidate compound, as well as monitor for any safety issues that might not be apparent from physical observations (e.g., hematologic, hepatic, or renal toxicities). Depending on the PK parameters of the candidate compound, patient follow-up could be necessary for days or even weeks after the initial exposure.

Although the majority of phase I clinical trials employ healthy volunteers, there are some notable exceptions. When the proposed therapeutic agent has a known degree of toxicity, such as potential cancer chemotherapies[38] or new antiviral agents for the treatment of HIV,[39] healthy volunteers are not used. In this instance, patients with the disease or condition are used in phase I clinical trials. Also, in many cases the phase I and phase II trials are merged in order to minimize the number of patients that are exposed to a candidate compound of unknown safety and efficacy. PK studies are also minimized for the same reason.

In general, phase I clinical trials, including patient follow-up after exposure, take approximately 1–1.5 years to complete. At a minimum, the studies must demonstrate that the candidate compound is safe for use in humans in order for a clinical program to proceed to phase II trials. PK studies conducted in the phase I trial, combined with the MTD and DLTs that are observed in this round of clinical study, are used to set the doses for phase II clinical trials.

PHASE II CLINICAL TRIALS

Once it has been demonstrated that a candidate compound can be safely studied in human subjects (i.e., successfully completed phase I clinical trials), phase II clinical trials can be initiated. Healthy volunteers are the norm in phase I clinical trials, but this is not the case in phase II clinical trials. Clinical trials in this stage of development are designed to obtain preliminary efficacy data to determine if the candidate compound has the potential to safely improve the lives of patients. All of the subjects in phase II studies are afflicted with the disease or condition of interest. A typical phase II study requires 100–300 patients and can require up to 2 years, depending on the nature of the indication under investigation. Candidate compounds targeting chronic disease or conditions such as arthritis require longer time periods. Irrespective of the length of the trial, patients are carefully chosen using strict inclusion and exclusion criteria in an effort to ensure a high degree of homogeneity in the patient population. This increases the likelihood of identifying a positive result if the drug is effective, while also minimizing variables that might obscure the results (i.e., produce a false negative). The question of whether or not the candidate compound will prove effective in the broader patient population is left to be addressed in phase III trials.

Another major goal of phase II clinical trials is to establish the dose (and dosing scheme) that will be employed in phase III studies if a positive outcome is achieved in phase II. Although phase I studies provide safety and pharmacokinetic data, they rarely provide information on the efficacy of a candidate compound. Fortunately, the PK and safety studies can be used to establish test doses that are high enough to theoretically provide a systemic concentration capable of supporting efficacy that are below the MTD. In practice, patients are typically divided into several cohorts so that multiple doses can be examined in the patient population. In one scenario, four doses might be chosen, one at the low end as a potentially ineffective dose, two intermediate doses, and a fourth, high dose that theoretically would not improve upon the effect of the dose just below it. Ideally, upon completion of these dose ranging trials, an efficacious dose will be selected for phase III clinical trials.

Although the primary goal of phase II clinical trials is the identification of a positive signal in the patient population, safety is still a high priority. In many cases, phase II clinical trials are the first time that patients in need of treatment will be exposed to the candidate compound (notable exceptions include cancer and HIV). It is possible that safety issues can occur in patients that would not have been apparent in the healthy volunteers in phase I. In addition, the cumulative patient exposure to a candidate compound is typically higher in phase II, leading to the possibility of additional side effects that might not occur in the shorter exposure windows of phase I trials. A phase II clinical trial for a chronic indication, for example, could be as long as 6 months. All of the patients must be carefully monitored to identify any potential safety risks that might become apparent as the candidate compound is used for longer and longer time periods.

The ability of a phase II clinical trial to arrive at successful conclusion is dependent not only on the candidate compound, but also the trial design and outcome measures that define success. The choice of trial design and outcome measure is highly dependent on the disease or condition that is being examined. A phase II clinical trial designed to determine whether or not an antibiotic is suitable for further clinical study (e.g., phase III clinical trials) will be very different from a phase II clinical trial designed to determine whether or not a candidate compound might be useful for the treatment of migraine headaches or cancer. The complexity of trial design and possible clinical endpoint is far beyond the scope of this text, but an understanding of some of basic trial designs and outcome measures can provide some perspective on the capabilities and limitations of each.

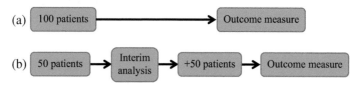

FIGURE 9.14   (a) In a simple single arm phase II study, all patients receive the clinical candidate. (b) In a staged, one arm, phase II study, one group of patients receives the candidate and the results are analyzed. If the interim analysis meets a predetermined goal, the remainder of the patients receive the candidate compound and the trial proceeds. If the predetermined goals are not met, the trial is terminated.

The simplest trial design is a single arm study (Figure 9.14(a)).[40] In this scenario, all of the patients are provided with the new treatment and monitored over time to determine whether or not the treatment is effective. This study design can be cost-effective when compared to more complex designs, as all of the resources are focused on a single patient group. The absence of a control group, however, prevents direct comparison with either a placebo or a standard of care patient group. This is a significant limitation, especially if it is unclear what effect the candidate compound will have on the disease state (e.g., a compound with a novel mechanism of action).

A single arm study can also be designed as a staged study in which a subset of the full group is treated with the candidate compound (Figure 9.14(b)). If the number of patients responding to the new treatment meets a predetermined goal, then the remaining subjects are recruited and the phase II trial continues to its planned end point. On the other hand, if the number of positive responders fails to meet the predetermined objective or a significant safety issue occurs, then the trial is halted. The staged trial design has the advantage of limiting the number of patients exposed to the candidate compound, as a "stopping rule" may end the trial early. This trial design can be advantageous if there is a risk of serious side effects, as it minimizes the number of patients exposed to potential new treatments. Evaluating an expensive candidate compound in a staged clinical trial also decreases the financial risk, as there is less capital outlay required to determine if additional expenses are warranted.

In many case, it is necessary to generate more data than can be provided in a single arm trial. If, for example, the level of response that will be elicited by a candidate compound is not well understood, incorporating a control group in the trial will provide a strong basis for comparison. A two armed, randomized phase II trial with one group receiving the candidate compound and a control receiving either a placebo or the standard of care can provide significant insight (Figure 9.15). The progression and results

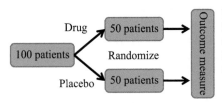

FIGURE 9.15 In a two armed, randomized phase II clinical trial, the patients receive either the drug substance or a placebo. Multiple treatment arms can be included if necessary.

from each arm of the trial can be used to optimize the design of phase III studies. It is also possible to monitor the rate of patient recruitment as the trial progresses. This will provide a sense of how long it might take to recruit a larger population of patients that will be required for phase III clinical trials, as well as any other logistical problems that may occur through the course of the trial. Although the patient population will not be large or diverse enough to definitively determine whether or not a new therapy is suitable for marketing, ideally it will be large enough to extrapolate the number of patient that will be required for the phase III trials. It is also worth noting that this trial design can be extended to include multiple treatment arms in order to determine the optimal dose for phase III trials. In addition, a staged trial design can also be employed in a similar manner to that described above (Figure 9.16).

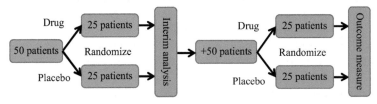

FIGURE 9.16 A two armed, randomized phase II clinical trial can be staged. Results from an interim analysis determine whether or not new patients enter the trial.

Determining whether or not a candidate compound is providing a benefit to the patients is measured using predefined outcome measures that are established when the trial design is approved. In some cases, the outcome may be directly tied to the disease or condition to be treated and evaluating the effect may be a simple matter of counting people. An increase in the number of patients surviving 1 year after a pancreatic cancer diagnosis when the candidate compound is added to the standard of care is a simple example of this type of outcome measure. Monitoring quantifiable attributes in patients (taking measurements) might also be used as an outcome measure. Changes in body weight in patients could be used as an outcome measure for a potential weight loss therapy, while changes in a patient's blood pressure could be used as an outcome measure for potential novel antihypertensive agents. Surrogate markers, a subset of biomarkers, such as cholesterol concentration[41] or viral load[42] can also be used as outcome measures. This type of outcome measure will be discussed in greater detail in Chapter 10.

Although a successful phase II clinical trial provides proof of principle, it is by no means a guarantee of success in phase III trials. There are many examples in the literature of candidate compound that produce a positive result in phase II that cannot be replicated in a larger, more heterogeneous patient population. A 2014 analysis of over 5800 clinical programs from 835 companies indicated that only ~60% of the compounds that produced positive results in phase II trials were also successful in phase III trials.[43] Given the expense associated with phase III clinical trials, a thorough examination of the phase II data should be undertaken to ensure that the data set is robust enough to warrant the expense of the next phase of development.

PHASE III CLINICAL TRIALS

Launching a phase III clinical trial represents a major commitment by the sponsor organization (almost always a pharmaceutical company), as the cost of this portion of the drug discovery and development process represents the largest financial commitment. In some case, as much as 90% of the costs associated with bringing a new drug to market are incurred during phase III clinical trials.[44] As a result, the successful conclusion of a phase II trial is not always enough to warrant initiating phase III studies. There are many reasons that an organization might choose to terminate a clinical program that have nothing to do with clinical trial results. A competitor drug may already be available that is more effective than the candidate compound. This would hamper the sponsor company's ability to gain market share and recoup the costs of developing a competing new therapy. The availability of a generic drug may also lead to a decision to abandon a clinical program rather than launch a phase III trial. In the absence of superiority over the cheaper drug (e.g., better efficacy, improved safety profile), market penetration for a new drug may be difficult to achieve. The expense of producing the candidate compound or formulation requirements may also be too high to warrant the cost of advancement. It is also possible that the length of time required for phase III clinical trials is too long relative to the remaining patent life for the candidate compound. Generic competition might be able to enter the market before the originating company would be able to recoup the cost of developing the candidate compound. There are many non-scientific reason why a candidate compound might be abandoned before a phase III trial is initiated. The decision to move forward or stop should be weighed carefully, given the enormous financial risk of moving forward.

If the decision is made to move forward, the nature of the phase III trial is heavily dependent on the purpose of the candidate compound. There are, however, some overarching goals that must be met in order for a phase III program to reach a successful conclusion (e.g., provide support for marketing the candidate compound). First and foremost, the phase III clinical trials, often referred to as pivotal studies, must confirm the efficacy and safety of the candidate compound in the broad patient population. This process generally requires 1000–3000 patients, possibly more, multiple clinical sites, an institutional review board (IRB) and they can last from 2.5 years to 5 years, depending on the nature of the disease and the patient population. In order to be considered a success, a phase III clinical program must produce at least two "adequate and well-controlled studies" that demonstrate both the safety and efficacy of the candidate compound. In other words, the studies must be robust enough that the end results can be extrapolated to the general population and the resulting information can be provided to the health care providers (e.g., physicians, nurses, pharmacists, etc.) in the form of a package insert.

In theory, a sponsor company could develop the data necessary to support a request for market approval (licensure) by conducting only two phase III clinical trials. In practice, however, many companies choose to run a third study. While this will add significant cost to the overall program, it is often viewed as a worthwhile investment. If two trials are conducted and it is determined that one of the two was not sufficiently supportive of the compound candidate's efficacy or safety, then a third trial would be required in order to develop the data necessary for marketing approval. There could also be procedural problems such as patient monitoring inconsistencies or data collection issues that result in a regulatory agency excluding a clinical trial from a New Drug Application (NDA) package. If only two studies are run, the sponsor organization would not have enough data to support licensure. A third trial would be necessary and would create a substantial delay in obtaining marketing authorization. This would reduce the time available to recoup the significant investment required to reach the market. In a sense, a third trial is an insurance policy against the unexpected.

Phase III clinical trials can be divided into some general categories based on their overall goal relative to the current standard of care. The goal of a superiority trial is, as the name implies, to demonstrate that the candidate compound is more effective in dealing with the targeted disease or condition. In this case, a competitor treatment may be run in parallel to the candidate compound so that a direct comparison may be made. Historical data may be available, but it is rarely suitable for inclusion in a licensure request (NDA). If there is no current treatment, a placebo control is used as a comparator group. In theory, a placebo control group could be used even when a treatment is available, but the ethics of not using at least the standard of care in the control group is questionable.

In some cases, phase III clinical trials are designed to demonstrate that a candidate compound is similar in efficacy to the current standard of care (an equivalency trial) or that it is not less effective that the current standard of care (a non-inferiority trial; the candidate compound could also be better). In these instances, a placebo control is generally not sufficient as a direct comparison with the standard of care is the desired end result. These types of studies are most often employed when the candidate compound's advantages over the standard of care are tied to something other than overall efficacy. A candidate compound that is expected to be safer, more convenient for the patient (e.g., once a week dosing versus daily dosing), or more cost-effective could be studied in an equivalency or non-inferiority trial.

The choice of outcome measures in a phase III clinical trial is a critical aspect of the trial design. They must be chosen very carefully and be well defined so that it is clear whether or not the trial has met its objective. In addition, the results must be capable of convincing regulatory officials and health professional that a change in the standard of care is warranted and that the candidate compound is suitable for market approval. Although the primary endpoints will vary based on the targeted condition or disease, they must be clinically relevant to both the researcher and the patient. Examples of primary endpoints include increased survival (e.g., cancer survival), shortened duration of an event (e.g., shorter recovery time from an infection), or changes in patient habits (e.g., smoking cessation). In some cases, it may be necessary to track more than one endpoint in order to demonstrate utility of a candidate compound, which necessitates a composite endpoint. Consider, for example, a trial designed to evaluate a potential new treatment for cardiovascular disease. In this case, there are multiple, relevant clinical events (e.g., myocardial infarction, stroke, acute coronary syndrome) that could be positively impacted by the new treatment. Individually, positive changes in a single outcome measure might not be statistically significant, but collectively they could demonstrate a benefit to patients. Importantly, only the first clinically relevant event is included in the results of a clinical trial, as treatment or disease management will generally shift to more adequately support the patient post-event. It would be difficult to distinguish between shifts in patient outcome that are based on the changes in patient management versus the candidate compound.

In practice, the choice of endpoints is, of course, dependent on the nature of the disease or condition to be treated. Objective endpoints such as radiological measures (X-rays, CT scans), physiological measurements (e.g., blood pressure, heart rate, lung capacity), and blood measurements (e.g., cholesterol concentration, white blood cell count) are especially useful when it is not possible to run a blinded study (i.e., the subjects and possibly the researchers do not know who is receiving the therapy). Subjective endpoints, such as pain sensation, mood modification, or other quality of life assessments can also serve as clinical trial endpoints, but they are less effective in non-blinded studies, as the subjects themselves may impact the results. A clinical trial for a new treatment for chronic pain, for example, could be significantly impacted if the patients knew that they were receiving the test compound and not a placebo. They might believe that they feel less pain as a result of the new therapy, irrespective of the actual utility of the candidate compound. In general, subjective measurements tend to add variability to study results that can mask the true results of a clinical trial. As a result, objective endpoints are preferred when possible.

In considering the overall design of a phase III clinical trial, it is important to establish whether or not all of the subjects of the trial will receive the test therapy. The simplest phase III clinical trial design is a parallel group trial in which each patient group receives only one treatment regimen (Figure 9.17). Patient management is simplified in this trial

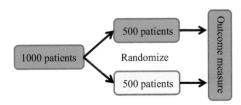

FIGURE 9.17 In a randomized, parallel group phase III clinical trial, patients receive only one treatment regimen. Multiple treatment groups can be established to incorporate placebo controls, standard of care controls, and multiple dose levels.

design, and it is possible run multiple treatment groups in parallel to the standard of care (or placebo). A crossover trial design can be a useful alternative (Figure 9.18). In this instance, the subjects of the trial are split

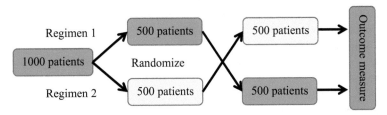

FIGURE 9.18 In a randomized, crossover phase III trial, the patients are divided into groups and assigned treatment regimens. After a set period of time, dosing regimens are switched, and the trial continues.

into two groups based on the order of treatments. In the first time period of the clinical trial, patients are randomly provided either the candidate compound or the standard of care. After a set time interval, the treatment groups are reversed, and the study is continued to its conclusion. As the trial proceeds, the patients are monitored to determine the impact of the therapeutic regimens. In this instance, since all of the patients receive both the new treatment and the standard of care, they are their own control group. As a result, fewer subjects are required for crossover trials. This trial design is an effective means of determining the utility of potential new therapies for chronic conditions in which symptomatic relief is the desired outcome (e.g., chronic pain, allergy suppression, etc.). Crossover trials are also useful in demonstrating that two therapeutic agents are bioequivalent.

There are some important limitations that should be considered before embarking on a crossover trial. First, the change of therapy should take place with minimal carryover effect of the first treatment regimen. Significant residual effects that carry over into the second phase of a crossover trial can make the results difficult to interpret. The amount of time required for the treatment regimens to "wash out" depends on the pharmacokinetic properties of the candidate compound and should be considered when determining the time lag between regimens at the crossover time point. It is also important to understand whether or not the disease or condition reverts back to its baseline level during the washout period, and to determine whether or not the treatment order has an impact on the results of the clinical trial. (Is patient improvement significantly different when treatment A is followed by treatment B versus when treatment B is followed by treatment A?) If these items are not easily addressed and accounted for, a standard, parallel trial may provide more meaningful results.

One of the fundamental drivers in the value and cost of a phase III clinical trial is the number of patients in each arm of the trial. In order for a phase III clinical trial to be supportive of a licensing effort for a candidate compound, the number of participants must be large enough to provide a statistically significant result. This is typically set at 5%. In other words, the chance of identifying an effect when one is not actually present is 5%. Viewed in a more positive light, there is a 95% chance that results observed in the course of the trial are real and not a result of random chance. The trial must also include enough patients such that its statistical power is high enough. Typically, the number of patients needs to be high enough that there is an 80–90% likelihood of detecting a difference between treatment groups if such a difference exists. The number of patients required to meet these criteria also depends on the size of the expected effect. As one would expect, larger effects are more easily detected and generally require smaller patient populations. None of these calculations are easily accomplished, as they require complex statistical analyses that are best accomplished in consultation with a statistician and one of several commercial software packages designed to estimate sample size requirements.[45]

Irrespective of the study size necessary to meet efficacy requirements, phase III clinical trials must also be large enough to provide an adequate safety package to support regulatory approval. Given the large number of disease states and conditions that require short term administration (e.g., cumulative exposure of less than 6 months), it is difficult to provide general guidance on the population size required to provide an adequate safety assessment. For longer term treatments that are designed to treat non-life-threatening conditions (e.g., cumulative exposure of greater than 6 months), however, there are some generally accepted guidelines. The number of patients exposed to the candidate compound at therapeutically relevant doses should be greater than 1500. Of this set, 300–600 should be studied for a minimum of 6 months, and at least 100 patients should reach the 1-year mark. If, however, the candidate compound is intended to treat a life-threatening disease, a debilitating condition, or a disease with a small patient population, a smaller number of subjects may be sufficient (Regulatory approval for this deviation from the norm would be required.). On the other hand, if animal studies, similar compounds, or other data indicate that a safety issue may exists, regulatory agencies may require a larger patient population in the safety database. Similarly, if the expected benefit is small (e.g., symptomatic improvement in the mild medical condition) or a safe alternative is already available, additional subjects may be required in order to establish an appropriate safety package for a new drug application.[46]

PHASE IV CLINICAL TRIALS

Once a candidate compound has successfully completed two adequate and well-controlled studies, a new drug application can be prepared and submitted to the appropriate regulatory body. If the regulators are satisfied with the information provided, then marketing approval will be granted. In many cases, however, there may be additional question about the candidate compound that, while not critical to initial approval, are considered important enough to warrant additional examination in a clinical setting. As a result, postmarketing surveillance studies, also referred to as phase IV clinical trials, are commonly required as a condition of marketing approval.

There are a number of possible objective of a phase IV clinical trial. In some cases, it may be necessary to determine the safety and efficacy of a compound relative to a competitor compound. These studies may be required by regulatory bodies, or they may be run proactively by the sponsor of the newer compound in an attempt to gain market share. Of course, if the new compound is shown to be less efficacious or not as safe as the established drug, the phase IV study results will likely lead to decreased market share, or possibly revocation of marketing rights.

Pharmacoeconomic studies designed to determine the value of the new therapy as compared to the standard of care have also entered the phase IV landscape over the course of the last few decades. In some countries (Canada, Finland, New Zealand, Norway, Sweden, Australia, and the U.K.), an analysis of this type is part of the approval process and can have an impact on whether or not a new drug will be subsidized by government health care programs. It is worth noting that this does not directly impact whether or not a candidate compound will be granted marketing approval. It only impacts who will pay for the new drug once it reaches the market.

Other objective of postmarketing surveillance studies include the identification of low-frequency adverse events, continued safety monitoring to better characterize known risks, gauging the potential for drug–drug interactions, establishing treatment guidelines for pediatric and geriatric populations, and of course, determining the "real-world" efficacy of the candidate compound. The patient population in a phase IV clinical trial is generally much larger and more heterogeneous than would be possible in a phase III study (Keep in mind that phase III studies are often double blinded, randomized trial with strict inclusion and exclusion criteria). Also, the strict control of dosing regimens provided by the research team is absent in most phase IV trials. These factors increase the variability in the drug's application (e.g., patients applying the medication late, incorrect dosing prescribed, etc.). If a drug fails to fulfill its anticipated goals in the broader population provided in a phase IV study, it might be removed from the market, especially if a more effective therapy is already available.

In some cases, there may be an opportunity to expand the use of an approved therapy. This may be driven by the need to develop new methods of administration or different dosing levels designed to treat a group of patients not served by the originally approved formulation of the candidate compound. It is also possible that diseases or conditions outside of the scope of the original clinical trial (new indications) might be positively impacted by the candidate compound. If a company is interested in increasing the utility of its candidate compound through any of these avenues, it will be necessary to run additional clinical trials in order to gain official marketing approval for the new indication, formulation, or dose level. Expanding the approved diseases and conditions that a candidate compound can be used to treat clinically is a common method of increasing the ongoing value of a drug (profitability).

ADAPTIVE CLINICAL TRIAL DESIGN[47]

As previously mentioned, clinical trials, especially phase III clinical trials, are the largest contributor to the overall cost of drug discovery and development. It should come as no surprise that there has been a concerted effort over the last few decades to identify more efficient methods of conducting clinical trials. The concept of adaptive clinical trial designs has been developed in response to the need to decrease the overall cost of new drug development. In an adaptive clinical trial, interim analysis time points are defined within the course of the overall study. When these time points are achieved, analysis of the data collected up until this point is used to determine how the remaining portion of the clinical trial will be conducted. In other words, the adaptive clinical trials are prospectively planned to include opportunities for modification of certain aspect of the trial (e.g., sample size, treatments, population, choice of outcomes, etc.) based on data collected during the trial itself.

The ability to monitor and change aspects of an ongoing clinical trial can have a significant impact on the time, costs, and risk to patients when compared to conventional clinical trials. In a conventional clinical trial, the methods and protocols are set prior to the start of the trial and are typically not altered as the trial proceeds. Although every effort is made to design efficient clinical trials, it is important to keep in mind that study designs represent the best estimate of various parameters (e.g., number of patients required, dosing schedule, variability in response rate, etc.) that may not be fully characterized at the beginning of the trial. If the estimates turn out to be inaccurate for any reason, the study could produce false negative or false positive results, both of which would be problematic. By incorporating an interim analysis step, adaptive trial designs allow the study to be modified based on data developed through the course of the study. If, for example, an interim analysis of the data determines that one of the study arms is having no impact on the patients (i.e., the dose is too low to produce an effect), this arm of the study could be terminated. Similarly, if it were determined that the study was highly successful, futile, or causing harm to the subjects, an interim analysis would provide an opportunity to end the study earlier than anticipated based on the original study design. Event rates can also be monitored to determine if the number of subjects is appropriate. If the event rate is higher than anticipated in the general population, it may be possible to recruit a smaller number of patients. On the other hand, if the event rate is lower, it may be necessary to recruit a larger than anticipated number of patients. These and other types of changes can significantly increase the efficiency of a clinical trial, decreasing the risk to patients and the overall cost of clinical development, provided the regulatory authorities are in agreement with the adaptive trial design.

QUESTIONS

1. Provide a definition of a clinical trial.
2. What are the general goals of phase I, II, and III clinical trials?
3. What is polymorphism?
4. Why is it often necessary to develop new methods of synthesis upon converting a lab scale preparation of a candidate compound to a commercial scale synthesis?
5. What are four types of delivery methods that can be used to administer a candidate compound?
6. What is the purpose of an enteric coating on a tablet?
7. What is an excipient?
8. What is the purpose of the "push chamber" in a two chambered osmotic pump delivery system?
9. What are the three major sections of an investigational new drug (IND) application?
10. What is the purpose of an institutional review board (IRB)?
11. What is allometric scaling?
12. What is a "3 + 3" phase I study?
13. What is the purpose of an interim analysis in a clinical trial?
14. What is the purpose of a stopping rule in a clinical trial?
15. Why would an organization choose to run a phase III clinical trial as an equivalency trial rather than attempt to show superiority over the standard of care?
16. What are the advantages of a cross-over study?
17. How are adaptive clinical trials different from standard clinical trials?

References

1. http://www.fda.gov/regulatoryinformation/legislation/federalfooddrugandcosmetic actFDCAct/default.htm.
2. http://www.fda.gov/ForConsumers/ConsumerUpdates/ucm322856.htm.
3. a. Kelly, K. *The History of Medicine: Early Civilizations, Prehistoric Times to 500 C.E*; Facts on File, Inc.: New York, 2009.
   b. Borchardt, J. K. The Beginning of Drug Therapy: Ancient Mesopotamian Medicine. *Drug News Perspect.* **2002**, *15* (3), 187–192. "History of ancient Medicine in Mesopotamia & Iran." Massoume Price, Iran Chamber Society, October 2001. http://www.irancha mber.com/history/articles/ancient_medicine_mesopotamia_iran.php. http://www. indiana.edu/~ancmed/meso.HTM.
4. Cyril P. Bryan, (translator). *The Papyrus Ebers; .* D. Appleton and Co, 1931.
5. Read, B. E. *Chinese Medicinal Plants from the Pen T'Sao Kang Mu*, 3rd ed.; Peking National History Bulletin, 1936.
6. Rush, B. *An Account of the Bilious Remitting Yellow Fever as it Appeared in the City of Philadelphia in 1793;* Dobson: Philadelphia, 1794.
7. Green stone, G. The History of Bloodletting. *BCMJ* **2010**, *52* (1), 12–14.
8. Dunn, P. M. Perinatal Lessons from the Past: James Lind (1716-94) of Edinburgh and the Treatment of Scurvy. *Arch. Dis. Child. – Fetal Neonatal Ed.* **1997**, *76* (1), F64–F65.

9. a. Morabia, A. Pierre-Charles-Alexandre Louis and the Evaluation of Bloodletting. *J. R. Soc. Med.* **2006,** *99* (3), 158–160.

   b. Louis, P. C. A. *Recherches sur les Effets de la Saignée;* De Mignaret: Paris, 1835.

10. Greenwood, M.; Yule, G. U. The Statistics of Anti-typhoid and Anti-cholera Inoculations and the Interpretation of Such Statistics in General. *Proc. R. Soc. Med., Sect. Epidemiol. State Med.* **1915,** *8,* 113–194.

11. Ferguson, F. R.; Davey, A. F. C.; Topley, W. W. C. The Value of Mixed Vaccines in the Prevention of the Common Cold. *J. Hyg.* **1927,** *26,* 98–109.

12. Marshall, G.; Blacklock, J. W. S.; Cameron, C.; Capon, N. B.; Cruickshank, R.; Gaddum, J. H.; Heaf, F. R. G.; Hill, A. B.; Houghton, L. E.; Hoyle, J. C.; Raistrick, H.; Scadding, J. G.; Tytler, W. H.; Wilson, G. S.; Hart, P. D. Streptomycin Treatment of Pulmonary Tuberculosis: A Medical Research Council Investigation. *Br. Med. J.* **1948,** *2* (4582), 769–782.

13. Smith, L. J. Types of Clinical Trials 107–122. In *From Drug and Biological Development from Molecule to Product and beyond;* Evens, R. P., Ed.; Springer Science: New York, 2007.

14. Carlton, R. A. *Pharmaceutical Microscopy,* 213–246. Springer Science: New York, 2011.

15. a. Bauer, J.; Spanton, S.; Henry, R.; Quick, J.; Dziki, W.; Porter, W.; Morris, J. Ritonavir: An Extraordinary Example of Conformational Polymorphism. *Pharm. Res.* **2001,** *18* (6), 859–866.

   b. Morisette, S. L.; Soukasene, S.; Levinson, D.; Cima, M. J.; Almarsson, O. Elucidation of Crystal Form Diversity of the HIV Protease Inhibitor Ritonavir by High-throughput Crystallization. *Proc. Natl. Acad. Sci. U.S.A.* **2003,** *100* (5), 2180–2184.

16. Devillers, G. Exploring a Pharmaceutical Market Niche & Trends: Nasal Spray Drug Delivery. *Drug Delivery Technol.* **2003,** *3* (3), 1–4.

17. Tacconelli, E.; De Angelis, G.; Cataldo, M. A.; Pozzi, E.; Cauda, R. Does Antibiotic Exposure Increase the Risk of Methicillin-resistant *Staphylococcus aureus* (MRSA) Isolation? A Systematic Review and Meta-analysis. *J. Antimicrob. Chemother.* **2008,** *61* (1), 26–38.

18. Fleming, A. Penicillin. *Nobel Lecture* **December 11, 1945**. http://www.nobelprize.org/nobel_prizes/medicine/laureates/1945/fleming-lecture.pdf.

19. a. Tiwari, G.; Tiwari, R.; Sriwastawa, B.; Bhati, L.; Pandey, S.; Pandey, P.; Bannerjee, S. K. Drug Delivery Systems: An Updated Review. *Int. J. Pharm. Invest.* **2012,** *2* (1), 2–11.

   b. Hillery, A. M.; Lloyd, A. W.; Swarbrick, J. *Drug Delivery and Targeting: For Pharmacists and Pharmaceutical Scientists;* Taylor and Francis: New York, 2001.

20. Cady, R. K.; Wendt, J. K.; Kirchner, J. R.; Sargent, J. D.; Rothrock, J. F.; Skaggs, H., Jr. Treatment of Acute Migraine with Subcutaneous Sumatriptan. *JAMA* **1991,** *265* (21), 2831–2835.

21. Doggrell, S. A. Zoledronate Once-yearly Increases Bone Mineral Density - Implications for Osteoporosis. *Expert Opin. Pharmacother.* **2002,** *3* (7), 1007–1009.

22. Black, D. M.; Cummings, S. R.; Karpf, D. B.; Cauley, J. A.; Thompson, D. E.; Nevitt, M. C.; Bauer, D. C.; Genant, H. K.; Haskell, W. L.; Marcus, R.; Ott, S. M.; Torner, J. C.; Quandt, S. A.; Reiss, T. F.; Ensrud, K. E. Randomised Trial of Effect of Alendronate on Risk of Fracture in Women with Existing Vertebral Fractures. Fracture Intervention Trial Research Group. *Lancet* **1996,** *348* (9041), 1535–1541.

23. Recker, R. R.; Barger-Lux, J. Risedronate for Prevention and Treatment of Osteoporosis in Postmenopausal Women. *Expert Opin. Pharmacother.* **2005,** *6* (3), 465–477.

24. Bauss, F.; Schimmer, R. C. Ibandronate: the First Once-monthly Oral Bisphosphonate for Treatment of Postmenopausal Osteoporosis. *Ther. Clin. Risk Manage.* **2006,** *2* (1), 3–18.

25. a. Weintraub, A. Pfizer's Exubera Flop. *Businessweek* **October 18, 2007**.

   b. Heinemann, L. The Failure of Exubera: Are We Beating a Dead Horse? *J. Diabetes Sci. Technol.* **2008,** *2* (3), 518–529.

26. Ameer, B.; Greenblatt, D. J. Lorazepam: a Review of its Clinical Pharmacological Properties and Therapeutic Uses. *Drugs* **1981,** *21* (3), 162–200.

27. Aulton, M. E., Taylor, K. M. G., Eds. *Aulton's Pharmaceutics: The Design and Manufacture of Medicines,* 4th ed.; Churchill Livingstone: Edinburgh, U.K, 2013.

28. Hussan, S. D.; Santanu, R.; Verma, P.; Bhandari, V. A Review on Recent Advances of Enteric Coating. *IOSR J. Pharm.* **2012,** *2* (6), 5–11.

29. a. Senel, S.; Hincal, A. A. Drug Permeation Enhancement via Buccal Route: Possibilities and Limitations. *J. Control. Release* **2001,** *72,* 1–3. 133–144.

    b. van Hoogdalem, E. J.; de Boer, A. G.; Breimer, D. D. Intestinal Drug Absorption Enhancement: An Overview. *Pharmacol. Ther.* **1989,** *44* (3), 407–443.

30. Merisko-Liversidge, E.; Liversidge, G. G.; Cooper, E. R. Nanosizing: a Formulation Approach for Poorly-water-soluble Compounds. *Eur. J. Pharm. Sci.* **2003,** *18,* 113–120.

31. Liversidge, G. G.; Cundy, K. C. Particle Size Reduction for Improve Ment of Oral Bioavailability of Hydrophobic Drugs: I. Absolute Oral Bioavailability of Nanocrystalline Danazol in Beagle Dogs. *Int. J. Pharm.* **1995,** *125* (1), 91–97.

32. a. Chen, H.; Khemtong, C.; Yang, X.; Chang, X.; Gao, J. Nanonization Strategies for Poorly Water-soluble Drugs. *Drug Discovery Today* **2011,** *16,* 7–8. 354–360.

    b. Sushant, S.; Archana, K. Methods of Size Reduction and Factors Affecting Size Reduction in Pharmaceutics. *Int. Res. J. Pharm.* **2013,** *4* (8), 57–64.

33. a. Shende, P.; Shrawne, C.; Gaud, R. S. Multi-layer Tablet: Current Scenario and Recent Advances. *Int. J. Drug Deliv.* **2012,** *4,* 418–426.

    b. Yadav, G.; Bansal, M.; Thakur, N.; Khare, S.; Khare, P. Multilayer Tablets and Their Drug Release Kinetic Models for Oral Controlled Drug Delivery System. *Middle-East J. Sci. Res.* **2013,** *16* (6), 782–795.

34. a. Herrlich, S.; Spieth, S.; Messner, S.; Zengerle, R. Osmotic Micropumps for Drug Delivery. *Adv. Drug Deliv. Rev.* **2012,** *64* (14), 1617–1627.

    b. Ghosh, T.; Ghosh, A. Drug Delivery through Osmotic Systems – an Overview. *J. Appl. Pharma. Sci.* **2011,** *1* (2), 38–49.

35. Code of Federal Regulations Title 21, Chapter 1, Subchapter D, Part 312, Subpart B, Section, 312.23.

36. Caldwell, G. W.; Masucci, J. A.; Yan, Z.; Hageman, W. Allometric Scaling of Pharmacokinetic Parameters in Drug Discovery: Can Human CL, Vss and T1/2 Be Predicted from *In-vivo* Rat Data? *Eur. J. Drug Metab. Pharmacokinet.* **2004,** *29* (2), 133–143.

37. Ivy, S. P.; Siu, L. L.; Garrett-Mayer, E.; Rubinstein, L. Approaches to Phase 1 Clinical Trial Design Focused on Safety, Efficiency, and Selected Patient Populations: A Report from the Clinical Trial Design Task Force of the National Cancer Institute Investigational Drug Steering Committee. *Clin. Cancer Res.* **2010,** *16,* 1726–1736.

38. Eisenhauer, E. A.; O'Dwyer, P. J.; Christian, M.; Humphrey, J. S. Phase I Clinical Trial Design in Cancer Drug Development. *J. Clin. Oncol.* **2000,** *18* (3), 684–692.

39. Chan-Tack, K. M.; Struble, K. A.; Morgensztejn, N.; Murray, J. S.; Gulick, R.; Cheng, B.; Weller, I.; Miller, V. HIV Clinical Trial Design for Antiretroviral Development: Moving Forward. *AIDS* **2008,** *22* (18), 2419–2427.

40. Ip, S.; Paulus, J. K.; Balk, E. M.; Dahabreh, I. J.; Avendano, E. E.; Lau, J. Role of Single Group Studies in Agency for Healthcare Research and Quality Comparative Effectiveness Reviews. Publication No. 13-EHC007-EF *Agency for Healthcare Research and Quality Publication* **January 2013.**

41. Vasan, R. S. Biomarkers of Cardiovascular Disease: Molecular Basis and Practical Considerations. *Circulation* **2006,** *113,* 2335–2362.

42. Strimbu, K.; Tavel, J. A. What Are Biomarkers? *Curr. Opin. HIV AIDS* **2010,** *5* (6), 463–466.

43. Hay, M.; Thomas, D. W.; Craighead, J. L.; Economides, C.; Rosenthal, J. Clinical Development Success Rates for Investigational Drugs. *Nat. Biotechnol.* **2014,** *32,* 40–51.

44. Roy, A. S. A. Stifling New Cures: The True Cost of Lengthy Clinical Drug Trials. In *Project FDA Report 5, Manhattan Institute for Policy Research;* 2012; pp 1–8.

45. a. Machin, D.; Campbell, M.; Fayers, P.; Pinol, A. *Sample Size Tables for Clinical Studies*, 3rd ed.; Wiley-Blackwell, 2009.

b. Noordzij, M.; Tripepi, G.; Dekker, F. W.; Zoccali, C.; Tanck, M. W.; Jager, K. J. Sample Size Calculations: Basic Principles and Common Pitfalls. *Nephrol., Dial., Transplant.* **2010,** *25* (5), 1388–1393.

46. *Guidance for Industry Premarketing Risk Assessment.* U.S. Department of Health and Human Services Food and Drug Administration Center for Drug Evaluation and Research (CDER), March 2005. http://www.fda.gov/downloads/regulatoryinformation/ucm126958.pdf.

47. Guidance for Industry: Adaptive Design Clinical Trials for Drugs and Biologics. www.fda.gov/downloads/drugs/guidancecomplianceregulatoryinformation/guidances/ucm201790.pdf.

10

Translational Medicine and Biomarkers

As the twentieth century came to a close and the twenty-first century was just beginning to unfold, drug discovery scientists found themselves equipped with tools and technologies capable of performing amazing feats of science. The entire human genome (and genomes of numerous other species) had been elucidated, and new fields such as genomics (the analysis of the structure and function of genes),[1] proteomics (the study of protein structure and function),[2] and metabolomics (the study of the metabolic "fingerprints" of cellular processes)[3] began to emerge and mature. The "omics" fields delivered a wealth of new information on the molecular basis of disease pathology and opened the door to a plethora of potential new therapeutic targets. At the same time, drug discovery scientists were developing the tools and methods necessary to assess the potential utility of candidate compounds against these new targets. There was an expectation that these advances would translate into new therapies for previously untreatable diseases such as Alzheimer's disease, Parkinson's disease, and ulcerative colitis, while also increasing the pace of identification of new therapies in general.

Despite the promise of increased efficiency in drug discovery and development based on newly developed biotechnological tools, the rate of drug approval and development costs has not improved. In fact, quite the opposite has occurred. The number of new drugs approved reached a high point of 51 in 1996, but dropped to 27 in 2000, and only 19 new drugs were approved in 2007.[4] At the same time, the cost of drug discovery and development has risen dramatically over the course of the last 40 years. In 1976, the average cost of developing a new therapy was approximately $137 million. By 1992, the cost had more than doubled to an estimated $318 million. The estimated costs continued to increase, reaching just over $800 million in 2000 and approximately $1.7 billion by 2011.[4a,5]

At the same time, a number of high-profile market withdrawals and late-stage clinical failures provided further indication that developing new therapies was becoming less efficient rather than more efficient. Hopes for a treatment for Alzheimer's disease, for example, were dashed when the γ-secretase inhibitor Semagacestat produced significant *declines* in cognitive function even though this enzyme has been definitively linked to β-amyloid plaque formation.[6] In a similar manner, Torcetrapib, a cholesteryl ester transfer protein (CETP) inhibitor was predicted to be "one of the most important compounds of our generation" by Jeff Kindler, Pfizer's chief executive on November 30, 2006.[7] Less than 1 month later, in December 2006, Pfizer announced that Torcetrapib increased mortality rates by nearly 60%, and that the $800 million spent on developing the compound would not be recouped.[7,8] Even compounds that successfully reached the market by modulating known drug targets have had major issues. Baycol® (Cerivastatin), for example, was brought to market to treat high cholesterol and cardiovascular disease by inhibiting HMG-CoA reductase, the same target as the widely successful drug Lipitor® (Atorvastatin). In 2001, however, Bayer AG withdrew it from market, as it caused an unacceptably high level of fatal rhabdomyolysis.[9] Similarly, Vioxx® (Rofecoxib), a selective cyclooxygenase-2 (COX-2) inhibitor for the treatment of chronic pain, was approved in 1999, had annual sale of $2.5 billion by 2003, but was withdrawn from the market by Merck in 2004. Phase IV clinical trial results definitively demonstrated that Vioxx® increased the risk of ischemic events.[10] Clearly, the expectations of rapid advances in therapeutic interventions that were envisioned to come with the opening of the age of "omics" were not being fulfilled.

To be fair, the various "omics" fields associated with drug discovery and development were not the only relatively new avenues of research that were not living up their expectations. High throughput screening, advanced molecular modeling, and high throughput chemistry (combinatorial chemistry) were all independently successful fields of research, but their arrival on the scene did not produce the expected increases in efficiency. As these events were unfolding in the scientific community, increasing pressure was mounting to rein in the growing costs of health care, especially with regard to the cost of marketed pharmaceuticals. It is against this backdrop that the field of translational medicine began to come into focus.

As the name implies, translational medicine incorporates a wide range of scientific disciplines and in a general sense refers to concerted efforts to apply new discoveries in basic science to the identification of new therapeutic modalities for patients. In other words, a "bench-to-bedside" approach to drug discovery in which there is a greater focus on linking discoveries in basic science labs to clinical data obtained in patients, improving clinical practices through the rapid adoption of best practices in the treatment of patients, and reapplying human clinical data to basic science research in order to foster additional advances. The National Institutes of Health divides the science of translational medicine into two distinct categories. "One is the process of applying discoveries generated during research in the laboratory, and in preclinical studies, to the development of trials and studies in humans. The second area of translation concerns research aimed at enhancing the adoption of best practices in the community. Cost-effectiveness of prevention and treatment strategies is also an important part of translational science."[11] These concepts have been divided into four subcategories designated T1 through T4.

In the T1 phase of translational medicine, scientific research is focused on understanding disease pathology on a molecular level and establishing links between molecular processes, animal models, and the human condition. The T2 phase is described as the study of the clinical impact of therapies with the end goal of providing the data necessary to support a change in clinical practice. These themes should sound familiar, as they are the end goals of discovery research through phase III clinical trials. As such, it should come as no surprise that translational medicine research incorporates a wide range of fields, from medicinal chemistry and molecular biology to *in vivo* pharmacology and clinical sciences. The majority of this chapter will focus on the T1 and T2 subcategories.

Research designated as "T3" is concerned with the real world impact of a new therapy after it has been approved for marketing. Does the drug live up to the results obtained in phase III clinical trials? Are there safety issues that appear as the drug is moved into a broader population? Are there important results that become apparent in clinical practice that would be relevant to the development of improved therapies and basic scientific research? All of this information is gathered in order to gain a better understanding of the clinical utility of a newly approved drug, to positively impact the next generation of therapies, and to support efforts in T4 research. This final area of translational medicine is primarily concerned with the incorporation of the findings of the previous phases of translational medicine into policies and procedures that positively change clinical practices. Creating a novel therapy capable of enhancing the lives of patients is, after all, of little consequence if clinicians and patients are unaware of them or if policies prevent their use.[12]

Of course, all of these themes predate the conception of translational medicine. The key piece that translational medicine provides is the overarching theme of creating definable links between the various aspects of the drug discovery and development process that can be objectively measured and correlated from one stage to the next. In principle, this would improve the overall efficiency of the drug discovery and development by adding a higher degree of predictability to the process. Consider, for example, a program directed towards the development of a treatment for HIV infection. A candidate compound that inhibit a key enzyme in the viral replication cycle, such as reverse transcriptase, could be assessed for clinical utility by determining its ability to prolong survival in HIV positive patients. In this scenario, it would be necessary to dose patients for an extended period of time in order to determine whether or not the candidate compound possessed the desired clinical utility. At the height of the AIDS crisis (mid-1980s to early 1990s), time was of the essence, as treatment options were severely limited. Only one drug, Retrovir® (AZT, Zidovudine, Figure 10.1(a)),[13] had reached the market by 1987. An alternative approach based on the clinical observations was developed in order to

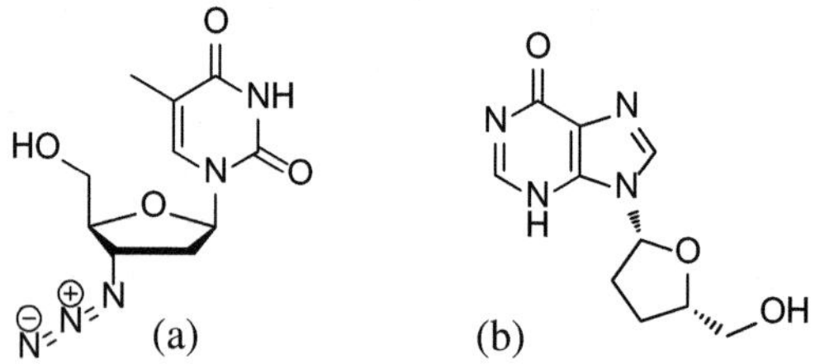

FIGURE 10.1   (a) Retrovir® (AZT, zidovudine) (b) Videx® (Didanosine).

increase the pace of clinical development. It had been observed that clinical progression of HIV infection was associated with a significant decline in CD4+ T-lymphocytes. Based on these observations, it was theorized that treatment with an effective anti-HIV therapy would lead to the restoration of normal CD4+ T-lymphocytes levels. Monitoring for positive changes in CD4+ T-lymphocytes levels could therefore be used in a clinical trial to demonstrate the efficacy of potential new therapies. In other words, CD4+ T-lymphocytes levels were used as a *biomarker* for disease progression (or lack thereof) in HIV infected patients. The reverse transcriptase inhibitor Videx® (Didanosine, Figure 10.1(b)) was successfully brought to market by Bristol Myers Squibb in 1991 as the second HIV therapy based in part on CD4+ T-lymphocytes level changes in patients.[14]

DEFINITION OF A BIOMARKER AND THEIR CLASSIFICATION

The term biomarker has been defined as "a characteristic that is measured and evaluated objectively as an indicator of normal biological processes, pathogenic processes, or pharmacological response to a therapeutic intervention."[15] This definition was provided by the NIH Biomarkers Definition Working Group in 2001 in an attempt to provide a framework for discussion on what constitutes a biomarker. An alternative definition that takes a more pragmatic view of the drug discovery and development process was provided by Lathia in 2002. According to this definition, a "biomarker is a measurable property that reflects the mechanism of action of the molecule based on its pharmacology, pathophysiology of the disease, or an interaction between the two. A biomarker may or may not correlate perfectly with clinical efficacy/toxicity, but could be used for internal decision-making within a pharmaceutical company."[16] Both of these definitions encompass a wide range of methods and techniques that span the length of the drug discovery and development process. Fortunately, it is possible to categorize biomarkers based on their purpose and the type of information that they provide. These categories include target engagement biomarkers, mechanism biomarkers, outcome biomarkers, toxicity biomarkers, pharmacogenomics biomarkers, and diagnostic biomarkers.

As the name implies, target engagement biomarkers provide insight into whether or not a candidate compound is interacting with the macromolecular target of interest. This type of biomarker can be effectively used to validate or invalidate the relationship between a disease and a specific biomolecular target, or explain why a compound fails to produce the expected result with a previously validated target. Consider, for example, a biomarker that definitively demonstrates that a candidate compound is interacting with a biomolecule that is hypothesized to be linked to the progression of a specific disease. If an *in vivo* response is also observed, then the biomarker has validated the link between the disease and the biomolecular target of interest. On the other hand, if the same candidate compound's interaction with the targeted biomolecule fails to provide a physiological response, this suggests that the targeted biomolecule is not a suitable target for the treatment of the disease or condition of interest. Imaging techniques designed to determine the location of radiolabeled compounds, such as PET and SPECT have been particularly useful in demonstrating that a candidate compound is interacting with GPCRs in the brain.[17] Additional details on imaging techniques and their practical application will be provided later in this chapter.

Mechanism biomarkers represent a second class of biomarkers. This class of biomarkers provides information on the physiological impact of a candidate compound. They measures changes in specific events theorized to be associated with the targeted disease or condition that occurs as a result of target engagement. Exemplary physiological events that could be measured include changes in enzymatic activity, gene expression, protein expression, behavioral changes in the subject, or even plasma concentration of specific chemicals. Blood glucose levels, for example, are an established biomarker for candidate compound efficacy in the treatment of diabetes,[18] while sleep induction is an easily detected indication of the efficacy of a candidate compound designed to treat insomnia.[19] It is important to keep in mind that mechanism biomarkers are not necessarily tied to efficacy, especially if the link between the targeted mechanism and the disease/condition of interest has not been established.

Outcome biomarkers, on the other hand, are biomarkers with a defined link to the disease or condition of interest that can be used as an indication of compound candidate efficacy. In some cases, such as changes in viral load or CD4+ lymphocytes in HIV infection, biomarkers are biochemically assessed. There are, however, physiological outcomes, such as blood pressure reduction (cardiovascular disease) or sleep induction (insomnia) that can serve as outcome biomarkers. In many cases, there is an overlap between mechanism biomarkers and outcome biomarkers. It is also important to be aware that outcome biomarkers can also be used to screen out candidate compounds that have undesired side effects. A candidate compound designed to treat a migraine headache that also induces sleep or sedation may have limited commercial utility. The early application of an outcome biomarker to detect potential problems can be very effective in limiting the forward momentum of flawed candidate compounds.

Toxicity biomarkers are similar to outcome biomarkers, but as their name implies, these biomarkers are associated with negative outcomes. Data developed through the application of a toxicity biomarker are generally perceived to be strong indication that the candidate compound has a serious flaw that must be considered before moving forward with further development efforts. QT prolongation and hERG channel blockade, for example, are well known physiological and biochemical toxicity biomarkers associated with torsade des pointes and sudden cardiac death[20] that are employed in almost every drug discovery and development program. Toxicity biomarkers that are associated with other forms of toxicity such as liver or kidney issues are also available, and their application early in a program via *in vitro* methods can provide an opportunity for a program to use medicinal chemistry tools to design out the toxicity. Identifying these kinds of issues prior to initiating advanced animal studies or human trials can be a very effective method of conserving resources and limiting patient exposure to potentially harmful candidate compounds.

Pharmacogenomic biomarkers are primarily used in clinical settings, and their main purpose is to provide an improved understanding of the target patient population, especially with regard to predicting which patients are likely to respond. Consider, for example a candidate compound that targets a specific variation of a biomolecule or physiological state, such as a mutation that activates or eliminates a specific gene. Identifying patients with this particular mutation would increase the likelihood of demonstrating utility in a clinical trial. At the same time, patient without the desired trait (e.g., genetic mutation) would not be admitted into the clinical trials, as it would already be apparent that they would be unlikely to benefit from exposure to the candidate compound. In this manner, patient risk is limited, clinical trial sizes are diminished, and overall program costs are lowered.

Diagnostic biomarkers are also used primarily in the clinical setting, but can also be used in animal models where appropriate. This class of biomarkers can be used to identify patients that are at risk for developing a particular disease or condition, provide insight into disease progression or regression, and in some cases, identify patients before the clinical manifestation of symptoms or outwardly apparent physiological changes. Knowledge of this type can be used to simplify clinical measurements, ensure proper targeting of clinical trials, and even as a method of screening patients for entry into a clinical program.[21] Viral load in HIV patients, for example, can be used as a biomarker for disease progression or staging, rather than using the occurrence of opportunistic infections (or lack thereof) as a measure of disease progression or regression. Similarly, concentration of human chorionic gonadotropin (hCG) can be used as biomarker to gate the admission of women into clinical trials. This particular biomarker is almost certainly the most widely recognized biochemical diagnostic biomarker, although it is not recognized widely by this name. This particular diagnostic biomarker is more widely recognized as the diagnostic component of over-the-counter pregnancy tests, and is capable of establishing pregnancy at a very early stage. Screening of female clinical trial participants using this diagnostic biomarker is a very effective method of limiting the likelihood that a pregnant woman and her unborn child might be exposed to a candidate compound with an unknown safety profile.

CHARACTERISTICS AND IMPACT OF BIOMARKERS

Although there are a wide variety of pharmacological, physiological, and biochemical end points that could be used as a biomarker, not all of them are suitable for application in a drug discovery and development program. As previously discussed, the role of a biomarker and translational medicine in general is to increase the efficiency of the identification of new therapeutic entities. As a result, biomarkers that are difficult to measure, associated with expensive or long experiments, or are difficult to reproduce are generally not suitable for a drug discovery and development program. Biomarkers of this type may provide important scientific information that could be useful to society as a whole, but their added cost would create an additional drag on a program rather than increase its efficiency. Ideally, biomarkers suitable for drug discovery and development program need to be easily measured, time efficient, quantitative, objective, and highly reproducible. Importantly, the results obtained from experiments using a biomarker should be useful for predicting results in the next stage of experimentations. In other words, *a biomarker should be translational in nature*. Clinical relevance and reliability across a heterogeneous patient population are also important hallmarks of an ideal biomarker.

Biomarkers have been developed using complex gene and proteins expression systems, biochemical tools to measure blood concentration of key molecules, and advanced imaging techniques such as computed tomography (CT) scanning, positron emission tomography (PET) techniques, and single-photon emission computed tomography (SPECT) imaging. To be clear, however, not all biomarkers are complex in nature. Blood pressure, heart rate, and even body temperature measurements would also be considered biomarkers based on the aforementioned definitions. These and other simple biomarkers have been available for a very long time. Most people can remember a time when their parents checked their temperature to determine whether or not a systemic infection was subsiding. However, the broad-based application of biomarker within the context of translational medicine to increase the efficiency of drug discovery and development dramatically increased at the beginning of the twenty-first century. Their primary purpose is to provide scientist with the data necessary to make more informed decision about whether or not a candidate compound (or hypothetical disease target) is worthy of further evaluation by providing data that can be used to predict performance in future studies. In theory, if scientists are better equipped to identify compounds with the desired properties using predictive assays and methods, then the compounds that move forward would be more likely to eventually reach the market. Similarly, the ability to predict which compounds possess undesirable properties (e.g., safety risks) would eliminate efficacious compound that would fail in clinical trials as a result of poor safety outcomes. In both cases, fewer compounds reach more advanced and more expensive assays, reducing the overall cost of the process.

In considering the use of biomarkers, it is important to understand that their use is not limited to clinical trials. Significant cost and time savings can be achieved by the judicious use of biomarker in the discovery/preclinical stage of the drug development process. Consider, for example, a program that has identified a set of 10 compounds with *in vitro* biochemical properties that suggest efficacy and the selectivity for the desired biochemical target. *In vivo* pharmacokinetic (PK) studies must be completed prior to launching *in vivo* efficacy studies. While it is certainly possible to run *in vivo* PK studies on all 10 compounds, there are several biomarker assays available that can be used to predict certain aspects of *in vivo* PK. *In vitro* microsomal stability studies[22] and permeability studies,[23] for example, can be used to identify compounds that are less likely to have the *in vivo* PK properties necessary to support efficacy. Compounds that are rapidly metabolized by microsomes are likely to be highly metabolized *in vivo*, while compounds that perform poorly in *in vitro* permeability assays are likely to have limited bioavailability. Although assay of this type are not perfect predictors of *in vivo* PK, they can be very effective in prioritizing compounds based on their likelihood having the desired PK properties. Ideally, this translates into fewer compounds moving into *in vivo* PK models, saving both time and money.

BIOMARKERS VERSUS SURROGATE END POINTS

It is important to keep in mind in any discussion of biomarkers and translational medicine that not all biomarkers are useful as a substitute for a demonstration of efficacy. In fact, only a small subset of biomarkers are recognized by regulatory bodies as sufficiently validated that they are accepted as indicators of efficacy in clinical trials. Biomarkers that meet these criteria are referred to as surrogate end points. The NIH Biomarkers Definition Working Group also provided a definition of this class of biomarkers, stating that a surrogate end point is "a biomarker that is intended to substitute for a clinical endpoint and is expected to predict clinical benefit or harm, or lack of benefit or harm, based on epidemiologic, therapeutic, pathophysiologic, or other scientific evidence."[15] In other words, in order for a biomarker to be useful as a surrogate end point, there must be clear and convincing scientific evidence (e.g., epidemiological, therapeutic, and/or pathophysiological) that the biomarker in question consistently and accurately predicts the clinical outcome. If this correlation does not exist, then the biomarker cannot be used as a surrogate end point in a clinical trial, irrespective of how objective or quantifiable it may be. There are very few biomarkers that meet these stringent criteria. Some examples include blood pressure changes, serum cholesterol concentrations, and lipid fraction as measure of cardiovascular disease risk[24]; CD4+ T-lymphocytes level and RNA viral load measurements for HIV progression[25]; and tumor size measurements in oncology studies, although there is an ongoing debate on the validity of this particular surrogate end point.[26]

Using surrogate end points and biomarkers in general during the prosecution of clinical programs provides a number of advantages that can translate into decreased patient risk, reduced costs, and shorter clinical trials. Consider, for example, a clinical trial designed to determine whether or not a candidate compound is capable of reducing the risk of heart attacks. In the absence of a surrogate end point, it would be necessary to recruit, treat prophylactically, and monitor a large set of patents for an extended period of time (months to years) while waiting for an infrequent event (e.g., heart attack) to occur. If, however, a surrogate end point were available, such as blood pressure changes or cholesterol concentrations, the picture becomes very different. Changes in these parameters would occur much more rapidly, substantially decreasing the length, and therefore both cost and patient risk, of a clinical trial. In addition, clinical trials of candidate compounds that failed to affect the surrogate end point (e.g., lower blood pressure or lower cholesterol) could be terminated at an earlier time point than a study designed to monitor clinical end points such as the occurrence of a heart attack or survival. The early termination of a clinical trial decreases patient risk by decreasing their exposure to failed candidate compounds. It also decreases the overall cost of drug discovery and development by allowing clinical efforts to be redirected away from a failed candidate compound at an earlier time point.

Although biomarkers that are not surrogate end points cannot be used as proof of efficacy, they can be used to increase the efficacy of clinical trials. A biomarker that is present only in certain patient populations, such as patients with HER2/neu-positive breast cancer, could be used as an entry criterion into a clinical trial designed to measure the clinical benefits of a candidate compound that modulates the target. Restricting the patient population in this manner increases the likelihood of success (assuming a positive result is possible), and decrease the risk to patients by preventing patient that could not benefit from the treatment (in this case, HER2/neu-*negative* breast cancer patients) from being exposed unnecessarily to candidate compounds that is unlikely to help them. Herceptin® (Trastuzumab), a monoclonal antibody for the treatment of HER2/neu-positive breast cancer, was brought to market using data based in part on this strategy.[27]

Safety-related biomarkers can also be very effective tools in clinical trials, as they can provide an early warning that a candidate compound has a potential problem. Consider, for example, a candidate compound under investigation for the treatment of osteoarthritis. Phase III clinical trials designed to determine the efficacy of a candidate compound for this condition would require a substantial number of people and long-term monitoring. Safety-related biomarkers that can be examined early in the trial could provide rational for ending a trial early if there is an indication that there may be problems with the candidate compound. If the candidate compound raised cholesterol concentrations in the blood or increased blood pressure in a subset of the patient population, it would be useful to know this early in the trial. If the risks are high enough, the trial might be ended early, or enrollment requirements may be altered so that patient at risk would not be allowed to enter the trial. In either case, the risk to patients is minimized and the overall cost of clinical trials is lowered as less time is required to terminate trials based on safety risks.

IMAGING TECHNOLOGIES

Although biochemical, physicochemical, and pharmacological biomarkers can be very useful, they are, at the end of the day, an indirect indication of what is truly occurring in the body. Elevated blood pressure and high cholesterol are certainly linked to an increased risk of cardiovascular disease, but recording changes in these biomarkers does not provide direct information on the condition of the cardiovascular system itself. Frequent measurements of blood pressure cannot on their own provide direct insight into the condition of the heart itself or indicate where blockages may be occurring. Similarly, human chorionic gonadotropin (hCG) can serve as a biomarker to indicate whether or not a woman is pregnant (e.g., at-home pregnancy tests via detection of hCG in the urine), but it provides no information on the overall progress of fetal development no matter how many times the test is administered.

Exploratory, invasive techniques are available to assess disease progression, and could be used to monitor the impact of a candidate compound on a wide array of disease states, but this would be impractical, expensive, and in most cases unethical. There are, however, a number of imaging technologies that make it possible to peer into the body in a non-invasive manner and create three-dimensional images of the inside of the body with a high degree of contrast (e.g., organs, bones, vasculature, and other systems distinguishable from each other). These methods have provided a wealth of understanding of how the body works, diagnostic techniques, and within the context of drug discovery and development, novel tools capable of monitoring disease progression and the impact of candidate compounds on therapeutic targets. The most commonly used imagining techniques are X-ray computed tomography (X-ray CT), positron emission tomography (PET), single-photon emission computed tomography (SPECT), magnetic resonance imaging (MRI), functional magnetic resonance imaging (fMRI), and ultrasound techniques.

X-ray computed tomography, more commonly referred to as X-ray CT or CAT scan (computed axial tomography scan), is perhaps the most commonly used imaging technology. This technique depends upon the differential attenuation of X-rays as they pass through different types of tissue. As with single X-ray images, highly mineralized tissues such as bones have a brighter intensity in this technique due the higher degree of attenuation of X-rays, while softer tissue (e.g., muscles, skin, vasculature, connective tissue) are illuminated to a lesser degree due to their lower attenuation of X-rays. In this imaging technique, a series of two-dimensional radiographic images are taken around a single rotational axis of the subject. The individual two-dimensional "slices," also referred to as tomograms, provide a cross-sectional view of the subject, much like removing a piece of bread from a loaf of bread provides a cross-sectional view of the loaf of bread. Three-dimensional images can be created by combining the two-dimensional "slices" using digital geometry processing, allowing scientist to non-invasively observe the inside of the body. In some cases, contrast agents, such as Hypaque® (Diatrizoic acid), and Ioxilan® (Oxilan) (Figure 10.2(a) and (b)) can be employed to enhance visualization of specific areas of the body based on their propensity to accumulate in specific regions.

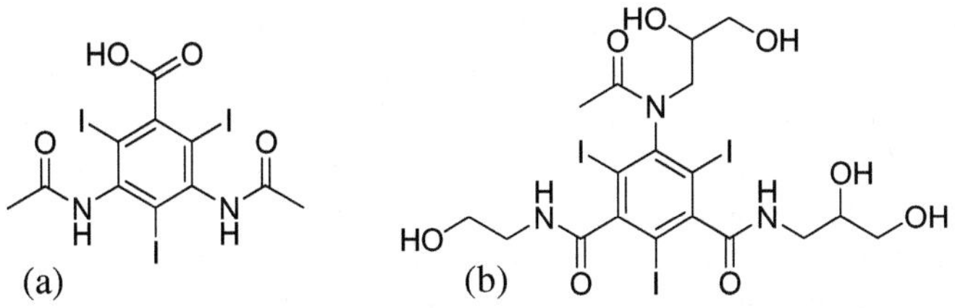

FIGURE 10.2   (a) Hypaque® (Diatrizoic acid) (b) Ioxilan® (Oxilan).

Positron emission tomography, also referred to as PET imaging is another useful tool capable of generating two- and three-dimensional images of a subject. This process takes advantage of the decay of specific types of radioactive isotopes, specifically those that undergo positron emission (also referred to as beta decay). As the radioisotope decays, it emits a positron (the antiparticle of an electron) that passes through the surrounding tissue until it encounters an electron. The encounter annihilates both particles and releases photons that travel in opposite directions. These photons can be detected with a photomultiplier tube. In a typical PET imaging experiment, the subject is treated with a ligand containing a suitable radioisotope, often referred to as a radioligand. This is followed by a short waiting period that allows the radioligand to reach its intended destination, and then the subject is slowly moved through a detector ring that creates "image slices" similar to those described in X-ray CT scanning. The "image slices" can then be assembled mathematically to produce a three-dimensional image of the subject. Importantly, radioligands are often designed to undergo specific, tight interactions with target biomolecules. As a result, specific pharmacological events or regions of the body will "light up" in a PET image. Amyvid® (Florbetapir, Figure 10.3(a)), for example, binds to β-amyloids

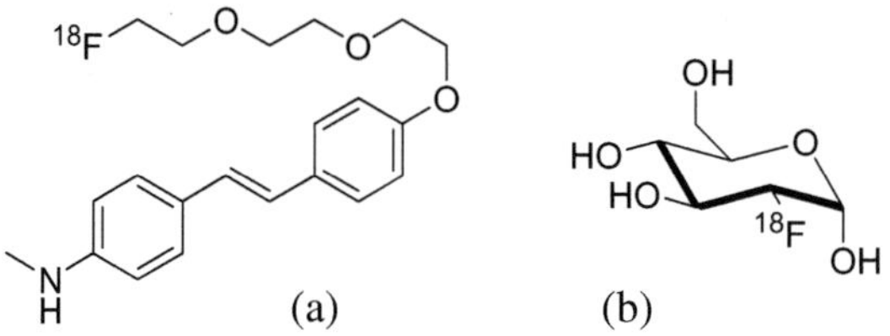

FIGURE 10.3   (a) Amyvid® (Florbetapir) (b) Fludeoxyglucose® (fluorodeoxyglucose, 18F-FDG).

and is used as a diagnostic tool for Alzheimer's disease,[28] while Fludeoxyglucose® (fluorodeoxyglucose, 18F-FDG, Figure 10.3(b)) acts a glucose mimic that accumulates in tumor cells and is a major clinical oncology diagnostic tool.[29]

Although PET imaging can be an effective tool in the drug discovery and development process, there are some limitations that must be considered. The list of radioisotopes that undergo positron emission is relatively small (Table 10.1). In addition, the radioactive half-life of these isotopes is very short (the longest, 18F, is only 110 min). As a result, PET radioligands have a short "shelf life" and must be used quickly once they are prepared. If too much time is allowed to elapse between creation of the radioligand and detection in an experiment, the PET signal will be below the threshold

TABLE 10.1   Positron-Emitting Radioisotopes

Isotopes	t½ (min)	Imaging time
^{18}F	110	~8 h
^{11}C	20	1.5 h
^{13}N	10	40 min
^{15}O	2	10 min

of detection due to radioactive decay. Once the positron emitting isotopes have been generated (using either a cyclotron or linear particle accelerator) they must be incorporated into a radioligand using the appropriate synthetic chemistry, purified, formulated, dosed in the subject, and all measurements must be taken before the signal fades due to radioactive decay. Needless to say, time is an important consideration in running PET experiments.

Single-photon emission computed tomography (SPECT) is similar to PET imaging in that it also requires a radioligand. In addition, as with PET imaging, a series of two-dimensional images are obtained and mathematically assembled to produce a three-dimensional rendering of the subject. There are, however, some important differences that should be considered. First and foremost, SPECT imaging systems are designed to detect gamma ray emissions. As a result, different radioisotopes are required for SPECT imaging techniques (Table 10.2)[30] and

TABLE 10.2   SPECT Radioisotopes

Isotope	t½	Isotope	t½
^{123}I	13.22 h	^{99m}Tc	6 h
^{131}I	8 days	^{111}In	2.8 days
^{177}Lu	6.6 days	^{67}Ga	3.2 days
^{186}Re	3.7 days	^{67}Cu	2.6 days

the radiochemical properties of these radioisotopes is the genesis of another significant difference between PET and SPECT. In general, the radioactive decay (expressed as t½) is slower in the radioisotopes necessary for SPECT imaging. This creates a larger time window of opportunity for imaging experiments, and substantially decreases the overall costs. As with PET imaging, radioligands with specific and tight binding to targeted biomolecules enables the visualization of specific regions

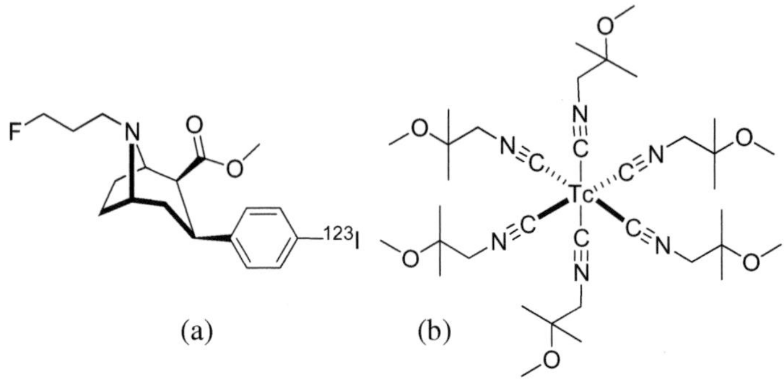

FIGURE 10.4 (a) DaTSCAN® (Ioflupane (^{123}I)) (b) Cardiolite® (Technetium (^{99m}Tc) sestamibi).

of the body or pharmacological events. DaTSCAN® (Ioflupane (^{123}I), Figure 10.4(a)), for example, is used in the diagnosis of Parkinson's disease based on its high affinity for the dopamine transporter (DAT) in the striatal region of the brain.[31] Cardiolite® (Technetium (^{99m}Tc) sestamibi, Figure 10.4(b)), on the other hand, incorporates a metastable Technetium atom and is routinely used for the visualization and diagnosis of cardiovascular problems.[32]

The merging of SPECT imaging techniques with antibody technology lead to the development of Prostascint (indium (^{111}In) capromab pendetide), a monoclonal antibody labeled with ^{111}In, for the identification of prostate cancer. In this case, the antibody specifically targets the prostate specific membrane antigen in prostate cancer cells, but not normal prostate cells.[33] It is worth noting at this point that SPECT imaging resolution is not as high as PET imaging. PET imaging is approximately 2–3 orders of magnitude more sensitive than SPECT imaging, and as a result, SPECT images typically have a lower image resolution than the corresponding PET images.[34]

In considering the application of both PET and SPECT imaging technologies, it is important to consider the nature and availability of an appropriate radioligand. As discussed above, PET ligands require special synthetic techniques, equipment, and dedicated equipment as a result of their radioactive nature. Similar limitations are also an issue with SPECT ligands, as they are also radioactive. There are, however, additional issues that must be addressed in order to develop a radioligand. It may not be possible to incorporate a radioisotope in a candidate compound that is selected for clinical study, as synthetic challenges are insurmountable. In this case, it may be possible to choose a structurally related compound that can serve the same purpose. The irreversible monoamine oxidase-B inhibitor Azilect® (Rasagiline, Figure 10.5(a)), for example, is useful for the treatment of early stage Parkinson's disease, but the synthesis of a

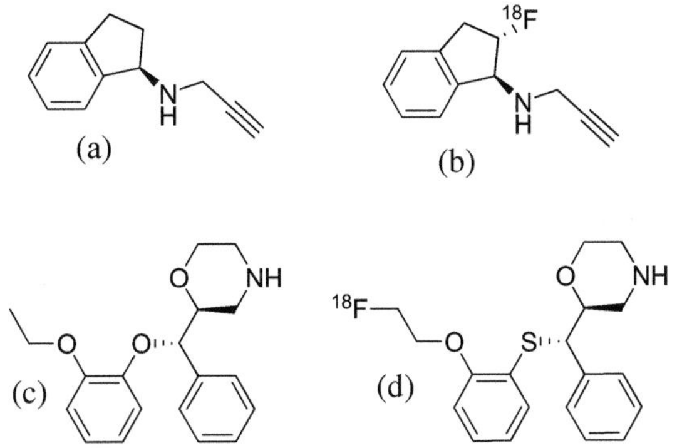

FIGURE 10.5 (a) Azilect® (Rasagiline) (b) [^{18}F] fluororasagiline (c) Edronax® (Reboxetine) (d) [^{18}F] fluororeboxetine.

PET ligand is not a viable option. In order to use Azilect as a radioligand it would be necessary to incorporate a ^{11}C into its structure, but the time required to synthesis ^{11}C labeled material based on the available synthetic chemistry is far too long for practical application (as mentioned above, the $t_{1/2}$ of ^{11}C is only 20 min). As an alternative, the ^{18}F labeled analog, [^{18}F] fluororasagiline (Figure 10.5(b)), was developed. The addition of a fluorine atom to Azilect had minimal impact on its overall properties, allowing it to serve as a surrogate for the clinical agent. New synthetic methods were required in order to prepare the radiolabeled compound, but the longer half-life of ^{18}F increased the time available for synthesis and PET imaging studies. In addition, incorporation of the radioisotope could be accomplished at the end of the synthetic pathway, greatly simplifying the radiochemistry requirements.[35]

In a similar manner, the norepinephrine reuptake inhibitor antidepressant Edronax® (Reboxetine, Figure 10.5(c)) was identified as a potential PET ligand useful for studying the norepinephrine transporter (NET). Edronax's low non-specific binding and high selectivity for NET over the serotonin transporter (SERT) and the dopamine transporter (DAT) provided an opportunity to study the brain distribution of NET. Unfortunately, much like Azilect, the synthesis of Edronax is not conducive to the incorporation of a radioisotope. As an alternative, a new radiolabeled analog of Edronax (Figure 10.5(d)) was identified as a viable alternative. The exchange of an oxygen atom for a sulfur atom had no impact on the NET/SERT/DAT selectivity, and synthetic methods capable of installing an ^{18}F in the final step of the synthetic sequence facilitated the development of this PET ligand. It is worth noting that in this case, alternative radioligands incorporating a ^{11}C were also developed.[36]

The importance of low non-specific binding of Edronax and its radioligand analogs should not be understated. Compounds that are target selective, but have a high propensity to undergo non-selective binding to plasma proteins such as albumin can present a significant challenge in an imaging experiment. Non-specific binding of the radioligand can substantially decrease the signal to noise ratio in a PET/SPECT imaging experiment. The radioligand would certainly bind to the intended biomolecular target and generate an observable signal, but it would also produce an observable signal in every non-specific binding event that it takes part in. If this non-specific binding is to albumin, then the radioligand would be nearly omnipresent in the body, as albumin is a major component of plasma. This could potentially drown out the signal from the macromolecule of interest. Expression levels of the biomolecular target also play a role in the signal to noise ratio in a PET/SPECT imaging experiment. Higher expression levels of the target macromolecule typically lead to better signal quality, while biomolecules with lower expression levels are more difficult to visualize.

The PK properties necessary for a successful radioligand are also not the same as those associated with therapeutic agents. Rapid uptake is highly desirable, especially in PET imaging agents where radioactive decay is rapid. In addition, a short pharmacokinetic half-life would decrease risk to subjects as a result of decreased exposure time to radioactive material. Pgp efflux can also be a substantial issue in the development and use of a radioligand. This is especially true if the targeted macromolecule is in the brain, as Pgp is heavily expressed in the blood–brain barrier. Finally, translation of the target between species should be high (e.g., the radioligand can be used to visualize the biomolecular target in animal models and humans). A radioligand that is useful in rats and dogs, but not in humans may provide some interesting information about the animal model, but it will be of limited predictive value in the human condition.

Magnetic resonance imaging (MRI) is an alternative technique that is also routinely employed to obtain detailed images of a subject. As the name implies, this imaging method is based on nuclear magnetic resonance techniques that are commonly employed in the study of organic and organometallic compounds. This method takes advantage of differential responses of hydrogen atoms to radio frequencies when they are in a magnetic field (other atoms can also be observed using this technique, but this discussion will be limited to hydrogen atoms). In brief, hydrogen atoms absorb energy when they are irradiated in the radio frequency range (60–1000 MHz) and move to an excited state. When the hydrogen atom relaxes back to the ground state, it releases energy that can be detected. The specific resonance frequency of any given hydrogen atom is dependent upon the nature of its immediate environment at the molecular level, and as a result, even small difference between two different hydrogen atoms can be detected and quantified. At the macroscopic level, differences in hydrogen atoms in various areas of the body can be mapped using sophisticated instrumentation designed to subject the body to a specific radio frequency in the presence of an oscillating magnetic field. The various tissues, organs, fluids, and other components of the body each have their own unique resonance frequency, which creates contrast between the various body components when an MRI image is generated. As with PET, SPECT, and X-ray CT, the data collected in this process can be used to generate three-dimensional images of the subject.

It is also possible to use contrast agents in MRI studies. Unlike PET and SPECT radioligands, however, the incorporation of radioisotopes is not necessary. MRI contrast agents, also referred to as shift reagents, contain paramagnetic atoms that can impact the resonance frequencies of nearby atoms. This creates an observable change in the MRI signals and can be especially useful if the contrast agent is differentially absorbed in the body. Dotarem® (Gadoteric meglumine, Figure 10.6), for example, is a Gadolinium chelating

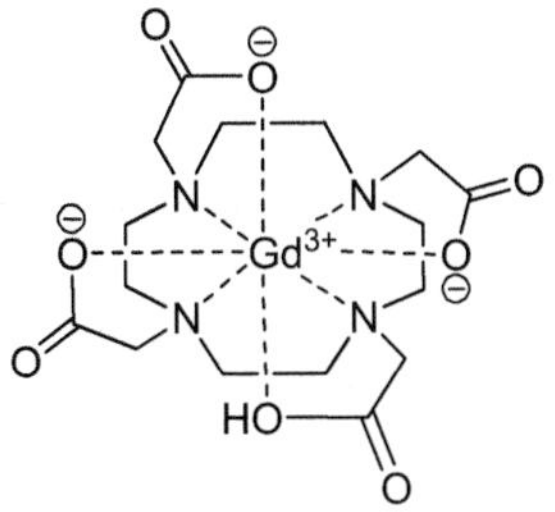

FIGURE 10.6   Dotarem® (Gadoteric meglumine).

agent that has been used to study cerebral pathologies, medullar pathologies, and vascular disease.[37] MRI has also been successfully applied to the study of Alzheimer's disease,[38] cardiovascular disease,[39] and multiple sclerosis.[40]

Functional magnetic resonance imaging (fMRI) is a variation on MRI technology that has provided substantial insight into brain function. This technique takes advantage of changes in cerebral blood flow that occur in concert with changes in neuronal activity. Blood-oxygen-level dependent (BOLD) contrast is the most commonly used fMRI procedure. In this method, differences in blood oxygenation are detected based on the magnetic properties of the iron atom in hemoglobin. Oxygenated hemoglobin (oxygen-rich blood) contains a diamagnetic iron atom, whereas the iron of deoxygenated hemoglobin (oxygen-poor blood) is paramagnetic. The two magnetic states of iron are distinct in their impact on MRI signals, and this difference can be exploited to create heat maps that describe amounts of oxygenated versus deoxygenated blood. Increased neuronal activity is associated with an increase in oxygen-rich blood. As a result, the aforementioned heat maps provide not only an indication of the oxygenation levels of blood in various areas of the brain, they also provide a heat map of brain activity across the various regions of the brain.[41] fMRI has been successfully applied in the study of depression,[42] pain,[43] as well as a wide range of CNS disorders.[44]

Ultrasound is perhaps the most recognized form imaging technology as a result of its common use monitoring fetal development through the course of pregnancy. First described for medical purposes in the 1940s by Dr. George Ludwig[45] and Dr. Jon Wild,[46] this technique depends on the generation, reflection, and subsequent detection of reflected sound waves. The typical sonography system generates a series of pulses in the ultrasound frequency range (typically 2–18MHz) at the surface of the skin. The sound waves are reflected back as echoes, the intensity and angle of which is dictated by the nature of the tissue encountered and its acoustical impedance. This technique can be used to visualize most regions of the body and does not require radioactive material, but image resolution is not as good as that available with other imaging techniques.[47] The relatively low cost of ultrasound as compared to the previously described techniques, however, has led to the wide application of ultrasound devices across the medical and clinical community.

THE PRACTICAL APPLICATION OF BIOMARKERS

Although there are a wide variety of biomarkers available to choose from, the application of any particular biomarker in a given program is at least in part dependent on the nature of the disease or condition under investigation. Biomarkers can be used to determine whether or not a candidate compound is capable of reaching the target of interest, provide an early indication of efficacy (or lack thereof), predict toxicity, and even identify patient populations that are more likely to respond to therapeutic agents. At the end of the day, the purpose of biomarkers and translational medicine is to increase the speed and efficiency of the identification of novel therapeutics. In order to accomplish this goal, the selection of biomarkers must be aligned with the goals of the program and the program scientist must be willing to objectively respond to the results even if the data support project termination. While this is not the most satisfying of outcomes, projects closed at an earlier stage expend less resources than those that are closed at a later point, allowing limited funds and time to be redirected to other priorities and projects. The importance of biomarkers in the drug discovery and development process is best demonstrated through a review of some real world examples of the impact of biomarkers. The following examples provide an illustration of programs whose key decision points would have been far more difficult to evaluate in the absence of biomarkers.

DPP-IV Inhibitors (Januvia®)

The development of inhibitors of the enzyme dipeptidyl peptidase IV (DPP-IV), a serine protease, for the treatment of type 2 diabetes is an interesting case. A number of biochemical biomarkers that provided translational bridges between *in vitro* screening results, animal models, and the human condition facilitated the discovery of new therapies such as Januvia® (Sitagliptin),[48] Onglyza® (Saxagliptin),[49] and Tradjenta® (Linagliptin) (Figure 10.7).[50] DPP-IV plays an indirect role in blood glucose

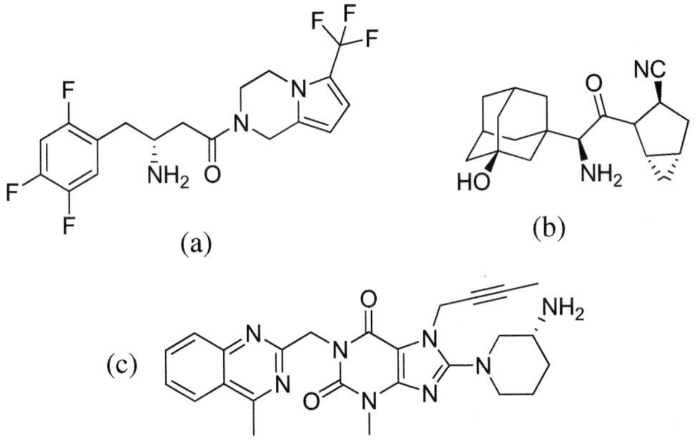

FIGURE 10.7 (a) Januvia® (Sitagliptin) (b) Onglyza® (Saxagliptin) (c) Tradjenta® (Linagliptin).

concentration, a key aspect in diabetes, through its ability to inactivate the incretin hormone glucagon-like peptide-1 (GLP-1). This hormone is released from the gut in response to food intake, stimulates insulin synthesis and secretion, and suppresses glucagon release.[51] Inhibiting DPP-IV would effectively slow the inactivation of GLP-1 (extend its half-life), prolonging its impact on glucose regulation.

When the aforementioned drugs were being developed, the link between DPP-IV, insulin, glucagon, glucose, and diabetes had not been firmly established. It was, therefore, necessary to develop animal models capable of predicting the impact of DPP-IV inhibitors on diabetes progression. Biomarkers played a key role in this process. Although blood glucose concentration could certainly be measured in animals, this assessment is not capable of demonstrating that a hypothetical DPP-IV inhibitor is actually engaging the target or that target engagement is actually a useful course of action. In the absence of an assay capable of demonstrating target engagement *in vivo*, the meaning of a negative result in a glucose tolerance test (a standard diabetes animal model) would not be clear. A negative result of this type could have been an indication that the hypothesis linking DPP-IV to diabetes progression was incorrect, that DPP-IV was not on the critical path of disease progression, or that the compound was simply not capable of reaching the target. The first two interpretations would have led to questioning of the utility of DPP-IV inhibitors as a whole for the treatment of diabetes, while the third conclusion would only be an indictment of the single candidate compound examined in the assay.

In order to pin down the link between DPP-IV, GLP-1, glucose, and diabetes progression, additional translational assays were necessary. The scientists at Merck, the company that developed Januvia® (Sitagliptin), chose to develop assays using GLP-1 plasma concentration and plasma DPP-IV activity as biomarkers.[48] Essentially, C57BL/6N male mice were treated with a candidate compound using conditions known to induce GLP-1 release (e.g., orally delivered glucose or dextrose), plasma samples were obtained, and the samples were analyzed for the presence of active GLP-1 and DPP-IV activity. If lower DPP-IV activity was detected in the plasma samples of the treated animals as compared to the control, this would indicate that the candidate compound had reached its intended target. Similarly, increased concentrations of GLP-1 in treated animal plasma versus control animals would be an indication that DPP-IV inhibition was, in fact, having an impact on GLP-1 concentration. Treatment of mice with Januvia® (Sitagliptin) produced a dose-dependent increase in active GLP-1 plasma concentrations, decreased DPP-IV activity, and improved glucose tolerance. Studying these biomarkers in human clinical trials provided an early indication of the clinical utility of Januvia® (Sitagliptin). Dose-dependent increases in active GLP-1 concentration, decreased DPP-IV activity, and improved glucose tolerance were observed in phase I clinical trials. These observations highlighted the importance of biomarkers in the identification of novel DPP-IV inhibitors for the treatment of diabetes and were supportive of further clinical study of Januvia® (Sitagliptin) for this purpose.[52] The FDA approved this drug for marketing in 2006.

Physiological Measurements as Biomarkers: Orexin Antagonists

As discussed earlier in this chapter, there are a number of physiological responses that can be used as biomarker to assist in the identification of novel therapeutic agents. Monitoring changes in blood pressure and heart rate are well-known and somewhat obvious biomarkers that are widely employed, but there are others that are less apparent. Consider, for example, the development of novel treatments for insomnia. If all types of sleep were equal, then identifying a compound capable of inducing sleep would be as simple as dosing animals with a candidate compound and monitoring for loss of consciousness and overall sleep duration. Of course, this simplistic view is not sufficient, as sleep is a complex phenomenon. Sleep is not simply the suppression of CNS activity. It consists of several stages, such as REM (rapid eye movement) sleep, non-REM sleep, slow wave sleep, and delta sleep. In addition, changes in the amount of time spent in each sleep stage (sleep architecture) can have a significant impact on memory consolidation and the restorative function of sleep.[53] In the absence of more sophisticated biomarkers, however, it would not be possible to distinguish between a candidate compound that induced normal sleep and one that globally suppresses CNS activity, such as the γ-aminobutyric acid (GABAa) system modulators Halcion® (Triazolam, Figure 10.8(a))[54] and Restoril®

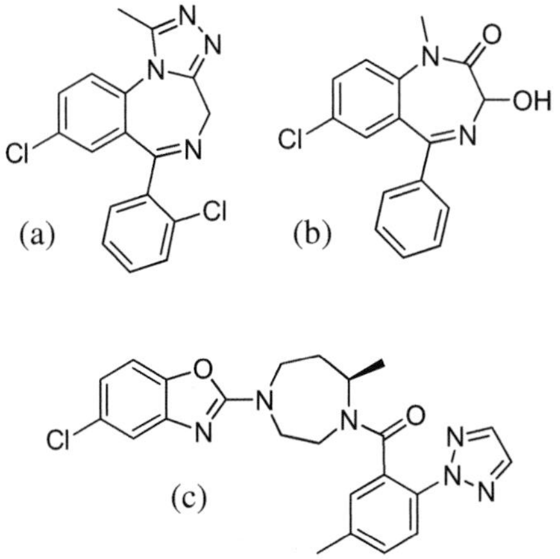

FIGURE 10.8 (a) Halcion® (Triazolam) (b) Restoril® (Temazepam) (c) Belsomra® (Suvorexant, MK-4305).

(Temazepam).[55] Determining the impact of candidate compounds on sleep architecture would also be impossible, as there would be no means of distinguishing between the different stages of sleep. These issues were addressed by Merck in a program designed to identify novel orexin receptor antagonists that culminated in the identification of Belsomra® (Suvorexant, Figure 10.8(c)), a potential treatment of insomnia.

The orexin receptor system is a critical regulatory component of the sleep/wake cycle. It is composed of two GPCRs, orexin receptor 1 (OX1R) and orexin receptor 2 (OX2R), and two associated ligands, orexin A and orexin B, and there is a high degree of structural conservation across multiple species.[56] In a normal sleep/wake cycle, orexin levels and activation of the corresponding receptor are maintained during wakeful periods, but the system is essentially inactive during sleep periods.[57] Genetically engineered mice that are unable to produce the orexin peptides display symptoms associated with narcolepsy (e.g., excessive daytime sleepiness, sleep fragmentation),[58] while knockout mice that are missing both OX1R and OX2R are acutely narcoleptic.[59] In human narcoleptics, there are substantially fewer orexin-producing neurons and low levels of orexin A during wakeful periods. These features are believed to be causative in symptomology.[60]

When taken together, these facts suggest that orexin receptor activity is required to maintain a wakeful state and that cessation of orexin receptor activity promotes sleep. The data also suggest that targeting this receptor with an antagonist would promote normal sleep rather than a global suppression of CNS activity (sedation). These observations do not, however, fully validate this approach. In order to accomplish this, it would be necessary to identify selective orexin receptor antagonists and, more importantly, develop an animal model capable of differentiating between normal sleep and sedation. The key to developing an appropriate animal model was the understanding that brain wave activity, eye movement, and skeletal muscle activity could be used as physiological biomarkers. Changes in each of these biological functions can be monitored using an electroencephalogram, an electrooculogram, and an electromyogram respectively (collectively referred to as polysomnography).[61] Normal sleep and global suppression of CNS activity produce significantly different results in these assays, providing a means for determining the impact of a candidate compound.

In the development of Belsomra® (Suvorexant), rats with implanted telemetry equipment designed to record electrocorticogram (an intracranial version of an electroencephalogram) and electromyogram signals were monitored to determine the amount of time each animal spent in the various stages of the sleep/wake cycle in the presence and absence of candidate compounds. A 30 mg/kg oral dose of Belsomra® (Suvorexant) produced a significant increase in both REM and delta stage sleep. Decreased wake time was also observed, and importantly, sleep architecture was consistent with normal sleep and rather than global sedation of CNS activity. Receptor occupancy measurements provided further confirmation that changes in sleep architecture were associated with a high level of drug/orexin receptor interaction.[62] Similar results were obtained with Belsomra® (Suvorexant) in dogs (1 and 3 mg/kg) and rhesus monkeys (10 mg/kg), highlighting the cross-species consistency of the sleep/wake cycle and the utility of these biomarkers.[63] These same biomarkers were effectively employed in human clinical trials using polysomnography to demonstrate the clinical efficacy of Belsomra® (Suvorexant) for the treatment of insomnia. Belsomra® (Suvorexant) was approved by the FDA for the treatment of insomnia in August 2014.

FDG PET Imaging Agent

2-deoxy-2-(^{18}F)fluoro-D-glucose (FDG, Figure 10.3(b)) was one of the first radioligands developed for PET imaging.[64] It was originally designed as a replacement for 2-deoxy-D-[^{14}C]glucose, a radioligand capable of imaging glycolysis via autoradiography, which requires the animal to be sacrificed as part of the process.[65] FDG is a substrate for the glucose transporter, so it is rapidly absorbed by cells, and it is also a substrate for hexokinase, the enzyme responsible for the first step in glucose metabolism. Hexokinase phosphorylates FDG, producing FDG-6-phosphate, but the absence of a hydroxyl group in the 2-position of this compound blocks further metabolism via the glycolysis pathway. As a result, FDG-6-phosphate accumulated in the cells.

Of course, all cells use glucose, so if all other things were equal, then the distribution of FDG-6-phosphate would be uniform across the body and PET imaging using this radioligand would not be very useful. There are, however, important differences between normal cells and malignant cells that provide an opportunity for PET imaging of several types of cancers. In many cases, the significantly greater energy demand of malignant cells is supported by upregulation of both glucose transporters (e.g., GLUT-1) and hexokinase.[66] This increases the uptake of FDG by certain types of malignancies relative to normal cells. The low levels of FDG-6-phosphate produced in normal tissues serves as a background for the higher levels of FDG-6-phosphate produced in cancerous tissues. These differences can be visualized using PET imaging techniques, providing a method of diagnosing and staging a number of different malignancies. This technique has been successfully applied to colorectal cancer, melanoma, lymphoma, and non-small-cell lung cancer.

It should also be clear that PET imaging agents such as FDG and other compounds that are selectively absorbed by malignant cells can be a very effective tool in the identification of new therapeutic agents. Tumor size and disease progression could be tracked over time in an appropriate animal model using a PET ligand. At the same time, animals can be treated with a candidate compound while monitoring for changes in tumor size and disease progression using the same PET ligand. If the candidate compound produces no change in the PET scan images when comparing treated versus untreated animals, then the candidate compound is not effective and should be discontinued for this particular indication. On the other hand, if tumor size decreases or disease progression is slowed/stopped, then further investigation of the candidate compound is reasonable. PET imaging methods can be applied at the *in vivo* animal model stage or in human clinical trials. In both cases, PET imaging techniques may provide insight into the utility of candidate compound well ahead of the traditional survival end points. Although these biomarkers are not sufficiently well-defined at this time to qualify as surrogate end points, the information derived from these experiments can be used to make educated decision on whether or not to continue pursuing a candidate compound and minimize patient risks associated with exposure to candidate compounds with limited safety profiles.

The Neurokinin 1 (NK$_1$) Receptor, Depression, and PET Imaging: The Aprepitant Story

There are certainly many cases in which PET and SPECT imaging are used as biomarkers to demonstrate that a candidate compound is engaging its intended target (e.g., receptor occupancy) so that a correlation between target binding and functional efficacy can be drawn. It is important to understand, however, the meaning of PET and SPECT imaging study that successfully demonstrates target engagement when *in vivo* efficacy is not observed. In this case, not only is the utility of the candidate compound called into question, the overall utility of the therapeutic target must also be reconsidered. Such is the case with Emend® (Aprepitant, Figure 10.9(a)), a selective neurokinin 1 (NK$_1$) receptor antagonist.[67] This drug is currently marketed as an antiemetic

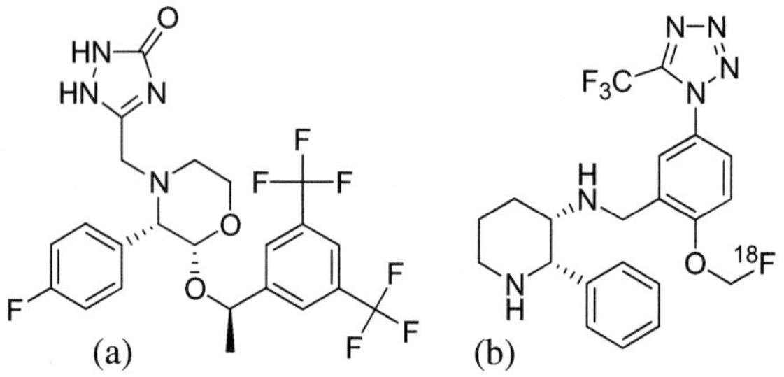

FIGURE 10.9   (a) Emend® (Aprepitant) (b) L-829,165-[18]F.

useful for the prevention of chemotherapy-induced and postoperative nausea and vomiting.[68] It was also clinically assessed for its ability to treat major depressive disorder.

Prior to the Emend® (Aprepitant) major depressive disorder clinical trials, a substantial body of literature provided preclinical evidence suggesting a link between NK$_1$ receptor activation by substance P and depression and anxiety. Physiological, neuroanatomical, and behavioral studies[69] detailing the impact of activation of the NK$_1$ by substance P lead the scientific community to hypothesize that excessive release of substance P in areas of the brain that control emotionality could produce a cascade of psychophysiological responses that would culminate in the symptomology associated with depression and anxiety. Emend® (Aprepitant), which at the time was known as MK-869, had been identified by Merck as a highly efficacious (EC$_{50}$ = 90 pM) and selective NK$_1$ antagonist. Pharmacokinetic studies (including blood–brain barrier penetration) indicated that sufficient exposure to the drug could be achieved in humans, and its safety profile was clean in preclinical models. Based on this information, Merck launched clinical trials designed to determine the efficacy of this compound for the treatment of major depressive disorder.

As part of these efforts, Merck also performed a series of PET imaging studies designed to determine the level of target engagement in the brain provided by clinical doses of this compound. In these experiments, patients were dosed initially with the L-829165-^{18}F (Figure 10.9(b)), a PET tracer ligand known to selectively bind to NK_1.[70] This was followed by treatment with the phase III clinical doses of Emend® used in the major depressive disorder trials. NK_1 receptor occupancy was then assessed via PET imaging based on displacement of the PET tracer ligand by the clinical candidate. The results of this study clearly demonstrated that up to 100% NK_1 receptor occupancy could be achieved with the clinical doses of Emend®, confirming target engagement. At the same time, however, clinical efficacy was *not* achieved. After 8 weeks of treatment, there was no significant difference between the treatment groups and placebo. In contrast, 20 mg of the antidepressant Paxil® (Paroxetine),[71] a selective serotonin reuptake inhibitor (SSRI), provides symptom relief after 8 weeks. The failure of the Emend® major depressive disorder clinical studies combined with the clear evidence of target engagement led to the conclusion that the NK_1 receptor is not a viable target for the treatment of depression. Although the PET imaging studies did not lead to a positive result, they did provide a better understanding of the limitations of NK_1 receptor antagonism and most likely prevented the further progression of additional NK_1 receptor antagonists into clinical studies. In the absence of the aforementioned PET imaging biomarker studies, it would have been difficult to determine if the failure of Emend® was compound-specific or target-related. Resources that could have been devoted to better therapeutic targets would have been wasted in pursuit of an effect that could never be achieved.

Identifying biomarkers that facilitate the characterization of candidate compounds can have a significant impact on the various stages of drug discovery and development. Biomarkers that are indicative of efficacy can shorten clinical trials and decrease the number of patients exposed to new compounds. Safety risks can also be mitigated by identifying biomarkers that are indicative of potential risks. Termination of clinical programs due to poor performance in biomarker studies, whether by virtue of efficacy issues or the identification of safety risks, allows resources to be quickly redirected to more fruitful areas, thereby adding efficiency to the drug discovery and development process. Bringing these tools together through the concepts of translation medicine in a "bench-to-bedside approach" can also improve the use of resources at the earliest stages of drug discovery. Biomarkers of efficacy in disease states can be used to determine the validity of hypothetical therapeutic targets. Rapid validation or invalidation of a potential therapeutic target provides an opportunity to redirect resources to areas more likely to bear fruit and away from "dry holes." Translation medicine and biomarkers are not always available, but modern drug discovery and development scientists who take advantage of these tools when they are available are likely to be a step ahead of those who do not.

QUESTIONS

1. What is the definition of translational medicine?
2. What are the two categories of translational medicine?
3. What is the focus of the four subcategories of translational medicine (T1–T4)?
4. What is a biomarker?
5. What are the major categories of biomarkers?
6. What qualities should an ideal biomarker possess?
7. What is a surrogate endpoint?
8. What are the benefits of using surrogate endpoints and biomarkers in a clinical trial?
9. What are the major types of imaging technology that can be used to peer inside the human body in a non-invasive manner?
10. Why is radioactive decay rate an important issue in PET imaging experiments?
11. Why is it sometimes necessary to develop an alternate synthetic pathway for a PET radioligand as compared to the non-radioactive compound of the same structure?
12. Why is low non-specific binding to proteins an important feature of PET and SPECT imaging?
13. What are the desired pharmacokinetic properties of a PET or SPECT radioligand?

References

1. a. "A Brief Guide to Genomics." National Human Genome Research Institute, National Institutes of Health, http://www.genome.gov/18016863.
2. a. James, P. Protein Identification in the Post-genome Era: The Rapid Rise of Proteomics. *Q. Rev. Biophys.* **1997,** *30* (4), 279–331.
   b. Anderson, N. L.; Anderson, N. G. Proteome and Proteomics: New Technologies, New Concepts, and New Words. *Electrophoresis* **1998,** *19* (11), 1853–1861.
3. a. Daviss, B. Growing Pains for Metabolomics. *The Scientist* **2005,** *19* (8), 25–28.
   b. Nicholson, J. K.; Lindon, J. C. Systems Biology: Metabonomics. *Nature* **2008,** *455* (7216), 1054–1056.

4. a. Paul, S. M.; Mytelka, D. S.; Dunwiddie, C. T.; Persinger, C. C.; Munos, B. H.; Lindborg, S. R.; Schacht, A. L. How to Improve R&D Productivity: The Pharmaceutical Industry's Grand Challenge. *Nat. Rev. Drug Discov.* **2010,** *9,* 203–214.

   b. Munos, B. Lessons from 60 Years of Pharmaceutical Innovation. *Nat. Rev. Drug Discov.* **2009,** *8,* 959–968.

5. a. DiMasi, J. A.; Hansen, R. W.; Grabowski, H. G. The Price of Innovation: New Estimates of Drug Development Costs. *J Health Econ.* **2003,** *22,* 151–185.

   b. *Research and Development in the Pharmaceutical Industry;* The Congress of the United States Congressional Budget Office, October 2006.

6. Extance, A. Alzheimer's Failure Raises Questions about Disease-modifying Strategies. *Nat. Rev. Drug Discov.* **2010,** *9,* 749–751.

7. Berenson, A. *Pfizer Ends Studies on Drug for Heart Disease;* The New York Times, December 3, 2006.

8. a. Nissen, S. E.; Tardif, J. C.; Nicholls, S. J.; Revkin, J. H.; Shear, C. L.; Duggan, W. T.; Ruzyllo, W.; Bachinsky, W. B.; Lasala, G. P.; Tuzcu, E. M. Effect of Torcetrapib on the Progression of Coronary Atherosclerosis. *N. Engl. J. Med.* **2007,** *356* (13), 1304–1316.

9. a. Furberg, C. D.; Pitt, B. Withdrawal of Cerivastatin from the World Market. *Curr. Controlled Trials Cardiovas. Med.* **2001,** *2,* 205–207.

   b. Psaty, B. M.; Furberg, C. D.; Ray, W. A.; Weiss, N. S. Potential for Conflict of Interest in the Evaluation of Suspected Adverse Drug Reactions: Use of Cerivastatin and Risk of Rhabdomyolysis. *J. Am. Med. Assoc.* **2004,** *292* (21), 2622–2631.

10. Karha, J.; Topol, E. J. The Sad Story of Vioxx, and What We Should Learn from it. *Cleveland Clin. J. Med.* **2004,** *71* (12), 933–939.

11. National Institutes of Health, RFA-RM-07–007. Clinical and Translational Science Award (U54), Part II: Full Text of Announcement, Section I: Funding Opportunity Description, Subsection 1: Research Objectives, Definitions. http://grants.nih.gov/grants/guide/rfa-files/RFA-RM-07-007.html.

12. a. Rubio, D. M.; Schoenbaum, E. E.; Lee, L. S.; Schteingart, D. E.; Marantz, P. R.; Anderson, K. E.; Platt, L. D.; Baez, A.; Esposito, K. Defining Translational Research: Implications for Training. *Acad. Med.* **2010,** *85* (3), 470–475.

   b. Khoury, M. J.; Gwinn, M.; Yoon, P. W.; Dowling, N.; Moore, C. A.; Bradley, L. The Continuum of Translation Research in Genomic Medicine: How Can We Accelerate the Appropriate Integration of Human Genome Discoveries into Healthcare and Disease Prevention? *Genet. Med.* **2007,** *9* (10), 665–674.

13. a. Fischl, M. A.; Richman, D. D.; Grieco, M. H.; Gottlieb, M. S.; Volberding, P. A.; Laskin, O. L.; Leedom, J. M.; Groopman, J. E.; Mildvan, D. The Efficacy of Azidothymidine (AZT) in the Treatment of Patients with AIDS and AIDS-related Complex. A Double-blind, Placebo-controlled Trial. *N. Engl. J. Med.* **1987,** *317* (4), 185–191.

   b. Brook, I. Approval of Zidovudine (AZT) for Acquired Immunodeficiency Syndrome. *J. Am. Med. Assoc.* **1987,** *258* (11), 1517.

14. a. Drusano, G. L.; Yuen, G. J.; Lambert, J. S.; Seidlin, M.; Dolin, R.; Valentine, F. T. Relationship between Dideoxyinosine Exposure, CD4 Counts, and P24 Antigen Levels in Human Immunodeficiency Virus Infection. A Phase 1 Trial. *Ann. Intern. Med.* **1992,** *116,* 562–566.

   b. Kahn, J. O.; Beall, G.; Sacks, H. S.; Merigan, T. C.; Beltangady, M.; Smaldone, L.; Dolin, R.; Lagakos, S. W.; Richman, D. D.; Cross, A.; et al. A Controlled Trial Comparing Continued Zidovudine with Didanosine in Human Immunodeficiency Virus Infection. *N. Engl. J. Med.* **1992,** *327* (9), 581–587.

15. Atkinson, A. J.; Colburn, W. A.; DeGruttola, V. G.; DeMets, D. L.; Downing, G. J.; Hoth, D. F.; Oates, J. A.; Peck, C. C.; Schooley, R. T.; Spilker, B. A.; et al. Biomarkers and Surrogate Endpoints: Preferred Definitions and Conceptual Framework. *Clin. Pharmacol. Ther.* **2001,** *69* (3), 89–95.

16. Lathia, C. D. Biomarkers and Surrogate Endpoints: How and when Might They Impact Drug Development? *Dis. Markers* **2002**, *18*, 83–90.

17. a. Schou, M.; Pike, V. W.; Halldin, C. Development of Radioligands for Imaging of Brain Norepinephrine Transporters *in Vivo* with Positron Emission Tomography. *Curr. Top. Med. Chem.* **2007**, *7* (18), 1806–1816.

    b. Paterson, L. M.; Kornum, B. R.; Nutt, D. J.; Pike, V. W.; Knudsen, G. M. 5-HT Radioligands for Human Brain Imaging with PET and SPECT. *Med. Res. Rev.* **2013**, *33* (1), 54–111.

    c. Lever, J. R. PET and SPECT Imaging of the Opioid System: Receptors, Radioligands and Avenues for Drug Discovery and Development. *Curr. Pharm. Des.* **2007**, *13* (1), 33–49.

18. a. Mu, J.; Woods, J.; Zhou, Y. P.; Roy, R. S.; Li, Z.; Zycband, E.; Feng, Y.; Zhu, L.; Li, C.; Howard, A. D.; et al. Chronic Inhibition of Dipeptidyl Peptidase-4 with a Sitagliptin Analog Preserves Pancreatic B-cell Mass and Function in a Rodent Model of Type 2 Diabetes. *Diabetes* **2006**, *55*, 1695–1704.

    b. Aschner, P.; Kipnes, M. S.; Lunceford, J. K.; Sanchez, M.; Mickel, C.; Williams-Herman, D. E. Effect of the Dipeptidyl Peptidase-4 Inhibitor Sitagliptin as Monotherapy on Glycemic Control in Patients with Type 2 Diabetes. *Diabetes Care* **2006**, *29* (12), 2632–2637.

19. a. Cox, C. D.; Breslin, M. J.; Whitman, D. B.; Schreier, J. D.; McGaughey, G. B.; Bogusky, M. J.; Roecker, A. J.; Mercer, S. P.; Bednar, R. A.; Lemaire, W.; et al. Discovery of the Dual Orexin Receptor Antagonist [(7R)-4-(5-chloro-1,3-benzoxazol-2-yl)-7-methyl-1,4-diazepan-1-yl][5-methyl-2-(2H-1,2,3-triazol-2-yl)phenyl]methanone (MK-4305) for the Treatment of Insomnia. *J. Med. Chem.* **2010**, *53* (14), 5320–5332.

    b. Winrow, C. J.; Gotter, A. L.; Cox, C. D.; Doran, S. M.; Tannenbaum, P. L.; Breslin, M. J.; Garson, S. L.; Fox, S. V.; Harrell, C. M.; Stevens, J.; et al. Promotion of Sleep by Suvorexant – A Novel Dual Orexin Receptor Antagonist. *J. Neurogenet.* **2011**, *25* (1–2), 52–61.

20. a. Sanguinetti, M. C.; Jiang, C.; Curran, M. E.; Keating, M. T. A Mechanistic Link between an Inherited and an Acquired Cardiac Arrhythmia: HERG Encodes the IKr Potassium Channel. *Cell* **1995**, *81* (2), 299–307.

    b. Sanguinetti, M. C.; Tristani-Firouzi, M. hERG Potassium Channels and Cardiac Arrhythmia. *Nature* **2006**, *440* (7083), 463–469.

21. Mayeux, R. Biomarkers: Potential Uses and Limitations. *NeuroRx J. Am. Soc. Exp. Neurother.* **2004**, *1*, 182–188.

22. Iwatsubo, T.; Hirota, N.; Ooie, T.; Suzuki, H.; Shimada, N.; Chiba, K.; Ishizaki, T.; Green, C. E.; Tyson, C. A.; Sugiyama, Y. Prediction of *in Vivo* Drug Metabolism in the Human Liver from *in Vitro* Metabolism Data. *Pharmacol. Ther.* **1997**, *73* (2), 147–171.

23. Hidalgo, I. J.; Raub, T. J.; Borchardt, R. T. Characterization of the Human Colon Carcinoma Cell Line (Caco-2) as a Model System for Intestinal Epithelial Permeability. *Gastroenterology* **1989**, *96* (3), 736–749.

24. Wittes, J.; Lakatos, E.; Probstfield, J. Surrogate Endpoints in Clinical Trials: Cardiovascular Diseases. *Statistics Med.* **1989**, *8*, 415–425.

25. Kanekar, A. Biomarkers Predicting Progression of Human Immunodeficiency Virusrelated Disease. *J. Clin. Med. Res.* **2010**, *2* (2), 55–61.

26. a. Mozley, P. D.; Schwartz, L. H.; Bendtsen, C.; Zhao, B.; Petrick, N.; Buckler, A. J. Change in Lung Tumor Volume as a Biomarker of Treatment Response: A Critical Review of the Evidence. *Ann. Oncol.* **2010**, *21*, 1751–1755.

    b. Zhao, B.; Oxnard, G. R.; Moskowitz, C. S.; Kris, M. G.; Pingzhen, W. P.; Rusch, V. M.; Ladanyi, M.; Rizvi, N. A.; Schwartz, L. H. A Pilot Study of Volume Measurement as a Method of Tumor Response Evaluation to Aid Biomarker Development. *Clin. Cancer Res.* **2010**, *16*, 4647–4653.

    c. Kogan, A. J.; Haren, M. Translating Cancer Trial Endpoints into the Language of Managed Care. *Biotechnol. Healthc.* **2008**, *5* (1), 22–35.

27. Hudis, C. A. Trastuzumab–mechanism of Action and Use in Clinical Practice. *N. Engl. J. Med.* **2007,** *357* (1), 39–51.

28. Carpenter, A. P., Jr.; Pontecorvo, M. J.; Hefti, F. F.; Skovronsky, D. M. The Use of the Exploratory IND in the Evaluation and Development of ^{18}F-PET Radiopharmaceuticals for Amyloid Imaging in the Brain: A Review of One Company's Experience. *Q. J. Nucl. Med. Mol. Imaging* **2009,** *53* (4), 387–393.

29. Som, P.; Atkins, H. L.; Bandoypadhyay, D.; Fowler, J. S.; MacGregor, R. R.; Matsui, K.; Oster, Z. H.; Sacker, D. F.; Shiue, C. Y.; Turner, H.; et al. A Fluorinated Glucose Analog, 2-fluoro-2-deoxy-d-glucose (F-18): Non-toxic Tracer for Rapid Tumor Detection. *J. Nucl. Med.* **1980,** *21*, 670–675.

30. Muller, C.; Schibli, R. Single Photon Emission Computed Tomography Tracer. *Recent Results Cancer Res.* **2013,** *187*, 65–105.

31. Antonini, A. The Role of 123I-iofl Upane SPECT Dopamine Transporter Imaging in the Diagnosis and Treatment of Patients with Dementia with Lewy Bodies. *Neuropsychiatric Dis. Treat.* **2007,** *3* (3), 287–292.

32. Leppo, J. A.; DePuey, E. G.; Johnson, L. L. A Review of Cardiac Imaging with Sestamibi and Teboroxime. *J. Nucl. Med.* **1991,** *32* (10), 2012–2022.

33. Manyak, M. J. Indium-111 Capromab Pendetide in the Management of Recurrent Prostate cancer. *Expert Rev. Anticancer Ther.* **2008,** *8* (2), 175–181.

34. Rahmima, A.; Zaidib, H. PET versus SPECT: Strengths, Limitations and Challenges. *Nucl. Med. Commun.* **2008,** *29*, 193–207.

35. Nag, S.; Lehmann, L.; Kettschau, G.; Heinrich, T.; Thiele, A.; Varrone, A.; Gulyas, B.; Halldin, C. Synthesis and Evaluation of [^{18}F]fluororasagiline, a Novel Positron Emission Tomography (PET) Radioligand for Monoamine Oxidase B (MAO-B). *Bioorg. Med. Chem.* **2012,** *20*, 3065–3071.

36. Zeng, F.; Jarkas, N.; Stehouwer, J. S.; Voll, R. J.; Owens, M. J.; Kilts, C. D.; Nemeroff, C. B.; Goodman, M. M. Synthesis, *in Vitro* Characterization, and Radiolabeling of Reboxetine Analogs as Potential PET Radioligands for Imaging the Norepinephrine Transporter. *Bioorg. Med. Chem.* **2008,** *16*, 783–793.

37. Fernandes, P. A.; Carvalho, A. T. P.; Marques, A. T.; Pereira, A. L. F.; Madeira, A. P. S.; Ribeiro, A. S. P.; Carvalho, A. F. R.; Ricardo, E. T. A.; Pinto, F. J. V.; Santos, H. A.; et al. New Designs for MRI Contrast Agents. *J. Computer-Aided Mol. Des.* **2003,** *17* (7), 463–473.

38. a. Grundman, M.; Sencakova, D.; Jack, C. R., Jr.; Petersen, R. C.; Kim, H. T.; Schultz, A.; Weiner, M. F.; DeCarli, C.; DeKosky, S. T.; van Dyck, C.; et al. Brain MRI Hippocampal Volume and Prediction of Clinical Status in a Mild Cognitive Impairment Trial. *J. Mol. Neurosci.* **2002,** *19*, 23–27.

    b. Jack, C. R., Jr.; Slomkowski, M.; Gracon, S.; Hoover, T. M.; Felmlee, J. P.; Stewart, K.; Xu, Y.; Shiung, M.; O'Brien, P. C.; Cha, R.; et al. MRI as a Biomarker of Disease Progression in a Therapeutic Trial of Milameline for AD. *Neurology* **2003,** *60*, 253–260.

    c. Fox, N. C.; Warrington, E. K.; Freeborough, P. A.; Hartikainen, P.; Kennedy, A. M.; Stevens, J. M.; Rossor, M. N. Presymptomatic Hippocampal Atrophy in Alzheimers Disease – A Longitudinal MRI Study. *Brain* **1996,** *119*, 2001–2007.

39. a. Choudhury, R. P.; Fuster, V.; Badimon, J. J.; Fisher, E. A.; Fayad, Z. A. MRI and Characterization of Atherosclerotic Plaque: Emerging Applications and Molecular Imaging. *Arterioscler. Thromb. Vasc. Biol.* **2002,** *22*, 1065–1074.

    b. Yuan, C. P.; Zhang, S. X.; Polissar, N. L.; Echelard, D.; Ortiz, G.; Davis, J. W.; Ellington, E.; Ferguson, M. S.; Hatsukami, T. S. Identification of Fibrous Cap Rupture with Magnetic Resonance Imaging Is Highly Associated with Recent Transient Ischemic Attack or Stroke. *Circulation* **2002,** *105*, 181–185.

40. Jacobs, L. D.; Beck, R. W.; Simon, J. H.; Kinkel, R. P.; Brownscheidle, C. M.; Murray, T. J.; Simonian, N. A.; Slasor, P. J.; Sandrock, A. W. Intramuscular Interferon-β-1a Therapy Initiated during a First Demyelinating Event in Multiple Sclerosis. *N. Engl. J. Med.* **2000,** *343*, 898–904.

41. Huettel, S. A.; Song, A. W.; McCarthy, G. *Functional Magnetic Resonance Imaging*, 2nd ed.; Sinauer Associates: Sunderland, Massachusetts, 2009.
42. Sheline, Y. I.; Barch, D. M.; Donnelly, J. M.; Ollinger, J. M.; Snyder, A. Z.; Mintun, M. A. Increased Amygdala Response to Masked Emotional Faces in Depressed Subjects Resolves with Antidepressant Treatment: An fMRI Study. *Biol. Psychiatry* **2001,** *50,* 651–658.
43. Wise, R. G.; Rogers, R.; Painter, D.; Bantick, S.; Ploghaus, A.; Williams, P.; Rapeport, G.; Tracey, I. Combining fMRI with a Pharmacokinetic Model to Determine Which Brain Areas Activated by Painful Stimulation Are Specifically Modulated by Remifentanil. *Neuroimage* **2002,** *16,* 999–1014.
44. Borsook, D.; Becerra, L.; Hargreaves, R. A Role for fMRI in Optimizing CNS Drug Development. *Nat. Rev. Drug Discov.* **2006,** *5,* 411–425.
45. Ludwig, G. D.; Struthers, F. W. *Considerations Underlying the Use of Ultrasound to Detect Gallstones and Foreign Bodies in Tissue;* Naval Medical Research Institute Reports, June 1949. Project #004 001, Report No. 4.
46. Wild, J. J.; Neal, D. Use of High-frequency Ultrasonic Waves for Detecting Changes of Texture in Living Tissues. *Lancet* **1951,** *257* (6656), 655–657.
47. Willman, J. K.; van Bruggen, N.; Dinkelborg, L. M.; Gambhir, S. S. Molecular Imaging in Drug Development. *Nat. Rev. Drug Discov.* **2008,** *7,* 591–607.
48. Kim, D.; Wang, L.; Beconi, M.; Eiermann, G. J.; Fisher, M. H.; He, H.; Hickey, G. J.; Kowalchick, J. E.; Leiting, B.; Lyons, K.; et al. (2R)-4-oxo-4-[3-(trifluoromethyl)-5,6-dihydro [1,2,4]triazolo[4,3-a]pyrazin-7(8H)-yl]-1-(2,4,5 Trifluorophenyl)butan-2-amine: A Potent, Orally Active Dipeptidyl Peptidase IV Inhibitor for the Treatment of Type 2 Diabetes. *J. Med. Chem.* **2005,** *48,* 141–151.
49. Augeri, D. J.; Robl, J. A.; Betebenner, D. A.; Magnin, D. R.; Khanna, A.; Robertson, J. G.; Wang, A.; Simpkins, L. M.; Taunk, P.; Huang, Q.; et al. Discovery and Preclinical Profile of Saxagliptin (BMS-477118): A Highly Potent, Long-acting, Orally Active Dipeptidyl Peptidase IV Inhibitor for the Treatment of Type 2 Diabetes. *J. Med. Chem.* **2005,** *48,* 5025–5037.
50. Eckhardt, M.; Langkopf, E.; Mark, M.; Tadayyon, M.; Thomas, L.; Nar, H.; Pfrengle, W.; Guth, B.; Lotz, R.; Sieger, P.; et al. 8-(3-(R)-Aminopiperidin-1-yl)-7-but-2-ynyl-3-methyl-1-(4-methyl-quinazolin-2-ylmethyl)-3,7-dihydropurine-2,6-dione (BI-1356), a Highly Potent, Selective, Long-acting, and Orally Bioavailable DPP-4 Inhibitor for the Treatment of Type 2 Diabetes. *J. Med. Chem.* **2007,** *50,* 6450–6453.
51. a. Holst, J. J. Glucagon-like Peptide 1 (GLP-1) a Newly Discovered GI Hormone. *Gastroenterology* **1994,** *107,* 1048–1055.
    b. Drucker, D. J. Glucagon-like Peptides. *Diabetes* **1998,** *47,* 159–169
    c. Deacon, C. F.; Holst, J. J.; Carr, R. D. Glucagon-like Peptide 1: A Basis for New Approaches to the Management of Diabetes. *Drugs Today* **1999,** *35,* 159–170.
    d. Livingston, J. N.; Schoen, W. R. Glucagon and Glucagon-like Peptide-1. *Annu. Reports Med. Chem.* **1999,** *34,* 189–198.
52. Herman, G. A.; Stein, P. P.; Thornberry, N. A.; Wagner, J. A. Dipeptidyl Peptidase-4 Inhibitors for the Treatment of Type 2 Diabetes: Focus on Sitagliptin. *Clin. Pharmacol. Ther.* **2007,** *81* (5), 761–767.
53. Iber, C.; Ancoli-Israel, S.; Chesson, A.; Quan, S. F. *The AASM Manual for the Scoring of Sleep and Associated Events: Rules, Terminology and Technical Specifications;* The American Academy of Sleep, Westchester: American Academy of Sleep Medicine, 2007.
54. Mamelak, M.; Csima, A.; Price, V. A Comparative 25-night Sleep Laboratory Study on the Effects of Quazepam and Triazolam on Chronic Insomniacs. *J. Clin. Pharmacol.* **1984,** *24* (2–3), 65–75.
55. Bixler, E. O.; Kales, A.; Soldatos, C. R.; Scharf, M. B.; Kales, J. D. Effectiveness of Temazepam with Short-intermediate-, and Long-term Use: Sleep Laboratory Evaluation. *J. Clin. Pharmacol.* **1978,** *18* (2–3), 110–118.

56. Tsujino, N.; Sakurai, T. Orexin/hypocretin: A Neuropeptide at the Interface of Sleep, Energy Homeostasis, and Reward System. *Pharmacol. Rev.* **2009,** *61* (2), 162–176.

57. a. Estabrooke, I. V.; McCarthy, M. T.; Ko, E.; Chou, T. C.; Chemelli, R. M.; Yanagisawa, M.; Saper, C. B.; Scammell, T. E. Fos Expression in Orexin Neurons Varies with Behavioral State. *J. Neurosci.* **2001,** *21,* 1656–1662.

   b. Lee, M. G.; Hassani, O. K.; Jones, B. E. Discharge of Identified Orexin/hypocretin Neurons across the Sleepwaking Cycle. *J. Neurosci.* **2005,** *25,* 6716–6720.

   c. Mileykovskiy, B. Y.; Kiyashchenko, L. I.; Siegel, J. M. Behavioral Correlates of Activity in Identified Hypocretin/Orexin Neurons. *Neuron* **2005,** *46,* 787–798.

58. Chemelli, R. M.; Willie, J. T.; Sinton, C. M.; Elmquist, J. K.; Scammell, T.; Lee, C.; Richardson, J. A.; Williams, S. C.; Xiong, Y.; Kisanuki, Y.; et al. Narcolepsy in Orexin Knockout Mice: Molecular Genetics of Sleep Regulation. *Cell* **1999,** *98,* 437–451.

59. Willie, J. T.; Chemelli, R. M.; Sinton, C. M.; Tokita, S.; Williams, S. C.; Kisanuki, Y. Y.; Marcus, J. N.; Lee, C.; Elmquist, J. K.; Kohlmeier, K. A.; et al. Distinct Narcolepsy Syndromes in Orexin Receptor 2 and Orexin Null Mice: Molecular Genetic Dissection of Non-REM and REM Sleep Regulatory Processes. *Neuron* **2003,** *38,* 715–730.

60. a. Thannickal, T. C.; Moore, R. Y.; Nienhuis, R.; Ramanathan, L.; Gulyani, S.; Aldrich, M.; Cornford, M.; Siegel, J. M. Reduced Number of Hypocretin Neurons in Human Narcolepsy. *Neuron* **2000,** *27,* 469–474.

   b. Crocker, A.; Espana, R. A.; Papadopoulon, M.; Saper, C. B.; Faraco, J.; Sakurai, T.; Honda, M.; Mignot, E.; Scammell, T. E. Concomitant Loss of Dynorphin, Narp and Orexin in Narcolepsy. *Neurology* **2005,** *65,* 1184–1188.

   c. Blouin, A. M.; Thannickal, T. E.; Worley, P. F.; Baraban, J. M.; Reti, I. M.; Siegel, J. M. Narp Immunostaining of Human Hypocretin (Orexin) Neurons: Loss in Narcolepsy. *Neurology* **2005,** *65,* 1189–1192.

61. Vaughn, B. V.; Giallanza, P. Technical Review of Polysomnography. *Chest* **2008,** *134* (6), 1310–1319.

62. Cox, C. D.; Breslin, M. J.; Whitman, D. B.; Schreier, J. D.; McGaughey, G. B.; Bogusky, M. J.; Roecker, A. J.; Mercer, S. P.; Bednar, R. A.; Lemaire, W.; et al. Discovery of the Dual Orexin Receptor Antagonist [(7R)-4-(5-chloro-1,3-benzoxazol-2-yl)- 7-methyl-1, 4-diazepan-1-yl][5-methyl-2-(2H-1,2,3-triazol-2-yl)phenyl]methanone (MK-4305) for the Treatment of Insomnia. *J. Med. Chem.* **2010,** *53,* 5320–5332.

63. a. Winrow, C. J.; Gotter, A. L.; Cox, C. D.; Doran, S. M.; Tannenbaum, P. L.; Breslin, M. J.; Garson, S. L.; Fox, S. V.; Harrell, C. M.; Stevens, J.; et al. Promotion of Sleep by Suvorexant – A Novel Dual Orexin Receptor Antagonist. *J. Neurogenet.* **2011,** *25* (1–2), 52–61.

64. Ido, T.; Wan, C. N.; Casella, V.; Fowler, J. S.; Wolf, A. P.; Reivich, M.; Kuhl, D. E. Labeled 2-deoxy-d-glucose Analogs: ^{18}F-labeled 2-deoxy-2-fluoro-d-glucose, 2-deoxy-2-fluoro-d-mannose and ^{14}C-2-deoxy-2-fluoro-d-glucose. *J. Labeled Compd. Radiopharm.* **1978,** *24,* 174–183.

65. Sokoloff, L.; Reivich, M.; Kennedy, C.; Des Rosiers, M. H.; Patlak, C. S.; Pettigrew, K. D.; Sakurada, O.; Shinohara, M. The [^{14}C]deoxyglucose Method for the Measurement of Local Cerebral Glucose Utilization: Theory, Procedure, and Normal Values in the Conscious and Anesthetized Albino Rat. *J. Neurochem.* **1977,** *28* (5), 897–916.

66. Smith, T. A. Mammalian Hexokinases and Their Abnormal Expression in cancer. *Br. J. Biomed. Sci.* **2000,** *57* (2), 170–178.

67. Hale, J. J.; Mills, S. G.; MacCoss, M.; Finke, P. E.; Cascieri, M. A.; Sadowski, S.; Ber, E.; Chicchi, G. G.; Kurtz, M.; Metzger, J.; et al. Structural Optimization Affording 2-(R)-(1-(R)-3,5-bis(trifluoromethyl)phenylethoxy)-3-(S)-(4-fluoro)phenyl-4-   (3-oxo-1,2,4-triazol-5-yl) methylmorpholine, a Potent, Orally Active, Long-acting Morpholine Acetal Human NK-1 Receptor Antagonist. *J. Med. Chem.* **1998,** *41* (23), 4607–4614.

68. Hargreaves, R.; Ferreira, J. C. A.; Hughes, D.; Brands, J.; Halle, J.; Mattson, B.; Mills, S. Development of Aprepitant, the First Neurokinin-1 Receptor Antagonist for the Prevention of Chemotherapy-induced Nausea and Vomiting. *Ann. N. Y. Acad. Sci.* **2011,** *1222,* 40–48.

69. a. Holmes, A.; Heilig, M.; Rupniak, N. M. J.; Steckler, T.; Griebel, G. Neuropeptide Systems as Novel Therapeutic Targets for Depression and Anxiety Disorders. *Trends Pharmacol. Sci.* **2003**, *24*, 580–588.

   b. Mantyh, P. W. Neurobiology of Substance P and the NK1 Receptor. *J. Clin. Psychiatry* **2002**, *63* (S11), 6–10.

   c. Rupniak, N. M. J. New Insights into the Antidepressant Actions of Substance P (NK1 Receptor) Antagonists. *Can. J. Physiol. Pharmacol.* **2002**, *80*, 489–494.

   d. Santarelli, L.; Gobbi, G.; Blier, P.; Hen, R. Behavioral and Physiologic Effects of Genetic or Pharmacologic Inactivation of the Substance P Receptor (NK1). *J. Clin. Psychiatry* **2002**, *63* (S11), 11–17.

70. Burns, H. D.; Gibson, R. E.; Hamill, T. G. *Preparation of Radiolabeled Neurokinin-1 Receptor Antagonists;* , 2000. WO 2000018403.

71. Johnson, A. M. Paroxetine: A Pharmacological Review. *Int. Clin. Psychopharmacol.* **1992**, *6* (S4), 15–24.

Organizational Considerations and Trends in the Pharmaceutical Industry

The number of scientific hurdles that must be overcome in order to convert scientific discoveries into clinically useful and commercially viable therapeutics is enormous by any measure. Unfortunately, there are a number of non-scientific issues that can, and often do, distract scientists and staff from their main goal of moving programs forward. The organizational structure of a company, for example, can either facilitate or impede program progress. Similarly, the evolutionary nature of project teams directly effects how and when scientists and support staff interact with each other as programs progress. The dynamics within this changing team will almost certainly have a significant impact on the chances of successfully reaching the market with a novel therapeutic and the careers of those involved in the process. Changes in the business climate can also influence the science of drug discovery and development. In some cases, changes to the industry occur slowly. The rise of contract research organizations and academic drug discovery research centers, for example, began just before the turn of the twenty-first century and their true impact on the pharmaceutical industry is still unfolding. In other cases, the impact is more immediate and jarring. Mergers and acquisitions, for example, almost always cause significant upheaval in the affected companies (and the lives of their employees). Finally, in smaller companies, gaining access to funding is a major issue that must be addressed, and understanding where to look for capital can mean the difference between continued scientific advancement and closure of the company. In short, a successful career in the pharmaceutical industry requires an understanding of not just the science of drug discovery and development, but also the ability to adapt to the changing requirements of the industry as a whole.

Basic Principles of Drug Discovery and Development
http://dx.doi.org/10.1016/B978-0-12-411508-8.00011-6

ORGANIZATIONAL STRUCTURES OF PHARMACEUTICAL COMPANIES

Although the previous chapters provide an overview of many important components of the drug discovery and development process, they do not provide an organizational framework for the various functions. Alone, none of the components will lead to the development of a clinically useful drug, but when brought together, the full process becomes greater than the sum of its parts. A major question that the pharmaceutical industry has faced over the decades, and indeed has as yet to fully answer, is how to most efficiently link these activities together. An organizational structure that allows the various functions to collaborate and communicate can have a substantial impact on the overall cost and time lines associated with bringing a new drug to market, just as they would impact the production costs and time required to assemble an automobile. While it is certainly possible to manufacture cars with installation of the air conditioning as the last step in the process, this is almost certainly less efficient than installing the various components of the air conditioning as the car is being assembled. In a similar manner, significant efficiency can be derived by establishing various working groups within a research and development organization and setting guidelines as to how and when they interact.

BUSINESS DIVISIONS INTERACTIONS

At the highest level, pharmaceutical company activities can be broadly grouped into three areas: discovery, development, and commercial. The first two areas, discovery and development, were broadly described in Chapter 1. Commercial organizations are generally responsible for ensuring the commercial viability of a new drug once it reaches the market (postapproval) and for the remainder of its marketable lifetime. Manufacturing, sales, and marketing operations fall within the purview of commercial groups. In most cases, these three areas are established as distinct working groups, each with their own management lines, productivity goals, and success metrics that define how well they have performed during a given time period. Each of these operating groups has a significant degree of autonomy from the others and operates independently on a number of levels.

It is important to realize, however, that in order for a company to operate efficiently, there must be clear lines of communication between divisions. In addition, each working group must be aware of the capabilities and limitations of the other groups. Consider, for example, the problems that might occur if a discovery organization spent time and resources pursuing a program to identify novel therapeutics for a disease that the development organization could not study clinically. Even if the discovery organization identified a potential novel therapeutic, the development organization would be unable to clinically validate the efficacy of the new treatment. In a similar sense, if the commercial group is unable to identify a patient population, pricing paradigm, and marketing strategy that will allow a novel therapy to generate the revenue necessary to support the future of the company, efforts in the discovery and development organizations are essentially wasted. Large-market diseases that could be highly profitable would certainly be attractive to commercial groups within a pharmaceutical company. In the absence of a viable scientific approach to produce novel treatments, however, efforts by the commercial organization to define the scope of the potential market, assess patient needs, and understand prescribing practices would be of limited value.

Ideally, each of the three major facets of a pharmaceutical company should be included in decisions regarding which programs, disease states, or therapeutic areas are within the scope of the organization's capabilities and interests. In addition, since the process of bringing a new drug to market is exceptionally long, continuous interaction between these organizational components is a must. Increased scientific understandings, the arrival of competitor therapeutics, or changes in patient demographics are just some of the forces that alter the value proposition of pursuing research and business strategies. Competitor compounds almost always have a significant impact on the commercial viability of a potential therapeutic, new scientific data could validate a previously unproven target (or disprove a target previously believed to be valid), and changes in patient demographics could complicate clinical and marketing strategies. In other words, the successful development of a novel therapeutic requires both rigorous scientific advancement of a program and constant, vigilant awareness of the scientific and competitive landscape.

THE DISCOVERY PROJECT TEAM EVOLUTIONARY CYCLE

It should be clear at this point that the process of bringing a new drug to market is an exceptionally complex endeavor. There is no way that a single individual could possess all of the expertise required to complete the task. Drug discovery and development is a team sport requiring a number of players with a wide array of skills and knowledge. Of course, the efforts of a collection of diverse scientists, technicians, and specialists must be coordinated in some manner in order to ensure that projects move forward. This is most often accomplished through the formation of a project team. The composition and leadership of this team is somewhat fluid in nature as a result of the many tasks required. Project needs and focus will change as a program advances, and as a result, the skill sets required for further advancement of the project will also change. In order to be effective, the composition and leadership of a project team should evolve as the project evolves.

In the early discovery stage of a project (prior to the identification of a lead series of compounds), project team membership is generally dominated by discovery scientists (e.g., *in vitro* pharmacologists, medicinal chemists) and led by a biologist, a medicinal chemist, or one of each who are experts in the molecular basis of the targeted disease/indication. In addition, experts in mid- to late-stage discovery efforts, such as *in vitro* ADME, pharmacokinetics, and *in vivo* pharmacology, should be a part of the project team. At this point in the process, the role of these individuals may be relatively small, as efforts will be focused on identifying lead classes of compounds. It is important, however, that they be aware of program activities so that they can prepare for the eventual transition of focus to animal models. In addition, if the target has as yet to be validated in an animal model, these individuals may be tasked with designing an appropriate animal model and validating the target. This would likely occur in parallel to the efforts to identify a lead series of compounds. Finally, in an ideal world, representatives from later-stage processes/organizations, such as safety, toxicology, clinical trials, and commercial will also be included. Although these individuals are not experts in early drug discovery, their input ensures that early stage programs are in-line with an organization's capabilities, goals, and business strategies.

As a project progresses through the various stages of discovery research (e.g., target discovery, lead discovery, lead optimization, early preclinical), the composition of the project team should change to match the needs of the program. If a project begins with target discovery (e.g., the identification and characterization of a novel therapeutic target), the project team will be dominated by *in vitro* biologists, as their skills and expertise will be required to identify, characterize, and develop assays for a novel therapeutic target. Medicinal chemistry participation, on the other hand, may be limited, as in the absence of a defined macromolecular target, developing chemical leads is challenging.

Once a novel therapeutic target has been characterized and *in vitro* screening assays have been established, the composition of the project team should shift to reflect the changing needs of the program. Scientists with expertise in identifying novel targets may be replaced with individuals skilled in running high-throughput assays in order to facilitate data acquisition. The number of medicinal chemists assigned to the program will increase as the need to interpret *in vitro* screening results increases. These scientists will also design and prepare new compounds based on hits identified in the first round of high-throughput screening. Experts in molecular modeling may also be added to the team in order to develop model systems capable of predicting which compounds will possess the desired *in vitro* activity. *In vitro* ADME scientists may also be brought on board in order to begin assessing compound properties as a prelude to eventual animal experiment (e.g., *in vivo* PK and efficacy). Team leadership typically remains unchanged at this point, and representation of late-stage processes/organizations is also typically maintained.

When a lead series is identified and the project moves into the lead optimization stage, the composition of the project team shifts again. Increased participation by medicinal chemists is generally required in order to facilitate improvements in the characteristics of project compounds. Process chemists may be added to the team in order to expedite the preparation of larger batches of materials necessary for *in vivo* PK and efficacy studies. The role of *in vitro* ADME scientists continues as the project team continues to evaluate compounds in an attempt to identify biologically relevant compounds that are also suitable for *in vivo* PK assessment (e.g., soluble, stable in microsome assays, permeable). Scientists specializing in *in vivo* experiments (e.g., *in vivo* PK, *in vivo* efficacy models, and formulation) become increasingly important to the project team, as the need to identify compounds with *in vivo* efficacy becomes a major focus of the program.

As the lead optimization stage of the project continues, ultimately headed toward a single clinical candidate, the composition of the project team shifts once again. Team leadership will likely transition to individuals from later-stage processes/organizations, especially once a small set of candidate compounds are identified as possible clinical candidates. Medicinal chemistry participation typically decreases as the need to identify new and improved compounds decreases (although there may be some efforts directed toward elaboration of a series of compounds for the purposes of patent enablement). At the same time, the role of process chemists and manufacturing specialists may rise as the need to prepare clinical supplies approaches. *In vitro* screening and ADME scientists also play a diminished role in the project team, while experts in *in vivo* safety, toxicology, and formulations become increasingly important. The need to characterize specific compounds in advanced *in vivo* efficacy, safety, and toxicology models will take center stage. By the time a single clinical candidate has been identified, the leadership roles of the project are typically transitioned to clinical scientists, manufacturing specialists, and commercialization experts.

Once a clinical candidate has been identified, the majority of discovery scientists on the project team are replaced by clinical scientists, regulatory experts, and manufacturing specialists. The discovery team leaders may be kept as part of the project team, so that they are available for consultation as needed, but other discovery scientists are generally not part of the process at this juncture. In some organizations, the discovery team will be tasked with identifying a backup compound that could move forward if the original lead compound fails in early clinical trials. Failure of a clinical candidate in phase I due to safety concerns, for example, might lead to advancement of a backup candidate. (A phase III study failure due to lack of efficacy, on the other hand, is likely to end pursuit of the clinical candidate and any backup compounds.) If a backup program is launched, the characteristics of the project team will be similar to that described for the lead optimization stage. In other cases, the discovery project team is disbanded and the resources (personnel) are reassigned to other discovery programs if there are additional needs within the organization. The development project team, on the other hand, continues to focus on the clinical candidate and advancing it through clinical trials as described in Chapter 9.

Although this is a relatively vague description of the evolution of a discovery project team, it demonstrates an important point. The role of an individual scientist in a discovery project team is not a permanent position. Scientists involved in drug discovery should expect to be a part of multiple, possibly overlapping, project teams through the course of their career. It is also important to understand that success and failure of a project both lead to the reassignment of scientists. When a project fails, resources, including project team members, are reassigned (hopefully, to positions within the same organization, but this is not always the case). Successful discovery projects also have limited use for discovery resources and personnel, as the project focus shifts toward development issues. As a result, resources and project team members are reassigned to other programs (again, hopefully, to positions within the same organization). In a large company, success in a program can be rewarded with advancement within the organization, but in smaller companies, the success of a discovery project team can lead to layoffs. Small companies with limited resources that successfully identify a clinical candidate may not have the ability (money) to support a clinical program and to launch a new discovery project. Since clinical programs require a different skill set, the discovery project team may be laid off so that the company can hire a development team capable of pushing their clinical candidate down the path to commercialization. In other words, in a small company, discovery scientists can be a casualty of their own success.

THE BUSINESS CLIMATE

While it is certainly true that pharmaceutical companies are heavily reliant on the scientific discoveries that lead to novel products and profitability, they do not operate in a vacuum. Changes in the overall business environment as a result of economic pressures, regulatory changes, or the political landscape, for example, can have a dramatic impact on how individual companies, and indeed the industry as a whole, functions. As discussed in Chapter 2, regulatory changes have dramatically altered the pharmaceutical industry through the creation of safety standards, efficacy standards, and increased access to generic drugs. Mergers and acquisitions have also played an ongoing role in the pharmaceutical industry, especially as the need to lower the cost of commercialization became more pronounced at the end of the twentieth century and the beginning of the twenty-first century. These same forces also contributed to two additional changes in how drugs are identified and commercialized. They led to the formation of contract research organizations and academic drug discovery centers. Whether or not these changes have had a positive or negative impact on the pharmaceutical industry as a whole is unclear, but it is clear that those employed in this sector will continue to be impacted by these kinds of changes. It is in the best interest of those wishing to pursue a career in drug discovery and development to have an understanding of these important, ongoing industry trends.

MERGERS AND ACQUISITIONS

Mergers and acquisitions are not unique to the pharmaceutical industry, but the level of this type of activity has increased substantially since the beginning of the twenty-first century. Between 2000 and 2009, over 1300 merger and acquisition (M&A) deals were consummated with a total value in excess of $690 billion. These deals played a major role in the loss of over 300,000 jobs in the US pharmaceutical industry in the same time frame.[1] Thousands of additional jobs in other parts of the world and in sectors that support the pharmaceutical industry were also negatively impacted by these deals. While the overwhelming majority of these deals involved smaller firms, large pharmaceutical companies were not immune from the feeding frenzy. A total of 46 deals with a value in excess of $2.0 billion occurred between 2000 and 2012, 16 of which topped $10.0 billion (Table 11.1). This dramatic increase in merger activity was driven in part by the need to replace billion dollar products such as Lipitor® (Atorvastatin), Cymbalta® (Duloxetine), Plavix® (Clopidogrel), and Singulair® (Montelukast) that were on the verge of losing patent protection (the "patent cliff"). It has been estimated that between 2011 and 2015, over $250 billion in sales of patent-protected drugs have become vulnerable to generic competition.[2]

Decreasing the overall costs associated with the research and development process was also a major factor leading to increased M&A activity. As discussed earlier in this text, the cost of bringing a single new drug

TABLE 11.1   Deals Valued Above $10 billion, 2000–2012

Year	Companies	Dollars in billion
2000	Pfizer–Warner Lambert	90
	Glaxo Wellcome–SmithKline Beecham	74
	Pharmacia and Upjohn–Monsanto	50
2001	Johnson & Johnson–Alza	12.3
2004	Sanofi Synthelabo–Aventis	62
	GE–Amersham	10.2
2005	Johnson & Johnson–Guidant	21
2006	Bayer AG–Schering AG	21
	J & J–Pfizer Consumer	16.6
2007	Schering–Plough–Organon	11
	AstraZeneca–Medimune	15.2
2008	Roche–Genentech	44
2009	Pfizer–Wyeth	68
	Merck–Schering–Plough	41
2010	Novartis–Alcon	51
2011	Gilead–Pharmasset	11

to market can reach $1.75 billion.[3] This high cost is seen by many as a leading factor in the high cost of prescription medications. The hepatitis C drug Sovaldi® (Sofosbuvir, Figure 11.1) developed by Gilead and approved by the FDA in 2013, for example, costs approximately $84,000 for a 12-week treatment. To be fair, the long-term costs associated with

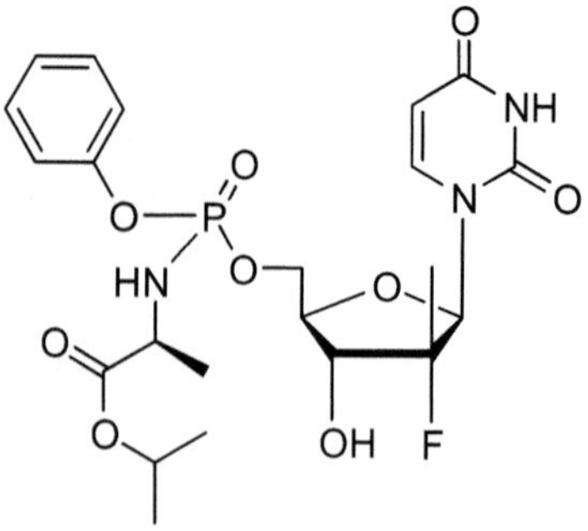

FIGURE 11.1   Sovaldi® (Sofosbuvir).

caring for hepatitis C patients is far in excess of the cost of this treatment. The 2014 annual costs hepatitis C patient care to the US health care system alone is over $30 billion and was expected to exceed $85 billion prior to the approval of this new drug.[4] This long-term view, however, is not often taken into account by patients, policy makers, and payers (e.g., insurance companies, governmental organizations) who are increasingly focused on lowering the cost of prescription medications. These issues are also largely ignored by stockholders, investors, and upper management of pharmaceutical companies who are largely focused on the profitability of their drug portfolio and the short-term value of their companies as a whole.

On the surface, mergers and acquisitions appear to be an attractive method of lowering the costs of identifying, developing, and commercializing new therapies. The surviving organization can eliminate duplicative efforts (i.e., two research programs focused on the same macromolecular target), redundant personnel (e.g., staff attached to eliminated programs or other areas of redundancy such as sales, human resources, or finance), and even R&D facilities that are no longer needed by the smaller surviving organization. These and other organizational restructuring events generally appeal to stockholders, investors, and management, as they produce an immediate impact on the "cost of doing business," but the true impact of mergers and acquisitions on R&D efficiency remains unclear.

Although monetary impacts can be quantified, there are substantial non-monetary effects of mergers and acquisitions that are not easily measured. Consider, for example, the elimination of redundant or undesired programs from a pair of merging companies. Two programs focused on the same biological target/disease state in competing organizations might lead to two marketed drugs in competition for market share, which might keep drug prices lower. In a merger (or acquisition) scenario, the two projects are viewed as redundant, and one will invariably be canceled in favor of the other. Resources are conserved in the postmerger organization, but one could argue that the efforts of the discarded program have been effectively thrown in the garbage and a 50% loss of potential clinical candidates has occurred. In a similar manner, one of the two merging companies may operate in areas that will not be of interest to the postmerger organization. Programs in the undesired area are likely to be terminated irrespective of their potential to produce marketable therapeutics. This is especially true if the business transaction is an acquisition rather than a "merger of equals."

The review process designed to determine which programs will continue within the emerging organization can also create drag on operational efficiencies that are not easily measured. A considerable amount of time and energy must be expended reviewing the combined project portfolios of each company to determine which ones will continue in the organization postmerger. While this is occurring, scientific productivity will almost certainly decrease as the program scientists must turn their attention toward proving that their program is superior to other programs vying for survival and away from scientific pursuits designed to advance the programs. The instinct for self-preservation may also cause key personnel to leave the organization in order to avoid being laid off if their program is selected for termination. Organizational memory and expertise may be sacrificed in the name of cost cutting. The true impact of these "human factors" is difficult to quantify. There can be little doubt, however, that fewer experiments can be designed and fewer discoveries can be made if there are fewer scientists and support staff available.

Small companies are often viewed as acquisition targets by established pharmaceutical companies, especially if the large company is nearing a "patent cliff" for one of its major products. Acquiring a small biotech operation can provide a pharmaceutical company access to late-stage programs and clinical candidates that have significantly less risk, thus improving the pipeline of the acquiring company (or merged entity). Many small and start-up companies are, in fact, focused on attracting the attention of a large pharmaceutical company in the hopes of being acquired rather than building capacity to form a fully integrated pharmaceutical company. The end goal for these organizations is not a commercial product, but rather mid-to-late phase 2 stage clinical candidates that they can use as "bait" to reel in a "big fish." In this scenario, the small company has substantially "de-risked" the program, and the owners/shareholders can reap a significant financial reward when the company is acquired. Of course, the acquiring company is unlikely to retain any of the staff of the small company once the deal is consummated, which is an important consideration for anyone interested in working in start-up biotechnology companies. The lifetime of a successful start-up biotechnology can be very short relative to the length of a career in the pharmaceutical industry.

It is worth noting at this point that large companies are not immune to the potential for major job cuts in the wake of a merger or acquisition. Perceived synergies or duplications in two merging R&D organizations can lead to significant decreases in the overall R&D program of the final entity, which in turn leads to loss of employment for those impacted. The acquisition of Wyeth by Pfizer, for example, led to a substantial decrease in R&D activities as measured by their reported R&D budgets. In 2008, the combined R&D budget for the two organizations was over $11 billion, whereas the postacquisition company reported an annual R&D budget of approximately $7 billion.[5] At the end of the day, M&A activity can lead to a decreased number of "shots on goal," which may translate into fewer novel therapeutics and in the longer term, decreased efficiency of R&D operations, even if short-term costs decrease.

In an attempt to determine the impact of M&A activity on the productivity of pharmaceutical companies, Munos et al.[3] examined changes in

FDA approval rates of new molecular entities (NMEs) at a series of companies that were involved in M&A activity. Their study included 10 large companies acquiring small companies, six companies involved in mergers of larger pharmaceutical companies, and 14 small biotech companies that acquired another company. Interestingly, they found that mergers of larger companies produced no significant changes in the rate of FDA approval of NMEs, while acquisitions by larger companies were quite disruptive. FDA approvals dropped by as much as 70%. Interestingly, small company acquisitions lead to better results, as FDA approval rates increased by up to 118% in the companies studied. Although this study is very small, it does suggest that the strategy of acquiring companies as a means of addressing productivity issues within the pharmaceutical industry may not be the solution the industry needs.

Irrespective of whether or not mergers and acquisitions can solve productivity issues in the industry, there is no question that the waves of mergers and acquisitions that occurred at the end of the twentieth century and the beginning of the twenty-first century caused a substantial decrease in the number of scientists working in industrial R&D laboratories. It has also decreased the industry's capacity to pursue both basic and applied science required for drug discovery and development, as fewer R&D employees can only lead to a decrease in the number of sustainable R&D programs within a given company. Of course, the industry's need for new and innovative therapeutics remains unchanged, and as a result the pharmaceutical industry has increased its reliance on two relatively new sources of R&D capacity and innovation. As the R&D staffing levels in pharmaceutical companies declined, contract research organization and academic drug discovery centers began to grow. The emergence of these two types of organizations has provided new employment opportunities for displaced industrial scientists, new sources of research capacity for both large and small pharmaceutical companies, and additional avenues for novelty and innovation that is required to sustain the pharmaceutical industry.

CONTRACT RESEARCH ORGANIZATIONS

Contract research organizations (CROs) first emerged in the pharmaceuticals industry in the early 1980s. At that time, CROs were primarily focused on clinical trial management, but this changed over the next few decades as large pharmaceutical companies became increasingly focused on cutting costs and improving efficiency in drug discovery and development.[6] By the early twenty-first century, nearly every aspect of the drug discovery and development process that were traditionally handled in-house by a fully integrated pharmaceutical company could be executed by a CRO. According to a 2012 analysis by Tufts University Center for the Study of Drug Development, the US pharmaceutical CRO market was valued between $32.5 and $39.5 billion. The corresponding global market was estimated to be worth $90 to $105 billion, and both were predicted to continue to grow in size and importance.[7]

Although the original growth of the CRO market was primarily a result of decreased R&D expenditures by fully integrated pharmaceutical companies (e.g., GlaxoSmithKline, Pfizer, AstraZeneca, Amgen, Roche), small and start-up biotechnology companies have also benefited from growth in this sector. The availability of "temporary resources," such as expertise in formulation, ADME, safety, toxicology, pharmacology, medicinal chemistry, clinical science, manufacturing, and many other fields have allowed smaller companies to access the talent necessary to run drug discovery and development projects without the need to build internal capacity. In some cases, this has led to the formation of "virtual" drug companies in which only a small number of scientists and business personnel manage projects that are fully executed by CROs. As a result, in modern drug discovery and development, scientists must be ready to interact across corporate and cultural boundaries, as the trend toward outsourcings is unlikely to abate.

Supplementing or even replacing internal resources by accessing resources through CROs, particularly those in lower cost markets such as India and China, has become common place across the pharmaceuticals industry, but it is not without inherent risks. It is important that everyone involved in the process be aware of these risks prior to establishing a working relationship with a CRO. Outsourcing aspects of the drug discovery process requires that a significant amount of control of the process be turned over to the CRO. This may provide a significant cost savings up-front, but the "client company" will no longer have control of day-to-day operations. Risks to quality control, scientific integrity, confidentiality of information, and intellectual property issues also exist and must be considered before a working relationship is established with a CRO. These and other risks can be minimized by conducting thorough due diligence prior to signing a deal with a CRO.

Routine, detailed communication between the CRO and the "client organization" can also decrease the risks of a project "running off the rails," especially, if something unexpected occurs. Consider, for example, a case in which a CRO is tasked with running *in vivo* efficacy studies on a set of 10 closely related compounds and the first three produce the desired results. If this information is submitted to the "client organization" in a timely manner, a decision might be made to shift gears and run additional studies on the three successful candidates rather than expend additional resources on the remaining seven. On the other hand, if the first three failed to produce the desired results, and the CRO keeps going with the remaining seven compounds without informing the client of the results in a timely fashion, then the "client organization" may miss out on an opportunity to conserve resources by canceling the remaining studies. High-quality communication is absolutely critical when working with a CRO. This is especially true when the two organizations are separated by multiple time zones. An 8–12 h time difference adds significant challenges to managing a relationship with a CRO, especially with regard to real-time communication of results.

Macroeconomic forces are also changing the value proposition of the CRO in the pharmaceutical industry. In the 1980s and 1990s, the cost of a full-time equivalent (FTE, the price paid for a single employee, including benefits) was substantially lower in developing economies such as India and China. Infrastructures and the availability of research materials, especially specialty chemical and biological reagents, however, were inconsistent, creating the potential for program delays. As the CRO market matured, infrastructure was established and suppliers set up more efficient means of getting research material to the CROs in developing regions. At the same time, however, economic growth in these regions caused the difference between a CRO FTE and a company employee to drop substantially. This has led some companies to reexamine their use of CROs in developing economies in favor of CROs in established economies.

The maturation of the CRO industry has also changed the corporate goals of many CROs. While the vast majority of CROs were established with the goal of acting solely as a service organization for the pharmaceutical industry, their aspirations shifted as they built infrastructure, capacity, and experience. Many CROs have begun to develop their own research programs in order to establish their own stable of products. Scynexis Inc., a small company in Durham, North Carolina, for example, was founded as a CRO in 2000. Over time, they built on the expertise and infrastructure required for CRO activities and added internal drug discovery programs to their operation. As of 2014, Scynexis Inc. had nine programs in its pipeline.[8]

CROs that have not developed internal programs have recognized the increased reliance of the pharmaceutical industry on CROs. They have leveraged this reliance to their benefit to increase their participation in financial rewards of successful programs. Milestone payments and profit-sharing clauses are not uncommon in CRO agreements. Still, other CROs have established long-term research contracts in which their scientists work at the pharmaceutical company's research center. Eli Lilly, for example, signed a 6-year deal with Albany Molecular Research Inc., (AMRI) that called for 40 AMRI medicinal chemists to work in Eli Lilly labs.[9]

Overall, the value equation of outsourcing research and development operations has changed as the CRO industry has matured. Although the immediate cost savings associated with a decreased internal workforce and the elimination of research sites may temporarily satisfy the short-term goals of shareholders and investors, the true impact of CROs on the efficiency of the R&D process remains unclear. It is clear, however, that CROs will continue to play an important role in the R&D process for the foreseeable future.

ACADEMIC DRUG DISCOVERY

There is no question that universities have played a key role in the evolution of our understanding of disease progression, the identification of potential therapeutic targets, and determining the molecular mechanism of drug action. Historically, the primary focus of academic laboratories has been basic research supported by grants provided by governmental bodies and non-profit/charitable organizations with an interest in scientific research. Grant funding is a highly competitive process with only a small percentage of proposals receiving funding (10%–15%). As a result, scientific findings produced by academic laboratories are often highly innovative. However, research programs focused on basic research rarely produce commercially viable products. The identification of a key enzyme in a critical biological process in an academic laboratory, for example, might provide an innovative target for future drug discovery efforts. It might even be hailed as a key milestone of scientific discovery, but identifying a target is a long, long way from having a compound suitable for human clinical trials.

To be fair, there was little motivation, at least within the United States, for academic researchers to pursue the applied research programs necessary to identify potential clinical candidates. Prior to the beginning of the twenty-first century, grant funding for drug discovery programs in the academic community was almost unheard of. Drug discovery was thought to be the purview of industrial organizations and outside of the capabilities of academic laboratories. In addition, prior to 1980, there was no method available to establish patent protection for federally funded scientific discoveries, including potential therapeutically useful compounds. Inventions developed using federal funds in the United States before 1980 were considered the property of the US government and they could only be licensed on a non-exclusive basis. As discussed earlier, in the absence of exclusivity, pharmaceutical companies have little motivation to commercialize new therapies. Once a compound had been approved for clinical use, other companies could launch generic competitors, making it next to impossible for the lead company to recoup the money required to fund clinical development. Of course, academic institutions whose primary goal is student education are not equipped to commercialize new clinical entities, so in the absence of an industrial partner, forward progression of an academic drug discovery program prior to 1980 was all but impossible.

As a result of these policies, the US government held over 28,000 patents by 1980 and less than 5% of these applications had been utilized for commercial purposes[10]

The landscape changed dramatically in 1980 with the passage of the Bayh–Dole Act, also referred to as the Patent and Trademark Law Amendment Act. This change in US law allowed non-profit organizations (and small businesses) to maintain ownership of inventions developed using federal funding. In addition, the rights to the invention could be assigned or licensed to another organization capable of commercializing the invention. In effect, the passage of this law gave academic laboratories and their associated universities and colleges the ability to profit from scientific endeavors funded by organizations such as the National Institutes of Health, provided certain conditions were met. Specifically, preference in licensing had to be given to small businesses, the balance of royalties and profit obtained from the invention (after expenses) had to be used for scientific research or education, and a portion of the royalty must be shared with the inventors of record. This last clause provided substantial financial incentive to academic scientists to increase their focus on applied sciences, as commercialization of their research would enrich them personally.[11] Given the significant upside potential of new drug discovery and development (potentially billions of dollars in annual sales), it should come as no surprise that there has been a significant increase in the number of academic scientists exploring the science of drug discovery since the Bayh–Dole Act became law. By 2001, 12 academic drug discovery centers had been founded in the United States, and by 2013 there were over 100 centers in the United States [12] and numerous academic drug discovery centers across the globe (Appendix 1).

Modern academic drug discovery centers are typically staffed by pharmaceutical industry veterans with the skills and experience necessary to execute drug discovery programs (e.g., target validation, hit-to-lead, lead optimization, pharmacokinetic analysis, *in vivo* efficacy studies). Many highly qualified scientists entered the academic arena as the wave of mergers and acquisitions of 2000–2013 forced them out of their industrial positions. Many of these organizations also received donations of high-cost equipment (e.g., screening robots, analytical equipment, etc.) as large pharmaceutical companies have undergone mergers and closed industrial research centers in an effort to become more competitive. As a result of this influx of talent and equipment, academic drug discovery centers have been able to incorporate scientific advances in areas that are common place in the pharmaceutical industry such as high-throughput screening, robotics, *in vitro* ADME, and molecular modeling. This has enabled them to successfully execute the majority of functions that are required for early drug discovery programs (target validation, lead identification, lead optimization, *in vivo* PK, etc.). Universities and colleges have also stepped up their support of these initiatives by providing financial support when possible, supporting patent activity, and increasing their technology transfer operations to enable licensing opportunities.

Of course, drug discovery programs operating in academic environments are not immune to the issues and challenges associated with the identification of novel therapeutics. There is no change in the requirements for safety and efficacy based on academic versus industrial origin of potential clinical candidates. This raises the question of what is the advantage of pursuing a drug discovery program in an academic environment as opposed to an industrial center. The advantage of academic drug discovery centers lies in their access to the innovative research being conducted in academic research laboratories. As mentioned earlier, the competitive nature of the grant funding process all but ensures that research efforts in academic centers is innovative in nature. Within the context of drug discovery and development, this often means that academic scientists are exploring previously unidentified potential drug targets. In developing an understanding of these new targets, academic scientists have the opportunity to determine the biological impact of modulating the activity of these macromolecules and possibly uncover novel strategies for the treatment of unmet medical conditions. Collaborating with academic drug discovery centers provides academic scientists with the ability to develop their findings into programs that would be of interest to an industrial partner. Full drug discovery programs, complete with medicinal chemistry, *in vitro* selectivity screening, *in vitro* ADME, *in vivo* PK, and *in vivo* efficacy testing can be established in an academic setting. Ideally, the availability of these resources in an academic setting will provide universities and colleges with the ability to convert their basic research programs into something that would be of interest to an industrial partner.

It is important to keep in mind, however, that there are some inherent downsides to the academic–industrial partnerships. All of these potential risks can be mitigated and academic drug discovery centers, such as the Moulder Center for Drug Discovery Research, the Vanderbilt Center for Neuroscience Drug Discovery, the Sanford–Burnham Medical Research Institute, and the Emory Chemical Biology Discovery Center, were established with this purpose in mind. As previously stated, the main purpose of these organizations is to provide academic scientists with a means of moving their research forward in a manner that would make their programs appealing to an industrial partner, but discussions of academic–industrial partnerships would be incomplete without at least mentioning possible roadblocks to success.

First and foremost, the primary goals of academic institutions and industrial organizations are not the same. While companies are solely interested in identifying their next commercial success, academic institutions are primarily focused on educating students. The discovery of novel therapeutics by academic scientists is certainly possible, but at the end of the day, this process will always be secondary to the education of students. In addition, publishing requirements in academic institutions are high. Faculty need to publish results in order to establish themselves as experts in their field and maintain grant funding for their research, and students need to publish their work in order to increase the likelihood of gaining employment once their education is complete. The need to publish research results can run countercurrent to the need to develop intellectual property to protect commercial opportunities if intellectual property and publication strategies are not properly managed. Strategically filed patent applications can protect the intellectual property developed in academic laboratories, while still providing academicians with the opportunity to publish important scientific findings. Ideally, the technology transfer departments in the academic institutions will be involved in this process. In practice, however, the majority of academic scientists are highly specialized in their area of expertise, have limited (if any) industrial experience, and are unaware of the intricacies of patent laws. As a result, there is a high likelihood that premature publication of potentially patentable intellectual property could occur if the proper precautions (i.e., solid communication channels) are not taken throughout the course of an academic/industrial collaboration.

Despite these potential issues, the identification of a number of important clinical compounds makes it clear that drug discovery is no longer the sole purview of industrial organizations. The HIV protease inhibitor Prezista® (Darunavir, Figure 11.2(a)) marketed by Janssen Therapeutics, for example, was the product of academic science. It was discovered by Professor Arun K. Ghosh and his colleagues at the University of Illinois, Chicago.[13] Similarly, the nucleoside reverse transcriptase inhibitor HIV therapy Emtriva® (Emtricitabine, Figure 11.2(b)), which is marketed by Gilead Sciences and listed in the World Health Organization's List of Essential Medicines, was originally identified at Emory University by Professor Dennis Liotta's research team.[14] Alimta® (Pemetrexed, Figure 11.2(c)) is an anticancer agent identified at Princeton University by Professor Edward C. Taylor and marketed by Eli Lilly,[15] and Pfizer's drug Lyrica® (Pregabalin, Figure 11.2(d)), which produced sales of $3.7 billion in 2011, was originally identified by Professor Richard Silverman's group at Northwestern University.[16] Each of these discoveries provided substantial financial benefits to the academic institutions and the scientists that originally invented the compounds. Northwestern University, for example, received a one-time payment of $700 million, as well as annual licensing fees for Lyrica® (Pregabalin).[17]

Given the large potential payout to the institution and inventors, it is not difficult to understand why academic institutions have increased their efforts to move their discoveries into the commercial arena. At the

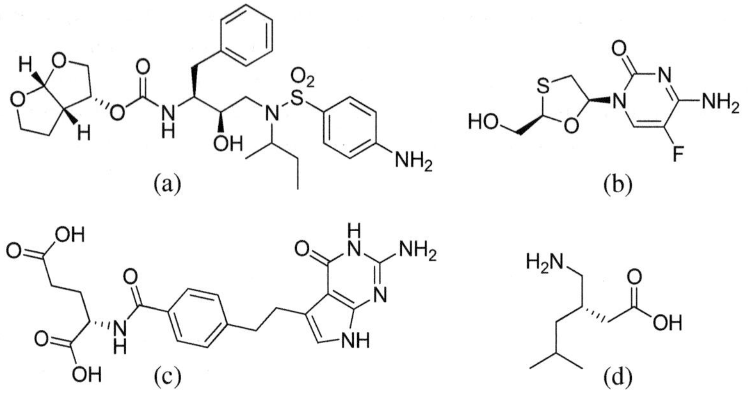

FIGURE 11.2   (a) Prezista® (Darunavir), (b) Emtriva® (Emtricitabine), (c) Alimta® (Pemetrexed), and (d) Lyrica® (Pregabalin).

same time, pharmaceutical companies have become far more interested in working with academic organizations to fill their pipelines with potential products. In the past, "of interest to an industrial partner" has been synonymous with "clinical candidate," but this is no longer the case. In fact, some pharmaceutical companies have been proactively seeking out interactions with the academic institutions. In 2012, for example, Merck established the California Institute for Biomedical Research (Calibr), an independent, non-profit academic institute dedicated to advancing the discoveries of academic scientists from around the world.[18] In a similar manner, Pfizer has established its "Centers for Therapeutic Innovation (CTI)" in Boston, New York, San Diego, and San Francisco with the expressed intent of collaborating with academic scientists.[19]

Other companies have taken a more hands-off approach. Eli Lilly's "Open Innovation Drug Discovery Program" provides academic scientists with the opportunity to submit compounds to an established set of *in vitro* screening assays. Eli Lilly provides biological screening data free of charge. If interesting results are identified, the academic scientists have the opportunity to engage the company to explore opportunities to move the program forward together or they can move forward on their own.[20] GSK's "Discovery Fast Track Challenge"[21] and "Discovery Partnerships with Academia (DPAc)"[22] also target academic scientists who believe their research has the opportunity to generate novel therapies for patients in need. In these programs, academic scientists submit a research proposal to a review board within GlaxoSmithKline that decides if the proposal is a match for their corporate interests and meets their scientific standards.

The long-term impact of academic drug discovery centers and academic–industry partnerships on the pharmaceutical industry remains unclear. It is very likely, however, that these organizations and arrangements will continue to thrive, as they serve the dual purpose of identifying novel scientific avenues for potential therapeutics and the training of future generations of drug discovery scientists. The pharmaceutical industry needs a constant supply of both in order to continue to thrive and grow.

FUNDING ISSUES

Although the scientific hurdles that must be surmounted in order to discover and develop a new drug are substantial, financial issues and the availability of funding can have a significant impact on the translation of interesting science into a commercially successful therapy. In large, well-established pharmaceutical companies, sales of existing products support R&D efforts to design the next generation of products. (Keep in mind, however, that once a patent expires, the profitability of a marketed therapeutic will decrease rapidly as generic versions become available.) Established companies also have access to capital markets (e.g., stock markets, debt securities markets) that can support company endeavors, so long as the investors are sufficiently satisfied with the overall performance of the company. Smaller companies, start-up biotechnology organizations, and academic investigators, however, generally do not have access to these markets, nor do they typically have marketed products bringing in revenue to support ongoing R&D efforts. Instead, these researchers and organizations rely on three main sources of financial support: (1) grant support, (2) angel investors, and (3) venture capital investors.

Grant funding is a major source of funding and support for early stage research, especially when that research is being pursued in an academic environment. Government organizations such as the National Institute of Health (NIH), the National Science Foundation, the European Research Council, and the Indian Council of Medical Research provide grant funding in amounts ranging from a few thousand dollars to several million dollars to successful grant applicants. In many cases, grant funds can only be awarded to not-for-profit organizations, but there are some programs that support small business, such as the NIH's Small Business Innovation Research (SBIR) program and Small Business Technology Transfer Program (STTR). Funding business operations through grant funding typically does not require any transfer of ownership in exchange for funding (it is "non-dilutive" funding), which makes it a very attractive means of launching a biotech start-up company. The success rate of grant applications, however, is low and has been trending lower since the beginning of the twenty-first century. In 2002, just over 30% of NIH grant applications received funding, but by 2013, the success rate had fallen to under 18%.[23]

While there are no definitive studies on the cause of this drop in funding rates, it is unlikely that the quality of grant applications has decreased over this time frame. Decreasing budgets as a result of economic changes and falling governmental budget and an increase in the number of applicants as a result of corporate downsizing (which has increased the number of scientists attempting to form their own companies or enter academia) are far more likely causes of this change in the overall success rate of grant applications. It is unclear whether or not this trend will be reversed in the future.

Angel investors are another source of funding that can be accessed by start-up and early stage biotechnology companies. These investors are generally affluent individuals that are considered "accredited investors" as defined by the laws of their countries. Usually, this means that they have a large annual income ($200,000+) and personal assets of over $1.0 million, although guidelines vary. The amount of money available is highly variable, and unlike the majority of grant funds, there are almost always strings attached. In return for their investment in the company, angel investors typically receive an ownership stake or convertible debt that will allow them to recoup their investment and make a substantial profit if the company is successful. It should come as no surprise that identifying individuals capable of making these kinds of investments is challenging at best, but there are some tools available to help scientists find angel investors who might be willing to invest in a biotech start-up company. Organizations such as Angel Capital Association (http://www.angelcapitalassociation.org/) and Gust LLC (https://gust.com/) have emerged to help facilitate creating links between start-up companies and angel investors. The competition to attract this kind of funding is very high and it is important to understand that in order to be successful, sound science must be coupled with a well-thought-out business strategy. Despite the benevolent sounding designation, angel investors are primarily guided by the ability to make a profit on their investments. In the absence of a sound business strategy, angel investors are unlikely to be interested in great science, irrespective of its potential real-world impact.

The third major source of funding for biotech start-ups and small companies is venture capital firms. These organizations can provide significant funding (multimillion dollar investments are the norm) to companies with limited operating history that are too small to raise funds in public market places (e.g., stock markets, bond offerings). Very few companies reach the point at which they are of interest to venture capitalists and their firms. The likelihood of attracting venture capital funding increases with each additional step toward commercialization, as the risk level decreases. The amount available also increases, as the financial requirements for sustaining clinical trials increase as drug candidates move through clinical trials.

The price of venture capital funding, however, is very high. In exchange for their investment, venture capital firms generally receive a significant ownership stake in the company (usually a controlling interest) and operational control of the organization. If the company is successful, the venture capitalists also receive a significant portion of the downstream financial windfall (based on their percentage of ownership and terms agreed upon when they made their investment in the company). Ideally, the company itself benefits from the support and business strategy knowledge of the venture capital firm. The founders of the company may be kept on board in some capacity to enable further advancement of the company's programs, but this is not always the case. As noted earlier, many company founders form companies with the intent of selling out to venture capital once the science and business has advanced into early clinical trials. (Most venture capital firms will only invest in programs that have reached clinical trials.) Employees of small biotechnology companies would be well advised to stay aware of the intentions of company owners with respect to venture capital plans, especially once a viable clinical candidate has been identified.

The discovery, development, and eventual commercialization of a novel therapeutic require a great deal more than scientific expertise. Internal corporate dynamics can, and often do, play a major role in determining whether or not a new drug will reach patients in need. Corporate cultures that are not conducive to the near constant information flow will be far less efficient and less able to move their programs forward. Navigating the changing business environment can also be a significant challenge. The pharmaceutical business climate is not static, and organization must adapt to survive as competitor compounds reach the market, patient populations shift in size and age, and new business models become available (e.g., CROs, academic drug discovery centers). For smaller companies, the added hurdle of funding research and development programs in an increasingly constrained financial environment must be overcome. Science alone cannot win the day, but when coupled with the right environment, resources, and organizational structures, new medications can be developed to help patients in need.

QUESTIONS

1. What are the three major facets of a pharmaceutical company?
2. What are some of the business climate factors that can impact a pharmaceutical company?
3. What are some of the non-monetary impacts of mergers and acquisitions?
4. What is the purpose of a contract research organization?
5. What are some of the potential risks of working with a contract research organization?
6. How did the Bayh–Dole act of 1980 impact academic and non-profit research communities?

References

1. Stone, K. "Big Pharma's Merger Mania." About.com. http://pharma.about.com/od/Big Pharma/tp/Big-Pharmas-Merger-Mania.htm.
2. Stovall, S. *Pharma Edges toward 'Patent Cliff'*; Wall Street Journal, June 15, 2011. http:// online.wsj.com/news/articles/SB10001424052702304186404576387073020214328.
3. Paul, S. M.; Mytelka, D. S.; Dunwiddie, C. T.; Persinger, C. C.; Munos, B. H.; Lindborg, S. R.; Schacht, A. L. How to improve R&D productivity: the pharmaceutical industry's grand challenge. *Nat. Rev. Drug Discov.* **2010,** *9,* 203–214.
4. Ward, A. *High Cost of Hepatitis C Treatment Pills Is Hard to Swallow*; The Financial Times, June 27, 2014.
5. LaMattina, J. *Drug Truths: A Holiday Gift from Merck's Ken Frazier*; December 20, 2011. http://johnlamattina.wordpress.com/2011/12/20/a-holiday-gift-from-mercks-ken-frazier/.
6. Schumacher, C. *CRAMS Outsourcing in Pharmaceutical and Chemical Research and Manufacturing, Part II*.StepChange Innovations GmbH, Science and Technology Blog; 2012. http://blog.stepchange-innovations.com/2012/08/crams-outsourcing-in-pharmaceutical-and-chemical-research-and-manufacturing-part-ii/#.VINxtslNcmQ.
7. Getz, K.; Lamberti, M. J.; Mathias, A.; Stergiopoulis, S. Resizing the Global R&D Contract Service Market. *Contract Pharma* **2012,** *14,* 54.
8. Based on information posted in the Scynexis Inc. corporate website. http://www.scynexis.com/pipeline/.
9. *AMRI Announces Collaboration Agreement with Lilly for In-sourced Chemistry Services*; Albany Molecular Research Inc, November 7, 2011. http://www.amriglobal.com/news_and_publications/AMRI_Announces_Collaboration_Agreement_With_Lilly_for_In-sourced_Chemistry_Services__207_news.htm.
10. *Technology Transfer, Administration of the Bayh-dole Act by Research Universities*; U.S. Government Accounting Office (GAO), May 7, 1978. Report to Congressional Committees.
11. a. 35 U.S.C. 18 Patent Rights in Inventions Made with Federal Assistance.
    b. 37 C.F.R. 401 – Rights to Inventions Made by Nonprofit Organizations and Small Business Firms under Grants, Contracts, and Cooperative Agreements.
12. Slusher, B. S.; Conn, P. J.; Frye, S.; Glicksman, M.; Arkin, M. Bringing together the academic drug discovery community. *Nat. Rev. Drug Discov.* **2013,** *12* (11), 811–812.
13. a. Ghosh, A. K.; Kincaid, J. F.; Cho, W.; Walters, D. E.; Krishnan, K.; Hussain, K. A.; Koo, Y.; Cho, H.; Rudall, C.; Holland, L.; Buthod, J. Potent HIV protease inhibitors incorporating high-affinity P2-ligands and (R)-[(hydroxyethyl)amino]sulfonamide isostere. *Bioorg. Med. Chem. Lett.* **1998,** *8* (6), 687–690.
    b. Vazquez, M. L.; Mueller, R. A.; Talley, J. J.; Getman, D. P.; Decrescenzo, G. A.; Freskos, J. N.; Bertenshaw, D. E.; Heintz, R. M. *Hydroxyethylamino Sulfonamides Useful as Retroviral Protease Inhibitors*; 1995. WO 9506030 A1.
14. a. Schinazi, R. F.; McMillan, A.; Cannon, D.; Mathis, R.; Lloyd, R. M.; Peck, A.; Sommadossi, J. P.; St Clair, M.; Wilson, J.; Furman, P. A. Selective inhibition of human immunodeficiency viruses by racemates and enantiomers of Cis-5-fluoro-1-[2-(hydroxymethyl)-1,3-oxathiolan-5-yl]cytosine. *Antimicrob. Agents Chemother.* **1992,** *36* (11), 2423–2431.
    b. Frick, L. W.; Lambe, C. U.; St John, L.; Taylor, L. C.; Nelson, D. J. Pharmacokinetics, Oral bioavailability, and metabolism in mice and cynomolgus monkeys of (2'R,5'S)-cis-5-fluoro-1-[2-(hydroxymethyl)-1,3-oxathiolan-5-yl] cytosine, an agent active against human immunodeficiency virus and human hepatitis B virus. *Antimicrob. Agents Chemother.* **1994,** *38* (12), 2722–2729.

15. a. Taylor, E. C.; Patel, H. H. Synthesis of pyrazolo[3,4-d]pyrimidine analogs of the potent antitumor agent N-[4-[2-(2-amino-4(3H)-oxo-7H-pyrrolo[2,3-d]pyrimidin-5-yl)ethyl]benzoyl]-l-glutamic Acid (LY231514). *Tetrahedron* **1992,** *48* (37), 8089–8100.

   b. Taylor, E. C.; Kuhnt, D. G.; Shih, C.; Grindey, G. B. *Preparation of N-[pyrrolo[2,3-d]pyrimidin-3-ylacyl]glutamates as Neoplasm Inhibitors;* 1993. US5248775.

16. a. Taylor, C. P.; Vartanian, M. G.; Yuen, P. W.; Bigge, C.; Suman-Chauhan, N.; Hill, D. R. Potent and stereospecific anticonvulsant activity of 3-isobutyl GABA Relates to *in vitro* binding at a novel site labeled by tritiated gabapentin. *Epilepsy Res.* **1993,** *14* (1), 11–15.

   b. Silverman, R. B.; Andruszkiewicz, R.; Yuen, P. W.; Sobieray, D. M.; Franklin, L. C.; Schwindt, M. A. *Preparation of GABA and L-glutamic Acid Analogs for Antiseizure Treatment;* 1993. WO 9323383A1.

17. Wang, A. L. *Northwestern University Leads Nation in Tech Transfer Revenue;* Crain's Chicago Business, October 29, 2012.

18. http://www.calibr.org/index.htm.

19. http://www.pfizer.com/research/rd_partnering/centers_for_therapeutic_innovation.

20. https://openinnovation.lilly.com/dd/about-open-innovation/how-does-open-innovation-work.html.

21. http://openinnovation.gsk.com/na/index.php.

22. http://www.dpac.gsk.com/.

23. http://report.nih.gov/nihdatabook/.

Intellectual Property and Patents in Drug Discovery

There is little doubt that the discovery and subsequent development of novel drugs has improved the quality of life for millions of people. Life expectancies have been extended well beyond what they would have been in the absence of the medical advances associated with the identification of novel therapeutic agents. It is also true, however, that the cost associated with the identification of useful and marketable therapeutic entities is staggering. As of 2011, it is estimated that a single new drug costs over $1.75 billion to discover and develop.[1] To provide a measure of comparison, the same amount of money could be used to buy 17 Boeing 737 jet aircrafts (based on 2012 prices on Boeing's Web site), purchase approximately 7000 homes (assuming $250,000 per home), 70,000 automobiles (average price $25,000), or provide for the raising 7000 children born in 2010 to the age of 18 years. These costs are primarily borne by companies that bring new products to market, the branded pharmaceutical companies, as they are responsible for gaining market approval for their new products.

Once a drug has been approved by regulatory agencies, competitor companies, such as generic drug companies, can enter the marketplace with copies of the drug at a substantially lower cost, as they can use an Abbreviated New Drug Application (ANDA). This process was established by the "Drug Price Competition and Patent Term Restoration Act of 1984," also known as the Hatch-Waxman Act. Unlike the full New Drug Application (NDA), the ANDA application package does not need to include preclinical (animal) or clinical (human) data to establish safety and efficacy. The generic companies need only show that their product is bioequivalent to the branded company's product. Bioequivalence can be established by measuring the pharmacokinetic properties of the generic copy of the branded drug in a small group of healthy volunteers (24–36). The generic company can use the pharmacokinetic data provided in this study (rate of absorption, bioavailability, etc.) to establish that their copy is able to deliver the same amount of active ingredients into a patient's bloodstream in the same amount of time as the innovator drug.[2] In other words, generic drug companies do not need to bear the cost of conducting clinical trials to demonstrate that their copy of the branded agent will produce the same result. This substantially decreases the costs associated with drug development. In addition, generic drug companies do not carry the costs associated with failed drug discovery efforts (9 out of 10 clinical trials ends in failure), further decreasing the costs of copying drugs that have already entered the marketplace.

If a generic company can produce the products that a branded pharmaceutical company brings to market at a fraction of the cost, why would any company invest in new drug development? Although one could argue that society as a whole benefits when companies invest in the identification of novel therapeutic entities, a substantial amount of profit must be generated in order for a company to be self-sustaining. In the absence of significant profits from successful drug discovery and development programs, pharmaceutical companies would be crushed by the financial weight of the process, especially in light of the high failure rate. Fortunately, the protection of intellectual property rights through patents provides an opportunity for companies to recoup the high cost of drug discovery by preventing others from entering the market while the patent is in force. Patents and the protection that they afford are the lifeblood of the pharmaceuticals industry. It is vitally important that those who choose to work in this industry have a basic understanding of the patent system.

In considering the role of patents in the drug industry, it is important to understand what a patent provides to the owners of the patent. In theory, patents provide incentive to invest and invent by providing the owners of the right to exclude other from using the invention described in the patent for a set period of time. In return, the owners/inventors of the patent provide the public access to the technology associated with the patent. This provides the public with access to information that might have otherwise been kept secret. Other companies and individual may develop additional intellectual property as a result of the information available when the patent is published, and once the patent expires, anyone can use the invention. In the pharmaceutical industry, the patent system prevents generic companies from launching copies of new drugs as they enter the market, allowing the innovator company time to recoup the substantial investment required and enough profit to make the overall risk worthwhile. Ideally, pharmaceutical companies seek to gain patents for their inventions across the globe so that they can sell their products across the widest range of countries possible, speeding the time to recoup investment and begin the cycle over again.

The process of gaining patent protection for an invention would probably be more efficient if there were single unified international system for the prosecution, issuance, and enforcement of intellectual property rights.

Unfortunately, no such system exists, and as a result, each country has its own set of rules and regulations that must be followed in order to achieve patent protection for an invention. Efforts to harmonize the process across various jurisdictions (countries) have produced some positive results, such as the Patent Cooperation Treaty (PCT) and the related application process which will be covered later in this chapter. It is absolutely critical, however, that those wishing to obtain patent protection for their inventions be aware of the laws and requirements of the countries in which they are pursuing patent protection. For the purposes of this text, emphasis will be placed on the laws and requirements for patent protection in the United States.

The United States Patent and Trademark Office (USPTO) administers the patent application process, working either directly with the inventors of record or an individual recognized as a representative of the inventor who is registered with the USPTO to represent others in the patent application process, a patent attorney or patent agent. While an inventor can represent himself/herself or a group of inventors submitting a patent application together, this is not generally recommended any more than representing oneself in a court of law. The application process is exceptionally complex, and it is highly recommended that interested parties work with a patent attorney or patent agent, both of which having been registered with the USPTO as experts in patent laws and procedures. For those interested, detailed information on the patent application process is provided by the USPTO in the form of the text entitled Manual of Patent Examining Procedure (MPEP).

PATENTABLE SUBJECT MATTER

Human ingenuity and innovation provide ample opportunity for the inventions or discovery of new science and technology that could be harnessed to support commercial ventures such as the pharmaceuticals industry. The discovery of something that has not previously been described in the public domain, however, does not guarantee that one can obtain a patent. There are specific requirements that must be met in order for a new invention or discovery to be considered patentable subject matter. First, the new invention or discovery must fall into one of four eligible invention categories. It must be a composition of matter, process, machine, or an article manufacture.[3] Compositions of matter are defined as "all compositions of two or more substances and all composite articles, whether they be the results of chemical union, or of mechanical mixture, or whether they be gases, fluids, powders, or solids." Small molecules, proteins, nucleic acids, as well as genetically engineered microorganism and animals all qualify as compositions of matter. Obtaining a composition of matter patents is a major focus of the pharmaceutical industry. A composition of matter patent covering a set of molecules, for example, provides the owner with sole owner-ship of the compounds irrespective of their use. Since small molecules are typically described using chemical structures rather than specific language, a markush structure is used to describe the scope of cover-age of a particular patent or patent application. A markush structure (Figure 12.1) is a general structure with variable units, often designated as R-groups, appended to the general structure. Each R-group has a specific definition that is provided in the patent. Proteins, antibodies,

The present invention is directed toward novel hydroxylated sulfamide derivatives, compounds of formula (I),

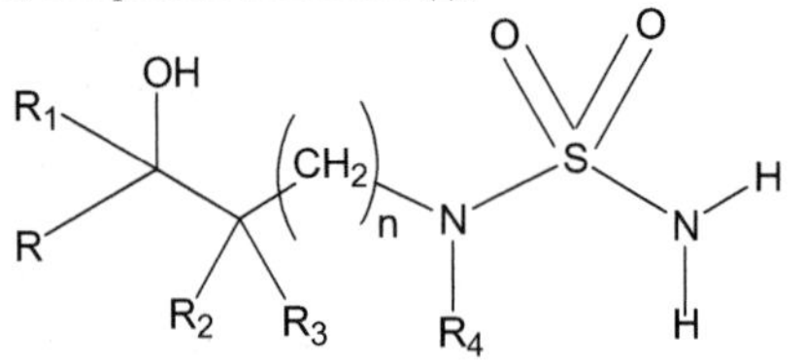

Including hydrates, solvates, pharmaceutically acceptable salts, prodrugs and complexes thereof, wherein:
R is selected from the group consisting of optionally substituted aryl, optionally substituted benzoisoxazole, and optionally substituted benzothiophene where R may be substituted by 0-5 moieties;
n is 1 or 2;
R^1 is selected from the group consisting of hydrogen and optionally substituted C_{1-6} alkyl;
R^2, R^3, and R^4 are each independently selected from the group consisting of hydrogen, halogen, and optionally substituted C_{1-6} alkyl; and
R^1 and R^4 are taken together with atoms to which they are bound to form an optionally substituted ring having from 5 to 7 ring atoms.

FIGURE 12.1   Markush structure from US patent 8609849.

nucleic acids, and other biological constructs can be described in a sim-ilar manner using the appropriate scientific coding sequence (e.g., amino acid sequence coding, nucleic acid sequence coding) with variable units incorporated as necessary to define the scope of the invention.

The second class of patentable subject matter is a process, which is defined as "an act, or a series of acts or steps." This is clearly a very broad definition that can apply to a wide range of activities within the phar-maceuticals arena. A synthetic procedure used to prepare a therapeutic agent falls into this category, as does a method of using a drug to treat a particular disease. Process inventions are not necessarily linked to com-position of matter inventions, which can lead to multiple, related patents. A drug could be protected by a composition of matter patent, a second patent could cover the method of use of the drug to treat a specific disease state, while a third patent could cover a synthetic method for making the aforementioned drug. In theory, these patents could be owned by separate organizations, leading to complications in the practice of each invention. These issues will be discussed separately.

The third category of patentable subject matter is a machine. Strictly speaking, a machine is defined as "a concrete thing, consisting of parts, or of certain devices and combination of devices." This definition includes mechanical devices or combinations of mechanical power and devices to perform a function or produce a defined result. Examples from the pharmaceuticals industry include an autoinjecting syringe or a pump system designed to regulate drug delivery based on patient feedback.

The fourth and final class of patentable subject matter, an article of manufacture is defined as "the production of articles for use from raw or prepared materials by giving to these materials new forms, qualities, properties, or combinations, whether by hand labor or by machinery." An article of manufacture can consist of multiple parts, but unlike machines, the interaction between the parts is generally static. In principle, something as simple as a hammer or mouse pad would qualify as an article of manufacture, as would a drug-coated stent, a transdermal patch containing a drug, or a tablet containing a specific drug and excipients.

Not every new and useful invention or discovery is eligible for patent protection. There are some areas of knowledge and invention that are specifically excluded from the patent system for a variety of different reasons. The patenting of "special nuclear material or atomic energy in an atomic weapon" is specifically barred by US patent law for reason that should be obvious.[4] In addition, although microorganism, cell-based material, and even animals can be patented in some circumstances, the patenting of "a human organism" is prohibited.[5]

Laws of nature, natural phenomena, or abstract ideas are also specifically excluded from patentability. This restriction prevents someone from gaining patent protection for the concepts such as Einstein's theory of relativity ($E = MC^2$), the law of gravity, or closer to the pharmaceuticals arena, the lock and key fit theory of enzyme activity. Each of these concepts was new at some time, but would not have been eligible for patent protection as they are each considered laws of nature. In a similar manner, new minerals discovered in the earth, new plants discovered in the wild, and even unchanged genes are not patentable as they are each considered to be natural phenomena. The issue of the patentability of isolated, unaltered gene was settled in the US Supreme Court in June 2013 when the court ruled against Myriad Genetics on a patent that described the breast cancer genes BRCA1 and BRCA2.[6] Importantly, this decision does not prevent the patenting of altered genes, human or otherwise, as the altered genes would no longer be considered natural phenomena. In a similar manner, genetically modified organisms, such as microorganisms designed to produce human proteins (e.g., insulin) or an animal that has been genetically modified to establish a disease model (e.g., the SOD1 mouse model of ALS) are patent eligible. The genetic modification required to produce the invention falls outside of the scope of products of nature.

The second requirement for an invention or discovery to be considered patentable subject matter is that it "must be useful or have a utility that is specific, substantial, and credible."[7] In other words, if one were to invent a previously unknown class of compounds, in the absence of a use for the compounds, they are not patentable subject matter. Insubstantial or non-specific uses such as using the new molecules as filler material for a landfill are not enough to qualify a discovery as patentable subject matter. There must be some real world, credible utility for a discovery in order for it to be considered patentable subject matter.

Within the context of the pharmaceuticals industry, utility is often the treatment or prevention of a disease or condition. If, for example, it can be demonstrated that a new class of compounds is capable of lowering blood pressure in patients with hypertension, then the class of compounds may be considered patentable subject matter. They have a specific, real world utility as antihypertensive agents. Of course, generating human data for a single compound requires substantial investment in time, resources, and funding. Requiring human data to demonstrate utility for the purposes of gaining patent protection for potential therapeutic agents would represent a substantial burden.

Fortunately, human data are not required in order for an assertion of utility of a compound or set of related compounds. In fact, the efficacy and safety of potential therapeutic agents in the human population is not a consideration in the determination as to whether or not an invention is considered patentable subject matter. These issues are the purview of other regulatory bodies (e.g., FDA) and the USPTO guidelines explicitly state that "there is no decisional law that requires an applicant to provide data from human clinical trials to establish utility for an invention related to treatment of human disorders."[8] *In vivo* animal studies with recognized disease models are also not required, although they certainly could be used to demonstrate utility. A demonstration of utility can be met with an *in vitro* assay that is reasonably correlated with the desired therapeutic utility of a compound, composition, or process.[9] If, for example, a new compound were shown to inhibit HMGCoA reductase, an enzyme known to be important in cholesterol biosynthesis, in an *in vitro* assay, the compound would be considered to have utility as a cholesterol-lowering agent. Additional data such as *in vivo* models or clinical data are not required, as the *in vitro* assay for HMGCoA reductase activity has been shown to correlate with cholesterol levels. It is worth noting at this point that there is no specific threshold set for biological activity that must be demonstrated in an *in vitro* or *in vivo* assay. A compound that has an IC_{50} of $100\,\mu M$ in an *in vitro* assay has demonstrated utility, as does a compound that has an IC_{50} of $1.0\,nM$ in the same assay, provided that there is a correlation between the *in vitro* assay and therapeutic application. In fact, it is not always necessary to have *in vitro* data at all. If it can be demonstrated that a compound is "structurally similar to a compound known to have a particular therapeutic or pharmacological utility,"[10] then an assertion of utility can be made. This presumption of structural similarity is the basis for claiming a wide range of compounds surrounding a smaller set of compounds with demonstrated *in vitro* activity in an assay. Of course, the definition of "structurally similar" is in the eye of the beholder, so great care must be taken in determining the breadth and scope that is appropriate for a particular patent application.

INHERENT PROPERTIES AND PATENTABILITY

It is often the case that there are multiple uses for a single material, some of which may not be recognized when the material becomes available to the public. The discovery of a new use, property, or scientific explanation for the function of a pre-existing material does not, however, make the material itself patentable as a composition of matter.[11] Consider, for example, the invention of the first ball. If the first ball was invented by Jim Bounce, he could gain patent protection for the ball, assuming that he could come up with a real world utility for the ball. If his brother, John Bounce were to discover that the ball was red, he would not be able to gain a new patent on the ball based on his determination of the color of the ball. The fact that the ball is red is an inherent property of the ball and does not change the fact that it was already part of the public domain when John Bounce determined its color. In a similar sense, compounds that are publically available are not patentable as compositions of matter if a new use is identified. Once a compound is available to the public (commercially available, published in the literature, etc.), discovery of an inherent property, such as activity in an *in vitro* assay, does not make the compound itself patentable as a composition of matter.

Continuing the red ball scenario, if Mark and Fred Parker, brothers and contemporaries of Jim Bounce, were to design a toy that incorporated the red ball as a component of the toy, then they could potentially receive a patent on the use of the red ball in the toy. Unlike the discovery of the color of the ball, the use of the red ball in the newly designed toy is not an inherent property of the ball. It is a new use or process that uses the red ball. The red ball itself is still no longer eligible for patenting as John Bounce already holds that patent, but the new use of the red ball can be patented. Of course, in order to sell the toy that uses the red ball, Mark and Fred Parker would need to obtain a license from John Bounce to use his ball. Similarly, John Bounce cannot sell his ball as part of Mark and Fred's toy without a license from them.

The same concepts of inherent properties and patentability apply to potential therapeutic agents. Newly identified small molecules and biologics are potentially patentable as new composition of matter provided they have a demonstrated utility. Those that are already part of the public domain are not patentable as new compositions of matter irrespective of the discovery of a new potential therapeutic utility. Once the compounds have been disclosed to the public, even if they are published in a paper or catalog with no known utility, a composition of matter patent application is not a viable option. If, however, it is discovered that a known compound is active in an *in vitro* assay that suggests utility as a therapeutic agent, then it is possible to obtain a patent on the use of the compound to treat the disease. In other words, if a commercially available compound is identified as an HMGCoA reductase inhibitor in an *in vitro* assay, it may be possible to gain patent protection for the use of the compound for the treatment of high cholesterol. The compound itself is not patentable as a composition of matter, but a patent application claiming the method of use of the compound for the treatment of high cholesterol is a viable path forward. A patent of this type would allow the owner to prevent others from selling the compounds as a cholesterol-lowering agent, but the compound could still be sold for other purposes not covered by the aforementioned patent.

In general, compositions of matter patents are preferred by the pharmaceuticals industry, as they give the owner the ability to prevent others from selling or using the material in the patent irrespective of the use. Process patents that cover the use of a material as a therapeutic agent against a particular disease only prevent others from making or selling the same material for the same use. If a different use for the same material is identified, the owners of the process patent cannot prevent the material in question from entering the market for a different use. Consider, for example, if one were to discover that aspirin was an effective treatment for toenail fungus. In theory, it would be possible to obtain a patent for the use of aspirin as a treatment for toenail fungus and prevent others from making or selling aspirin for this particular use. Other companies could not market aspirin as a treatment for toenail fungus, but they could still market it as a pain reliever. Consumers could then choose to adopt the off-label use of aspirin for the treatment of toenail fungus. This would severely limit the financial return for the investment required to gain marketing approval from the FDA for the newly identified use of an already existing therapeutic agent.

NOVELTY AND THE PRIOR ART

Issues of utility and patentable subject matter consideration are important, but they are not the only factors that determine whether or not an invention or discovery may be protected by a patent. The subject of a patent application must also be novel. In order for something to be considered novel, it cannot be "known or used by others in this country (U.S.) or patented or described in a printed publication in this (U.S.) or a foreign country, before the invention thereof by the applicant for a patent."[12] Essentially, this means that if someone other than the patent applicants has already filed a patent application or published the material anywhere in the world, a patent cannot be obtained in the U.S. Further, if the material is available for use in the U.S. or if the material has been offered for sale by others prior the submission of a patent application, then it can no longer be protected by a patent. All of these actions make the subject matter part of the "prior art" and act as a statutory bar against the issuance of a patent on the subject matter.

The breadth and scope of the prior art that is available at the time a patent application is filed is exceptionally important, as it will be a major factor in determining whether or not something is eligible for patent protection. As such, it is important to understand what is considered part of the prior art. In a general sense, the prior art refers to the sum total of knowledge available to the public prior to the date of a particular patent application. Any information, knowledge, or material that is part of the prior art can be used as evidence indicating that the subject of a patent application is either not novel or would be considered obvious to someone skilled in the relevant art at the time of the invention. Patents, patent applications, and scientific journal articles are common sources of prior art material, but they are not the only sources. Printed publications, whether they are in a scientific journal, a newspaper, trade publication, or even a Ph.D. thesis placed on a shelf in a university library[13] are all considered part of the prior art on the day that they become available to the public. Electronic publications and information housed in publicly accessible databases, also become part of the prior art on the date they are available to the public. Even NIH grant applications become part of the prior art once they are available to the public under the Freedom of Information Act.[14] In addition, posters presentation, whether they are part of a nationally recognized meeting of a scientific organization such as the American Chemical Society or a small college or university's efforts to showcase the talent of their students, also become part of the prior art on the date of the presentation. A poster that is only displayed for a few days is considered prior art, irrespective as to whether copies are provided, as the contents of the poster have been made available to the public.[15] Oral presentations can also be considered part of the prior art "if written copies are disseminated without restriction."[16] In short, if something is available to the public, even for a limited time, it is part of the prior art. Importantly, material that is only distributed internally within an organization and are intended to remain confidential are not considered part of the prior art.[17]

OBVIOUSNESS AND THE PRIOR ART

Even if an invention is not directly described in the prior art, it still may not be patentable. In addition to being novel, inventions must also be non-obvious. Specifically, if the subject matter of a patent application would be obvious to one of ordinary skill in the art to which the application pertains, then the invention is not considered patentable.[18] This, of course, raises the question of how to define what is obvious to one of ordinary skill in the art. Current guidelines in the MPEP state that making an object portable, applying aesthetic changes (e.g., changes in size or color), rearranging the order of parts or steps in a sequence, or automating a manual activity would all be considered obvious changes, rendering the subject ineligible for patent protection.[19] The adaptation of a known *in vitro* screening assay to a high-throughput platform, for example, would be considered an obvious adaption of the known screening technique. In other words, the HTS variant of a known *in vitro* screening paradigm is not patentable by virtue of the adaptation of the screen to an HTS platform. Similarly, changing the color of a pill used to treat a disease does not make the pill patentable simply because it is a different color.

If these were the only issue to consider in determining whether or not an invention would be considered obvious, the situation would be fairly simple. There are, however, two additional criteria that are used to determine if something is obvious relative to the prior art. In some instance, the purification of a pre-existing material could be considered an obvious change, preventing patent protection.[20] Consider, for example, a publically available compound that is useful as a therapeutic agent when used in 78% purity. Purifying the material to 95% purity and finding that it still performs the same function (acting as a therapeutic agent for a particular condition) is considered an obvious change. The 95% pure product is not once again patentable just because it is now available in a purer form. (Keep in mind that the method of purification might be patentable, but not the compound itself.) On the other hand, if it were found that the 95% pure material possessed a utility that the 78% pure material did not, then it might be possible to claim that purification lead to a non-obvious change in properties. This could potentially make the purified material patentable.

The distinction between obvious and non-obvious changes becomes even more complex when one considers combinations of prior art. If multiple prior art references can be combined to produce a result that would have been expected by one of ordinary skill in the art at the time of the invention, then the invention may be considered obvious. It is not necessary that the prior art suggests the modification or combination of knowledge to arrive at the invention at issue. Additionally, there must be a reasonable expectation of success based on the prior art available at the time. If these conditions are met, then an invention may be considered obvious, and therefore not patentable. These conditions are somewhat vague and one could argue that obviousness, much like beauty, is in the eye of the beholder. The concept of obviousness is best understood through the examination of examples.

Consider, for example, the hypothetical scenario in which (1) has been disclosed by an inventor in a patent application as useful for the treatment of malaria (Figure 12.2). If a different inventor were to file a separate patent

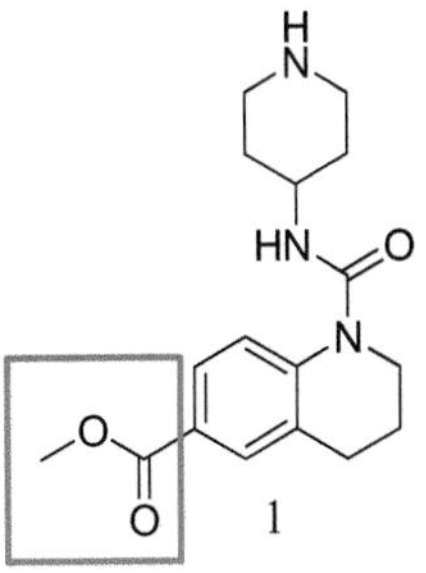

FIGURE 12.2 If compound (1) is known in the prior art as a compound useful for the treatment of malaria, changing the methyl ester of (1) to an ethyl ester (2) is likely to be viewed as an obvious change. If (2) is an antiviral agent, obviousness with respect to (1) is less likely.

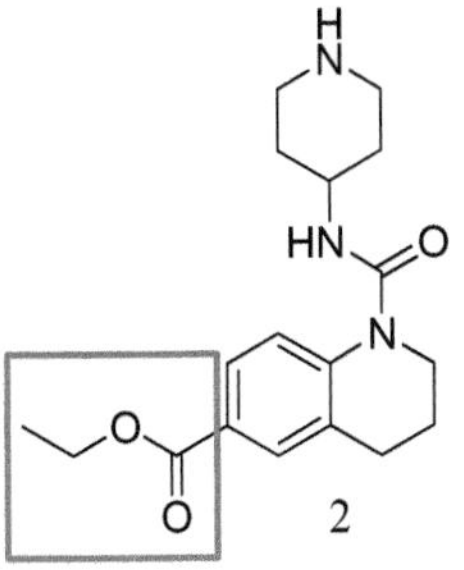

application describing (2) as being useful for the treatment of malaria after the first patent was published (became part of the prior art), it is likely that (2) would be considered obvious. The only difference between (1) and (2) is the addition of a methylene to the ester side chain. A medicinal chemist of ordinary skill would reasonably expect this difference to have little consequence in the antimalarial activity of the two compounds. On the other hand, if a third inventor were to determine that (2) was useful as an antiviral agent, then a patent describing the method of using (2) as an antiviral agent might not be considered obvious. Although the first patent described the use of (1) as an antimalarial agent, antiviral activity is a significantly different application. In the absence of other prior art suggesting the use of (1) as an antiviral agent, the first patent application is not likely to be enough to argue that the use of (2) as an antiviral agent would be obvious.

Suppose a fourth inventor filed a patent application describing compound (3) as an antimalarial agent after the publication of the first patent (Figure 12.3). One could argue that the original patent on (1) makes the antimalarial activity of (3) obvious as the overall structure is intact. On the other hand, one could also argue that it is not obvious that the antimalarial activity would be preserved when the ester group is removed from the scaffold. In this scenario, whether or not the leap from (1) to (3) is obvious will depend at least partly on the inventor's (or their representative's)

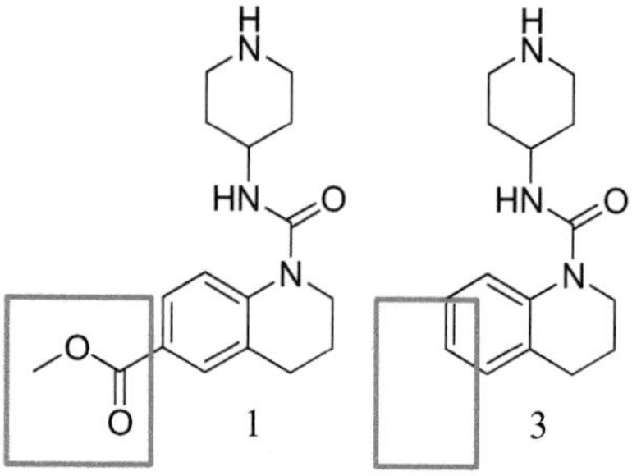

FIGURE 12.3   Whether or not (3) would be considered obvious to one skilled in the appropriate art in light of (1) depends on the context. If (1) and (3) share the same utility, then obviousness is likely to be an issue. On the other hand, if (1) and (3) have substantially different utility, obviousness is less likely to be a concern in a patentability determination.

ability to effectively argue to the USPTO why (3) is different enough from (1) as to be unexpected based on the prior art available at the time.

Finally, consider the discovery of (4). If this compound were described in a patent application as an antimalarial agent after the publication of (1) and (3), would it be considered obvious based on the other patent applications (Figure 12.4)? Maybe, maybe not. It could be

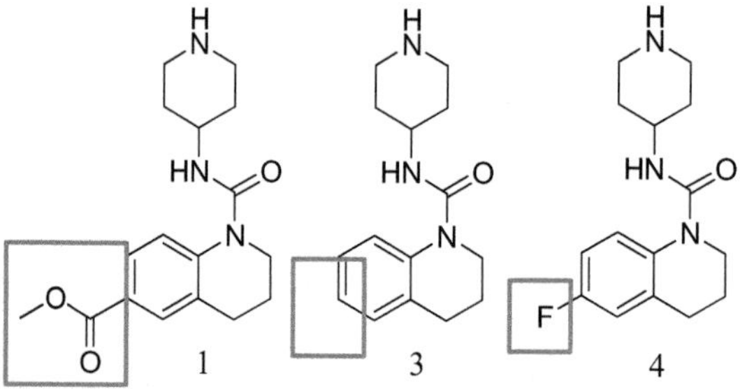

FIGURE 12.4   If (1), (3), and (4) share the same *in vitro* antimalarial activity, but only (4) is capable of providing *in vivo* efficacy, an argument could be made that prior art describing (1) and (3) as antimalaria agents is not enough to block patentability based on obviousness issues.

argued that since (1) and (3) all have antimalarial properties a person of ordinary skill in the practice of medicinal chemistry would reasonably expect that (4) would share the same features. On the other hand, what if it was known that (1) and (3) have antimalarial properties, but are incapable of eliminating a malaria infection in an animal model, while (4) cleared the condition? In this scenario, it could be argued that the *in vivo* activity of (4) is not anticipated based on (1), and (3), making the invention of (4) a non-obvious discovery (e.g., patentable). Whether or not something is obvious based on the prior art is a complex determination that requires expert analysis, and is best done in conjunction with experts in patent law.

INVENTORSHIP

One of the most contentious issues in the patent process is identifying the inventors of record for a patent application. Designating the correct inventors for a patent application is of critical importance, as the consequences of an incorrect list of inventors can be very serious. If an inventor is left off a patent application, either by mistake or with intent, it is possible that any patent resulting from the application could be deemed unenforceable. In a similar manner, if individuals that do not qualify under the legal definition of an inventor of record are included in the list of inventors of a patent application, either by mistake or with intent, it is possible that any patent resulting from the application could be deemed unenforceable. The designation of who is an inventor for a particular invention and the associated patent application is a legal determination that must be made by strictly adhering to the letter of the law. Failure to do so can lead to loss of all patent rights for the entire invention.

The legal rules for identifying the inventors of record for a particular application require that inventors must always be a person and not a corporation. Corporations can certainly own patents, provide salaries, supplies, and infrastructure in support of the inventive process, but only people can be inventors of record. In order to be considered an inventor of record, an individual must contribute to at least one claim in the application. To be clear, there is no requirement that an individual perform the work necessary to enable the invention. It is only necessary that an individual conceive of one aspect of at least one claim of a patent application in order to be considered an inventor of record. In contrast, performing the work required to reduce an invention to practice does not guarantee that a person will be considered an inventor of record. Who is and who is not an inventor of record depends entirely on what the patent claims as the invention.

Consider, for example, a group of scientist, Susan, Mark, John, and Beth. The four scientists are working on developing a novel class of antibacterial agents. Susan designs a novel class of compounds that she believes will be useful for the treatment of Gram-positive bacterial infection. Susan then asks Mark to prepare a list of compounds. Mark diligently prepares the compounds after devising a synthetic strategy, and hands the compounds to John. John, an expert in bacterial infection, realizes that the compounds might also be useful for the treatment of Gram-negative infections. John conveys his thoughts to Susan, and instructs his collaborator Beth to screen the compounds that Mark prepared against Gram-positive and Gram-negative bacteria. The tests are successful, and the four scientists decide that they should file a patent application to protect their intellectual property. They present their discoveries to Allen, their manager who oversees the antibacterial efforts in their company. Who are the inventors of record?

Since the compounds are novel, then the person that *conceived* of the compounds, Susan, is an inventor of record for a composition of matter patent that claims the compounds themselves. On the other hand, even though Mark worked very hard to make the compounds, in this scenario, he did not contribute to the conception of the compounds. In this instance, Mark is *not* an inventor of record on a patent application that claims the compounds in question. Interestingly, if the patent application were to also claim the methods of preparation of the compounds that Mark developed, then Mark would be an inventor of record. What about John and Beth? If the patent application claims a method of using the compounds to treat only Gram-positive infections, then John and Beth are not inventors of record. They may have been responsible for executing the screens to determine if the compounds would be effective, but they did not contribute to conceiving that the novel compounds would be useful for the treatment of Gram-positive infections. As mentioned earlier, performing the work to enable an invention does not equate to inventorship.

If the same patent application were to also claim a method of using the compounds to treat Gram-negative infections, then John is an inventor of record for this aspect of the invention. John conceived of this utility for the compounds designed by Susan and prepared by Mark. The last scientist, Beth, did not contribute to the conception of the compounds, the methods of preparing the compounds, or their use as treatments for either Gram-positive or Gram-negative bacterial infections. Even though Beth may have done exceptional work in screening the compounds, she is not an inventor of record. Inventorship requires contribution to the conception of at least one claimed aspect of an invention.

What about Allen, the manager of antibacterial research? He championed the program, garnered support from upper management, coordinated research presentations, and successfully convinced the company to file a patent to protect the intellectual property that his team developed. Despite any hard work that Allen may have done to advance the program in the corporate setting, he did not contribute to any of the claimed inventions. Allen may be the manager of antibacterial research, but he is not an inventor of record and should not be included no matter how good of a manager he may have been.

In this case, Allen and Beth simply do not qualify as inventors as defined by the MPEP. In addition, since inventorship is not transferable, Susan, John, and Mark cannot simply attribute some of their inventive efforts to Allen and Beth. The list of inventors on a patent application must be limited to only the people that contributed to the conception of at least one claim in the application. Failure to list the correct inventors, whether too many or too few people are listed can lead to invalidation of the related patents.

ASSIGNMENT AND OWNERSHIP

Although inventorship is not transferable, patent applications and patents have the attributes of personal property. Much like a house or a car, they can be sold to an individual or company. In addition, the owners of a patent can license part or all of the invention to a third party (or multiple third parties) as they see fit. This, of course, raises the question of who owns a patent application and what rights are they afforded? In the absence of documents filed with the USPTO, ownership (assignment) of a patent or patent application defaults to the inventors of record. In most cases, the inventors work for a company that has agreed to pay them in exchange for ownership of the intellectual property that they develop during the course of their employment. In other words, the employees have agreed to assign their rights for the patents and patent applications to the company in exchange for their compensation package. This transfer of ownership is recorded with assignment documents that must be recorded with the USPTO. To be clear, when an inventor assigns his rights to an invention to a company, then the company owns the patent or patent application.

If there is only one inventor and he assigns his rights to a patent to a single company, then only the company has the right to prevent others from using the invention in the patent in the US. What happens if there are two inventors of record? Although one might expect that each inventor would own the rights to only the material that they contributed to, this is not the case. The two inventors have equal ownership and rights to all aspects of the patent and neither can prevent the other from practicing the invention in the U.S. Also, each is free to assign their rights to another individual or company without the approval of the other inventor. In principle, two inventors on the same patent could each assign their rights to two different companies, and then both companies would be able to "make, use, offer to sell, or sell the patented invention within the United States, or import the patented invention into the United States, without the consent of and without accounting to the other" company.

If a patent covering a drug invented by two people is assigned to two different companies, one by each inventor, then both companies would be able to pursue market approval for the drug without the consent or approval of the other. In theory, if two companies both hold an assignment to the patent, and one company received FDA approval to market a compound covered by the patent, the second company could file an ANDA for the same product as soon as they were ready. The second company would not have to endure the expense of clinical trials and FDA approval born by the first company, giving the second company a significant financial advantage. Given the costs associated with new drug development, pharmaceutical companies generally require that they are exclusively assigned the rights of all of the inventors of record in order to ensure that they have complete ownership of a patent.

CLASSIFICATION OF PATENTS AND PATENT APPLICATIONS

Patents and patent applications can be divided into three basic categories, utility, plant, and design. Utility patents are issued for "any new and useful process, machine, manufacture, or composition of matter, or any new and useful improvement thereof."[21] This is the most common type of patent employed within the pharmaceuticals industry and it can be employed to protect "compositions of matter" for the enforceable lifetime of the patent. Small molecules, biologics (e.g., proteins, antibodies, nucleic acids), and even genetically engineered organisms (e.g., bacteria, animals) are compositions of matter for the purposes of a patent. Utility patents can also protect many other commercially important aspect of therapeutic agents such as a method of use (e.g., the use of a compound to treat a specific disease), a method of drug delivery (e.g., extended release formulations of a drug, type of pills), or a method of manufacturing a drug (e.g., a specific synthetic route, specialized manufacturing techniques). In general, utility patents are enforceable for 20 years beginning on the date of the earliest filed application. There are, however, some provisions for extension if regulatory delays occur within the USPTO or other governmental bodies (such as the FDA).[22] Over 540,000 utility patent applications were filed in 2012 in the U.S. alone. This represents over 92% of the total patent application filed with the USPTO.[23] Given the importance of this type of patent, this chapter will focus primarily on utility patents and the associated applications.

The second general class of patents is design patents. These patents protect "new, original and ornamental design for an article of manufacture." In other words, while a utility patent protects the way an article is used and works, a design patent protects the way an article looks. Since the design itself is inseparable from the article to which it is applied it cannot exist alone. Although outside of the world of pharmaceuticals, consider, for example, the personal computer. The personal computer has been available for purchase since the mid-1970, and Apple Inc. certainly has its share of utility patents covering a variety of aspect of its computer technology. In addition, however, Apple Inc. has received numerous design patents covering the appearance of the iMac computer, protecting the unique look of the product. This prevents other companies from marketing personal computers using independent technology with the same appearance as the iMac. Unlike utility and plant patents, design patents are enforceable for 14 years from the date the patent is granted by the USPTO[24] and extensions based on regulatory delays are not available.[25] This type of patent is far less common than utility applications. Less than 33,000 design patent applications were submitted to the USPTO in 2012,[26] and very few design applications are filed by the pharmaceuticals industry.

Plant patents are the third class of patents and they cover the invention or discovery of asexually reproducible "distinct and new variety of plant, including cultivated sports, mutants, hybrids, and newly found seedlings, other than a tuber propagated plant or a plant found in an uncultivated state."[27] This class of patent is primarily used to protect the rights of plant breeders, but in principle could be used in the pharmaceuticals industry. A plant designed to produce a therapeutically useful entity could be protected by a plant patent. Similar to utility patents, plant patents are enforceable for 20 years beginning on the date of the earliest filed application, and extension based on regulatory delays are possible.[28] Less than 1200 plant patent applications were filed with the USPTO in 2012.[29]

IMPACT OF OVERLAPPING PATENTS

Although any given invention can only be covered by one patent, it is possible to have multiple, overlapping patents that protect a commercially important compound. In theory, a compound could be protected by a patent, its method of manufacturing could be covered by a second patent, and a third patent could describe an extended release formulation. Each of these patents would be enforceable for 20 years from the earliest application date. Consider, for example, a patent on a compound designed to treat a disease that was initially applied for in 2010. If there are no extensions, this patent would be enforceable until 2030. While it would not be possible to obtain a second patent on the compound itself, if a new use for the compound were identified, such as the ability to use the compound to treat a new disease, a separate patent could be obtained for this method of use. If the method of use application was filed in 2015, a patent awarded based on this application would be enforceable through 2035. If a patentable extended release version was developed and applied for in 2020, then the associated patent would be enforceable until 2040 if it were granted. Although the second and third patents do not protect the therapeutic entity itself, they do provide protection for potentially valuable aspects of the compound. The commercial lifetime of many therapeutically useful compounds has been extended by stacking patents in this manner.

PATENT APPLICATIONS AND THEIR CONTENTS

While there are several different types of patent application, the three most common applications are provisional application,[30] non-provisional applications, and Patent Cooperation Treaty (PCT) application. In all cases, only one invention is allowed per application. Although not required, many applicants choose to begin the process by filing a provisional application if they are filing a utility patent application (plant and design applications cannot be filed provisionally). When this application is filed, it sets the priority date for the material embedded within the application, defining the material available to the public (the prior art) before the invention was filed with the USPTO. Contrary to popular belief, a provisional application requires the same level of details as that required by all other application. The only exception is that a specific claim set defining the scope and breadth of the invention is not required. In other words, a provisional application must contain an enabling disclosure that allows one skilled in the relevant art to make and use the invention described in the patent application. Failure to include the proper level of detail can result in loss of patent rights. In addition, provisional applications have a 12-month lifetime. Failure to follow up on a provisional application within 12 months will result in automatic abandonment of the application with no recourse. Given that there is very little difference in the requirements for a provisional application and other types of applications why would one choose to file a provisional application rather than move forward with the full process?

There are a couple of advantages to filing provisional application. First, the application is not available to the public during its 12-month pendency, and unlike other types of applications, provisional applications are not reviewed for patentability and are never published. Non-publication prevents potential competitors from gaining access to the technology embedded within the application. In addition, since a priority date is established, the owners of the application can show the material to potential investors without concern regarding creating a prior art disclosure, provided they do not allow the application to become automatically abandoned as mentioned above. The cost of a provisional application is also much lower than other application types, making it easier for smaller companies to afford the initial steps of the patent process. Finally, since the application is never published, if an application is abandoned after 12 months, the inventors can refile the application without concern that their original application has become part of the prior art that might prevent them from gaining patent protection. Of course, if anyone has published material in between the abandoned first application and the second provisional filing, that material may have an impact on the patentability of the second provisional application.

As mentioned above, a provisional application has a finite lifetime of 12 months. If no further action is taken, the provisional application is abandoned and the patent application process ends with no review and no issued patent. In order to continue the process, a second application must be filed before the 12-month clock winds down. This second application is filed as either a non-provisional application[31] or a PCT application,[32] and is said to claim the priority date of the original provisional application. The importance of this is that material that has entered the public domain during the pendency of the provisional application will not have an impact on patentability decisions for a non-provisional or PCT application claiming priority to a previously filed provisional application.

A non-provisional application, also referred to as a regular application, can be filed for plant, design, and utility related inventions. When the non-provisional application covers a utility related invention, then the non-provisional application can claim the benefit of a priority date of a previously filed provisional application, but this is not required. Unlike provisional applications, non-provisional applications must contain a full set of claims describing in a distinct and definite manner the nature of the invention to be covered by the patent applicant. No new material can be added to the application once it has been filed with the USPTO and the application must cover only one invention. In addition, non-provisional applications are reviewed by the USPTO for patentability and are published by the USPTO during their pendency, typically 18 months after the earliest priority date.

It is important to understand at this point that a non-provisional application is a U.S. patent application only and has no status outside of the U.S. Successful prosecution of an application of this type would provide coverage for the invention in the U.S. only. In order to obtain patent protection in other countries, it is necessary to file an application in those jurisdictions and then pursue the patent process to its ultimate conclusion in each jurisdiction. Contrary to popular belief, there is no such thing as a world patent. There is, however, an international system that allows an inventor to file a patent application in large number of countries using a single application, rather than filing an application in each country of interest. The Patent Cooperation Treaty (PCT) governs this process, and it is administered by the World Intellectual Property Organization (WIPO). As of July 2013 the PCT has 148 contracting countries, meaning that a single application submitted to a WIPO receiving office could designate up to 148 countries for patent prosecution. Much like the US non-provisional patent application, a PCT application must contain a full set of claims describing in a distinct and definite manner the nature of the invention to be covered by the applicant. Although not required, PCT applications can claim the benefit of the filing date of a previously filed provisional application, provided the provisional application has not been abandoned (expressly or automatically). This claim of priority provides the same benefit described above for a US non-provisional application. In addition, once the application is filed, no new material can be added to the application and it is published by WIPO, typically 6 months after the filing date (if a provisional application is in place, this would be 18 months from the original filing date).

Once the filing is complete and accepted by designated receiving office, it is reviewed. The application is reviewed for patentability by the International Searching Authority (ISA), which provides a patentability opinion based on the available prior art, but WIPO does not provide a final decision on patentability of an invention. This part of the process occurs when the application reaches the national stage in each designated country. At this point, a U.S. national stage application is handled in the same manner as a non-provisional filing. If the application process begins with a U.S. provisional application, then the application will generally enter the U.S. national phase of the process 30 months after the original filing date (Figure 12.5).

| U.S. Provisional Application | PCT or US Non-Provisional Application | Publication of Application | ISA Report and patentability opinion | National Phase Patent Application |

| 0 Months | 12 Months | 18 Months | 24 Months | 30 Months |

FIGURE 12.5 The U.S. patent application process often begins with a U.S. provisional patent application. Once the provisional application is filed, there is a 12-month deadline for filing a follow-up Patent Cooperation Treaty (PCT) or U.S. non-provisional application. Failure to file a follow-up application will result in loss of the priority date associated with the original provisional filing. Eighteen months after the provisional application is filed, the corresponding PCT or U.S. non-provisional patent application is published and becomes part of the prior art. Between 18 and 24 months, the International Search Authority (ISA) will review the patent application and issue a written report and a patentability opinion, and at 30 months the patent application enters the national phase of the process. The time line for the issuance of a patent is highly dependent on the content of the patent, the nature of the prior art, and the interaction between the applicant and the national patent office. As a result, it is not possible to provide a time line for these events.

As mentioned above, there is a 12-month gap between the filing of a provisional application and a subsequent filing of either a U.S. non-provisional application or PCT application. This 12-month window is often viewed as a time period during which the invention can be perfected or elaborated upon. This is partially correct. Supportive data, such as additional example compounds and their methods of preparation, can be included in a follow-up non-provisional or PCT application based on a provisional application. It is important to understand, however, that the added material must fall within the scope of the original provisional application in order to share the benefit of the original filing date. If, for example, new compounds are added to PCT application claiming priority to a provisional application, then in order for the new examples to share the priority date of the original filing, they must be within the scope of the originally claimed set of compounds. If they are outside of the scope of the material originally described in the provisional application, they can still be included in the follow-up application, but the added material will have a priority date of the follow-up application, not the provisional filing. This means that material that entered the public domain in the 12 months leading up to the follow-up application may be available as prior art in a patentability decision on the newly added material. Great care should be taken in determining whether or not to add additional material to a patent application at the 12-month filing decision.

As mentioned above, a patent application is restricted a single invention, but the guidelines as to what constitutes a single invention are not clear-cut. There are often situations in which an inventor files an application believing that it contains a single application, but the patent office disagrees. In these situations, the patent office will indicate the number of inventions they believe are present in the application, request that the inventor choose one to continue to pursue (a restriction requirement), and provide the inventors with the opportunity to file separate applications covering the other inventions designated by the patent office. The new applications are referred to as divisional applications.[33] Although they are submitted after the original filings, they share the priority dates of the original application, so the amount of prior art available to review in a patentability decision remains unchanged. Importantly, new material cannot be added at this juncture.

The continuation in part application, also referred to as a CIP application, is another common type of application, but it is only available in the U.S. This application allows inventors to protect improvements made to the originally submitted invention through the submission of a patent application that significantly repeats the material included in the original non-provisional application, but includes new material not disclosed in the original filing. The new material will receive a new filing date, but material from the original filing will retain its filing date for the purposes of patentability determinations. Outside of the U.S., this type of application is not available. Great care should be taken in determining if a CIP is appropriate for a given situation versus filing a separate application on the additional material that could garner broader coverage for the intellectual property in question.

There are several other types of applications available in the patent process, such as continuation applications, substitute applications, reissue applications, and continued prosecution applications. Drug discovery and development scientists, however, are generally only peripherally involved in these types of applications. As such, they will not be discussed in this text. Those interested in additional information on the various types of patent applications are encouraged to consult an intellectual property expert or Chapter 200 of the MPEP.

CONTENTS OF A PATENT APPLICATION

Although there are a number of different types of applications, the structure of an application is fairly consistent irrespective of the nature of the application. In all cases, a patent application must contain "a written description of the invention, and of the manner and process of making and using it, in such full, clear, concise, and exact terms as to enable any person skilled in the art to which it pertains, or with which it is most nearly connected, to make and use the same, and shall set forth the best mode contemplated by the inventor of carrying out his invention."[34] Stated in another way, a patent application must contain an adequate written description of the invention that explains why the invention is novel and support for a claim of a credible and substantial real world utility. This section of the patent application is generally referred to as an enabling disclosure. While the enabling disclosure is assumed to be adequate unless proven otherwise, it must contain enough detail such that one of ordinary skill in the art (the area of expertise related to the application) can understand and use the invention without undue experimentation. Failure to include an adequate written description and an enabling disclosure in a patent application can lead to loss of all patent rights. In addition, it is not possible to add this information after the application has been filed. This information must be included when the application is filed.

The rules surrounding the enabling disclosure are somewhat vague, as they must apply to any and all inventions that make their way to the USPTO. Within the context of drug discovery, however, it is possible to establish some general guidelines that are useful as guideposts in determining whether or not an application contains an adequate written description and enabling disclosure. Consider, for example, a patent application that describes a novel class of compounds that are useful for the treatment of atrial arrhythmias. Since the compounds are novel, there would be no information in the prior art on how to make the compounds or any demonstration as to how their biological activity could be demonstrated to provide evidence of utility. If a patent application contains only the structures of the new compounds and a table describing the compounds activity in an assay, it will not be considered to have provided an enabling disclosure. In this situation, there is no information provided regarding how the compounds could be prepared and no information as to how one might identify compounds with the desired utility. A medicinal chemist might be able to figure out how to make the compounds disclosed in the application, but patent law requires that a patent application provide enough information that the aforementioned medicinal chemist could make them "without undue experimentation." In other words, the patent application must include detailed information describing how the compounds can be prepared. The same requirements hold for the preparation of proteins, nucleic acids, genetically modified organisms, or any other type of novel material that is described in a patent application. Of course, if the material is already described in the prior art, then this information can be incorporated into the patent application to meet these requirements.

In a similar manner, claims of utility for the aforementioned compounds suitable for the treatment of arrhythmia must be supported by an enabling disclosure. In order to claim a method of using a group of compounds to treat arrhythmia, it is not enough to simply state that they are useful for this purpose. A person skilled in cardiovascular biology might be able to design methods for screening the compounds to determine whether or not they are useful in this disease state, but patent applications that do not include an adequate description of how this might be accomplished run afoul of the "without undue experimentation" requirement. In order to meet this requirement, a patent application must include detailed information on how to determine that the compounds in question are useful for the treatment of arrhythmia. As mentioned earlier in this chapter, an *in vitro* or *in vivo* assay can support a claim of utility, but an application must describe how these experiments can be performed, not just the results themselves. Data do not appear in a vacuum any more than novel compounds can appear out of thin air.

Another aspect of an enabling disclosure is that it must contain the best mode of practicing the invention at the time the application is filed.[35] Consider, for example, a patent that claims a new compound useful for the treatment of a disease. If the inventors have designed two separate methods for preparing the compound, and one of the two methods is better than the other, it is not sufficient to only include the inferior method of preparation. Inventors are required to provide the public with the best methods available when the patent application is filed in order to be eligible for patent protection under U.S. law. Interestingly, there is no requirement that the best mode be identified as such in a patent application, only that it be present in the application. It can even be represented as a preferred range of conditions or group of reagents if necessary. Also, if new and improved methods are developed after an application is filed, it is not necessary to update the application. In fact, the new method of preparation could be filed as a separate patent altogether.

The enablement requirements do not require that the invention work perfectly, only that the applicant provides the necessary information to prove that he/she is in "possession of the invention." In simpler terms, a method that can be used to prepare a novel compound in 1% yield is no less valid than a method that provides 95% yield of the same novel compound. If the best mode available when an application is filed provides only a 1% yield, this is sufficient for the purposes of filing composition of matter on the novel compound. The commercial viability of an invention does not impact whether or not an application contains an enabling disclosure based on the knowledge available when the patent application is filed.

Another critical section of a patent or patent application is the claims section. While the body of the patent contains information about the invention, the claims are found at the end of the patent or patent application and define what will be protected by the patent itself. Material that is presented in a patent or patent application, but is not included in the claims that are eventually approved by the USPTO are not protected by the patent. When a patent application is filed, the applicants provide a list of claims covering what they believe should be protected by the patent. (Provisional applications are a notable exception, as claims are not required if provisional applications.) Although it is possible that the USPTO agrees and issues a patent on all of the claims, in many cases, the patent examiners at the USPTO will reject portions of the claims based on the prior art available when the application was filed. Ideally, a dialog between the applicant and the USPTO patent examiners will produce a finalized set of claims that will be protected under U.S. patent laws. The final claims are often a subset of the original claims. Anything not included in the final claims set is not afforded protection under U.S. patent laws, but is available to the public as a prior art document by virtue of its publication in the patent application process. Ultimately, the claims in a patent application define the breadth and scope of protection provided to the owners of a patent. Of course, the claims must be supported by the written description of the application.

The claims themselves are required to have certain characteristic. First and foremost, they must "particularly point out and distinctly define the metes and bounds of the subject matter that will be protected by the patent grant."[36] The claims must be distinct and definite. In other words, someone reading that patent must have a clear understanding of exactly what is protected by the patent. Consider, for example, a patent claim that describes "a bicycle whose wheelbase is between 58% and 75% of the height of the rider." Since there is no standard definition of the height of a rider, it is not possible to define the scope of the invention with this claim. It is indefinite. In a similar manner, describing a method of dosing a drug based on the weight of a patient is also indefinite as there is no set standard defining the weight of a patient. Using terms such as similar, relatively, and superior are often associated with indefinite claims that are not allowed by the USTPO, as they are subjective in nature. Claims must be definite in their nature so that the public is clearly informed of the boundaries of patent protection afforded by the patent.

The claims themselves can be grouped into three distinct types: independent, dependent, and multiple dependent claims. An independent claim stands on its own and does not depend on any other claim to define its meaning. A markush claim describing a set of compounds can serve as an independent claim. The first claim of a patent application is always an independent claim, and it is also the broadest claim in the patent application. The second type of claim is a dependent claim. As the name implies, these claims does not stand on their own, as a portion of their scope is defined in an earlier claim. In a patent directed toward a set of chemical compounds, for example, an independent claim could be followed by a dependent claim that describes a subset of compounds described in the first claim (Figure 12.6).[37] The third type of claim, a

1. A compound of formula (**I**):

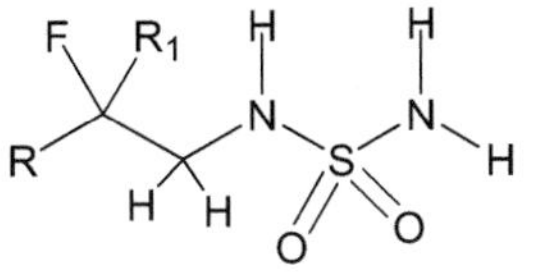

wherein:
R is selected from the group consisting of optionally substituted aryl, optionally substituted benzoisoxazole, and optionally substituted benzothiophene where R may be substituted by 0-5 moieties; and
R^1 is selected from the group consisting of hydrogen, fluorine, and optionally substituted C_{1-6} alkyl; or a pharmaceutically acceptable salt form thereof.

2. The compound according to claim 1 wherein R is phenyl, 2-fluorophenyl, 3-fluorophenyl, 4-fluorophenyl, 2,4-difluorophenyl, 2,5-difluorophenyl, 2,6-difluorophenyl, 2-chlorophenyl, 2-bromophenyl, 2-trifluoromethylphenyl, 3-trifluoromethylphenyl, 2,6-dichlorophenyl, 2,4-dichlorophenyl, 2-methylphenyl, 2-ethylphenyl, 2-methoxyphenyl, 2-chloro-6-fluorophenyl, 2-chloro-4-fluorophenyl, 2-fluoro-6-methoxyphenyl, 4-fluoro-2-methoxyphenyl, 2-chloro-6-methoxyphenyl, benzo[b]thiophen-3-yl, or benzo[d]isoxazol-3-yl.

FIGURE 12.6   The first claim provides a general structural class of compounds (a markush set) that the inventors wish to claim as their invention. The second claim is a dependent claim that provides a narrower definition of compounds that fall within the scope of the invention. Dependent claims must be narrower in scope that the claims upon which they depend.

multiple dependent claim, is similar to a dependent claim, but it derives its scope from two or more claims that are listed earlier in the patent claims set.

Claims construction, the writing of a claims set for a patent application, is a critical aspect of any patent application. As mentioned earlier, the claims of a patent define what is protected by the patent, so it is critical that the claims be written properly. While this may seem like a simple task, it is not. The language must be precise and follow the guidelines of the MPEP to the letter in order for the claims to stand up to the scrutiny of examination by the USPTO. Something as simple as the use of the word "or" when the word "and" is required can create a serious issue. Consider, for example, a group of limitations on a claim such as the definition of a substituent on a benzene ring. Defining an R-group appended to the ring as being "selected from the group consisting of Cl, F, and Br" is not the same as defining the R-group as "being selected from a group consisting of Cl, F, or Br." In the first definition, the group of allowed substituents is set without question. The R-group must be one of the three halogens listed. In the second definition, the use of the word "or" instead of "and" makes the nature of the limitation unclear. In contrast, a claim stating that "R is Cl, F, or Br" is considered acceptable, as the limitations are clear to the reader, even though the word "or" is used in this context.

In a similar manner, the terms "consisting of" and "comprising of" are very different within the context of a claim. Describing something as "optionally" involved in a process, methods, or other type of invention also has a specific meaning in the patent arena. The language rules of claims drafting are exceptionally complex and although not stated earlier, the same rules apply to the rest of the application.

Writing a proper patent application requires an in-depth knowledge of patent laws and technical knowledge of the invention itself. While inventors will have an in-depth knowledge of their invention, they often do not have the expertise required to draft a patent application that will survive the rigorous review of a patent agency. The inventors also may not recognize the full scope of what they can claim as their invention. There are special skills and knowledge that allow inventors to recognize new technologies, and in a similar manner, there are specific skills and knowledge that allow patent agent and patent attorneys to design patents that protect the rights of the patent owners. As mentioned earlier, patents are the lifeblood of the pharmaceuticals industry. Failure to gain patent protection for a new drug through the filing of proper patent application can make all of the research and development efforts worthless. In the absence of patent protection, very few organizations will move forward with a drug development program. The cost is simply too great.

For more information on the patent prosecution process, readers are encouraged to consults the USPTO Web site (www.USPTO.gov) and the MPEP (8th edition, version 9, released August 2012). Consultation with an expert in patent law is highly recommended in matters of intellectual property and possible prior art disclosures.

QUESTIONS

1. What are the four classes of patentable subject matter?
2. What types of inventions and discoveries cannot be patented?
3. What requirements must be met in order for an invention or discovery to be eligible for patent protection?
4. What is the meaning of the term "prior art"?
5. If it is discovered that a compound has a previously unreported utility, can a new patent be obtained on the compound? Why or Why not?
6. Scientist A conceives of a small molecule, its potential use in treating malaria, and designs a synthesis for the small molecule. Scientist A directs scientist B to prepare the compound, and the compound is screened for activity in a malaria assay by Scientist C. Who are the inventors of record for the small molecule?
7. Can inventorship be transferred between individuals and how does an individual qualify as an inventor of record on a patent application?
8. What is an assignment and how does it differ from inventorship?
9. What is the lifetime of a provisional patent application and what happens once this lifetime is exceeded?
10. Are provisional applications published? If so, when?
11. Are PCT patent applications published? If so, when?
12. Is a PhD thesis in a university library considered part of the prior art?
13. What level of detail must be provided in a patent application in order for it to be considered properly enabled?
14. What is a Markush structure?

References

1. Paul, S. M.; Mytelka, D. S.; Dunwiddie, C. T.; Persinger, C. C.; Munos, B. H.; Lindborg, S. R.; Schacht, A. L. How to improve R&D productivity: the pharmaceutical industry's grand challenge. *Nat. Rev. Drug Discov.* **2010,** *9,* 203–214.
2. http://www.fda.gov/drugs/developmentapprovalprocess/howdrugsaredevelopedan dapproved/approvalapplications/abbreviatednewdrugapplicationandagenerics/defau lt.htm.
3. MPEP 2104, 35 U.S.C. 101.
4. MPEP 706.03(b), 42 U.S.C. 2014.
5. MPEP 2105, Leahy-smith America Invents Act (AIA), Public Law112-29, sec. 33 (a), 125 Stat. 284.
6. Association for Molecular Pathology, et al. v. Myriad Genetics, Inc., et al.
7. MPEP 2104, 35 U.S.C. 101.
8. MPEP 2107.3 section IV.
9. MPEP 2107.3 section I.

10. MPEP 2107.3 section II.
11. MPEP 2112.
12. MPEP 2132, 35 U.S.C. 102 (a)1.
13. MPEP 2108.01, section I.
14. Freedom of Information Act (FOIA), 5 U.S.C. 552.
15. MPEP 2108.01, section IV.
16. MPEP 2108.01, section III.
17. MPEP 2108.01, section II.
18. MPEP2141, 35 U.S.C. 103.
19. MPEP 2144.04, sections I through VI.
20. MPEP 2144.04, section VII.
21. MPEP 2104, 35 U.S.C. 101.
22. MPEP 2700, 35 U.S.C. 154.
23. USPTO 2012 statistics, http://www.uspto.gov/web/offices/ac/ido/oeip/taf/us_stat.htm.
24. MPEP 1500, 35 U.S.C. 171.
25. MPEP 2710, 35 U.S.C. 154.
26. USPTO 2012 statistics, http://www.uspto.gov/web/offices/ac/ido/oeip/taf/us_stat.htm.
27. MPEP 1601, 35 U.S.C. 161.
28. MPEP 2700, 35 U.S.C. 154.
29. USPTO 2012 statistics, http://www.uspto.gov/web/offices/ac/ido/oeip/taf/us_stat.htm.
30. MPEP 201.04-201.09, 35 U.S.C. 111 (b), and 37 C.F.R. 1.53 (c).
31. MPEP 201, 35 U.S.C 111 (a), and 37 C.F.R. 1.53 (b).
32. MPEP 1800.
33. MPEP 201.06.
34. MPEP2161, 35 U.S.C. 112 First Paragraph.
35. MPEP 2165.
36. MPEP 2171, 35 U.S.C. 112, Second Paragraph.
37. Smith, G. R.; Brenneman, D. E.; Reitz, A. B.; Zhang, Y.; Du, Y. *Novel Fluorinated Sulfamides Exhibiting Neuroprotective Action and Their Method of Use* WO2012074784. , 2012.

13

Case Studies in Drug Discovery

The previous chapters of this text each emphasized a portion of the drug discovery and development process focusing on important theoretical aspects, tools, and techniques. Understanding and appreciating this material provides a firm foundation for those planning on pursuing a career in the pharmaceutical industry. There are, however, numerous important lessons that can be learned by examining the practical application of these concepts. The history of the pharmaceutical industry is filled with descriptions of successful programs and failed projects, and it is incumbent upon each generation of scientists to take heed of the lesson of the past to ensure that novel, safe therapeutics are discovered in an increasingly efficient manner. As the early twentieth-century philosopher George Santayana stated "Those who cannot remember the past are condemned to repeat it." Although he was almost certainly referring to geopolitical conflicts, this sentiment can just as easily be applied to the pharmaceutical industry. Routine review of modern scientific literature is strongly recommended for anyone pursuing a career in this challenging and complex arena, as every drug discovery program is unique and lessons can be learned from each. In an effort to provide some insight into what can be learned from both successful and failed drug discovery and development programs, several cases studies have been selected from the literature. Although the full details of each program are no doubt quite interesting, specific aspects of each case have been selected to provide real world examples of the process and emphasize important key learnings. Those who are interested in the full details of these programs are strongly encouraged to seek additional information in the modern literature.

TAMIFLU: FROM MECHANISM OF ACTION TO MARKETED DRUG

The symptoms of influenza virus infection, commonly referred to as the flu, were first described by Hippocrates approximately 2400 years ago.[1] While the symptoms of this infection are similar to those of many other respiratory infections, it is clear that influenza has been and continues to be a public health issue. The pandemic outbreak of Spanish flu in 1918–1920 led to the deaths of at least 20 million people (some estimates are as high as 40 million),[2] and despite significant advances in medical technology, the 2009 pandemic swine flu outbreak led to the death of more than 500,000 people (based on computer modeling of disease epidemiology).[3] Despite the importance of influenza infection, prior to 1999, only a few medications such as a Symmetrel® (Amantadine)[4] and Flumadine® (Rimantadine)[5] were available for clinical use. This changed with the approval of Tamiflu® (Oseltamivir) in 1999 (Figure 13.1).[6] The identification of this

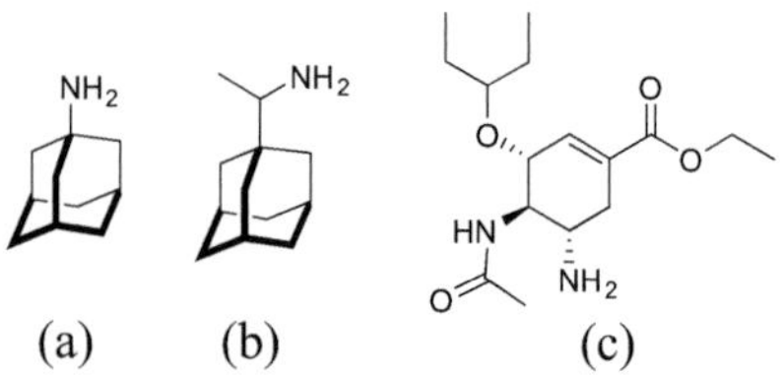

FIGURE 13.1   (a) Symmetrel® (Amantadine), (b) Flumadine® (Rimantadine), and (c) Tamiflu® (Oseltamivir).

compound demonstrates how an understanding of the underlying molecular mechanism, molecular modeling, and the use of prodrugs can lead to important advances in drug therapy.

The origin of Tamiflu® (Oseltamivir) actually begins with the identification of 2,3-didehydro-2-deoxy-N-acetylneuraminic acid (Neu5Ac2en, Figure 13.2(a)), an inhibitor of viral neuraminidase (also referred to as

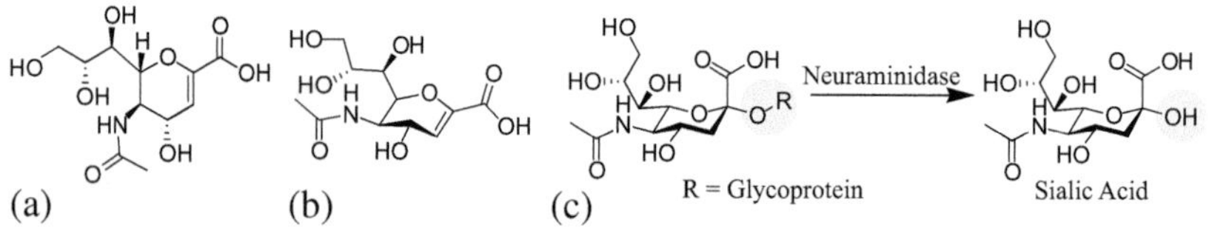

FIGURE 13.2   (a) Neu5Ac2en two-dimensional representation and (b) Neu5Ac2en three-dimensional representation. (c) Cleavage of sialic acid from glycoprotein by neuraminidase.

sialidase).[7] Neuraminidase is a critical enzyme in the life cycle of influenza, as viral replication is dependent on its ability to catalyze the cleavage of sialic acid from surface glycoproteins (Figure 13.2(c)).[8] Blocking this process prevents viral replication, thereby decreasing the impact of influenza infection. In addition, the amino acid sequence of the catalytic site is highly conserved across the various strains of influenza, which makes this enzyme a good target for antiviral therapy. Although it has been clearly demonstrated that Neu5Ac2en is an inhibitor of neuraminidase and that it binds to the active site of the enzyme, a comparison of the three-dimensional structures of this compound and the enzyme's substrate (sialic acid-containing glycopeptide) indicates that there are significant differences. Sialic acid exists primarily in a chair-like configuration, while the three-dimensional configuration of Neu5Ac2en is significantly distorted (Figure 13.2(b)) relative to sialic acid. This raises the question of how do compounds with very different overall shapes fit into the same binding site? This question also offers an important lesson in compound design.

The answer lies in an understanding of the configuration sialic acid glycopeptides in the active site (Figure 13.3). In isolation the three-dimensional

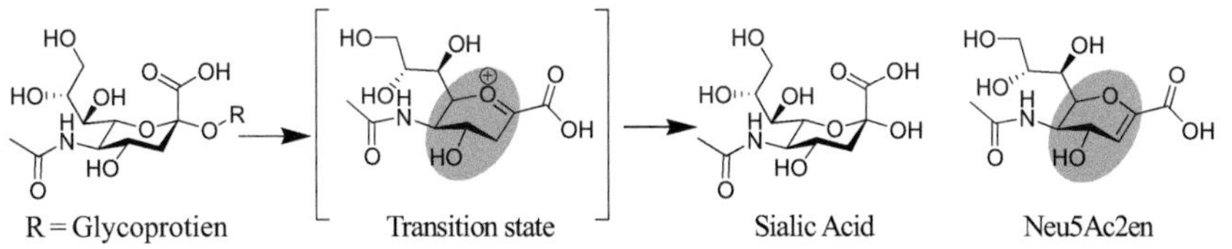

R = Glycoprotien   Transition state   Sialic Acid   Neu5Ac2en

FIGURE 13.3   Cleavage of sialic acid from a glycoprotein by neuraminidase is believed to proceed through a transition state in which the sialic residue's configuration is significantly distorted. The neuraminidase inhibitor Neu5Ac2en ring configuration is very similar to the configuration of the hypothetical transition state (both highlighted in green) which allows it to occupy the catalytic site and inhibit the enzyme.

structure of the sialic acid portion of the substrate glycopeptide is chair-like, but in the active site of the enzyme, this is not the case. X-ray crystal structure studies of enzyme/substrate complexes have clearly demonstrated that the sialic acid residue of the substrate is significantly distorted by three arginine residues located in the catalytic site.[9] These distortions enable the catalytic activity of the enzyme by bringing the substrate to a higher energy state that is closer in configuration to the theoretical transition state. At the same time, the higher energy configuration of the substrate is very similar to the configuration of Neu5Ac2en, and it is this fact that allows Neu5Ac2en to acts as neuraminidase inhibitor.[10] The realization that the substrate is significantly distorted in the active site was critical to the eventual discovery of Tamiflu® (Oseltamivir), and is an example of an important point in drug discovery and development. In designing novel therapeutics, it is important to consider not just the structure of a ligand, but also the structure of the ligand in the presence of its intended target. Significant structural distortions may occur in the binding site as a consequence of the local environment of the catalytic sites. These changes in configuration may be critical to the underlying molecular mechanism of the targeted biomolecule and it may be possible to exploit these changes in developing novel ligands.

While Tamiflu® (Oseltamivir) was eventually identified by scientists at Gilead using the structural observations provided by X-ray crystal structure data, it was not the first neuraminidase inhibitor to reach the market. Relenza® (Zanamivir), a potent neuraminidase inhibitor with an IC_{50} of

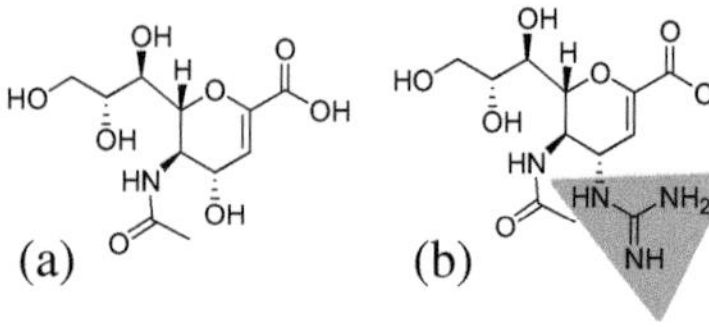

FIGURE 13.4 (a) Neu5Ac2en, (b) Relenza®
(Zanamivir) contains a guanidine residue (green)
in place of an alcohol. This change increases
potency by a factor of 1000.

(a)   (b)

10 nM was approved for clinical use first (Figure 13.4). This compound is approximately 1000 times more potent than Neu5Ac2en, but despite this increase in potency, considerable controversy surrounded its FDA approval for the treatment of influenza. In fact, the FDA advisory committee voted against approval of this drug 13-4 on the grounds that it did not work. The FDA leadership overruled the decision of the advisory committee and approved the drug based on positive results of a single clinical trial (at the time, other clinical trials failed to indicate efficacy over the placebo).[11]

Although this drug validated neuraminidase as a therapeutic target for treatment of viral infections, it also serves as a reminder that target potency is not the only issue to consider in the development of novel therapeutics. While Relenza® (Zanamivir) is capable of blocking viral replication (thereby providing therapeutic relief to patients), its method of delivery is somewhat unconventional. Rather than delivery via a tablet or pill, this drug is administered via inhalation of the powder. The reason for this unusual route of administration is directly tied to its physicochemical properties. The compound's highly polar side chains, particularly the triol and guanidine moieties, support tight interactions with the enzyme's catalytic site. At the same time, however, they also contribute to Relenza's® (Zanamivir's) high polar surface area (TPSA = 198) and low cLogP (cLogP = −5.7). These properties support high aqueous solubility, but also make Relenza's® (Zanamivir's) cell permeability very low. As a result, its oral bioavailability is extremely low, and it is ineffective as an orally delivered medication. In developing an orally delivered therapy, physicochemical properties can be just as important as target potency. The scientists who developed Relenza® (Zanamivir) were almost certainly aware of this, but they also recognized the viability of alternate forms of drug delivery. By keeping an open mind with respect to drug delivery, they were able to develop a clinically important therapy and begin to recoup their investments.

Returning to the development of Tamiflu® (Oseltamivir), scientist at Gilead hypothesized that the core ring structure of Neu5Ac2en and Relenza® (Zanamivir) could be replaced with a cyclohexene ring (Figure 13.5). They reasoned that the cyclohexene ring would preserve the overall configuration of the core ring system, while also providing novel chemical space for further exploration (keep in mind that patentability is a key consideration in developing novel therapeutics). The cyclohexene scaffold was selected as a platform for investigation and a wide range of neuraminidase

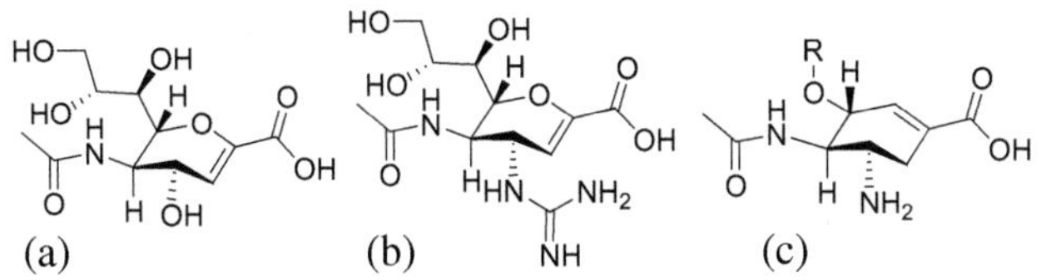

FIGURE 13.5   (a) Neu5Ac2en, (b) Relenza® (Zanamivir), and (c) the cyclohexene core of Tamiflu® (Oseltamivir). Modification of the R-group increased enzyme inhibition by a factor of 6300.

inhibitors, including GS-4701, were identified. An examination of the impact of structural changes in this series provides significant insight into the impact that even small structural changes can have on biological activity. Consider, for example, the difference in neuraminidase inhibition as a function of the nature of the R group of (Figure 13.5(c)). The free alcohol (R=H) is a marginal inhibitor with an IC_{50} of 6300 nM. The corresponding methyl ether is nearly twice as potent, but is still a relatively weak inhibitor of neuraminidase (IC_{50}=3700 nM). Interestingly, exchanging the methyl ether for an n-propyl ether produces a 20-fold increase in activity (IC_{50}=180 nM), and an additional 180-fold increase in binding potency is achieved with the 3-pentyl ether (GS-4701, IC_{50}=1.0 nM). In total, the relatively minor change in structure from methyl ether to 3-pentyl ether produced a 3700-fold increase in neuraminidase inhibition, indicating that the nature of the ether moiety is critical to neuraminidase binding efficiency. It also serves to highlight the fact that even small structural changes can have a dramatic impact on biological properties.[12]

Despite its potency, GS-4701 was not the molecule that eventually became the multi-billion dollar product known as Tamiflu® (Oseltamivir). Overall, the structure of GS-4701 is very different from Relenza® (Zanamivir). The most notable difference in structure included elimination of the guanidine moiety and the triol side chain, but these changes were not sufficient to establish acceptable oral bioavailability. Relenza's® (Zanamivir's) oral bioavailability in rats was only 3.7%, and GS-4701 demonstrated only a marginal improvement (4.3% bioavailability in rats). As with Relenza® (Zanamivir), the lack of oral bioavailability observed with GS-4701 can be explained by its high polar surface area (TPSA=102) and low cLogP (cLogP=−1.84). In order to address these issues, the ethyl ester prodrug GS-4104, which eventually became Tamiflu® (Oseltamivir, Figure 13.6(b)), was developed. The decrease in polar surface area (TPSA=90.6) and increase in cLogP (cLogP=1.16) were sufficient to allow reasonable bioavailability in rats (35%), mice (30%), dogs (73%), and most importantly humans (75%).[13] The prodrug is significantly less potent than the active agent (GS-4701), but it is rapidly converted to GS-4701 by esterases (enzymes that hydrolyze

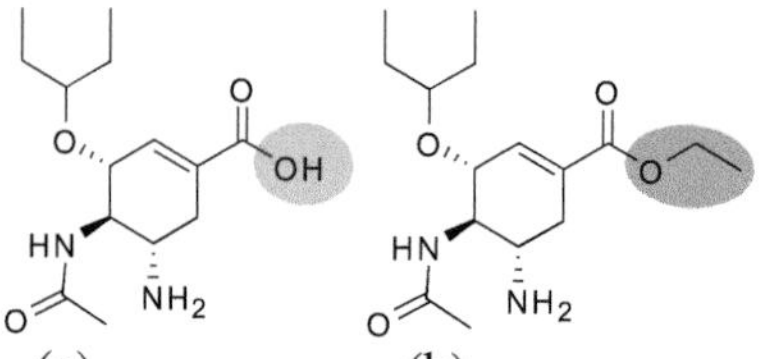

FIGURE 13.6   (a) GS-4701. (b) Tamiflu® (Oseltamivir). Replacing the acid of GS-4701 (orange)
with an ethyl ester (green) increased the bioavailability in rats by a factor of 10.

(a)   (b)

esters to the corresponding acid and alcohol), which, in turn, leads to
efficacy against influenza infection. Thus, the final lesson of the Tamiflu®
(Oseltamivir) story is that prodrugs can be a powerful tool in the development of novel therapeutics. If the scientist responsible for the development
of Tamiflu® (Oseltamivir) had not recognized this, they would have missed
out on a multibillion dollar opportunity.

HISTONE DEACYLASE INHIBITORS: PHYSICOCHEMICAL OPTIMIZATION VIA STRUCTURAL CHANGE

The vast majority of drug discovery programs eventually reach a point
at which an interesting series of compounds has been identified, but further advancement is blocked by physicochemical issues. As previously discussed, in many cases, a careful analysis of structure activity relationships
and structure property relationships can provide insight into structural
changes that will "fix" the physicochemical properties while maintaining
the desired properties of the series. Program scientists at Roche faced this
type of challenge in their efforts to develop novel inhibitors of histone
deacetylase-1 (HDAC-1). This zinc-dependent enzyme removes acetyl
groups from the amino side chain of lysine residues of histones (proteins
that play a key role in DNA coiling) and has been implicated in the progression of hepatocellular carcinoma (HCC, the most common form of
liver cancer).[14] Initial efforts by program scientists led to the identification
of selective HDAC-1 inhibitor typified by (1) (Figure 13.7). Although this
compound demonstrated the desired level of target selectivity, compound
clearance was far too rapid and bioavailability was far too low.

In order to address these issues, a series of strategic structural modifications were applied to the lead compound (Figure 13.7). Although many
compounds were prepared, it is instructive to consider the impact of key
structural changes as the program moved forward. Removal of the morpholine ring of (1) as seen in (2), for example, led to a substantial increase in
bioavailability and improvement in compound clearance, but compound
solubility decreased by a factor of 6. Although this first modification did
not solve all of the issues, it did demonstrate that the morpholine ring was
not a required feature of the pharmacophore. In addition, it established that

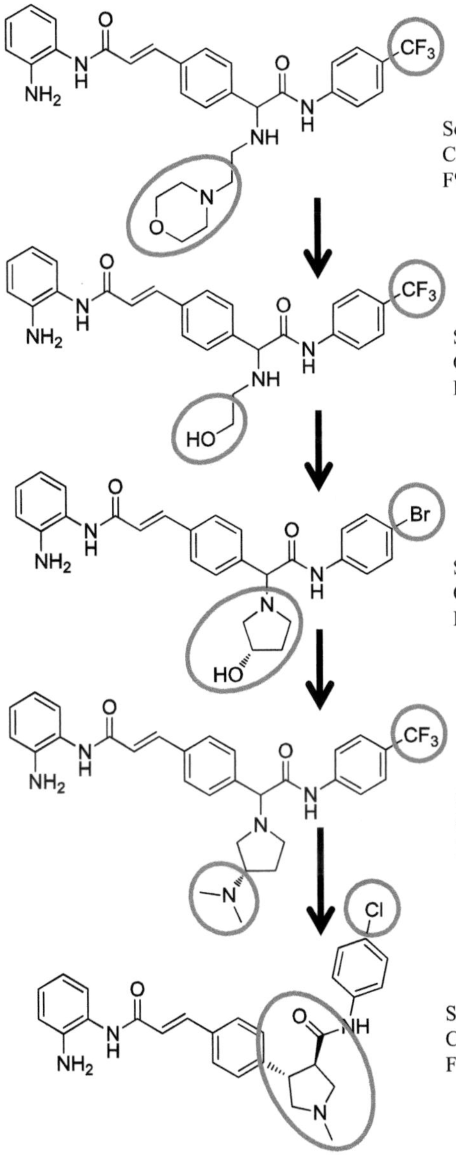

(1)
Sol.: 32 μg/mL
CL: 122 mL/mg/kg
F%: 11%

(2)
Sol.: 5 μg/mL
CL: 49 mL/mg/kg
F%: 71%

(3)
Sol.: 5 μg/mL
CL: 8mL/mg/kg
F%: 13%

(4)
Sol.: 610 μg/mL
CL: 67 mL/mg/kg
F%: 21%

(5)
Sol.: 136 μg/mL
CL: 39 mL/mg/kg
F%: 100%

FIGURE 13.7 Structural changes incorporated into a series of HDAC inhibitors (**1**–**5**, changes highlighted in red) provided improvements in physicochemical properties.

an alcohol could be tolerated in this region of the HDAC-1-binding pocket and suggested that other polar side chains might also be possible. Reclosing the side chain to form a hydroxyl pyrrole (**3**) improved *in vivo* clearance in the mouse, but solubility and bioavailability did not improve as much as desired. Solubility was significantly improved, however, when the alcohol was replaced with a dimethyl amine (4), a substituent known to impart significant solubility when tolerated. Bioavailability also improved, although it did not return to the level observed with (2). For reasons that are beyond the scope of this discussion, progression toward an eventual clinical candidate required a "scaffold hope" to a related series of compounds that eventually culminated with (5). Molecular modeling and HDAC-1 homology model docking studies demonstrated that the class of compounds typified by (5) would occupy the same space in the HDAC-1-binding site and as a result, most of the SAR data collected on the first series could be applied to the new series. In addition, the lead compound in this new series (5) is sufficiently soluble in water, metabolically stable, highly bioavailable, and was viewed as a viable candidate for clinical study.[15]

The progression from a compound incapable of further progression (1) to a viable clinical candidate (5) required only a few relatively small changes in the overall structure of the lead compound. Although in hindsight, it may be easy to rationalize why these changes worked, it probably was not as clear that they would be effective during the course of the program. Keeping an open mind about possible structural changes in a lead scaffold can be an important part of the drug discovery process. Tunnel vision with respect to chemical structure is a significant risk as programs move forward.

HIV PROTEASE INHIBITORS: CHEMICALLY COMPLEX MIRACLE DRUGS

The global HIV/AIDS epidemic that began in the early 1980s was one of the greatest health care challenges that the world faced in the twentieth century. The first clinical cases were diagnosed in 1981.[16] It quickly became clear that none of the drug regimens available at the time were useful for the treatment of this deadly viral infection. Government agencies, academic research labs, and, of course, the pharmaceutical industry launched research campaigns focused on determining the cause of HIV/AIDS, identifying potential therapeutic targets, and developing new drugs capable of arresting disease progression. In March of 1987, Retrovir® (azidothymidine, AZT), a reverse-transcriptase inhibitor, became the first drug approved for the treatment of HIV/AIDS.[17] Although this drug was certainly a watershed moment in the history of HIV/AIDS, it did not eliminate the problem. Even at the highest recommended doses Retrovir® (azidothymidine, AZT) does not completely block HIV replication. Low levels of viral replication still occur and prolonged treatment eventually leads to the development of AZT-resistant mutants.[18]

In an effort to deal with the emergence of drug-resistant HIV/AIDS, a number of companies and research institutions began working on alternative targets, including HIV protease, an enzyme responsible for the proteolytic cleavage of key proteins required for the production of viral particles. HIV protease inhibitors were the second class of anti-HIV drugs to become available to patients in need, and their impact on the course of the HIV/AIDS epidemic was profound. In 1995, before the introduction of the first HIV protease inhibitors, the estimated annual death rate in the United States was 50,628. Two years after the introduction of this new class of antiviral agents, the annual death rate in the United States had dropped by 60% to 18,851, and the number of AIDS-related fatalities has continued to decline (Figure 13.8).[19]

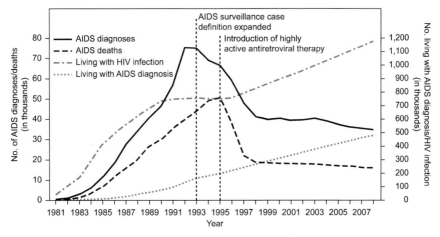

FIGURE 13.8 Annual reporting of AIDS diagnoses, deaths, and patients living with AIDS as reported by the Center for Disease Control and Prevention in the Morbidity and Mortality Weekly Report. The introduction of HIV protease inhibitors and HAART in the early to mid-1990s led to a dramatic decrease in AIDS-related fatalities. *Source: Center for Disease Control and Prevention Morbidity and Mortality Weekly Report, "HIV Surveillance United States, 1981-2008" June 3rd, 2011.*

The staggering decline in AIDS-related mortality was the result of both the availability of the HIV protease inhibitors and the introduction of the triple therapy treatment regimen, also known as highly active antiretroviral therapy (HAART).[20] In this protocol, HIV patients are treated with a protease inhibitor in combination with two additional antiviral agents (non-protease inhibitors). The combined presence of three different antiviral agents caused a decrease in HIV viral load in patients to nearly undetectable levels, restored CD4 T-cell counts to normal levels, and revitalized the immune system so that it can appropriately respond to pathogens.[21]

Although combination therapy was possible prior to the clinical use of HIV protease inhibitors, combinations of previously available antiviral drugs (reverse transcriptase inhibitors such as AZT) were unable to suppress viral replication for extended periods of time, and patients eventually succumbed to opportunistic infections. The development of HIV protease inhibitors was a key turning point in the battle against HIV and AIDS.

The importance of the HIV protease inhibitors is undeniable, and their impact on the treatment of HIV/AIDS is remarkable. This class of drugs is among the most structurally complex clinically available small molecule therapeutics. The fact they are on the market at all is a testament to the skill and creativity of the chemist, chemical engineers, and associated scientist responsible for the original preparation and eventual manufacture of these highly important drugs. Consider, for example, Crixivan® (Indinavir), the most potent HIV protease inhibitor that was developed by Merck and approved for clinical use by the FDA in March of 1996. This compound contains 5 chiral centers, making it one of 32 possible isomers with its overall chemical structure. Crixivan® (Indinavir) is the only isomer of the 32 possible combinations that is active as an HIV protease inhibitor. In addition, the laboratory scale synthesis of the Crixivan® (Indinavir) family of compounds is exceptionally complex (Figure 13.9) and it does not take advantage of naturally occurring chiral

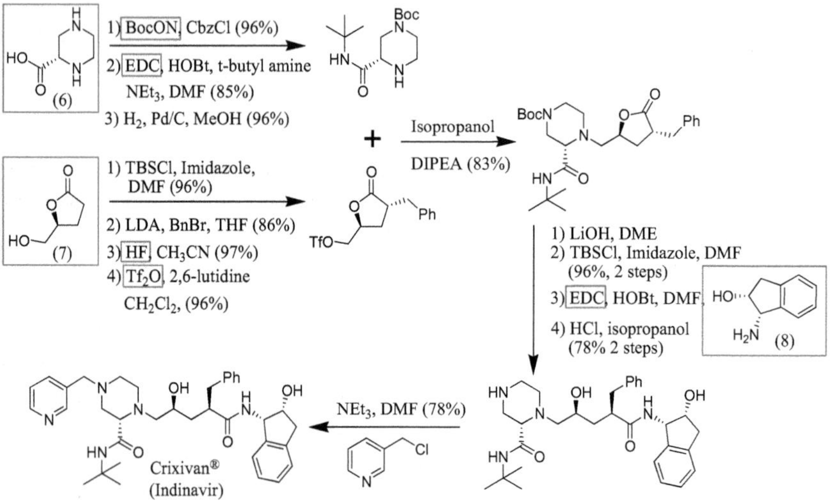

FIGURE 13.9 A laboratory scale synthesis of Crixivan® (Indinavir). Although this synthesis was useful for the preparation of small quantities of drug, the reagents and intermediates highlighted in red are not suitable for manufacturing commercial quantities of Crixivan® (Indinavir).

compounds (e.g., amino acids, sugars, the "chiral pool") to facilitate its synthesis. All of the chiral centers are established using enantioselective chemistry (i.e., chemistry that provides only one of two possible enantiomers). For perspective, the vast majority of small molecule therapeutics (>90%) contain two or fewer (often fewer) chiral centers, and many derive their stereochemistry from naturally occurring sources of chirality. It is also important to keep in mind that many other potentially useful compounds were prepared as the program moved toward its ultimate conclusion. No one knew ahead of time which molecule would end out being the marketed therapeutic agent. As a result, expedience in preparing screening samples was valued over potential scalability of the chemistry.[22]

The discovery level synthesis of Crixivan® (Indinavir) met the goal of producing enantiopure compound, but it was not suitable for manufacturing scale processes. Many of the reagents employed in this process, such as 1-ethyl-3-(3-dimethylaminopropyl)carbodiimide (EDC), trifluoromethanesulfonic anhydride (Tf$_2$O), 2-(tertbutoxycarbonyloximino)-2-phenylacetonitrile (BocON), and hydrofluoric acid (HF) can be effectively employed in a laboratory to prepare milligram to gram scale quantities of a compound, but using them to prepare multi-ton quantities required for global clinical use is simply not practical. In order to move forward with manufacturing scale synthesis, at least five steps in the laboratory synthesis needed to be replaced with chemistry suitable for multi-ton production methods. In addition, although compounds (**6**), (**7**), and (**8**) (Figure 13.9) could be purchased as starting points, the cost would have been prohibitive. An alternative route to each of these materials would also be necessary.

The need to develop alternative synthetic strategies that can be scaled to multi-ton manufacturing capacity is a common issue in the pharmaceuticals business. To be clear, however, this is not the result of poor planning by the discovery scientists as they pursue their projects. Discovery scientists are required to sort through hundreds, if not thousands of candidate compounds in an attempt to find the one compound that has a chance of becoming a marketed therapeutic. As a result, the medicinal chemists charged with preparing candidate compounds are focus on expediency of compound production. Optimizing synthetic protocols for individual compounds is simply not an efficient use of time when it is not clear which compounds will advance and which will fall be the wayside. Once the clinical candidate has been selected, however, the game changes, and a new set of skills are required to identify a commercially viable synthetic route. The process chemists, chemical engineers, and manufacturing specialist are often unsung heroes whose expertise is vital to the ultimate success of a drug candidate.

In the case of Crixivan® (Indinavir), the scientist at Merck were aware of the limitations of the laboratory scale synthesis, but they were also aware of the significant benefits to HIV/AIDS patients (and to the company itself) that would be provided if they could determine how to mass produce the compound. Their efforts led to an almost entirely new, manufacturing friendly process that is vastly different from the originally reported route

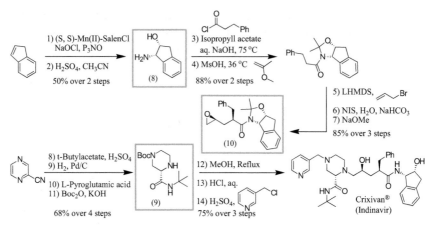

FIGURE 13.10 The commercial synthesis of Crixivan® (Indinavir). Key intermediates are highlighted in green.

(Figure 13.10). Scalable methods to prepare enantiopure intermediates **(8)** and **(9)** were developed to eliminate the need to purchase expensive starting material (thereby decrease the overall cost of goods). In addition, **(7)** was eliminated from the process entirely, as it was determined that the chiral centers derived from this starting material could be installed in an enantioselective manner. Steps **(5)**–**(7)** installed the chiral centers originally derived from **(7)**. The chiral centers of **(8)** drive the stereoselectivity of this chemistry, leading to the preparation of enantiopure **(10)**. The left **(9)** and right halves **(10)** of the molecules are then stitched together (step 13), and the final piece, the 3-pyridinylmethyl side chain, is added using a new procedure as well (although the reagent itself, 3-(chloromethyl)pyridine, is still the same). It is also important to note that reagents not compatible with manufacturing scale processes (EDC, Tf₂O, BocON, and HF) were eliminated or replaced, further simplifying the overall process and decreasing the total overall cost.[23]

On paper, these changes seem relatively straightforward, but in practice the road from bench-top synthesis to manufacturing plant production was long and difficult. In late 1993, early prototype production runs required almost 4 months to prepare batches of Crixivan® (Indinavir) and the overall process was inefficient (less than 15% overall). By November of 1994, the new process (Figure 13.10) had been developed and could produce clinical-grade material in 6 weeks, but the infrastructure to support full scale commercialization was not in place. At the time, Merck could produce enough material to support clinical trials (approximately 300 patients in early trials and 2000 in late stage trials), but if the drug proved to be successful, hundreds of thousands of patients would need daily supplies of the medication. At the time, each patient was taking six 400 mg pills daily (876 g annually). If 100,000 patients (a conservative estimate at the time) were to be prescribed the medication, annual supply requirements of drug substance, assuming no failed batches or recalls, would be 87,600 kg. In addition, supply shortages could not be tolerated, as patients losing access to the medication would present an opportunity for the development of drug-resistant variants of HIV. In February of 1995, more than 1 year before FDA approval and well before final clinical results would be available Merck took the risk of authorizing the construction of production lines capable of meeting the expected high demand for Crixivan® (Indinavir).

Of course, the gamble paid off for Merck, as Crixivan® (Indinavir) became an exceptionally important drug, but it was not without a great deal of effort on the part of the process and manufacturing teams. More than 400 people were involved in moving the process from the bench to full scale manufacturing, and by November of 1996, less than 9 months after FDA approval, more than 90,000 patients were using Crixivan® (Indinavir).[24] It is often the case that the scientist who identify a clinical candidate and the physicians who run the clinical trials are given credit for the launch of new and improved therapies, but the role of process and manufacturing teams such as those involved in the commercialization of Crixivan® (Indinavir) should not be overlooked. The clinical efficacy of a novel therapeutic agent is of little real world value if it cannot be commercialized, and this requires a robust manufacturing process. The scientist and staff responsible for establishing a manufacturing process are essential to the drug discovery and development process, as demonstrated by the efforts required to convert the Crixivan® (Indinavir) laboratory synthesis to full scale manufacturing.

NITROFURANTOIN: A SURPRISINGLY SUCCESSFUL DRUG

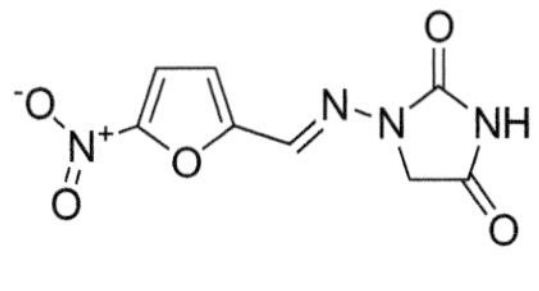

FIGURE 13.11   Nitrofurantoin.

In 1953, nitrofurantoin (Figure 13.11), which is known commercially as Macrobid®, Macrodantin®, and Furadantin®, was introduced for the treatment of urinary tract infections. At the time, no one could have predicted that it would eventually be included in the World Health Organization's List of Essential Medicines. Despite the development of a wide range of modern antibiotics, nitrofurantoin remains a front-line treatment of uncomplicated urinary tract infections. In addition, despite decades of clinical experience, reports of bacterial resistance remain rare. This is in stark contrast to the major classes of antibiotics (e.g., penicillins, quinolones), each of which has experienced the development of resistant strains as a result of widespread clinical application. Given the low level of antibiotic resistance that has developed since its original commercial introduction, it might seem logical to apply this medication to other types of bacterial infections, but this is not possible. Nitrofurantoin is uniquely capable of treating urinary tract infection, but other bacterial infections are not effectively treated with this drug.

In order to understand how this drug has retained its position in the medicine cabinet, one must examine both its mechanism of action and its pharmacokinetic properties. In the case of nitrofurantoin, there are multiple mechanisms of action that kill invading bacteria. Studies have shown that this drug kills bacteria via DNA damage, RNA damage, protein damage, and inhibition of the citric acid cycle.[25] The fact that it acts through multiple mechanisms explains the lack of resistance development. Resistance develops as a result of natural selection and mutation. In order for bacteria to develop resistance to penicillin, only one mechanism of action must be overcome. In the case of nitrofurantoin, however, mutations that confer resistance to at least the four mechanisms listed must all occur in a single organism. The odds of this occurring are so small that it almost never occurs.

Interestingly, the antibacterial activities of nitrofurantoin are driven by a structural feature that would be avoided in modern drug discovery programs, an aryl nitro group. This particular functionality is avoided in most modern drug discovery programs as it is a known risk factor for carcinogenicity, mutagenicity, and teratogenicity. In both humans and bacteria, aryl nitro groups are activated by the enzyme nitroreductase, which converts the nitro group to a nitroso group, a reactive functionality that damages DNA, RNA, and proteins via nucleophilic reactions and redox chemistry.[26] When these events occur in a bacteria, the organism dies and the patient is happy (Figure 13.12(a)), but if these events occur in cells of the patient, the results can be catastrophic (e.g., cancer, toxicity, Figure 13.12(b)). How is it

FIGURE 13.12   (a) Bacterial Nitroreductase converts nitrofurantoin to the corresponding nitroso compound (red) which kills bacteria via DNA damage, RNA damage, and proteins damage. (b) Human Nitroreductase also converts nitrofurantoin to the corresponding nitroso compound (red) which is a potential carcinogen, mutagen, and teratogen.

that a compound known to produce potential carcinogens, mutagens, and teratogens can remain on the market, especially in light of the strong focus on safety in the modern era of drug discovery?

The answer to this question is relieved by examining the pharmacokinetic properties of nitrofurantoin. When delivered orally, a 100 mg dose is rapidly removed after initial absorption. Approximately 75% of the dose is metabolized on first pass, while the remaining ~25% is excreted into the urinary tract as unchanged drug. As a result, the peak plasma concentration of a 100 mg dose of nitrofurantoin is less than 1 µg/mL and tissue penetration is negligible in all parts of the body except the urinary track. Drug concentration in the urinary tract, on the other hand, can exceed 200 µg/mL. This is in excess of the concentration required to kill invading bacteria.[27] Essentially, nitrofurantoin's pharmacokinetic properties prevent it from reaching other parts of the body, limiting its utility to urinary tract infections and preventing damage associated with its mechanism of action. Tissue distribution, metabolism, and excretion are key to the decades of success that this drug has seen. Interestingly, it is unlikely that this drug would have been developed in modern drug discovery programs. It would almost certainly been tossed aside based on possible risk associated with the aryl nitro group.

SELDANE® (TERFENADINE) VERSUS ALLEGRA® (FEXOFENADINE): METABOLISM MATTERS: SAFETY

The science of drug discovery and drug development is littered with examples of promising compounds and even marketed drugs that have encountered serious safety issues. In the case of marketed therapeutics, the sponsor companies often withdraw the drug from the market even if this is not required by regulators. The sudden and unexpected loss of revenue can radically change the fortunes of the sponsor company. At the same time, the removal from the market of a previously successful drug can provide an opportunity for the development of novel therapies, especially if a related compound with similar properties can be identified. Such is the case with the non-sedating antihistamine Allegra® (Fexofenadine, Figure 13.13(b)).

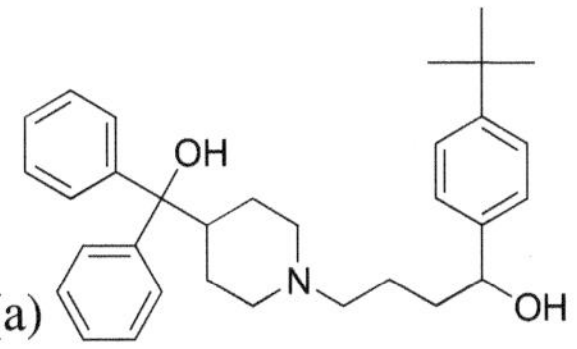

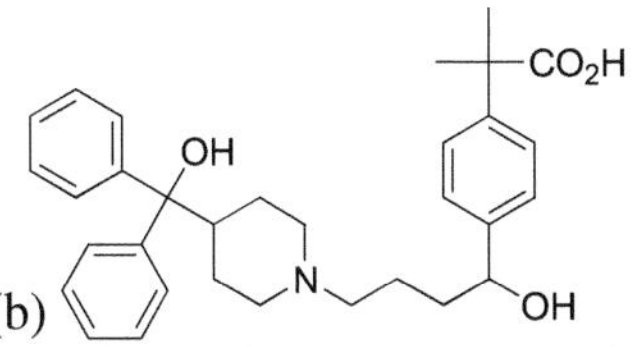

FIGURE 13.13   (a) Seldane® (terfenadine) and (b) Allegra® (Fexofenadine).

   The identification and eventual commercialization of Allegra® (Fexofen-adine) actually begins with the once successful drug Seldane® (Terfenadine, Figure 13.13(a)).[28] When it was originally approved by the FDA in 1985, Seldane® (Terfenadine) was viewed as a breakthrough drug, as it was the first non-sedating antihistamine approved for the treatment of allergic rhi-nitis (also known as "hay fever").[29] Hoechst Marion Roussel (now Sanofi) estimated that by 1990 this drug had been used by more than 100 million patients globally,[30] and by 1996 annual sales of Seldane® (Terfenadine) had reached $440 million. This compound was viewed as a huge success, but as discussed in earlier chapters, a serious flaw in this product led to its even-tual removal from the market. In 1990, the FDA issued a report on the risk of ventricular arrhythmia in patients using Seldane® (Terfenadine) con-comitantly with some macrolide antibiotics and ketoconazole.[31] By 1992, the FDA upgraded the warning associated with this drug to a "black box warning" describing the risk of ventricular arrhythmia, ventricular tachy-cardia, torsades de pointes, and sudden cardiac death when this drug was used in patients also taking ketoconazole and macrolide antibiotic, particu-larly erythromycin. A "black box warning" is the strongest warning that the FDA applies to a drug and is indicative of substantial medical risk or life-threatening side effects.[32] In 1997 the FDA recommended that all products containing Seldane® (Terfenadine) be removed from the market.

   The rapid fall of Seldane® (Terfenadine) marked a turning point in the industry with regard to drug safety studies and at the same time pre-sented a previously unknown pharmaceutical company the opportunity to reach the market with a drug that would eventually become a billion dollar blockbuster. The appearance of serious cardiovascular issues after the widespread use of Seldane® (Terfenadine), despite a relatively clean safety profile in clinical trials, was a surprise to both regulators and the manufacturer. Eventually, it was determined that the risk of cardiovascu-lar adverse events is tied to Seldane® (Terfenadine) metabolism, specifi-cally metabolism by CYP450 3A4, and the hERG channel, a voltage-gated potassium channel critical to maintaining ventricular rhythm.[33] Under normal circumstances, Seldane® (Terfenadine) is rapidly metabolized by CYP450 3A4 to the corresponding carboxylic acid (Figure 13.14). Although

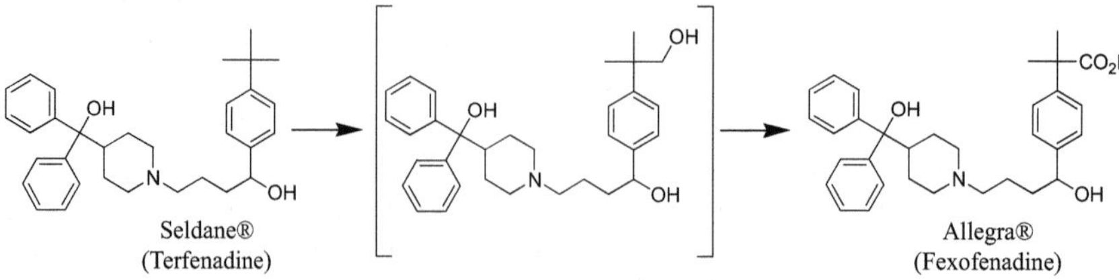

FIGURE 13.14   Seldane® (Terfenadine) is actually a prodrug that is metabolically converted to the biologically active compound Allegra® (Fexofenadine) by CYP450 3A4. Normally, this conversion is rapid, but in the presence of CYP450 3A4 inhibitors, dangerously high plasma concentrations of Seldane® (Terfenadine) can lead to dangerous cardiovascular side effects.

the scientists at Hoechst Marion Roussel (now Sanofi) were not aware of it at the time, it was later determined that Seldane® (Terfenadine) is actually a prodrug and the corresponding carboxylic acid metabolite Allegra® (Fexofenadine) is the true biologically active compound. At the same time, Seldane® (Terfenadine) is a potent hERG channel blocker ($IC_{50} = 10 \, nM^{34}$), but its rapid removal from the systemic circulation by CYP450 3A4 metabolism prevents it from achieving a significant plasma concentration. Therefore, under normal circumstances, Seldane® (Terfenadine) does not impact cardiac function, as it is rapidly eliminated. If, however, CYP450 3A4 activity is inhibited by other drugs, such as ketoconazole and erythromycin, plasma concentrations of Seldane® (Terfenadine) can increase rapidly, leading to adverse cardiovascular effects (e.g., ventricular arrhythmia, ventricular tachycardia, torsades de pointes, and sudden cardiac death).[35] In other words, it is the combination of suppression of CYP450 3A4 metabolism (which alters pharmacokinetic parameters of the drug) and potent hERG blockade of the parent compound that produce the potentially deadly effects observed with Seldane® (Terfenadine).

As the commercialization of Seldane® (Terfenadine) was coming to an end, the small biotech start-up company Sepracor (now Sunovion) was actively attempting to take advantage of the discovery of the active metabolite of Seldane® (Terfenadine). The scientist at Sepracor correctly hypothesized that this active metabolite, which was eventually marketed as Allegra® (Fexofenadine), would possess all of the positive attributes of the parent compound (e.g., non-sedating antihistamine, good PK properties), but not the cardiovascular risk profile. They based this hypothesis on the observation that Allegra® (Fexofenadine) had no activity as a hERG channel blocker (keep in mind that relationship between the hERG channel and cardiovascular risk had not been firmly established at the time). This hypothesis, of course, turned out to be correct. Sepracor secured a patent on the metabolite and then licensed the compound to Hoechst Marion Roussel (now Sanofi) in 1993. By 2004, annual sales of Allegra® (Fexofenadine) reached $1.87 billion,[36] and the drug was approved for over the counter sales in 2011, further broadening the market.[37]

The discovery and development of Seldane® (Terfenadine) and Allegra® (Fexofenadine) teach some important lessons. First, the safety of a therapeutic agent is controlled by more than the physical properties of the therapeutic agent itself. Metabolites can create significant safety risks which must be evaluated as candidate compounds are considered for further progression in the drug discovery and development process. Second, and perhaps more importantly, the pharmacokinetic profile of one drug can be significantly impacted by the presence of a second drug. Although this concept is well known to modern drug discovery scientists (and is the basis for screening candidate compound for CYP450 inhibition), it was novel science when this story was unfolding. Finally, understanding the role of metabolites in biological activity can be the key to identifying new therapeutic agents. In this case, Hoechst Marion Roussel's missed out on a major opportunity by failing to understand that their compound was actually a prodrug. Sepracor, on the other hand, was able to exploit this opportunity and develop a novel therapeutic agent that eventually became a blockbuster product.

CLARITIN® (LORATADINE) VERSUS CLARINEX® (DESLORATADINE): METABOLISM MATTERS: PHARMACOKINETICS

The rise and fall of Seldane® (Terfenadine) caught the attention of a number of pharmaceutical companies and regulatory agencies. The identification of novel, non-sedating antihistamines was a hotly pursued area in the 1980s and early 1990s. Schering–Plough (acquired by Merck) identified a compound that was eventually marketed as Claritin® (Loratadine, Figure 13.15),[38]

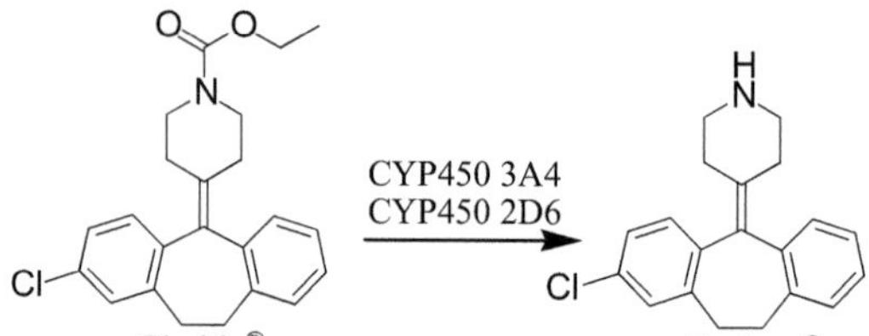

FIGURE 13.15 Claritin® (Loratadine) is converted to the active metabolite Clarinex® (Desloratadine) by CYP450 3A4 and CYP450 2D6.

but the road to market approval was longer than anticipated. They submitted an NDA to the FDA in 1986, but the drug was not approved for commercialization until 1993, nearly 77 months later. When the approval process began, the FDA was reluctant to approve another non-sedating antihistamine that did not appear to provide any additional benefits over the competition. This position changed radically when the problems with Seldane® (Terfenadine) began to come to the surface. Peak annual prescription sales of Claritin® (Loratadine) exceeded $2.0 billion, and over the counter sale was approved in 2002, the same year the patent covering this drug expired.[39]

Unlike its predecessor, Claritin® (Loratadine) has remained on the market, but once again the failure of the innovator company to fully appreciate the metabolic fate of their compound provided an opportunity for another company to enter the market with a competitor compound. In this case, the scientists at Sepracor identified an active metabolite of Claritin® (Loratadine) that was produced by CYP450 3A4 and CYP450 2D6. This metabolite, which was eventually brought to market as Clarinex® (Desloratadine, Figure 13.15), has substantially improved pharmacokinetic properties over the parent compound. As indicated in Table 13.1, both the C_{max} and T_{max} of

TABLE 13.1   PK Properties of Claritin® (Loratadine) and Clarinex® (Desloratadine)

PK Parameter	Claritin® (Loratadine)	Clarinex® (Desloratadine)
C_{max} (mg/L)	17	16
T_{max} (h)	1.2	1.5
$AUC_{0-\infty}$ (mg h/L)	47	181
$T_{1/2}$ (h)	6	13.4

the parent and metabolite are comparable. The metabolic half-life ($t_{1/2}$) of the metabolite, however, is more than twice that of the parent compound, and this translates into a significantly higher overall exposure (AUC) for the metabolite.[40] Sepracor obtained a patent on the metabolite, licensed it to Schering–Plough in 1997, and the FDA approved its commercial sale in 2002 (the same year that Claritin® (Loratadine) became available over the counter). Clarinex® (Desloratadine) became a blockbuster drug for Schering–Plough and Sepracor (by 2006, annual sales reached \$722 million[41]). This case is another stark reminder of the importance of understanding the metabolic fate of therapeutic agents. In this case, Schering–Plough missed a major opportunity to be the sole owner of a major therapy. Sepracor, on the other hand, was able to capitalize on this oversight and grab a "piece of the pie" in a major market area. The moral of the Claritin® (Loratadine)/Clarinex® (Desloratadine) and the Seldane® (Terfenadine)/Allegra® (Fexofenadine) stories is to pay close attention to the metabolic fate of candidate compounds. The potential risks and rewards can be substantial.

MPTP: PARKINSON'S DISEASE IN A BOTTLE

There are certainly many lessons to be learned by reviewing case studies of drug discovery and development programs, but there is also a great deal of useful information and insight that can be gained from an examination of the science developed in related fields. Consider, for example, the history of 1-methyl-4-phenyl-1,2,3,6-tetrahydropyridine, also referred to as MPTP (Figure 13.16(a)). This story actually begins with the identification of

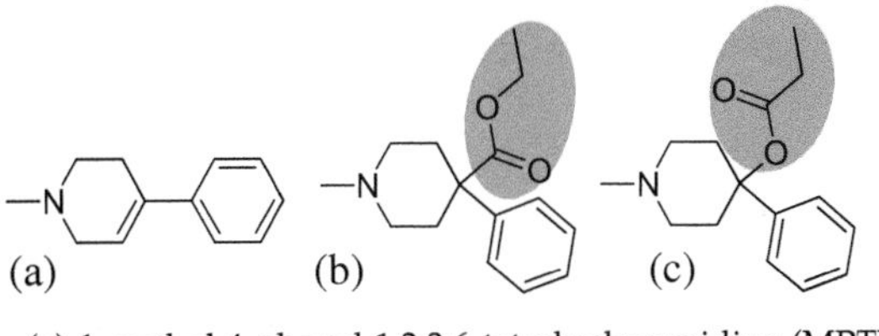

FIGURE 13.16   (a) 1-methyl-4-phenyl-1,2,3,6-tetrahydropyridine (MPTP), (b) Demerol® (Meperidine), and (c) Desmethylprodine (MPPP).

Demerol® (Meperidine, Figure 13.16(b)), a synthetic μ-opioid receptor agonist first prepared in 1932.[42] This morphine analog has been successfully used to treat moderate to severe pain, but carries with it a risk of addiction and abuse. It should come as no surprise that its distribution is highly regulated, it has been used for "recreational purposes," and that possession of this material without a valid prescription can results in serious legal trouble. A related compound, Desmethylprodine (MPPP, Figure 13.16(c)), was identified by scientists at Hoffmann-La Roche in the 1940s.[43] This compound is also a potent μ-opioid receptor agonist with analgesic properties, is highly regulated, but is not currently used clinically. The only significant difference between Demerol® (Meperidine) and MPPP is the inversion of the ester functionality (highlighted in green, Figure 13.16). Although this change may seem minor, the biological properties of the two compounds are radically different. Demerol® (Meperidine) is a commercially successful drug, while MPPP is an exceptionally dangerous material. The dangers of MPPP were identified as the result of the unfortunate efforts of recreational drug users in the 1970s and 1980s.

Although drug enforcement laws currently outlaw the use of a wide range of commercially available material and related compounds for recreational purposes, these laws were not as strictly established in the 1970s and 1980s. Possession and recreational use of drugs such as Demerol® (Meperidine) could lead to significant jail time, but this was not the case with the so-called "designer drugs." Designer drugs were analogs of known pharmaceutical agents that could be used for recreational purposes, but technically were not illegal at the time, as they were not part of the definition of illegal substances. In 1976, a chemistry graduate student, Barry Kidston, decided to take advantage of this loop-hole in the law. He was aware of the literature reports regarding MPPP that described its potency as a μ-opioid receptor agonist and guessed that it would provide the same effects without the legal issues. Unfortunately, Barry's attempt to skirt the law came with dire, unpredicted consequences. He was successful in his efforts to prepare MPPP and began using it for recreational purposes. Within a short period of time, he began to develop the symptoms of Parkinson's disease (e.g., shaking, rigidity, slowness of movement, difficulty walking). Treatment with L-dopa, the standard of care of Parkinson's disease patients, alleviated some of his symptoms, but was not curative. Later, in 1982, another group of drug addicts acquired and used MPPP, and they also suffered severe consequences. These drug users also developed Parkinson's disease symptoms, and in some cases total paralysis. Although it took several years of research, the neurologist Dr. J. William Langston in conjunction with collaborators at the National Institutes of Health eventually identified MPPP as the culprit.[44] This conclusion, however, raises an interesting question. How is it possible that the minor difference between Demerol® (Meperidine) and MPPP (the inversion of an ester) can lead to such drastic consequences?

The answer to this question can be found in an examination of differences in metabolism and drug distribution. Demerol's® (Meperidine) metabolic pathway leads to eventual excretion of metabolites via the kidneys (Figure 13.17). In this case, the parent compound can be de-esterified,

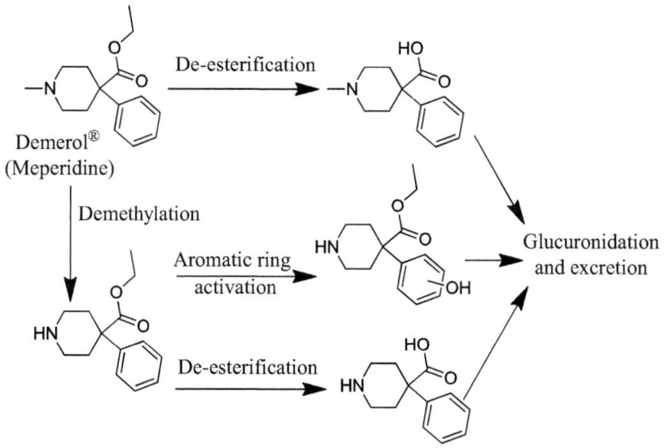

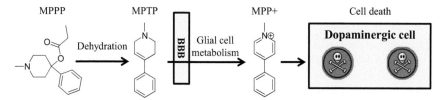

FIGURE 13.17   Demerol® (Meperidine) is metabolized to compounds that are safely removed from the body. The metabolites eventually undergo glucuronidation and are excreted from the body.

glucuronidated, and then excreted into the urine. Alternatively, removal of the N-methyl group can be followed by activation of the aromatic ring (e.g., hydroxylation) or de-esterification. Glucuronidation of these metabolites is then followed by excretion into the urine and elimination. Essentially, all of these pathways lead to the removal of Demerol® (Meperidine) from the systemic circulation.[45]

FIGURE 13.18   MPPP is converted to MPTP *in vivo*, which passes through the BBB. Once inside the brain, glial cells absorb the MPTP and convert it to MPP⁺. This compound enters the dopaminergic cells via dopamine transporters, killing these specialized cells.

The metabolic fate of MPPP, however, is quite different (Figure 13.18). In this case, the inverted ester opens a different metabolic pathway that leads to the production of 1-methyl-4-phenyl-1,2,3,6-tetrahydropyridine, more commonly referred to as MPTP. In the absence of further metabolic activity, this compound is non-toxic, but it is highly lipophilic and readily passes through the blood brain barrier. Once inside the brain, MPTP undergoes further metabolic conversion in glial cells. The enzyme Monoamine oxidase B (MAO-B) converts MPTP into the cationic species 1-methyl-4-phenylpyridinium (MPP$^+$). Diffusion across the blood brain barrier is no longer possible, as the positive charge prevents passage through the lipophilic barrier, and as a result MPP$^+$ cannot exit the brain. Unfortunately, MPP$^+$ is a substrate for the dopamine transporters (DATs) of the dopaminergic cells found in the substantia nigra region of the brain. Uptake of MPP$^+$ into these specialized cells leads to cell death, eliminating the brain's ability to produce dopamine, a key chemical required for coordination and movement. As the number of dopaminergic cells in the substantia nigra decreases, Parkinson's disease symptoms begin to appear, eventually leading to paralysis. In this case, the combination of altered metabolic pathways and brain distribution leads to the severe consequences associated with MPPP.[46]

Although the unfolding of the MPPP story is not technically a drug discovery story, there are a number of important lessons that can be gleaned from these tragic events. First, as discussed earlier in this chapter, metabolites can play a key role in the safety of candidate compounds. In this case, a simple structural change produced dramatic and severe changes in the physiological impact of the two compounds. Second, and perhaps more importantly, compound distribution can significantly impact compound safety. Consider, for example, how this story would have changed if MPPP were incapable of crossing the blood brain barrier. If MPPP were not able to cross the blood brain barrier, glial MAO-B would not have produced MPP$^+$, uptake of MPP$^+$ by dopaminergic cells would not have been possible, and Parkinsonian symptoms would not appear. Candidate compounds that do not reach the intended target cannot provoke the desired biological response, and at the same time compounds with the potential to elicit a negative response (side effect) can only do so if they reach the associated "anti-target."

Finally, species differences can be critical. In an effort to unravel the mystery behind the impact of MPPP and MPTP, a number of research teams examined the impact of these compounds on a variety of rodent species. None of the studies recapitulated the disastrous impact of these compounds that was observed in humans. Although it was not known at the time, rodents are not nearly as susceptible to these compounds as humans. Squirrel monkeys, on the other hand, are just as vulnerable as humans and Dr. J. William Langston's work with these monkeys and MPTP was crucial to understanding the development of Parkinson's disease in the addicts described above.[47] The critical lesson here is that animal safety models are not always predictive of human events. Consider the terrible consequences that could have come to pass if a company had attempted developed MPPP as a new pain killer with only rodent safety studies as an indication of expected safety in humans. Virtually all of the healthy phase 1 clinical trial volunteers would have quickly developed Parkinson's disease. Of course, clinical development of the compound would have been immediately terminated, but that would have been little comfort to healthy volunteers of the phase 1 studies. Extensive safety studies must be completed before any candidate compound is given to humans in order to minimize the likelihood of an unexpected, tragic outcome.

BUPROPION AND METHYLPHENIDATE: IMPROVING PERFORMANCE VIA FORMULATION CHANGES

While it is often possible to address weakness in the pharmacokinetic profile of candidate compounds by apply structural changes to design new analogs, there are times when this is not the optimal solution. Consider, for example, an approved therapeutic agent that requires three times per day dosing as a result of a shorter than desired metabolic half-life. In some cases, altering the drug's formulation can decrease the number of doses required for efficacy. Of course, changes in drug formulation will not alter the rate at which the compound is metabolized, but there are some formulations that can alter the time course of drug delivery. A number of extended release or sustained release mechanisms, such as multilayered tablets and osmotic pump systems, are capable of delivering drugs to patients over a longer period of time than a simple tablet. These, and similar technologies, have been effectively employed to increase the utility and improve patient outcomes with a number of important therapeutic agents. The clinical utility of bupropion and methylphenidate, for example, were greatly enhanced as a result of formulation changes designed to modify the drug delivery time course.

Bupropion (Figure 13.19) was originally identified in 1969 by scientists at Burroughs Wellcome (now GlaxoSmithKline, GSK). It was approved

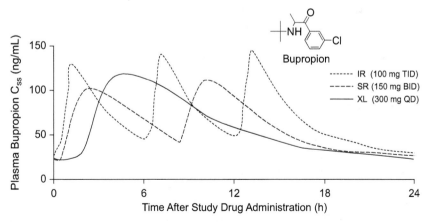

FIGURE 13.19 Comparative plasma concentration curves of Wellbutrin IR® (immediate release), Wellbutrin SR® (Sustained Release), and Wellbutrin XL® (Extended Release). *Source: This figure was published in Jefferson, W.J.; Pradko, J.F.; Muir, K.T. Bupropion for Major Depressive Disorder: Pharmacokinetic and Formulation Considerations. Clin. Ther. 2005, 27 (11), 1685–1695, Copyright Elsevier, 2005.*

by the FDA for the treatment of depression and marketed as Wellbutrin® in 1985. The maximum dosage approved was 600 mg three times per day, but a significant number of patients developed epileptic seizures, and the drug was withdrawn from the market in 1986. An examination of clinical data and drug pharmacokinetics demonstrated that the risk of seizure was highly dose dependent, and Wellbutrin® was relaunch in 1989 with a lower maximum dose (450 mg). While many patients were able to successfully resume drug therapy at the lower doses, patients who required higher doses for symptom relief were no longer able to use Wellbutrin®.[48]

Although it would have been possible to develop a new clinical candidate devoid of seizure risk, an alternate path forward was selected. In this case, it was hypothesized that peak plasma concentrations achieved with the highest doses were responsible for the increased risk of seizures, and that if a new dosing regimen could be designed to "smooth out" these peaks, then the risk of seizures would be eliminated. In other words, this new hypothetical formulation would allow plasma concentrations of the drug to reach therapeutically useful levels, but they would not reach concentrations required to cause seizures. This led to the development of a sustained release formulation of bupropion, Wellbutrin SR®, as a twice daily medication (FDA approved in 1996[49]), and an extended release formulation, Wellbutrin XL®, as a once daily medication (FDA approved in 2003[48]). Although modern drug discovery and development program routinely consider extended release formulations such as embedding the drug in slow dissolving polymer matrix or an osmotic pump system, this technology was a relatively new at the time. These new formulations provided greater control of plasma concentration when compared with the immediate release tablet and, as expected, lowered the peak plasma concentrations of bupropion (Figure 13.19).[50] This, in turn, allowed the drug to be used in a wider range of patients, simplified the dosing regimen, and effectively extended the patent life of the Wellbutrin® franchise for GSK. Patents covering the use of bupropion in conjunction with extended release technologies were filed well after the original bupropion patents, providing additional patent protection for both Wellbutrin SR® and Wellbutrin XL® (the patent covering Wellbutrin XL® expires in 2018).

Changes in the formulation of methylphenidate also provided improved patient outcome, while avoiding the process of developing an entirely new chemical entity. Methylphenidate was originally prepared in 1944 by Leandro Panizzon, a scientist at Ciba–Geigy Pharmaceutical Company (now Novartis). It was identified as a stimulant and was brought to market in 1954 as Ritalin® for the treatment of "chronic fatigue, lethargy, depressive states, disturbed senile behavior, psychosis associated with depression and narcolepsy." The most common use for methylphenidate, however, is the treatment of children with attention deficit disorder (ADD) and attention deficit hyperactivity disorder (ADHD). It is considered the standard of care for children with ADD or ADHD and has been successfully used for this purpose for over 50 years.[51]

Despite its commercial success under the name Ritalin®, it was not a perfect drug for the treatment of children. Its safety profile and efficacy profile were not an issue, but the three times daily administration was viewed as a problem by some. Although this may not seem like a significant issue to many, ask any parent about the difficulties of giving their child medicine and they will tell you how challenging this can be. Add the complication of ADD or ADHD to the picture, and the difficulty only increases. The dosing regimen is, of course, a result of methylphenidate's PK profile (Figure 13.20), which produces peak

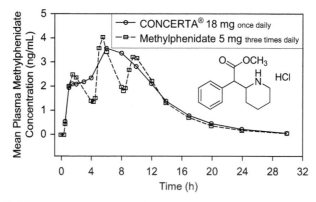

FIGURE 13.20 Comparison of the mean plasma concentration of drug observed with immediate release methylphenidate and Concerta® extended release formulation. *Source: Reproduced with the permission of Janssen Pharmaceuticals, Inc.*

plasma concentrations approximately 2h after ingestion of immediate release tablets (Ritalin®). Plasma concentration of the drug then falls, and additional doses are required to maintain efficacy during day time hours (Night time coverage is generally not necessary). Scientists at Janssen Pharmaceuticals recognized the opportunity to improve the utility of methylphenidate by developing an alternative formulation of this important drug. They designed an extended release delivery system that delivered efficacious plasma concentrations of the drug over a 12-h period, enabling once-daily dosing (Figure 13.20).[52] Their product was approved by the FDA in 2000 and marketed as Concerta®. Safety and efficacy were essentially the same as the original formulation, but the once-a-day dosing system greatly simplified the lives of children receiving the medication and their caregivers. In this case, improved performance generated by a change in formulation provided Janssen Pharmaceuticals with a blockbuster product based on an old medication. This case highlights the importance of the application of newer technologies to established medications. Significant advances in treatment and new products may be available to those astute enough to recognize the opportunities.

SELECTIVE INHIBITION OF CYCLOOXYGENASE-2: THE IMPACT OF AN INADEQUATE WRITTEN DESCRIPTION

In the late 1980s and early 1990s, a series of discoveries to led to the identification of two closely related enzymes, cyclooxygenase-1 (COX-1) and cyclooxygenase-2 (COX-2). The two enzymes have markedly different purposes. COX-1, also referred to as prostaglandin–endoperoxide synthase 1 (PGHS-1), promotes the production of the mucus lining of the stomach and plays an important role in protecting the stomach from its acidic contents. On the other hand, COX-2, also referred to as prostaglandin–endoperoxide synthase 2 (PGHS-2), plays an important role in inflammatory responses and pain sensation. Non-steroidal anti-inflammatory drugs (NSAIDS) such as aspirin, ibuprofen, and acetaminophen prevent pain and inflammation by inhibiting COX-2, but they also inhibit COX-1.[53] This lack of selectivity can lead to gastrointestinal side effects, possibly producing serious consequences. In April of 2000, the University of Rochester was issued a patent, US 6048850,[54] that claimed the following:

1. A method for selectively inhibiting PGHS-2 activity in a human host, comprising administering a non-steroidal compound that selectively inhibits activity of the PGHS-2 gene product to a human host in need of such treatment.
2. The method of claim 1 in which the compound inhibits the enzymatic activity of the PGHS-2 gene product, and has minimal effect on enzymatic activity of PGHS-1.
3. The method of claim 1 in which the activity of PGHS-1 is not inhibited.
4. The method of claim 3 in which the compound is a non-steroid anti-inflammatory drug.

Essentially, this patent claimed *any* compound that could selectively inhibit COX-2 versus COX-1 and their use as a drug. As noted in Chapter 12, an awarded patent gives the owner the right to prevent others from using the claimed invention, so the University of Rochester sued companies that were marketing selective COX-2 inhibitors. The University of Rochester asserted that marketing drugs such as Celebrex, a selective COX-2 inhibitors marketed by Pfizer, represented an infringement of their patent and that the University of Rochester deserved compensation for the use of their patent. Of course, the companies marketing selective COX-2 inhibitors disagreed, as they had also received patents for specific, novel compounds and their use for the treatment of pain via the COX-2 pathway. The case eventually made its way to the federal court system. The courts decided that the University of Rochester patent was not valid, as it did not contain an adequate written description of the invention. The patent in question provided assays that could be used to identify compounds that had the desired activity and selectivity. It did not, however, describe any small molecules, peptides, proteins, etc. that were capable of selectively inhibiting COX-2 over COX-1. In the absence of examples, working or hypothetical, the patent did not contain the information necessary for someone of ordinary skill in the relevant art to make and use the invention. In other words, the patent did not contain an enabling disclosure, making it invalid. These decisions were upheld on appeal, and an application to the Supreme Court for review was denied without comment.

This case highlights just one of many reasons why patent protection could be deemed insufficient and the potential disastrous consequences. The University of Rochester spent approximately $10 million attempting to enforce a patent application that they believed would give them a portion of a multi-billion dollar therapeutic market. Unfortunately, their patent coverage was not nearly as complete as they believed it to be. The importance of a well written patent cannot be understated. In the absence of patent protection, it would be nearly impossible to recoup the massive costs associated with bringing a new therapy to market. Patents are the life blood of the pharmaceuticals industry, and individuals interested in a career in the pharmaceuticals industry (or as academic drug discovery scientists) should have an understanding of at least the basic principles of patent law.

The pharmaceuticals industry is an ever changing arena that challenges its participants to maintain knowledge of a wide range of science, business practices, regulatory factors, and legal issues. As discussed earlier in this text, it would not be possible for any single individual to possess all of the skills and expertise (not to mention the equipment and finances) necessary to bring new therapeutic entities to market. Drug discovery and development is a complex, team driven endeavor in which each participant needs to have expertise in a few areas, a basic understanding of many areas, and the capacity to work and learn in a team setting. As illustrated in this chapter, the lesson of the past can provide a great deal of insight into the potential paths to success and pitfalls that should be avoided. The best drug discovery and development scientist are typically those who learn not just from their contemporaries, but also study the triumphs and failures of the industry. While those who do not remember the past are doomed to repeated, those who fail to learn from the past are less likely to find success in their future, especially in the pharmaceuticals industry.

QUESTIONS

1. Why is Tamiflu® (Oseltamivir) sold as a prodrug?
2. Nitrofurantoin is an antibiotic marketed for the treatment of urinary tract infections, but it is known to be highly mutagenic, highly teratogenic, and highly carcinogenic. Why is this possible?
3. Seldane® (Terfenadine) was the first nonsedating antihistamine. Launched in 1985, it was removed from the market in 1997 even though it had been successfully used by over 100 million patients. Why was it removed from the market, and how did this event change the drug discovery process?
4. What does the development of Allegra® (Fexofenadine) teach about the importance of metabolism?
5. What does the development of Claritin® (Loratadine) and Clarinex® (Desloratadine) teach about the importance of metabolism?
6. Desmethylprodine (MPPP) is a potent μ-opioid receptor agonist that is not safe for clinical use? Why and how is this relevant to drug discovery and development as a whole?
7. Why were extended release and sustained release mechanisms important in the development of Bupropion?
8. What is the possible impact of a poorly written patent application?

References

1. Martin, P.; Martin-Granel, E. 2500-Year Evolution of the Term Epidemic. *Emerg. Infect. Dis.* **2006,** *12* (6), 976–980.
2. Mills, C. E.; Robins, J. M.; Lipsitch, M. Transmissibility of 1918 Pandemic Influenza. *Nature* **2004,** *432* (7019), 904–906.
3. Dawood, F. S.; Iuliano, A. D.; Reed, C.; Meltzer, M. I.; Shay, D. K.; Cheng, P. Y.; Bandaranayake, D.; Breiman, R. F.; Brooks, W. A.; Buchy, P.; et al. Estimated Global Mortality Associated with the First 12 Months of 2009 Pandemic Influenza a H1N1 Virus Circulation: A Modelling Study. *Lancet Infect. Dis.* **2012,** *12* (9), 687–695.
4. Maugh, T. H. Amantadine: An Alternative for Prevention of Influenza. *Science* **1976,** *192* (4235), 130–131.
5. Wintermeyer, S. M.; Nahata, M. C. Rimantadine: A Clinical Perspective. *Ann. Pharmacother.* **1995,** *29* (3), 299–310.
6. Kaiser, L.; Wat, C.; Mills, T.; Mahoney, P.; Ward, P.; Hayden, F. Impact of Oseltamivir Treatment on Influenza-Related Lower Respiratory Tract Complications and Hospitalizations. *Arch. Intern. Med.* **2003,** *163* (14), 1667–1672.
7. a. Meinal, P.; Bodo, G.; Palese, P.; Schulman, J.; Tuppy, H. Inhibition of Neuraminidase Activity by Derivatives of 2-deoxy-2,3-dehydro-N-acetylneuraminic Acid. *Virology* **1974,** *58* (2), 457–463.
8. von Itzstein, M. The War against Influenza: Discovery and Development of Sialidase Inhibitors. *Nat. Rev. Drug Discov.* **2007,** *6* (12), 967–974.
9. Varghese, J. N.; McKimm-Breschkin, J. L.; Caldwell, J. B.; Kortt, A. A.; Colman, P. M. The Structure of the Complex between Influenza Virus Neuraminidase and Sialic Acid, the Viral Receptor. *Proteins: Struct. Funct. Genet.* **1992,** *14,* 327–332.

10. Bossart-Whitaker, P.; Carson, M.; Babu, Y. S.; Smith, C. D.; Laver, W. G.; Air, G. M. Three-dimensional Structure of Influenza A N9 Neuraminidase and its Complex with the Inhibitor 2-deoxy 2,3-dehydro-*N*-acetyl Neuraminic Acid. *J. Mol. Biol.* **1993,** *232* (4), 1069–1083.

11. Heneghan, C. J.; Onakpoya, I.; Thompson, M.; Spencer, E. A.; Jones, M.; Jefferson, T. Zanamivir for Influenza in Adults and Children: Systematic Review of Clinical Study Reports and Summary of Regulatory Comments. *Br. Med. J.* **2014,** *348,* g2547.

12. a. Kim, C. U.; Lew, W.; Williams, M. A.; Wu, H.; Zhang, L.; Chen, X.; Escarpe, P. A.; Mendel, D. B.; Laver, W. G.; Stevensd, R. C. Structure-Activity Relationship Studies of Novel Carbocyclic Influenza Neuraminidase Inhibitors. *J. Med. Chem.* **1998,** *41,* 2451–2460.

    b. Kim, C. U.; Lew, W.; Williams, M. A.; Liu, H.; Zhang, L.; Swaminathan, S.; Bischofberger, N.; Chen, M. S.; Mendel, D. B.; Tai, C. Y.; et al. Influenza Neuramini-dase Inhibitors Possessing a Novel Hydrophobic Interaction in the Enzyme Active Site: Design, Synthesis, and Structural Analysis of Carbocyclic Sialic Acid Analogues with Potent Anti-Influenza Activity. *J. Am. Chem. Soc.* **1997,** *119,* 681–690.

13. Li, W.; Escarpe, P. A.; Eisenberg, E. J.; Cundy, K. C.; Sweet, C.; Jakeman, K. J.; Merson, J.; Lew, W.; Williams, M.; Zhang, L.; et al. Identification of GS 4104 as an Orally Bioavailable Prodrug of the Influenza Virus Neuraminidase Inhibitor GS 4071. *Antimicrob. Agents Chemother.* **1998,** *42* (3), 647–653.

14. a. Rikimaru, T.; Taketomi, A.; Yamashita, Y.; Shirabe, K.; Hamatsu, T.; Shimada, M.; Maehara, Y. Clinical Significance of Histone Deacetylase 1 Expression in Patients with Hepatocellular Carcinoma. *Oncology* **2007,** *72,* 69–74.

    b. Lu, Y. S.; Kashida, Y.; Kulp, S. K.; Wang, Y. C.; Wang, D.; Hung, J. H.; Tang, M.; Lin, Z. Z.; Chen, T. J.; Cheng, A. L.; et al. Efficacy of a Novel Histone Deacetylase Inhibitor in Murine Models of Hepatocellular Carcinoma. *Hepatology* **2007,** *46,* 1119–1130.

15. Wong, J. C.; Tang, G.; Wu, X.; Liang, C.; Zhang, Z.; Guo, L.; Peng, Z.; Zhang, W.; Lin, X.; Wang, Z.; et al. Pharmacokinetic Optimization of Class-Selective Histone Deacetylase Inhibitors and Identification of Associated Candidate Predictive Biomarkers of Hepato-cellular Carcinoma Tumor Response. *J. Med. Chem.* **2012,** *55* (20), 8903–8925.

16. John, E., Bennett, J. E., Dolin, R., Blaser, M. J., Dolin, G. L., Eds. *Mandell, Douglas, and Bennett's Principles and Practice of Infectious Diseases,* 7th ed.; Churchill Livingstone, Elsevier: Philadelphia, PA, 2010; p 169.

17. Brook, I. Approval of Zidovudine (AZT) for Acquired Immunodeficiency Syndrome. A Challenge to the Medical and Pharmaceutical Communities. *J. Am. Med. Assoc.* **1987,** *258* (11), 1517.

18. Jeffries, D. J. Zidovudine Resistant HIV. *Br. Med. J.* **1989,** *298,* 1132–1133.

19. Center for Disease Control and Prevention Morbidity and Mortality Weekly Report. *HIV Surveillance—United States, 1981–2008,* June 3, 2011.

20. a. Ho, D. D. Time to Hit HIV, Early and Hard. *N. Engl. J. Med.* **1995,** *333,* 450–451.

    b. Hammer, S. M.; Katzenstein, D. A.; Hughes, M. D.; Gundacker, H.; Schooley, R. T.; Haubrich, R. H.; Henry, W. K.; Lederman, M. M.; Phair, J. P.; Niu, M.; et al. A Trial Comparing Nucleoside Monotherapy with Combination Therapy in HIV-Infected Adults with CD4-Cell Counts from 200 to 500 Per Cubic Millimeter. AIDS Clinical Trials Group Study 175 Study Team. *N. Engl. J. Med.* **1996,** *335,* 1081–1090.

    c. Gulick, R. M.; Mellors, J. W.; Havlir, D.; Eron, J. J.; Gonzalez, C.; McMahon, D.; Richman, D. D.; Valentine, F. T.; Jonas, L.; Meibohm, A.; et al. Treatment with Indi-navir, Zidovudine, and Lamivudine in Adults with Human Immunodeficiency Virus Infection and Prior Antiretroviral Therapy. *N. Engl. J. Med.* **1997,** *337,* 734–739.

21. Zuniga, J. M., Whiteside, A., Ghaziani, A., Bartlett, J. G., Eds. *A Decade of HAART: The Development and Global Impact of Highly Active Antiretroviral Therapy;* Oxford University Press Inc.: New York, 2008.

22. Dorsey, B. D.; Levin, J. R. B.; McDaniel, S. L.; Vacca, J. P.; Guare, J. P.; Darke, P. L.; Zugay, J. A.; Emini, E. A.; Schleif, W. A.; Quintero, J. C.; et al. L-735,524: The Design of a Potent and Orally Bioavailable HIV Protease Inhibitor. *J. Med. Chem.* **1994**, *37* (21), 3443–3451.

23. Reider, P. J. Advances in AIDS Chemotherapy: The Asymmetric Synthesis of Crixivan®. *Chimia* **1997**, *51*, 306–308.

24. Tanouye, E. Medicine: Success of AIDS Drug Has Merck Fighting to Keep Up the Pace. *The Wall Street Journal* **November 6, 1996**, A1.

25. McOsker, C. C.; Fitzpatrick, P. M. Nitrofurantoin: Mechanism of Action and Implications for Resistance Development in Common Uropathogens. *J. Antimicrob. Chemother.* **1994**, *33* (Suppl. A), 23–30.

26. a. Letelier, M. E.; Izquierdo, P.; Godoy, L.; Lepe, A. M.; Faúndez, M. Liver Microsomal Biotransformation of Nitro-aryl Drugs: Mechanism for Potential Oxidative Stress Induction. *J. Appl. Toxicol.* **2004**, *24*, 519–525.

    b. Neumann, H. G. Monocyclic Aromatic Amino and Nitro Compounds: Toxicity, Genotoxicity and Carcinogenicity, Classification in a Carcinogen Category; *The MAK-Collection Part I: MAK Value Documentations*, Vol. 21; Wiley-VCH: Weinheim, Germany, 2005.

27. Cunha, B. A. Nitrofurantoin—Current Concepts. *Urology* **1988**, *32* (2), 67–71.

28. Sorkin, E. M.; Heel, R. C. Terfenadine. A Review of its Pharmacodynamic Properties and Therapeutic Efficacy. *Drugs* **1985**, *29* (1), 34–56.

29. Masheter, H. C. Terfenadine: The First Nonsedating Antihistamine. *Clin. Rev. Allergy* **1993**, *11*, 5–34.

30. Thompson, D.; Oster, G. Use of Terfenadine and Contraindicated Drugs. *J. Am. Med. Assoc.* **1996**, *275* (17), 1339–1341.

31. a. Pulmonary-Allergy Drugs Advisory Committee. *Proceedings of the Pulmonary-Allergy Drugs Advisory Committee;* Food and Drug Administration, Public Health Service, US Dept of Health and Human Services: Rockville, MD, 1990.

    b. Honig, P. K.; Wortham, D. C.; Zamini, K.; Connor, D. P.; Mullin, J. C.; Cantilena, L. R. Terfenadine-Ketoconazole Interaction. Pharmacokinetic and Electrocardiographic Consequences. *J. Am. Med. Assoc.* **1993**, *269*, 1513–1518.

32. Marion Merrell Dow Inc.. *Important Drug Warning;* Marion Merrell Dow Inc.: Kansas City, MO, July 1992.

33. Sanguinetti, M. C.; Tristani-Firouzi, M. hERG Potassium Channels and Cardiac Arrhythmia. *Nature* **2006**, *440* (7083), 463–469.

34. Guo, L.; Guthrie, H. Automated Electrophysiology in the Preclinical Evaluation of Drugs for Potential QT Prolongation. *J. Pharmacol. Toxicol. Methods* **2005**, *52*, 123–135.

35. a. Jurima-Romet, M.; Crawford, K.; Cyr, T.; Inaba, T. Terfenadine Metabolism in Human Liver. *In vitro* Inhibition by Macrolide Antibiotics and Azole Antifungals. *Drug Metab. Dispos.* **1994**, *22*, 849–857.

    b. Woosley, R. L.; Chen, Y.; Frieman, J. P.; Gillis, R. A. Mechanisms of the Cardiotoxic Actions of Terfenadine. *J. Am. Med. Assoc.* **1993**, *269*, 1532–1536.

    c. Monahan, B. P.; Ferguson, C. L.; Killeavy, E. S.; Lloyd, B. K.; Troy, J.; Cantilena, L. R. Torsades de Pointes Occurring in Association with Terfenadine Use. *J. Am. Med. Assoc.* **1990**, *264*, 2788–2790.

36. a. *Teva and Barr Announce Launch of Generic Allegra® Tablets by Teva Under Agreement with Barr;* . Teva Pharmaceuticals press release, September 6, 2005.

    b. *Barr Granted Summary Judgment on Three Patents in Allegra(R) Patent Challenge;* . Barr Pharmaceuticals press release, July 1, 2004.

37. FDA Prescription to Over-the-Counter (OTC) Switch List, http://www.fda.gov/AboutFDA/CentersOffices/OfficeofMedicalProductsandTobacco/CDER/ucm106378.htm.

38. Kay, G. G.; Harris, A. G. Loratadine: A Non-sedating Antihistamine. Review of its Effects on Cognition, Psychomotor Performance, Mood and Sedation. *Clin. Exp. Allergy* **1999**, *29* (S3), 147–150.

39. Hall, S. S. *The Claritin Effect; Prescription for Profit*. The New York Times, March 11, 2001.

40. Zhang, Y. F.; Chen, X. Y.; Zhong, D. F.; Dong, Y. M. Pharmacokinetics of Loratadine and its Active Metabolite Descarboethoxyloratadine in Healthy Chinese Subjects. *Acta Pharmacol. Sin.* **2003**, *24* (7), 715–718.

41. Smith, A. *Big Pharma Teaches Old Drugs New Tricks: Drugmakers Hunt for New Patents on Old Blockbusters to Try and Postpone the Inevitable: Generic Competition.* CNNMoney.com, March 21, 2007.

42. Hori, G.; Gold, S. Demerol in Surgery and Obstetrics. *Can. Med. Assoc. J.* **1944**, *51* (6), 509–517.

43. Ziering, A.; Lee, J. Piperidine Derivatives; 1,3-dialkyl-4-aryl-4-acyloxypiperidines. *J. Org. Chem.* **1947**, *12* (6), 911–914.

44. a. In *The Case of the Frozen Addicts: How the Solution of a Medical Mystery Revolutionized the Understanding of Parkinson's Disease*; Langston, J. W., Palfreman, J., Eds. IOS Press: Amsterdam, The Netherlands, 2014.

    b. Langston, J. W.; Ballard, P.; Tetrud, J. W.; Irwin, I. Chronic Parkinsonism in Humans Due to a Product of Meperidine-Analog Synthesis. *Science* **1983**, *219* (4587), 979–980.

45. a. In *Goodman and Gilman's the Pharmacological Basis of Therapeutics*, 9th ed.; Hardman, J. G., Limbird, L. E., Molinoff, P. B., Ruddon, R. W., Goodman, A. G., Eds. McGraw-Hill: New York, NY, 1996; p 542.

    b. Chan, K.; Kendall, M. J.; Mitchard, M. Quantitative Gas-Liquid Chromatographic Method for the Determination of Pethidine and its Metabolites, Norpethidine and Pethidine N-oxide in Human Biological Fluids. *J. Chromatogr.* **1974**, *89* (2), 169–176.

46. a. Pifl, C.; Giros, B.; Caron, M. G. Dopamine Transporter Expression Confers Cytotoxicity to Low Doses of the Parkinsonism-inducing Neurotoxin 1-methyl-4-phenylpyridinium. *J. Neurosci.* **1993**, *13* (10), 4246–4253.

    b. Burns, R. S.; Markey, S. P.; Phillips, J. M.; Chiueh, C. C. The Neurotoxicity of 1-methyl-4-phenyl-1,2,3,6-tetrahydropyridine in the Monkey and Man. *Can. J. Neurol. Sci.* **1984**, *11* (1), S166–S168.

    c. Markey, S. P.; Johannessen, J. N.; Chiueh, C. C.; Burns, R. S.; Herkenham, M. A. Intraneuronal Generation of a Pyridinium Metabolite May Cause Drug-induced Parkinsonism. *Nature* **1984**, *311* (5985), 464–467.

47. a. Langston, J. W.; Forno, L. S.; Rebert, C. S.; Irwin, I. Selective Nigral Toxicity after Systemic Administration of 1-methyl-4-phenyl-1,2,5,6-tetrahydropyridine (MPTP) in the Squirrel Monkey. *Brain Res.* **1984**, *292* (2), 390–394.

    b. Irwin, I.; Langston, J. W. Selective Accumulation of MPP+ in the Substantia Nigra: A Key to Neurotoxicity? *Life Sci.* **1985**, *36* (3), 207–212.

48. Fava, M.; Rush, A. J.; Thase, M. E.; Clayton, A.; Stahl, S. M.; Pradko, J. F.; Johnston, J. A. 15 Years of Clinical Experience with Bupropion HCl: From Bupropion to Bupropion SR to Bupropion XL. *Prim. Care Companion J. Clin. Psychiatry* **2005**, *7* (3), 106–113.

49. Dunner, D. L.; Zisook, S.; Billow, A. A.; Batey, S. R.; Johnston, J. A.; Ascher, J. A. A Prospective Safety Surveillance Study for Bupropion Sustained-Release in the Treatment of Depression. *J. Clin. Psychiatry* **1998**, *59* (7), 366–373.

50. Jefferson, W. J.; Pradko, J. F.; Muir, K. T. Bupropion for Major Depressive Disorder: Pharmacokinetic and Formulation Considerations. *Clin. Ther.* **2005**, *27* (11), 1685–1695.

51. Lange, K. W.; Reichl, S.; Lange, K. M.; Tucha, L.; Tucha, O. The History of Attention Deficit Hyperactivity Disorder. *ADHD Atten. Def. Hyp. Disord.* **2010**, *2* (4), 241–255.

52. Concerta® (Methylphenidate HCl) Extended Release Tablet Prescription Package Insert.

53. Dubois, R. N.; Abramson, S. B.; Crofford, L.; Gupta, R. A.; Simon, L.; Van De Putte, L. B. A.; Lipsky, P. E. Cyclooxygenase in Biology and Disease. *FASEB J.* **1998**, *12* (12), 1063–1073.

54. Young D. A.; O'Banion M. K.; Winn V. D. Method of Inhibiting Prostaglandin Synthesis in a Human Host. US 6048850, **2000**.

Answers to Questions in Textbook by Chapter

1. The three major phases of drug discovery are (1) target discovery, (2) lead discovery, and (3) lead optimization. (See Figure 7.)
2. The four major phases of drug development are (1) preclinical, (2) proof of concept, (3) full development, and (4) registration and launch. (See Figure 7.)
3. The lead optimization cycle begins with the identification of a lead structure ("hit") in a relevant biological assay. New analogs with structural modifications are prepared and screened in the biological assay. If the assay results improve, then the changes are kept and the cycle is repeated. If the changes are detrimental, then they are discarded and the cycle is repeated. This process continues until a candidate compound with the desired properties is identified. (See Figure 16.)
4. A screening cascade, also referred to as a screening tree, is a series of experiments designed to decrease the number of candidate compounds as a discovery program proceeds toward a clinical candidate. At each stage of the cascade, criteria, also referred to as "gates," are established that determine whether or not a test compound proceeds to the next level of screening. In principle, the number of compounds that pass through each gate will decrease toward the bottom of the screening tree, limiting the number of compounds that need to be studied in more complex, time-consuming, and costly experiments designed to identify in vivo active, safe candidate compounds. (See Figure 18.)
5. Compound selectivity is an important aspect of drug discovery because there is often substantial overlap between the macromolecular target of interest and many other biomolecules. In cases where sufficient selectivity is not achieved between the targeted macromolecule and structurally related biomolecules, a biochemical response may be elicited from the interaction with both the desired target and unintended targets. These unintended interactions have the potential to create unwanted side effects that could hamper future development of a candidate compound.

6. The term in vitro ADME is a reference to a series of screening experiments designed to predict the absorption, distribution, metabolism, and excretion (ADME) of a candidate compound. They were generally performed early in a program in order to ensure that candidates reaching the drug development pipeline are "druglike" in nature.

7. A compound's in vitro ADME profile includes metabolic stability, plasma stability, aqueous solubility, Pgp efflux susceptibility, solution stability, CYP450 inhibition, bioassay solubility, blood–brain barrier penetration, and permeability. (See Figure 19.)

CHAPTER 2

1. Paul Ehrlich's experiments that determined the affinity of trypan red, trypan blue, methylene blue for biological tissues led him to postulate the existence of "chemoreceptors." He further postulated that the interaction of chemicals around the cell with "chemoreceptors" could produce a biological response by the cells. Finally, he theorized that the "chemoreceptors" of cancer cells and infectious organisms are different from those of the host and that these differences could be exploited to produce a therapeutic benefit for patients.

2. The Wistar rat that was introduced in 1906 marked the first effort to produce a "pure strain" animal model for scientific testing and medical research. Over 50% of all laboratory rat strains are descendants of the original Wistar rat colony. It is also one of the most commonly employed rat strain in modern medical research.

3. The SCID mouse is a strain of mice first identified at the Fox Chase Cancer Center. The acronym stands for severe combined immune deficient, and SCID mice are severely deficient in B- and T-lymphocytes. This renders them highly vulnerable to infectious diseases and unable to reject implanted foreign tissues. The later point makes SCID mice very useful for the study of cancer. These mice are distinct from nude mice, a hairless mouse that is also immunocompromised as they are congenitally athymic (no thymus). In the absence of a thymus, nude mice are unable to generate mature T-lymphocytes, limiting their ability to mount an immune response.

4. Transgenic animal models are animals whose genome has been altered using genetic engineering techniques to add a new gene. Transgenic animal models are developed through a combination of selective breeding and genetic manipulation. A gene construct suitable for insertion into an organism's DNA is prepared and then inserted into a fertilized egg via microinjection. The altered embryos

are then implanted into a suitably pseudopregnant female and carried to term. After birth, genetic profiling is employed to identify offspring that are carriers of the transgene. Identification of transgene positive progeny is then followed by selective breeding to further the germ line. (See Figure 9.)

5. A knockout animal model is a genetically engineered animal model in which the expression of a gene has been suppressed by either replacing the gene with a new, nonfunctional sequence of DNA or the insertion of additional genetic material into a gene that disrupts its expression. (See Figure 11.)

6. High throughput chemistry refers to the application of parallel synthesis, robotics, polymer-supported chemistry, and multicomponent reactions to the preparation of libraries of chemically related compounds. Chemical libraries containing thousands of unique compounds have been prepared using these methods.

7. Recombinant DNA refers to DNA strands that are artificial in nature. It is produced through the application of a series of enzymes capable of building, degrading, and modifying DNA chains (DNA ligases, exonucleases, terminal transferases, reverse transcriptases, and restriction enzymes). The selected application of these enzymes provides a means of carving out a specific DNA sequence from one set of genes and inserting them into a different set of DNA strands. (See Figure 20.)

8. Transfection technology is used to insert recombinant DNA into cells to generate cell lines that overexpress the product of the inserted DNA. (See Figure 21.)

9. In the polymerase chain reaction process, DNA is denatured at high temperatures, cooled, and then replicated through the action of DNA polymerase. After only 30 rounds of denaturing, cooling, and replication, a single DNA strand is amplified in number to over 1 billion copies. Taq polymerase is important to this process, because it is stable at the elevated temperatures necessary to denature a DNA double helix. Prior to the discovery of this enzyme, fresh DNA polymerase had to be added in each round of the process, making the process costly and time-consuming. (See Figure 22.)

10. Hybridoma cells are the result of the fusion of an antibody-producing B-cell with a myeloma cell. The resulting hybridoma cell can be isolated, and cloned to provide a stable cell line capable of producing a single antibody (a monoclonal antibody). (See Figure 23.)

11. Receptor construct fusion proteins are biologic therapeutic agents that are composed of a protein receptor segment and an immunoglobulin structure. The protein receptor portion provides selectivity for the target of interest, while the immunoglobulin structure provides metabolic stability.

12. In 1937, the S. E. Massengill Company launched a product containing sulfanilamide, an antibiotic, raspberry flavoring, and diethylene glycol, a substance now known to be toxic. No safety studies were performed, as none were required by law. One month after sales began, over 100 people had died after using the mixture and the product was recalled and destroyed. The medication was confiscated and destroyed not because of its dangerous properties, but rather because it was mislabeled as an "elixir" even though it contained no alcohol. The company denied any responsibility, as safety studies were not legally required at that time. The Food, Drug, and Cosmetic Act of 1938 requires pharmaceutical companies to prove the safety of their new products through animal safety studies prior to receiving marketing approval. In addition, this law requires manufacturers to submit an application for marketing approval to the FDA before new products could be brought to market. The application is also known as a new drug application (NDA).

13. Prior to the thalidomide disaster, the placenta was believed to be a perfect barrier that protected an unborn child from toxic chemicals. The birth defects made it clear that this was not true. Separately, it was recognized that single enantiomer of a compound could have different biological properties.

14. The Kefauver-Harris Amendment of 1962 gave the FDA virtually complete authority over drug approval and marketing. It required new drug candidate be proven to be safe and effective prior to market approval, and mandated a review of the safety and efficacy of drugs brought to market between 1938 and 1962. Nearly 40% were found to be ineffective and stripped of marketing approval. This law also required that clinical trial design be approved by the FDA, mandated informed consent for clinical trial participants, and stipulated that known side effects be disclosed to the public. In addition, good manufacturing practices (GMP) requirements were put in place and the FDA was given the authority to access company control and production records to ensure the quality of the final product. Advertising of prescription drugs was placed under strict regulation by the FDA and generic drugs could no longer be marketed as "breakthrough" therapies.

CHAPTER 3

1. See Figure 10
   a. Oxidoreductases—Catalyzes redox chemistry, employs cofactors.
   b. Transferases—Catalyzes functional group transfer.
   c. Hydrolases—Catalyzes the hydrolysis of a chemical bond.

    d. Lyases—Catalyzes bond cleavage other than hydrolysis or oxidation, often forming double bonds or rings.

    e. Isomerases—Catalyzes the structural rearrangement of isomers.

    f. Ligases—Catalyzes joining of large molecules with new bond.

2. Noncovalent interactions

    a. Hydrophobic interaction—A bonding interaction that occurs between hydrophobic side chains as they fold together to create hydrophobic pockets within proteins that exclude water. These interactions are characterized as Van der Waals interactions.

    b. Electrostatic/salt bridge—A bonding interaction that occurs between a positively charged amino acid side chain and a negatively charged side chain of an amino acid.

    c. Hydrogen bond—A bonding interaction that is the result of dipole–dipole interactions between polarized hydrogen atoms of a hydrogen bond donor and a lone pair of electrons from a hydrogen bond acceptor.

    d. π-stacking—A bonding interaction between two aromatic systems in which the π-orbitals of the aromatic systems interact. The interaction can be either face to face or face to edge (T-shaped) and the strength of the interaction is highly distance dependent.

    e. π cation interaction—A bonding interaction created by the interaction of an aromatic system with a positively charged amino acid side chain (e.g., protonated lysine side chain). Distance, angle of interaction, and the electron density of the aromatic system involved will impact the strength of this interaction.

3. The three methods of enzyme inhibition include (1) competitive inhibition in which an inhibitor reversibly blocks the active site, (2) irreversible inhibition in which an inhibitor covalently binds to the active site, and (3) allosteric inhibition in which an inhibitor binds to an allosteric binding site, altering the active site, preventing substrate binding. (See Figure 17.)

4. The three key structural features of a GPCR are (1) a series of seven transmembrane segments designated TM-1 to TM-7, (2) a carboxy terminus is on the cytoplasmic side of the cell membrane, and (3) an amino terminus is on the extracellular side of the membrane.

5. The two major signaling pathways for GPCRs are the cyclic AMP pathway and the phosphatidylinositol signaling pathway (IP3).

6. The ion channels listed are defined as follows:

    a. Ligand-gated ion channels are activated when an agonist interacts with a specific binding site on the channel.

    b. Voltage-gated ion channel are ion channels that open and close as a result of changes in membrane potential produced as electrical currents move through biological systems. They have no natural ligand.

    c. Temperature-gated ion channel are ion channels that open and close based on thermal thresholds.

    d. Mechanosensitive-gated ion channel are ion channels that are activated by mechanical deformations of membranes such as increased tension or changes in curvature.

7. A channelopathy is a disease or condition caused by improper function of an ion channel.

8. A passive transport system is a transport system that facilitates diffusion across a biological membrane. Solutes in high concentration on one side of the membrane bind to the transporter protein causing a conformational change that transports the solute through the membrane to an area of lower concentration. (See Figure 42.)

9. An active transport system is a transport system that uses energy to transport solutes across a biological membrane irrespective of solute concentrations on each side. They often move solutes against a concentration gradient (from areas of low concentration to areas of high concentration).

CHAPTER 4

1. An IC_{50} is defined as the concentration that provides 50% of the maximal response of the compound under the assay condition employed. In the case of an enzyme, it is the concentration at which 50% of the enzyme activity is blocked. (See Figure 1.)

2. An EC_{50} is defined as the concentration that produces 50% of the maximum effect produced by the test compound under the assay condition employed. The natural ligand's efficacy is set at 100%, but it is important to understand that the maximum effect of the natural ligand and a given test compound may not be the same. (See Figure 2.)

3. Agonists, antagonists, and inverse agonists are defined as follows: (See Figure 3.)

    a. Agonist—A compound that elicits the same functional biological response as the natural ligand.

    b. Antagonist—A compound that binds to a GPCR blocking the natural ligand, but eliciting no biological response.

    c. Inverse agonist—A compound that binds to a GPCR and blocks basal activity. In many cases, they are also antagonists.

4. High receptor reserve in a cell line can influence in vitro screening results. If the receptor reserve represents a 10-fold increase relative to normal expression levels, an agonist may appear to be 10-fold more potent in the test system than it would be in the normal cell line (the target cells).

5. Streptavidin and biotin form a very strong, stable, noncovalent interaction with a dissociation constant in the femtomolar range. The interaction is virtually inert under the conditions of a biological screen. Biotin can be attached to a surface of a microtiter plate and streptavidin will bind to the biotin, attaching it to the surface of the microtiter plate. If a second protein that is tagged with biotin is added, it will bind to additional binding sites of streptavidin, linking the biotin-tagged protein to the surface of the microtiter plate. (See Figure 6.)

6. Scintillation proximity assays take advantage of the ability of some compounds to emit light (scintillate) when they are in the presence of ionizing radiation. Compounds containing β-particle-emitting atoms (e.g., ^3H, ^{14}C, ^{33}P, and ^{35}S) or Auger electrons-emitting atoms (e.g., ^{125}I) are employed in a screening system in conjunction with a scintillant such as polyvinyltoluene, polystyrene, yttrium silicate (YSi), and yttrium oxide (YOx). In a standard radioligand assay, membranes or cells containing a target of interest are incubated with a mixture of a candidate compound and a radiolabeled ligand for a set period of time. After incubation, filtration and washing is employed to separate the membranes or cells from any unbound ligands, and then a scintillant is added to induce light emission from the resulting dry material. Compounds that compete with the radioligand for a binding site will displace it from the molecular target, and free radioligand will be washed away in the filtration step. This will decrease scintillation in a quantifiable manner, and the change in signal intensity provides direct insight into the strength of binding interaction between the candidate compound and the macromolecular target. Scintillation proximity assays have been developed to limit the amount of radioactive waste generated in this type of assay. (See Figure 8.)

7. An ELISA assay is an enzyme-linked immunosorbent assay in which an antibody is linked to an enzyme whose activity generates a quantifiable signal. There are several variations, but the simplest form is a direct ELISA in which a surface coated with an antigen is treated with a compatible antibody that has been covalently linked to an enzyme. When a suitable substrate is added, a color change is produced. The intensity and rate of the color change are a function of the activity of the enzyme and the amount of substrate present. The presence of an enzyme inhibitor will change the rate and intensity of the color change in a quantifiable manner.

8. Fluorescence resonance energy transfer assays (FRET assays) and time-resolved fluorescence resonance energy transfer assays (TR-FRET assays) both take advantage of the fact that when a fluorescent donor molecule absorbs energy, it can transfer the energy to a nearby

fluorescent acceptor molecule that will then fluoresce at a lower wavelength. The intensity of the lower wavelength emission can be quantified to determine the impact of candidate compound on a biological system. In a standard FRET assay, the emission of lower wavelength energy is immediate, and background fluorescence of other material (e.g., plastic, cellular components) can interfere with assay readouts. In TR-FET assay systems, specially designed organic scaffolds containing lanthanide elements that have an extended fluorescent decay rate are employed. The decay rates of these specialized fluorescent acceptor molecules are long enough that background emission is no longer an issue in signal detection. (See Figure 20.)

9. In a reporter gene assay system, expression of target gene is tied to expression of a nonnative gene that will produce a measurable signal upon expression. (See Figure 28.)

10. A label-free assay system is an assay system that measures changes in physical or chemical properties of cellular systems, particularly cell monolayers without the use of radiolabeled fluorescent tags, or other artificial detection systems. Change in cell volume, pH, refractive index, membrane potential, electrical impedance, refractive index, and optical properties have been employed as means of determining the impact of test compounds on cellular function.

CHAPTER 5

1. The American Chemical Society's division of medicinal chemistry defines medicinal chemistry as "the application of chemical research techniques to the synthesis of pharmaceuticals." A broader and more accurate description, however, is that medicinal chemistry applies knowledge derived from synthetic chemistry, biochemistry, pharmacology, physiology, and molecular biology in an effort to understand the biological impact of candidate compounds. It is an interdisciplinary science whose primary goal is the identification of therapeutically useful compounds. (See Figure 1.)

2. The term structure–activity relationship refers to the relationship between the structure of a series of candidate compounds and their biological activity at a target biomolecule. Understanding the impact of structural changes on biological activity provides medicinal chemists with opportunity to improve potency at the target.

3. Chirality can have a significant impact on biological activity, as pairs of enantiomers cannot occupy the same physical space. As a result, simple changes such as the inversion of a chiral center in a biologically active compound can result in a significant decrease in biological activity. (See Figures 5–7.)

4. Quantitative structure–activity relationships refer to the mathematical correlation of biological activity with structural changes in a series of compounds and using this information to predict the biological activity of new candidate compounds.

5. A pharmacophore is the subset of the atoms and functionalities within the molecular framework of a candidate compound that are required for biological activity.

6. The auxophore is the subset of atoms and functionality within the molecular framework of a candidate compound that are not required for biological activity. Changes to these features of the compound will have minimal impact on biological activity at the biomolecular target of interest.

7. Identifying an initial set of hits in a drug discovery program, irrespective of their source and methods employed, is generally the beginning of an iterative process. Structural analogs containing specific structural changes are prepared and assessed for biological activity. Favorable changes are maintained, unfavorable changes are discarded, and a new round of compounds is prepared. Each successive round of synthesis and biological assessment builds on the previous round with the ultimate goal of optimizing biological activity. (See Figure 22.)

8. Fragment-based drug design is a means of developing SAR data starting with low molecular weight molecules, typically in the 100–250 AMU range. This is much smaller than most drug molecules, and as a result binding affinities are low. Sensitive techniques such as X-ray crystallography, protein NMR, or surface plasmon resonance (SPR) are used to measure binding affinities (usually greater >100 μm). Fragments found to bind to the target of interest can either be "grown" by adding additional molecular features, or two fragments believed to occupy adjacent sites can be linked together to form a more potent molecule as a result of synergistic binding of the combination. (See Figure 20.)

9. The two general approaches to determine structure–activity relationships are physical screening of compound libraries and virtual screening of compound libraries. (See Figure 14.)

10. The two general approaches to virtual screening are docking a series of ligands into a binding site of a biomolecule and ligand-based design. In the docking method, a model of the target of interest is created using X-ray crystal data, possibly through the preparation of a homology model, and a virtual library of compounds are "docked" into a proposed binding site. The model is used to score the docked compounds based on their theoretical binding energies. In ligand-based drug design, a known ligand is used as a molecular template for potential new ligands. Features such as lipophilic

potential surfaces, electrostatic potential surfaces, area suitable for hydrogen bonding, lipophilic regions, and a wide array of molecular properties are defined. New ligands are then compared for their fit to the molecular properties of the known ligand in an effort to identify potential new ligands. (See Figures 15–17.)

11. A bioisostere is a functional group or collection of atoms that can be substituted into a candidate compound for a different functional group or collection of atoms without significant impact on biological activity at the target of interest. (See Figures 23–27 and Table 1.)

CHAPTER 6

1. Methods of increasing solubility include adding polar groups, adding ionizable groups (e.g., amines, acids), decreasing molecular weight, and decreasing planarity. (See Figures 5–9.)

2. Compound (2) should be more soluble as a result of the pyridine ring in the lower right side of the molecule, especially at acidic pH.

3. As lipophilicity in a series of compounds increases, solubility in aqueous media will decrease. A comparison of a series of compounds is often accomplished through a comparison of their cLogP values.

4. The aromatic ring in the upper left side of (1) is replaced with a cyclohexene ring in (2). The resulting decreased planarity of (2) leads to improved solubility. This change decreases the number of aromatic rings available for pi-stacking in a solid matrix, decreasing the stability of the solid. (See Figure 9.)

5. The five major methods of moving a compound across a cellular barrier are (1) passive diffusion, (2) active transport, (3) endocytosis, (4) efflux, and (5) paracellular transport. (See Figure 10.)

6. P-glycoproteins (Pgps) are efflux pumps that can prevent a compound from crossing a biological barrier. They are capable of acting on a broad range of substrates and are highly expressed in the blood–brain barrier.

7. In order for a compound to move across a biological barrier, it must disassociate from the water molecules that surround it, pass through the lipophilic region of the cellular barrier, and then resolvate with water on the other side of the cellular barrier. ClogP and TPSA are measures of lipophilicity and polarity respectively, and changes in these will have a significant impact on solubility. Compounds with a clog in a range of 1–3 are often permeable through a cellular barrier. If clogP is higher than 3, the compound is generally more soluble in lipids and may become trapped in the lipophilic section of a cellular barrier. Compounds with clogP below 1 are very polar and therefore, less able to desolvate from water making them less able to enter the

lipophilic region of the cellular barrier. In a similar fashion, high TPSA values are indicative of highly polar compounds that will have difficulty desolvating and entering the nonpolar region of a cellular barrier. (See Figures 13–15.)

8. The "free drug hypothesis" states that plasma protein bound compounds are incapable of passive diffusion and paracellular transport. Only "free" compounds, those that are not bound to plasma proteins, can cross biological barriers and gain access to tissues and organs outside of the bloodstream. (See Figure 22.)

9. In general, phase 1 metabolism reactions modify the molecular structure of a compound via processes such as oxidation or dealkylation, often leading to the production of a material that possesses a "handle" that could be utilized as an attachment point for a polar group that would facilitate excretion by the kidneys. Phase 2 metabolism takes advantage of the presence of potential points of attachment, such as hydroxyl groups, by adding polar groups to molecules, such as glucuronic acid or glutathione. These reactions are referred to as conjugation reactions.

10. There are essentially three methods of addressing metabolic instability through structural modifications of candidate compounds, remove, replace, and restrict.

11. The difference in stability is a result of the presence of the two fluorine atoms on the ethyl side chain on the left aromatic ring. The placement of these atoms in (2) most likely blocks metabolic oxidation of the compound (probably incorporation of a hydroxyl group) that would happen at this position in the absence of these two atoms. (See Figure 35.)

12. Enterohepatic recirculation occurs as follows. Compounds and their metabolites are excreted into the gall bladder along with bile fluid produced by the liver. The gall bladder releases bile fluids into the small intestine, providing an opportunity for reabsorption of the compounds in the bile fluid back into the systemic circulation. (See Figure 44.)

13. A compound with a high volume of distribution is most likely highly distributed into the lipophilic regions of the body and has a low concentration in the systemic circulation.

14. The two independent parameters that determine the in vivo half-life of a compound are its volume of distribution and its clearance.

15. The clearance of a compound by an organ or tissue is defined by the blood flow into the organ or tissue (Q) and the extraction rate (Er) of a compound by the organ or tissue.

16. If two compounds have the same clearance, but one compound has a fivefold greater in vivo half-life than the other, this suggests that the volume of distribution of the two compounds is also separated

by a fivefold window. The compound with the longer half-life most likely has a volume of distribution that is fivefold higher than the other.

17. If a compound under investigation has an oral bioavailability of greater than 100%, this suggests that a metabolic pathway has been saturated at the dose provided. Increases in compound dose will produce larger than expected increases in exposure. This situation is referred to as nonlinear pharmacokinetics. (See Figure 54.)

CHAPTER 7

1. The term therapeutic index refers to the ratio of the dose of a compound required to elicit a desired biological response and the dose required to elicit a negative response (side effect).
2. Homologous animal models are those that have the same causes, symptoms, and treatment options available for humans. They are the rarest and most difficult animal model to achieve.
3. An isomorphic animal model has the same symptomology as the human condition, and treatment options are generally the same. However, the root cause of the disease or condition in the animal model is not the same as that observed in humans.
4. Predictive animal models are common when a disease or condition is poorly understood. The animal model itself may have little or no obvious resemblance to the human condition, but there are facets of the model that allow researchers to use it as a predictive tool. The impact of potential therapeutic intervention in the model can be correlated to how humans will respond to the same compound. The model displays the treatment characteristics of the disease or condition and is said to have predictive validity.
5. More than one animal is required as a result of a number of factors. These include differences within the animal population selected (e.g., genetic difference that occur within a population of otherwise homogenous animals such as those described above), inaccuracies inherent to the means of measuring the "signal," and the intensity of the "signal" relative to the untreated population (a control is always necessary).
6. In the absence of treatment, a mouse will prefer to stay in the enclosed regions of an elevated plus maze. A compound that increases the amount of time a mouse spends in the open section is thought to have antianxiolytic efficacy. (See Figure 2.)
7. In the absence of treatment, a mouse will spend more time exploring a new object (relative to a known object already present in the chamber) that is placed in the testing chamber. Candidate compounds that cause a mouse to decrease the amount of time it

spends exploring a new object (relative to a known object already present in the chamber) suggest that the candidate compound is having a negative impact on memory formation. Both objects appear new to the mouse, even though only one is new. (See Figure 3.)

8. The rotarod performance test employs a rotating horizontal cylinder suspended above the cage floor. Under normal conditions, rodents placed on the rotating rod will attempt to stay on the rod for as long as possible. Neurodegenerative diseases and candidate compounds that impact balance, coordination, motor skills, and wakefulness (sedatives) will have a negative impact on the length of time a rodent is able to stay on the rod. Candidate compounds that delay the progression of neurodegenerative diseases can be studied in this model. In addition, this model has been used to identify compounds that cause sedation. (See Figure 6.)

9. Prolonged exposure to mineralocorticoids such as deoxycorticosterone acetate (DOCA) produces hypertension in several useful animals like rats, dogs, and pigs.

10. Although this model provides a means of identifying compounds that lower cholesterol and LDL, atherosclerotic plaque formation and disease progression is not the same as it is in humans. Differences in plaque structure and disease progression make these animal models unsuitable for identifying compounds capable of altering these facets of the disease.

11. A Kaplan–Meier survival curve plots the percentage of animals surviving against time and is used to compare the impact of candidate compounds on survival. It is frequently used to report results in animal models of infectious diseases and cancer. (See Figure 17.)

12. There are several limitations of animal models of infectious diseases. First, species-specific interactions between microbial ligands and host receptors are a significant part of the disease process that is often not replicated in an animal model of human infectious disease. Second, there may be differences between the clinical and laboratory strains of the infectious agents. Finally, transmission pathways can play an important role in disease progression, but very few animal models take this into account.

13. In the xenograft model of tumor progression, a tumor is established in an immunocompromised mouse using either patient-derived tumor cells or a stable tumor cell line. Once the tumor is established, treatment with a candidate compound or vehicle is initiated. Changes in the tumor size are monitored to determine the efficacy of candidate compounds. (See Figure 18.) In contrast, in an allograft model of tumor progression, an immune-competent mouse is used and mouse tumor cells are implanted instead of humans. In the allograft model, tumor cells are from the same species as the model animal, whereas in the xenograft model, tumor cells are from a species different from the model species.

CHAPTER 8

1. The abbreviation NOEL (no observed effect level) refers to the highest dose or exposure of a compound at which no toxicity is observed. In a similar manner, the abbreviation NOAEL (no adverse effect level) refers to the maximum dose or exposure level of a compound that can be employed without creating unmanageable toxicity.

2. The maximum tolerable dose (MTD) is the highest dose of a compound that can be employed without creating unmanageable toxicity.

3. Mechanism-based toxicity, also referred to as target-based toxicity, refers to the toxicity associated with biological target of interest. Musculoskeletal syndrome associated with inhibition of matrix metalloproteinase when they are used to treat arthritis is an example of this type of toxicity.

4. The hERG channel is a key component of cardiac rhythm, and blockade of this channel by candidate compounds can lead to ventricular arrhythmia, Torsades de pointes, and sudden cardiac death. Compounds capable of blocking the hERG channel are rarely moved forward in drug discovery programs. In addition, if a compound produces a metabolite that is capable of blocking the hERG channel, then the parent compound may represent a hERG risk. (See Figures 3 and 17.)

5. A hapten is a molecule that can produce an immune response, but only when it is bound to a carrier protein. The hapten reacts with a carrier protein and forms a covalent bond. The new hapten–protein adduct is no longer recognized by the immune system as "self," which causes the immune system to respond as if foreign material has entered the body. Antibodies are generated and an allergic response is initiated. The severity of the response can be as simple as a skin rash or as life threatening as anaphylactic shock. Penicillin allergy is a well-known example. (See Figure 22.)

6. Acutely toxic compounds produce undesirable effects after a single dose of a compound, whereas chronically toxic compounds, require prolonged or repeated exposure in order to elicit untoward effects.

7. The MTT human hepatotoxicity assay is used to assess the cytotoxic potential of candidate compounds. It monitors the rate of conversion of MTT to formazan in a cell culture. Compounds that are cytotoxic decrease the rate of production of formazan from MTT.

8. The AMES assay is used to identify potentially carcinogenic compounds. In the AMES assay, cells genetically engineered to be incapable of producing histidine (such as *Salmonella typhimurium*) are cultured in the presence of candidate compounds and a limited

supply of histidine. Once the histidine is depleted, only cells with mutations that restore histidine synthesis pathways will survive. Candidate compounds that promote mutations (mutagenic compounds) will lead to an increase in cell survival in the absence of a histidine-rich media. (See Figure 10.)

9. The micronucleus assay is used to identify compounds with the potential to cause chromosomal damage. In this assay, Chinese hamster ovarian cells (CHO cells) are grown in the presence of a candidate compound for a set period of time and then exposed to cytochalasin B. This blocks cytokinesis, leading to a buildup of binucleated cells. The presence of micronuclei indicates that a candidate compound can cause chromosomal damage. (See Figure 11.)

10. The comet assay detects compounds that are capable of causing DNA strand breaks. In this assay, cells are incubated with a candidate compound for a defined time period. A single cell is embedded in a gel electrophoresis matrix (agarose matrix) lysed, and then an electric field is across the gel. If the candidate compound causes DNA strand breaks, staining of the gel upon completion of the experiment will produce a comet-shaped image as a result of differential rates of migration of the full DNA as compared to the DNA fragments produced by strand breaks. (See Figure 13.)

11. Drug–Drug interactions occur when the presence of one drug impacts the metabolism of a second drug, leading to toxicity that is not seen when the drugs are used independently of each other. If, for example, a compound is normally metabolized to a safe product, but the presence of a second compound blocks the normal metabolic pathway of the first drug, forcing the first drug into a different metabolic pathway that leads to the formation of a toxic by-product. Alternatively, blockade of the normal metabolic pathway of one compound by the presence of a second compound could lead to higher than normal exposure to the first compound, possibly leading to concentrations above the safety margin for the first compound. (See Figure 14.)

12. Displacement of [^{3}H]-dofetilide from its binding site on the hERG channel by candidate compounds can be used to assess the level of hERG risk-associated candidate compounds. However, this assay will only identify compounds that bind to the dofetilide binding site. There are other binding sites on the hERG channel that compounds can bind to and block this channel. This assay is not capable of assessing for hERG risks mediated by these alternative binding sites. (See Figure 18.)

13. Major cardiovascular parameters that are routinely examined in safety studies include heart rate, blood pressure, contractile force, and ejection fraction.

14. The rotarod assay can be used to determine whether or not a compound has a risk of causing sedation.
15. Teratogenic compounds are capable of interfering with the normal growth and development of a fetus. Teratogen exposure can lead to malformations, growth retardation, functional deficits (mental or physical), and even death of the embryo/fetus.

CHAPTER 9

1. Within the context of the modern pharmaceutical industry, a clinical trial can be defined as a biomedical or behavioral experiment conducted on population of humans that is designed to answer key questions about a potential therapeutic agent. The studies are designed to generate safety and efficacy data that can be presented to the appropriate regulatory bodies in order to gain market approval.
2. Phase 1 studies define safety margins and the pharmacokinetic profile of a candidate compound. Phase 2 studies are pilot studies designed to provide an initial assessment of efficacy and safety in the target patient population. Phase 3 studies are broad efficacy and safety studies designed to determine the risk to benefit ratio of a candidate compound in the target population.
3. Polymorphism refers to the ability of a solid material, such as a candidate compound, to exist in more than one crystalline form.
4. It may be necessary to develop new chemistry to move the synthesis of a candidate compound from the laboratory to manufacturing scale for a number of reasons. Reagents and solvents employed in the laboratory may not be compatible with manufacturing scales. Solvents such as ether and methylene chloride, for example, are avoided in commercial scale production due to health and safety risks. Some reagents, like phosgene, are simply too dangerous to use on a commercial scale. The price of reagents may also be an issue if the original starting material or reagents are costly to procure or produce. Additionally, it may not be practical to achieve reaction conditions on a commercial scale in the manner that is possible in a laboratory setting. Laboratory scale reactions are easy to run at −78 °C, but manufacturing scale processes cannot be run in this manner. There are other additional issues described in the text of this chapter.
5. Candidate compounds can be delivered to a subject orally (PO), intravenously (IV), intraperitoneally (IP), subcutaneously (SC), via intramuscular injections (IM), transdermally, or via intranasal delivery.

6. An enteric coating is used to allow a tablet to survive the acidic environment of the stomach and arrive in the intestines unchanged. This polymeric coating is chemically inert with respect to the acidic environment of the stomach, but when it reaches the basic environment of the intestines, the coating dissolves and the contents of the tablet are released for absorption.

7. An excipient is any material that is added to a pill, tablet, or other delivery mechanism that is not the active pharmaceutical ingredient. Examples include fillers, disintegrants, solution binders, dry binders, glidants, lubricants, and antiadherents.

8. In a two-chambered osmotic pump system, the two chambers are separated by a movable wall. The "push chamber" is surrounded by a semipermeable membrane. Water enters the push chamber as a result of osmotic pressure causing the material inside the "push chamber" to expand. This action pushes the movable wall, forcing the drug substance out of the other chamber of the osmotic pump system. (See Figure 12.)

9. The three section of an investigational new drug application are (1) the animal pharmacology, safety, and toxicology study data section, (2) the chemistry, manufacturing, and controls (CMC) section, and (3) and the clinical trial protocols including investigator information section.

10. An institutional review board (IRB), also referred to as an independent ethics committee or ethical review board, is a committee that is tasked with ensuring that clinical trials are run in an appropriate manner. They are formally charged with the tasks of monitoring, reviewing, and approving all aspects of the clinical programs. This includes any changes to protocols that may be required as a result of new data that emerge through the course of the trials. Risk to benefit analysis may also be performed by the IRB as the clinical trials move forward in order to determine whether or not it is safe to continue the trials.

11. Allometric scaling is a mathematical calculation used to estimate interspecies dosing changes based on known PK parameters and the NOAEL (no observed adverse effect level) determined in animal studies. It is used as a guide in determining initial dosing levels in human clinical trials.

12. In a standard "3 + 3" phase 1 clinical trial, patients are divided into groups of three and given a single dose of the drug. If no adverse reactions are observed, the dose is doubled. This doubling of the dose is continued until adverse events occur, which defines the MTD for a single dose exposure. (See Figure 13.)

13. An interim analysis in a clinical trial is a point in the trail at which the data collected to date is analyzed to determine if the clinical trial should continue or be terminated. Incorporating an interim analysis in a clinical trial serves to limit the number of patients exposed to a candidate compound of unknown utility and safety.

14. A "stopping rule" is part of a staged clinical trial. It is a predetermined set of rules that define when a clinical trial must be stopped as the risk to patients outweighs the potential benefits. (See Figures 14 and 16.

15. An organization might choose to run a phase 3 clinical trial as an equivalency trial rather than a superiority trial when the potential benefit of the candidate compound over the candidate of care is not related to its efficacy. Typical examples include a more convenient dosing strategy or a better safety profile.

16. In a crossover study, all of the patients receive both the new treatment and the standard of care and as a result, they are their own control group. In addition, fewer subjects are required for crossover trials, so fewer patients are exposed to a candidate compound of unknown utility and safety.

17. Unlike standard clinical trials, adaptive clinical trials include interim analysis time points within the course of the overall study. When these time points are achieved, analysis of the data collected up until this point is used to determine how the remaining portion of the clinical trial will be conducted. In other words, the adaptive clinical trials are prospectively planned to include opportunities for modification of certain aspect of the trial (e.g., sample size, treatments, population, choice of outcomes, etc.) based on data collected during the trial itself.

CHAPTER 10

1. The term translational medicine refers to the concerted effort to apply new discoveries in basic science to the identification of new therapeutic modalities for patients. It requires the incorporation of a wide range of scientific disciplines and is often referred to as a "bench to bedside" approach to drug discovery and development.

2. The two categories of translational medicine are (1) the process of applying discoveries generated during research in the laboratory, and in preclinical studies, to the development of trials and studies in humans and (2) enhancing the adoption of best practices in the community.

3. In the T1 phase of translational medicine, scientific research is focused on understanding disease pathology on a molecular level and establishing links between the molecular processes, animal

models, and the human condition. The T2 phase is described as the study of the clinical impact of therapies with the end goal of providing the data necessary to support a change in clinical practice. Research designated as "T3" is concerned with the real world impact of a new therapy after it has been approved for marketing. Finally, the T4 subcategory of translational medicine is primarily concerned with the incorporation of the findings of the previous phases of translational medicine into policies and procedures that positively change clinical practices.

4. A biomarker is a measurable property that reflects the mechanism of action of the molecule based on its pharmacology, pathophysiology of the disease, or an interaction between the two. A biomarker may or may not correlate perfectly with clinical efficacy/toxicity, but could be used for internal decision-making within a pharmaceutical company.

5. The major categories of biomarkers are target engagement biomarkers, mechanism biomarkers, outcome biomarkers, toxicity biomarkers, pharmacogenomics biomarkers, and diagnostic biomarkers.

6. An ideal biomarker should be easily measured, time-efficient, quantitative, objective, and highly reproducible. Importantly, the results obtained from experiments using a biomarker should be useful for predicting results in the next stage of experimentations. In other words, a biomarker should be translational in nature.

7. According to the NIH Biomarkers Definition Working Group, a surrogate endpoint is "a biomarker that is intended to substitute for a clinical endpoint and is expected to predict clinical benefit or harm, or lack of benefit or harm, based on epidemiologic, therapeutic, pathophysiologic, or other scientific evidence." In other words, a surrogate endpoint is a biomarker for which there is clear and convincing scientific evidence (e.g., epidemiological, therapeutic, and/or pathophysiological) that the biomarker in question consistently and accurately predicts the clinical outcome.

8. Using surrogate endpoints and biomarkers in general during the prosecution of clinical programs provides a number of advantages that can translate into decreased patient risk, reduced costs, and shorter clinical trials.

9. The most commonly used imagining techniques are X-ray computed tomography (X-ray CT), positron emission tomography (PET), single photon emission computed tomography (SPECT), magnetic resonance imaging (MRI), functional magnetic resonance imaging (fMRI), and ultrasound techniques.

10. Radioactive decay rate is an important issue in PET imaging experiments, as the rate of radioactive decay limits the "shelf-life"

of the imaging agents. PET imaging requires a positron emitting radioisotope. The most commonly used radioisotope of this type is fluorine, whose half-life is only 110 h. This limits the time available for synthesis, purification, and use of an imaging agent. In order to maximize the time available for clinical application of a PET ligand, the radioactive isotope is almost always put in place in the last synthetic step. This also decreases the amount of radioactive waste produced during synthesis. It is also worth noting that positron emitting elements must be prepared using either a cyclotron or linear particle accelerator. In practice, these types of equipment are housed in the same facility as the PET imaging equipment in order to minimize the time lag between preparation of a suitable isotope and clinical application.

11. It may be necessary to develop an alternate synthetic pathway for a PET radioligand as compared to the nonradioactive compound of the same structure when it is not possible to incorporate the radioisotope in the last step of the synthetic protocol. In some cases, the synthetic challenges that must be overcome to incorporate an appropriate radioisotope into a compound may be insurmountable. In this instance, an analog with similar properties that is more easily tagged with a radioactive element is used instead.

12. If a PET or SPECT radioligand does not have low nonspecific binding to proteins, then the signal to noise ratio in a PET or SPECT experiment would be low. The radioligand would bind to its intended target and a signal could be observed, but it would also produce an observable signal at every nonspecific binding event that it takes part in. If this nonspecific binding is to albumin, then the radioligand would be nearly omnipresent in the body, as albumin is a major component of plasma. This could potentially drown out the signal from the macromolecule of interest.

13. A PET or SPECT radioligand should have rapid uptake is highly desirable, a short pharmacokinetic half-life (to decrease risk to subjects as a result of decreased exposure time to radioactive material), and Pgp efflux should be minimized. This is especially true if the targeted macromolecule is in the brain, as Pgp is heavily expressed in the blood–brain barrier.

CHAPTER 11

1. The three major facets of a pharmaceutical company are discovery, development, and commercial.
2. Business climate issues that can have an impact on the pharmaceutical company include economic pressures, regulatory

changes, and changes in the political landscape. Changes to safety standards, efficacy standards, and increased access to generic drugs have had a tremendous impact on the pharmaceutical industry, as have the formation of contract research organizations.

3. Nonmonetary impacts of mergers and acquisitions include the elimination of redundant or undesired research and development programs. The discontinued programs may have been independently viable, but having two programs focused in the same area or a program outside of a company's desired product portfolio would cause it to be eliminated. Considerable time and energy is required to evaluate programs in the merged organization, taking time and resources away from scientific efforts focused on bringing new drugs to market. In addition, key thought leaders and experienced personnel may leave the merged organization in an attempt to find a more secure working environment, potentially draining organizational memory and expertise.

4. A contract research organization (CRO) is a company that provides research and development services/expertise to other companies, but typically does not have internal discovery and development pipelines of their own. In theory, using a CRO to conduct research and development activities could increase the efficiency of the drug discovery and development process, although this has yet to be proven.

5. The Bayh-Dole Act, also referred to as the Patent and Trademark Law Amendment Act, changed US law to permit nonprofit organizations (and small businesses) to maintain ownership of inventions developed using federal funding. In addition, the rights to the invention could be assigned or licensed to another organization capable of commercializing the invention. In effect, the passage of this law gave academic laboratories, universities, colleges, and nonprofit organizations the ability to make profit from scientific endeavors funded by organizations such as the National Institutes of Health, provided certain conditions were met. Specifically, preference in licensing had to be given to small businesses, the balance of royalties and profit obtained from the invention (after expenses) had to be used for scientific research or education, and a portion of the royalty had to be shared with the inventors of record. This last clause provided substantial financial incentive to academic scientists to increase their focus on applied sciences, as commercialization of their research would enrich them personally.

CHAPTER 12

1. In order for an invention or discovery to be eligible for patent protection, it must be a composition of matter, process, machine, or an article manufacture. Compositions of matter are defined as "all compositions of two or more substances and all composite articles, whether they be the results of chemical union, or of mechanical mixture, or whether they be gases, fluids, powders or solids." The term "process" in this instance refers to "an act, or a series of acts or steps." This category is very broad in scope and includes synthetic methods necessary to produce a chemical compound or a method of using a compound to treat a disease. The third category of patentable subject matter, a machine, is defined as "a concrete thing, consisting of parts, or of certain devices and combination of devices." The fourth and final class of patentable subject matter, an article of manufacture is defined as "the production of articles for use from raw or prepared materials by giving to these materials new forms, qualities, properties, or combinations, whether by hand labor or by machinery." An article of manufacture can consist of multiple parts, but unlike machines, the interaction between the parts is generally static.

2. The utilization of "special nuclear material or atomic energy in an atomic weapon" is specifically barred by US patent law, as are patents covering "a human organism." Laws of nature, natural phenomena, or abstract ideas are also specifically excluded from patentability.

3. In order for an invention or discovery to be considered patent-eligible, it must fit into one of the categories of patent-eligible subject matter and it "must be useful or have a utility that is specific, substantial, and credible." In addition the invention or discovery must be novel and nonobvious.

4. In a general sense, the prior art refers to the sum total of knowledge available to the public prior to the date of a particular patent application. This includes patents, patent applications, scientific journals, newspapers, trade publication, PhD thesis (on the day that they become available to the public), poster presentations to the public, and electronic publications and information housed in publicly accessible databases. NIH grant applications also become part of the prior art once they are available to the public under the Freedom of Information Act. In addition, oral presentations can also be considered part of the prior art "if written copies are disseminated without restriction." In short, if something is available to the public, even for a limited time, it is part of the prior art. Importantly, material that is only distributed internally within an organization and are intended to remain confidential are not considered part of the prior art.

5. The discovery of a previously unreported utility for a known compound does not permit the discoverer of this utility the ability to obtain patent protection for the compound in question. The previously unreported utility is considered an "inherent property" of the previously known compound, and the discovery of an inherent property of material that is part of the prior art does not make it eligible for patent protection. It might be possible, however, to obtain a patent on the use of the compound.

6. In this scenario, only scientist A is an inventor of record of the small molecule. Since scientists B and C did not conceive of the compound, they are not part of the inventive process that led to the conception of the small molecule. Performing science at the direction of others does not make an individual an inventor of record.

7. Inventorship cannot be transferred between individuals. In order to be considered an inventor of record, an individual must contribute to at least one claim in a patent application.

8. Within the context of the patent process, the term assignment refers to the ownership rights of a patent or patent application. In the US, assignment documents are filed with the USPTO to record the ownership of patents and patent applications. In the absence of a recorded assignment with the USPTO, patents and patent applications are assigned to the inventors of record. Unlike inventorship, an assignment of rights to a patent or patent application can be transferred to individuals or entities other than the inventors of record. Also, while corporation cannot be inventors of record, they can own patent assignments.

9. A provisional patent application has a 12-month lifetime. Upon expiration of this deadline, a provisional patent application is considered automatically abandoned if no follow-up application has been filed that claims the provisional application as a priority document. Since the provisional patent application is never published, it does not enter the prior art.

10. Provisional patent applications are never published.

11. A PCT patent application is published 6 months after it is filed. In practice, a provisional application is typically filed, followed by a PCT application 12 months after the provisional application, and publication of the PCYT application occurs 6 months later, or 18 months after the original provisional application was filed.

12. A PhD thesis becomes part of the prior art as soon as it is available to the public, even it is only available in a university library.

13. A patent application must provide enough detail so that one skilled in the art can reproduce and use the invention without undue experimentation. If a patent application described new compounds, then the patent application must include detailed information describing how the compounds can be prepared. The

same requirements hold for the preparation of proteins, nucleic acids, genetically modified organisms, or any other type of novel material that is described in a patent application. In addition, patent applications must include detailed information on how to determine that the compounds in question are useful. Biological data alone is not enough.

14. A markush structure is a general chemical structure with variable units, often designated as R-groups, appended to the general structure. Each R-group has a specific definition that is provided in the patent or patent application. (See Figure 1.)

CHAPTER 13

1. Tamiflu® (Oseltamivir) sold as a prodrug, because the active agent, GS-4701, has a very low oral bioavailability. The presence of the free acid in GS-4701 limits its ability to penetrate biological barriers. Capping the acid as an ethyl ester significantly increases its oral bioavailability, and the ester prodrug is rapidly cleaved in the systemic circulation to release the active agent, GS-4701.

2. Nitrofurantoin can be used clinically as a result of its pharmacokinetic profile. When delivered orally, a 100 mg dose is rapidly removed after initial absorption. Approximately 75% of the dose is metabolized on first pass, while the remaining ~25% is excreted into the urinary tract as unchanged drug. Peak plasma concentration of a 100 mg dose of nitrofurantoin is less than 1 µg/ml and tissue penetration is negligible in all parts of the body except the urinary tract. Drug concentration in the urinary tract can exceed 200 µg/ml, which is well above the concentration required to kill bacteria. Essentially, nitrofurantoin's pharmacokinetic properties prevent it from reaching other parts of the body, limiting its utility to urinary tract infections and preventing damage associated with its mechanism of action.

3. Seldane® (Terfenadine) was removed from market as a result of its propensity to promote ventricular arrhythmia, ventricular tachycardia, Torsades de pointes, and sudden cardiac death in certain situations. Seldane® (Terfenadine) is a potent hERG blocker, but it is rapidly metabolized by CYP3A4, preventing it from achieving significant concentration in the systemic circulation. In the presence of a compound that blocks CYP3A4 metabolism (e.g., ketoconazole and erythromycin), however, systemic concentration of Seldane® (Terfenadine) can increase rapidly, leading to substantial risk of ventricular arrhythmia, ventricular tachycardia, Torsades de pointes, and sudden cardiac death. The drug was removed from the market as a result of this risk, and this event led to broad screening

of candidate compounds for inhibition of CYP3A43 and blockade of hERG.

4. Allegra® (Fexofenadine) was developed in order to compete with Seldane® (Terfenadine) and eventually replaced it. Although the makers of Seldane® (Terfenadine) did not realize it at the time, the actual active agent in Seldane® (Terfenadine) is Allegra® (Fexofenadine). By understanding the metabolism of a drug, scientists at Sepracor (now Sunovion) were able to identify, patent, and license a new therapeutic agent to Hoechst Marion Roussel (now Sanofi). It is important to be aware of the metabolites that are produced from a candidate compound. The metabolites may also be biologically active and useful therapeutic agents under the correct circumstance.

5. Claritin® (Loratadine) was brought to market as a nonsedating antihistamine by Schering-Plough (acquired by Merck). This drug eventually reached the overthecounter market and has been very successful. The scientists at Sepracor (now Sunovion), however, identified an active metabolite with better pharmacokinetic properties and eventually brought this compound to market as Clarinex® (Desloratadine). Understanding the metabolism of Claritin® (Loratadine) provided an opportunity to develop a multibillion dollar drug, which provides a clear indication of the value of understanding the metabolism of candidate compounds.

6. Desmethylprodine (MPPP) is not safe for clinical use as it is metabolized to very toxic material. MPPP is converted to MPTP in vivo, which passes through the BBB. Once inside the brain, glial cells absorb the MPTP and convert it to MPP+. This compound enters the dopaminegic cells via dopamine transporters, killing these specialized cells. This leads to the rapid development of Parkinson's disease. This is an example of how important it is to understand how and where compounds are metabolized. A candidate compound may appear safe in predictive assays, but the safety of the metabolites can be just as important as the safety of the candidate compound itself.

7. Bupropion was originally available as an immediate release 600 mg tablet, but a significant number of patients developed epileptic seizures. When the drug was reformulated into both extended release and sustained release formulations, therapeutically useful plasma concentrations could be achieved, but peak plasma concentration were substantially lower than those observed with the immediate release tablets. This significantly reduced the risk of seizures associated with this drug, allowing more patients to use the medication.

8. A poorly written patent application may not provide the protection necessary to prevent others from using the invention described in the application and recoup the costs associated with drug discovery and development.

Subject Index

Note: Page numbers followed by "b", "f" and "t" indicate boxes, figures and tables respectively.

Drug Index

Note: Page numbers followed by "f" and "t" indicate figures and tables respectively.